flora of
·EASTERN·SAUDI·ARABIA·

J A M E S · P · M A N D A V I L L E

FLORA OF EASTERN SAUDI ARABIA

Number 1 in the Series, Studies in the Flora of Saudi Arabia
Sponsored by the National Commission for Wildlife Conservation and
Development, Riyadh, Kingdom of Saudi Arabia

flora of
·EASTERN·SAUDI·ARABIA·

JAMES·P·MANDAVILLE

Routledge
Taylor & Francis Group

LONDON AND NEW YORK

First published 1990 by Kegan Paul International

2 Park Square, Milton Park, Abingdon, Oxon OX14 4RN
711 Third Avenue, New York, NY 10017, USA

Routledge is an imprint of the Taylor & Francis Group, an informa business

First issued in paperback 2016

Publisher's Note
The publisher has gone to great lengths to ensure the quality of this reprint
but points out that some imperfections in the original copies may be
apparent. The publisher has made every effort to contact original copyright
holders and would welcome correspondence from those they have been
unable to trace.

British Library Cataloguing in Publication Data
A catalogue record for this book is available from the British Library

ISBN 13: 978-0-7103-0371-4 (hbk)
ISBN 13: 978-1-138-97438-8 (pbk)

CONTENTS

Preface and Acknowledgements

Foreword

PART I — INTRODUCTION

PART II — THE FLORA

PREFACE AND ACKNOWLEDGEMENTS

I needed a book like this years ago when I was struggling to learn the east Arabian flora with unkeyed lists and from the synopses in the old family copy of Post's *Flora of Syria, Palestine and Sinai*. That exercise progressed finally to the aim of writing a practical manual for identifying the plants of Saudi Arabia's Eastern Province. I hope it may assist to some extent in the development and conservation of natural resources in these lands that have been my home for 40 years.

Effective use of such a tool requires some basic knowledge of botanical terminology and field practice, and this is presupposed on the part of the reader. One of the main purposes, however, in including a fair number of color plates with the text was to make it useful also to the non-specialist wishing to identify desert wildflowers in our area.

I have tried, at the cost of some neatness, to avoid taxonomic absolutism. Arguing that the student or general user is better not troubled with persisting problems of species delimitation or nomenclature is ineffectual; anyone with minimal powers of observation will discover these soon enough. Areas needing further work are therefore pointed out where they lie.

For support which made possible bringing this work to publication, I thank the National Commission for Wildlife Conservation and Development, Kingdom of Saudi Arabia. I am particularly obliged to the Commission's Managing Director, His Royal Highness Prince Saud Al-Faisal, for providing the Foreword, and to His Excellency Professor Abdulaziz Abuzinada, the Secretary-General, for sponsorship arrangements. Dr Abdulrasheed Nawwab provided valuable coordination assistance.

I owe a great deal to many people for technical help. Certainly first is Dorothy Hillcoat, now retired from the herbarium of The Natural History Museum, London, who over the years was so generous with her knowledge of Arabian plants. She has been, at great cost to her own research time, tutor to two generations of Arabian botanical collectors and writers. The herbarium staff at the Royal Botanic Gardens, Kew, assisted by naming some specimens during early phases of the field work. More recently Dr T. A. Cope there advised on the nomenclature of grasses and offered much helpful advice on the treatment of this, our largest family. Also at Kew Charles Jeffrey helped with some points in the Compositae, and A. Radcliffe-Smith identified several *Euphorbias*.

Ian Hedge at the Royal Botanic Garden, Edinburgh, freely offered the benefit of his broad experience with the Middle East flora during later work. He read and commented most helpfully on draft treatments of the Labiatae, Cruciferae and Capparaceae, fielded numerous queries, and saved me from error on many occasions. Jennifer Lamond and Rosemary King, also at Edinburgh, helped with reviews or queries in, respectively, Umbelliferae and Caryophyllaceae.

Dr V. Botschantzev at the Botanical Institute of the Academy of Sciences of USSR, Leningrad, reported on many of our saltbushes. Prof. H. Freitag of the University of Kassel also checked some chenopod specimens and offered highly valuable advice on nomenclature in this group. Among other specialists whose help in reviewing specimens increased the value of this work were Dr B. R. Baum at the Biosystematics Research Institute, Ottawa *(Tamarix)* and Prof. M. Nabil El Hadidi, University of Cairo *(Zygophyllaceae)*. Dr S. A. Chaudhary, Curator of the National Herbarium at the Agricultural and Water Research Center, Riyadh provided several unique collection records as well as helpful discussion. Valuable comments on some specimens were also provided by Dr H. W. Lack at Vienna *(Picris)*, Dr H. Hosni at Cairo *(Tribulus)*, Dr Abdul Ghafoor at Karachi (some aspects of *Anthemis*), Prof. J. Léonard of the Belgian National Botanical Garden *(Prosopis)*, Prof. H. Scholz at Berlin *(Bromus)* and Dr Ilkka Kukkonen at Helsinki *(Cyperus)*. The Saudi Arabian Oil Company (formerly the Arabian American Oil Company) allowed reproduction of weather data and provided the base for the distribution maps. Needless to say, and particularly because I have not in every instance followed all the advice of some of these helpful people, responsibility for errors rests wholly with myself.

Finally I would offer remembrance to my mother, Maxine Ruth (Newcomb) Mandaville, who in those earliest years taught us how nature is so wonderfully diverse.

BY HIS ROYAL HIGHNESS PRINCE SAUD AL-FAISAL
Managing Director,
National Commission for Wildlife Conservation and Development
Kingdom of Saudi Arabia

THE KINGDOM OF SAUDI ARABIA is committed to economic self-development in support of its citizens' aspirations for a continued full, peaceful and productive life. The nation's extensive hydrocarbon reserves are the primary natural resources for industrial growth, but great progress is also being achieved in other sectors, including agriculture.

Yet all of these activities, along with a burgeoning human population, carry potential for causing adverse, sometimes irreversible, effects on another natural resource: plants, wild animals and their habitats. These life-support systems are vital to our well-being and are a long-established part of our cultural milieu, commemorated in tradition and literature, and constituting another kind of riches for the people.

Saudi Arabia, in working to conserve this wildlife heritage, wants to build upon the experience of the international community as well as upon a comprehensive analysis of its own needs and priorities. The primary instrument and coordinating body of the national conservation movement was established in 1986 with the founding of the National Commission for Wildlife Conservation and Development.

The Commisson's work has forged ahead on two main fronts. The first has been to take urgent measures to protect wildlife in immediate danger of extinction and to learn, quickly, about the needs of these threatened species. Wildlife research centers are thus already in operation. In recognition that long-term conservation also requires habitat preservation, broad tracts of natural lands have also been set aside for surveillance and protection. The second main sphere of activity comprises field and laboratory research and publication of the results so as to catalogue and understand the Kingdom's wild organisms and their life histories. Plants and animals cannot be conserved effectively until we know what taxa we have and how they relate to each other; the ecologist must be able to identify, accurately, what he finds. The Commission is pleased to sponsor publication of the present book as a tool in this important endeavor.

This flora is the culmination of 25 years' field work by its author in Saudi Arabia's Eastern Province. These eastern reaches of our country include the sites of the Kingdom's oil and gas production centers, other world-scale industrial complexes, and expanding urban populations; they are thus of conservation survey priority. Plants support all higher life. We hope that the *Flora of Eastern Saudi Arabia* will be one of many contributions to a better understanding of the Kingdom's flora which the Commission will be fostering. We are confident that it will be as useful to botanists, other specialists, visitors, and the general public as we know it will be to our own staff.

PART I — INTRODUCTION

1. HISTORY OF BOTANICAL INVESTIGATIONS IN EASTERN SAUDI ARABIA

Early Travelers: Modern botanical exploration in northeastern Arabia had tentative beginnings in that golden age of Asian species description celebrated in 1867 with the first volume of Boissier's *Flora Orientalis*. Plant collecting began here as an incidental pursuit of British travelers who were exploring and describing the fringes of the Indian Empire. English thus became the first language of east-Arabian botany, and London was for long its principal herbarium.

Lieutenant Colonel Lewis Pelly, who as British Political Resident for the Gulf crossed into what is now the Eastern Province of Saudi Arabia from Kuwait territory on 21 February 1865, was the first to bring plant specimens from this region to European herbaria. A secondary purpose of his diplomatic mission to Riyadh, as related in his official report of 1866, was to study the geography of the area, 'collecting at the same time, such natural specimens as one might be able, en route.' Dr W. H. Colville, Civil Surgeon of the Residency at Bushire, traveled as medical officer of Pelly's party and was apparently assigned responsibility for the scientific studies.

Pelly and his group left Kuwait on 18 February, traveling on camels and entering the present Eastern Province north of Abraq al-Kabrit. From here their direction was southwest, across the Wari'ah ridge and on to the edge of the Summan at the wells of Wabrah. From Wabrah a generally southerly course took them across the Summan and the Dahna sands to Rumayhiyah, near Rumah on the threshold of Najd.

Pelly's return route from Riyadh, which led by way of al-Hasa to the Gulf port of al-'Uqayr opposite Bahrain Island, apparently followed the well known Darb Mazalij to al-Hufuf. His plant collecting — and it is not clear whether Pelly himself or Dr Colville did most of this field work — was restricted almost entirely to the Kuwait-Riyadh portion of the journey. The collection totaled 60 specimens, and of these about 25, as indicated by the data published from his sheets, could have been gathered within the present Eastern Province. The specimens were deposited at Kew, and J. D. Hooker provided a list of names published with Pelly's report to the Government of India (Pelly 1866).

It was not until 58 years later, when R. E. Cheesman undertook his expedition to al-Hasa and Yabrin, that there were any certain records of further botanical collecting in the Eastern Province. Although the earlier 1900s saw the first systematic exploration of the area by Europeans, precedence in this was given to mapping, and political conditions made scientific work extremely difficult.

Colonel S. G. Knox, British Political Agent at Kuwait who around 1907 collected plants with Sir Percy Cox in Kuwait territory, made inland trips to ar-Ruq'i and Hafar al-Batin in 1906 and south to Nita' in 1908. He did not, apparently, bring back specimens from our area. Knox's successor, Captain W. H. I. Shakespeare, was the son of a former Indian forestry officer and according to Blatter (1933) collected plants at Kuwait that were sent to the British Museum. His extensive travels in the future Eastern Province between 1909 and 1914, however, did not seem to include any botanical work.

In early 1912, while al-Hasa province was still under Ottoman occupation though soon to be reunited with the Saudi state, the young Danish geographer Barclay Raunkiaer crossed eastern Arabia from Kuwait to Riyadh and then back to al-'Uqayr.

His father was Christen Raunkiaer (1860 – 1938), an internationally recognized botanist still well known today to ecology students for his 'life form' classification of vegetation (see chapter 6), and during his Arabian journey the son carried plant collecting equipment provided by the University of Copenhagen Botanical Gardens. The suspicions of his escorts, however, prevented his collecting specimens, and his contribution to botany at that troubled moment in Arabian history was limited to a few general remarks on the vegetation contained in his account of the trip (Raunkiaer 1913). It is difficult, in these mechanized and more ordered times, to appreciate how knowledge of a land's plant life was once intelligence of high significance to the animal transport of armies. To the practical desert mind of that period, such surveys raised only visions of colonialist ventures.

Explorers and Collectors, 1917 – 1950: H. St. John Philby, the British political officer, explorer and later adviser to the King of Saudi Arabia, entered the Arabian scene through the Eastern Province port of al-'Uqayr on a political mission in 1917. Philby, determined to excel in all phases of Arabian knowledge, made biological specimen collecting an essential part of all his major expeditions. This ambition, coupled with a personal bent for natural history dating back to his boyhood, the opportunities of his Arabian journeys, and his prompt publication of results, all led to significant advances in knowledge of the biology of Arabia.

Philby's contribution to botany was not as significant as that to zoology — particularly ornithology — but his plant collections in the Rub' al-Khali in 1931 and the 'Asir mountains in 1936 provided some of the earliest data for these regions. Philby also had more 'feel' for vegetation than most other early Arabian explorers. Able to make maximum use of his guides' knowledge of terrain, landmarks and place names, he soon appreciated that among the desert Arabs important keys to this lore lay in the composition of vegetation and its zonal boundaries.

Philby entered Arabia in 1917 with equipment to collect plants as well as other specimens. He seems to have reserved these resources, however, until embarking on the previously unexplored part of the season's itinerary south of Riyadh toward Wadi ad-Dawasir; specimens sent by him to Blatter in Bombay after this trip all appear to have been gathered in central Arabia. Nor did he collect plants in the Eastern Province during any of his several subsequent crossings out of Najd to Kuwait and Iraq, although his geographical notes contain useful references by Arabic name to plants and vegetation he encountered (Philby Papers).

Philby's 1918 collection of plants from central Arabia reached Ethelbert Blatter, S. J. in Bombay as that botanist was preparing for press the first volume of his *Flora Arabica*, the first attempt at a Peninsula-scale flora of Arabia (Philby Papers). Blatter, Professor of Botany at St. Xavier's College, Bombay, published his flora at Calcutta in six parts from 1919 to 1936 as Volume VIII of *Records of the Botanical Survey of India*.

Aimed at including 'all the information available regarding the systematic botany of Arabia', Blatter's work was premature, undertaken at a time when major parts of the Peninsula were still unseen by scientific observers or collectors. Based almost entirely on specimens already at Kew and the British Museum, it was nevertheless a useful summary of the state of botanical knowledge up to the 1920s. It listed species by family, recording all known specimens without keys or descriptions, and consisted largely of records from southern Arabia, which had seen British collecting activity and from Yemen, the part of Arabia with the longest record of scientific study by Westerners.

2

For the present Eastern Province of Saudi Arabia Blatter had only 19 records, of which three were specimens collected by Cheesman and the rest by Pelly.

Eastern Saudi Arabia, meanwhile, had seen another expedition by a British collector — one more directly oriented toward natural history: Major R. E. Cheesman, personal secretary to the British High Commissioner in Iraq, Sir Percy Cox. Cox, long an active patron of the natural sciences in Arabia, had arranged permission from King 'Abd al-'Aziz Al Sa'ud for Cheesman to make a collecting trip in 1923 – 24. Cheesman, primarily a field zoologist, brought back with many bird, insect and mammal specimens a collection of 23 plant species that were listed with notes by R. Good and C. Norman of the British Museum (Natural History) in their appendix to Cheesman's book (1926).

Cheesman in 1921 had explored by ship the Arabian coast on the Bay of Salwah south of al-'Uqayr. In 1923 – 24 he landed at al-'Uqayr and traveled inland to the al-Hasa Oasis where he made collecting excursions. From there he moved south along the eastern flanks of al-Ghawar to Yabrin, the major objective of the trip and as yet unvisited by Westerners. Cheesman's plant collection was limited in scope but included specimens of several community dominants that were thus correctly identified for the first time. An unusual discovery was the European swamp orchid, *Orchis laxiflora*, growing in the al-Hasa reed swamps. It has not been found again in Eastern Arabia.

The turn of the decade saw interest in east Arabian scientific travel focusing on an almost entirely unknown part of the country, the Rub' al-Khali. Philby and Bertram Thomas, the Adviser to the Ruler of Muscat and Oman, were vying for first honors in a crossing of the southern sands.

Thomas in his 1929 – 1930 season had explored the southern fringe of the sands around Wadi Muqshin and al-'Ayn. In December 1930 he launched his full crossing following a south-to-north route into Qatar along the 51st Meridian. He collected insects, reptiles and a few mammals, but apparently no plants, although his route descriptions (1932) contain references to vegetation by vernacular names. Like other travelers able to communicate beyond rudiments with his Bedouin guides, he soon became conscious of the major vegetation zones so important to nomads and to any traveler depending on animal transport.

Thomas' victory in this race was a bitter disappointment to Philby who then, still driven to excel, aimed his route the following season into even more remote quarters. His early-1932 trip took him through Yabrin to an intersection of Thomas' route at Farajah and then kept to new territory to the south, making a wide triangular circuit just south of the 20th Parallel. He left the sands in the west near Wadi ad-Dawasir after overcoming great hardships across a wide waterless sector. The list of his plant specimens, provided by J. Ramsbottom as an appendix in Philby's book describing the journey (1933), includes some 43 names. These are given without locality, and some clearly came from areas outside the Rub' al-Khali proper. Philby's major contribution to botany lay not in the few specimens he brought back, but in the frequent descriptions of vegetation contained in his minutely detailed route notes. These, until very recently, provided the only botanical data for some parts of the Rub' al-Khali.

The following years up to the time of World War II — the early period of oil exploration and discovery — saw little botanical activity in eastern Saudi Arabia. It was at this time, however, that Mrs (later Dame) Violet Dickson of Kuwait began her extensive Kuwait plant collections for Kew in work that provided the first comprehensive floristic list for northeastern Arabia. Dame Violet's numerous field

3

outings with her husband, Colonel H. R. P. Dickson, on several occasions touched on the northern plains region of the Eastern Province, and their trips in 1942 and 1947 allowed her to collect along routes between Kuwait and Dhahran and from Dhahran to al-Hasa. She published a first Kuwait list in 1938 (Dickson 1938) and later a more fully annotated list in book form with sketches (1955). This includes a number of records from eastern and central Saudi Arabia. B. L. Burtt and Patricia Lewis (1950, 1952, 1954) began a detailed taxonomic account of Dame Violet's specimens in *Kew Bulletin*, but this work unfortunately was terminated after publication of three installments covering 23 families.

In other parts of Arabia, meanwhile, British-led regional cooperation in anti-locust campaigns led to the first ecological field work in the Peninsula aimed at gaining information about this economic scourge of the Middle East. A wealth of data on plant life as well as the associated fauna was soon being gathered.

Saudi Arabia was represented at the Fourth International Anti-Locust Conference in Cairo in 1936, but it was the threat of wartime famine that led to the first anti-locust field work in eastern Saudi Arabia. Directed from the Middle East Supply Centre in Cairo, a series of five annual cool-season campaigns was carried out in Arabia under the direction of Desmond Vesey-Fitzgerald. A small mobile unit with mixed British and Indian personnel operated in Oman and eastern Saudi Arabia in 1942 – 1943. That season set the pattern for three succeeding expeditions and included a large motor convoy from Cairo to Baghdad, Basra, Kuwait and Dhahran. The campaign was remodeled on civilian lines with the closure of the Middle East Supply Centre in October 1945, and Arabian headquarters was established at Burayman near Jiddah (Uvarov 1951). The Saudi Arabian Government took over anti-locust operations in succeeding years and continued to play a major role in international efforts to study and combat the locust threat.

Hundreds of plant specimens — including some from the Eastern Province — were collected by anti-locust ecologists and entomologists, and many were studied and identified by Dorothy Hillcoat of the British Museum (Natural History)'s Department of Botany. Vesey-Fitzgerald, on the basis of these findings, provided the first general description of the main plant communities of eastern Arabia (Vesey-Fitzgerald 1957). Except for this and two papers on western Arabia by the same author, little was published of the specimens and data gathered during these wide-ranging wartime expeditions across the Peninsula.

Yet another contribution of the anti-locust program was providing official purpose and entry for the last camel-riding European explorer of southern Arabia, Wilfred Thesiger. Much of his travel was in the Rub' al-Khali borderlands, in territories belonging to the then-Aden Protectorates, Oman and the present United Arab Emirates. Faithful to his sponsors, he brought back plant specimens and made notes on the vegetation and possible locust habitats. A. H. G. Alston and Dorothy Hillcoat of the British Museum (Natural History) identified many of his plants, which were listed by vernacular and scientific names in his travel accounts (1946, 1948, 1949, 1950). These contain valuable botanical information, although his data for the strict area of this flora is sparse. Most useful for our purposes are his descriptions of the eastern Rub' al-Khali, crossed in 1946 – 1947, and the far southwest, in 1947 – 1948.

Some other locust campaigners wrote accounts of Arabian travel, and a few — such as D. G. Bunker's (1953) description of the southwestern corner of the Rub' al-Khali

— include some notes on vegetation. Popov (n.d.) and Popov and Zeller (1963) wrote unpublished papers describing plant life in adjoining parts of Arabia. For the most part, however, such efforts in our eastern territory remain scientifically undocumented.

Beyond Mid-Century: Botanical field work by Saudi Arabian universities and Government agencies such as the Ministry of Agriculture and Water Resources was underway by the 1960s. Some results have been published, dealing primarily with the central and western parts of the country. Government hydrological surveys in the Eastern Province have involved some plant collecting and ecological work by consulting firms, leading to such studies as the 1969 report by Italconsult on grazing resources and water in eastern Saudi Arabia. Little of this consultants' data has been made publicly available, although in at least one case it has stimulated a publication by academic collaborators: De Marco and Dinelli's floristic list (1974).

More recent individual collectors residing in or visiting the Eastern Province have made significant contributions of eastern plant specimens. Neil Munro, of Hunting Technical Services Ltd, has deposited specimens at the National Herbarium, Riyadh. C. Parker, a tropical weeds specialist, collected weeds in eastern Saudi Arabia in the early 1970s. Dr S. A. Chaudhary also made collections in agricultural centers of the Province. Dr A. C. Podzorski traveled through the Province in 1982 as part of a wider Saudi Arabian visit to collect specimens for Riyadh, Edinburgh, and Lund. Kenneth Naylor collected while at Dhahran in the early 1980s. Sheila Collenette subsequently made several collecting visits. There have doubtless been other visitors or residents who have gathered specimens, at least on a small scale.

The integration of field data into broad-scale botanical publications was beginning to get underway by the 1970s. Professors A. M. Migahid and M. A. Hammouda's *Flora of Saudi Arabia*, published in Riyadh in 1974 (1st ed.) and 1978 (2nd ed.), was designed to be Kingdom-wide in scope. It provided a first local reference for student use, although the basis for its attribution of taxa to the Eastern Province was somewhat unsure. More recently, a growing number of botanical papers by national workers as well as Government planning for projects of wider scope promise a bright future for this field of scientific endeavor in Saudi Arabia.

2. GEOGRAPHICAL AREA OF THE FLORA: TOPOGRAPHY AND SUBUNITS

Geographical Limits: The geographical coverage of this flora (Map 2.1) approximates the political limits of the Eastern Province, Saudi Arabia. Except in the extreme north it also encompasses a more or less natural floristic unit. Arabia west of the central Dahna, as well as Qatar, the southern Gulf coast and Oman, lie largely in Sudanian floristic territory, with floras differing in several respects from that of our area. The Rub' al-Khali sands, while included here formally within the nebulous east Arabian borderlands of the Sudanian zone, also has plant associations contrasting with those of its bordering wadis and plains to the east, south, and west. The total areal coverage of this book is approximately 605,000 square kilometers, of which the Rub' al-Khali portion accounts for nearly 400,000.

General Topography: Northeastern Arabia east of the Dahna sand belt is divisible into two broad topographic provinces, each providing somewhat different environments for plant life. These take the form of two north-south concentric arcs, both of which

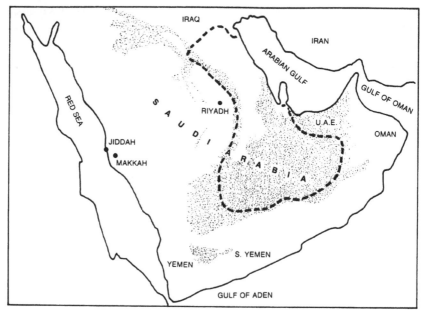

Map 2.1. The geographical area covered by this flora, shown enclosed by the dashed line. The outline indicates the approximate limits of field work on which the systematic account is based; it does not represent political boundaries.

parallel the Dahna and join it in reflecting broad patterns of geological structure in central and western Arabia. The eastern one, lying along the Gulf coast and about 100 kilometers wide near the middle, may be called the Coastal Lowlands. This is backed at higher elevations by the second, the Summan Plateau. The boundary between them is marked by a more or less distinct topographic step which in some parts takes the form of steep escarpments (Plate 1). North of 27° 30'N, as the Coastal Lowlands merge into the Northern Plains, and south of 23° S, near the northern limits of the Rub' al-Khali, this escarpment fades out leaving a less distinct boundary. Differences in soils, however, are still significant. Centrally, around the latitude of al-Hasa Oasis, this simplified picture is complicated by Summan-like terrain at Shadqam and neighboring structures.

Geologically, this whole territory is characterized by surface exposures of Tertiary sediments, largely carbonates, with significant areas covered by eolian sands and late Tertiary to Quaternary gravel sheets. Eastward, particularly in the coastal zones, evaporites including salt, marl and gypsum deposits strongly affect local vegetation potential. Powers et al. (1966) provide a detailed account of the rock units with an interpretation of the geological structure of this area.

These two broader provinces are subdivided here, for the purpose of describing the distribution of individual plant species, into eight subregions. These are derived by dividing the Coastal Lowlands north-to-south into three parts, termed the Northern Plains, the Central Coastal Lowlands, and the South Coastal Lowlands; the Summan and Dahna into northern and southern sectors; and adding the Rub' al-Khali (Map 2.2). This arrangement, apart from providing geographic units of convenient size, was chosen to reflect more or less natural groupings of plant distributions.

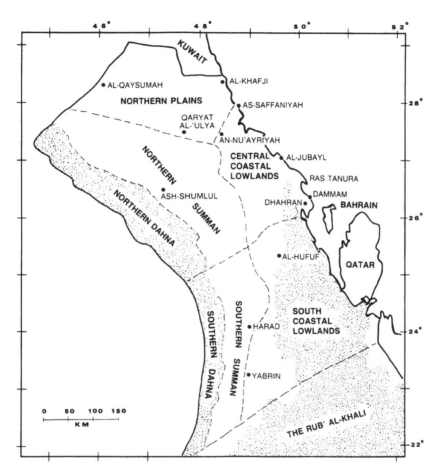

Map 2.2. Topographic subregions of eastern Saudi Arabia, showing the units used in terrain descriptions and for grouping specimen records of each species.

The Northern Plains: The Northern Plains, bounded for the purpose of our studies by the great linear depression of al-Batin on the northwest, in fact extends well across this line toward ash-Shu'bah and the Iraqi border. It partakes of some of the characteristics of the Central Coastal Lowlands in the far east but is marked, except along its southeastern edge, by its very low relief and an absence of sand terrain. For thousands of square kilometers it is a nearly dead-flat plain rising very gradually from sea level in the east to about 400 meters elevation in the west (Plate 2). Broad parts of it are made up largely of late Pliocene or early Pleistocene alluvial deposits from the Wadi ar-Rumah – al-Batin drainage system, with surfaces strewn with lag gravels and cobbles brought all the way from the igneous and metamorphic exposures of the Arabian Shield. These ancient flows also led to the deposition of a considerable body of silts, which figure prominently in the soils here and attract plant communities adapted to that substrate.

The presence of the Haloxyletum and some other saltbush communities indicate that soils here may be somewhat saline. In the southeast, however, relatively well drained

7

loamy sands prevail, leading to the establishment of an exceptionally well developed Rhanterietum. The concentration of gravels into 'desert pavements' restricts plant life in some central and northern areas. Water resources are limited, the few hand-dug wells being very deep (40 – 60 m). The Northern Plains extend through Kuwait in the north and are bounded on the south at the contact of the al-Batin alluvial fan with the underlying Summan rocklands. In the southeast the boundary is more subjective, placed near an-Nu'ayriyah where relief increases.

The Central Coastal Lowlands: This region, extending from Ras as-Saffaniyah on the northern coast to just south of Dhahran in the southeast, is bounded on the west by the escarpment of the Summan. It is basically a coastal plain, rising slowly inland with a gradient of approximately 1 in 1000. Relief is moderate, and the surface is generally covered by sands, although limestone exposures in the form of ridges, domes and minor escarpments are frequent.

A distinctive terrain type here, particularly in coastal areas, is the *sabkhah* (pl. *sibākh*), or highly saline flat with crusted surface. These are apparently derived in some cases from incursions of the sea during earlier geologic times. A *sabkhah* surface is level, being determined by the capillary rise of shallow ground water. Above this level, wind deflation removes the drying sediments; below it, they are fixed by moisture and evaporative concretion. Some *sabkhahs* may have surface areas of hundreds of square kilometers, and they are nearly always devoid of higher plant life although zoned halophytes may be found around their edges. Even shallow sand cover or hummocks on the *sabkhah* surface, however, can provide the improved drainage required for the establishment of plants.

Nearly barren unstabilized dunes, representing a northern extension of the Jafurah sands through Abqaiq toward al-Jubayl, cover parts of this region in the southeast (Plate 3). Greater areas are covered with fixed sands and better vegetation. Sand-covered coastal lowlands in the central part of this region provide the habitat preferred by wild or naturalized date palms (see systematic section, *Phoenix dactylifera*), and these — apart from some large *Tamarix* shrubs — provide the only uncultivated tree forms. Unlike the Northern Plains, the Central Coastal Lowlands have numerous shallow hand-dug wells which provide relatively good if brackish water. Soils are very limited in development, ranging from ill-drained saline formations, barren rock and moving dune sands — all relatively sterile — to more fertile loamy silt-sands in the northwest.

The Saudi Arabian Gulf islands are classed as part of this topographic unit.

The South Coastal Lowlands: This region is similar in many respects to the Central Coastal Lowlands, although with less relief. It is set apart largely on floristic grounds because of its noticeably greater penetration by Sudanian elements. It also differs climatically, lying largely below the 50 mm isohyet and thus having an overall poorer development of vegetation. Much of this region has fairly deep sand cover, heaviest in the Jafurah. Where stabilized, these sands bear shrubs (Plate 4), although these are more widely spaced and less well developed than in the north. Like the Northern Plains, it is underlain by broad gravel deposits, these radiating fan-like from the Wadi as-Sahba channel. The gravel flats are largely barren of vascular plants or support only very wide-spaced and stunted shrub associations.

This region is bounded on the west by the higher relief of the Ghawar anticline — site of the world's largest oil field — and on the southwest by the escarpment of Summan

8

Yabrin. On the south it grades imperceptibly into the Rub' al-Khali around the 23rd Parallel.

The Northern Summan: Bounded on the north and east by the Northern Plains and the Central Coastal Lowlands, and on the west by the Northern Dahna, these rocky uplands provide generally poorer conditions for plant growth than do their neighboring zones. They are characterized by broad exposures of limestones, and soils are shallow or non-existent. Except in marginal areas of the east and southeast, there is little relief beyond scattered limestone knolls, and it is only in the occasional silt-floored basins among these that any dense vegetation may be found (Plate 5). This, however, in a favorable rainy season may be exceedingly lush, with lawn-like islands of continuous annual cover around scattered shrubs of *Ziziphus nummularia*.

The rocky surface of the country leads to the frequent formation of long-lasting pools after moderate to heavy rains. Vertical, cave-like solution cavities and sinks may be found in the west-central parts. Shrub cover is generally stunted and sparse, consisting of degraded saltbush associations led by *Haloxylon* or *Anabasis*. *Ziziphus* is frequently seen in the scattered basins. Because of the contracted nature of the more favorable silt basin vegetation, this tends to be heavily exploited by herdsmen, and it often suffers from overuse and trampling. Under these conditions *Astragalus spinosus* and other overgrazing indicators increase.

The Southern Summan: Topographically similar to its northern counterpart, the Southern Summan is marked by its more arid climate, lower species diversity, and greater frequency of Sudanian penetrants. As defined here it includes a major part of the Ghawar anticline southwest of al-Hasa Oasis, although the Summan as a proper name strictly only applies to regions further west, beginning with al-Bidah. South of the gap formed by the Wadi as-Sahba breakthrough, the escarpment behind the Yabrin depression extends the Southern Summan's eastern boundary south to about the 23rd Parallel. Beyond this the escarpment dwindles, but the edge of Summan terrain is still apparent for another degree south, to where it merges with the gravel sheet of Abu Bahr.

The Northern Dahna: The Dahna sand belt, which bounds the overall area of our flora on the west, is one of the major topographic features of the Arabian Peninsula (Plate 6). It provides a natural boundary between central and eastern Arabia and links the Peninsula's two major sand deserts: the Great Nafud in the north and the Rub' al-Khali. Its sands, consisting of orange-red, iron oxide-coated grains, are visibly distinct from those of the coastal lowlands. They are akin, rather, to the dune zones of the central Peninsula, with which they merge in the far northwest of our area.

The Dahna consists of several parallel longitudinal sand ridges, which may be separated from one another by sand sheets or by bands of rocky or gravelly ground. The ridges are composed of merged barchans or smaller linear forms, and in the north massive 'star dunes' sometimes lie on them. All these red sands are characterized by a distinctive plant community led by codominant *Artemisia monosperma* and *Calligonum comosum*. Along the Dahna's eastern edges, in a transition zone with red sand shallowly overlying the Summan limestones, there is a band of *Rhanterium*. After good rains the Dahna, particularly in the north, supports a wide variety of annuals; it is considered a good grazing zone throughout the winter and spring. Access to grazing here was formerly limited by the absence of ready water supplies; today this is overcome by the Bedouins' use of tank trucks to transport water from drilled wells.

The Southern Dahna: This subregion is divided arbitrarily from its northern counterpart in the vicinity of the old Riyadh-Dammam road crossing near Khurays. It shares in the same plant community but supports distinctly fewer annual species. Its fringes include a greater proportion of Sudanian derivatives such as *Rhazya,* and its southern extremity is taken over by Rub' al-Khali dominants such as *Cornulaca arabica.*

The Rub' al-Khali: This region of some 500,000 square kilometers, with its extremely arid environment, endemic community dominants and very restricted species diversity, comprises a natural subunit in the vegetation of eastern Arabia. About 80 percent of it, or nearly 400,000 square kilometers, lies within the coverage area of this flora. Described as the largest area of continuous sand cover in the world, the Rub' al-Khali is structurally a basin with a long axis falling from southwest to east-northeast. Its sands are underlain in the west by gravels derived from ancient flows of Wadi ad-Dawasir. Further downslope the floor grades into evaporites with marls and *sabkhahs* exposed amid the dune massifs of the east and northeast.

Dune structures here are spectacular and of immense scale by world standards, ranging from parallel linear forms scores of kilometers long to rounded 'sand mountains' up to 250 m high (Plate 7). These major forms, selected largely by local wind regime and sand supply, are composed of, or accompanied by, the more usual primary dune elements: barchans, domes, sheets and smaller longitudinal types.

The sand-covered and greater part of this huge hyper-arid region supports a largely continuous although very diffuse cover of wide-spaced, drought-adapted shrubs. Only in the smooth rolling sand sheets of the central northwest, the salt-crusted floors of the east, and in the west-bordering gravel flats does one see sizeable areas completely devoid of vegetation. Because of the severe climate regime, however, it is not uncommon for many parts to experience a total failure of rains for several years or more, leading to a complete local destruction of vegetation over hundreds or thousands of square kilometers. See Mandaville (1986) for a fuller description of conditions in this region.

3. THE CLIMATE OF EASTERN SAUDI ARABIA

Causal Factors and General Characteristics: Eastern Saudi Arabia has a subtropical desert climate, arid basically because of its geographical situation in global patterns of atmospheric circulation.

Heated air rising in the equatorial regions moves northward and southward in the upper atmosphere to descend in bands of territory lying roughly along the 30th parallels of north and south latitude, or the poleward margins of the Trade Wind belts. These regions thus receive generally stable descending air which is adiabatically warmed as it loses altitude and consequently dried. This leads to an almost complete dispersal of cloud and an absence of rain except when this pattern is disturbed by incursions of rare storm centers from outside. It also leads to the formation over these desert territories of semi-permanent high pressure zones, or anticyclones, with divergent circulation that further surpresses local precipitation. The Trade Winds of Arabian latitudes represent the low-level return of this circulation toward the lower pressures nearer the equator. In the northern hemisphere, the Trades are generally northeasterlies. In northeastern Arabia their direction becomes north to northwest as a result of locally dominant pressure patterns over the Gulf and the Asian land mass to the east.

The southern parts of the Arabian Peninsula lie along the southern margin of this desert climate province. In summer, with the annual northward shift of the Intertropical Convergence Zone, they may be touched by precipitation from the rising air masses of this belt. This, with the monsoon circulation of the Indian Ocean, brings regular summer rains to a narrow coastal fringe of the Peninsula in Dhufar and increases chances for local summer convectional rains further inland as far as the southern Rub' al-Khali.

The whole area, except the southern Rub' al-Khali, has a Mediterranean climate regime in the sense of exhibiting a clear division into hot and cool seasons with rainfall confined almost exclusively to the cool months of October to April. The winter rains are associated with 'Mediterranean depressions', or low pressure storm centers entering from the Mediterranean region, that move east and southeast across the Arabian Peninsula. Particularly in spring, these may bring squall lines and thunderstorms with brief torrential rains and gale-force winds. Rain totaling 80 mm fell in one 24-hour period at Abqaiq in December 1955 (Arabian American Oil Company, n.d.). In midwinter there tend to be prolonged cloud build-ups with longer and more beneficial rains. These depressions are gradually dissipated as they move across Arabia, and the probability of rain thus decreases to the southeast. They do not normally carry precipitation as far as the Rub' al-Khali, but there are exceptions, such as the apparent broad frontal development with widespread rains experienced across the central and southwestern sands in February, 1982.

The low rainfall and high temperatures of our area certainly class it as a desert by virtually any of the many definitions of the term. Both in climate and vegetation types, a transect from its northern limits into the central Rub' al-Khali has much in common with a line leading from the northern into the central Sahara.

Arid Climate Classification: More than a dozen different indices have been devised to measure and classify the degree of aridity of deserts. One of the more widely used of these, that of Thornthwaite, uses calculated potential evapotranspiration in a water balance formula and was followed by P. Meigs in his maps prepared for the UNESCO arid zones research program (McGinnies, Goldman and Paylore 1968). In this scheme, as revised by Meigs in 1960, the climate of eastern Saudi Arabia north of the 24th Parallel is classed as 'arid' and that of the east-central Rub' al-Khali as 'extremely arid', the last classification being reserved for areas experiencing at least 12 consecutive months without any precipitation. Another way of comparing climates, one of particular interest because of its already wide application in other regions, is the climate diagram of Walter (1979). This is a composite graph of monthly rainfall and mean temperatures that gives a very useful impression of the seasonal march of aridity in desert lands. The ratio of the temperature and rainfall scales is chosen so that the area between the curves provides an approximate indication of water balance excess or deficit (Fig. 3.1).

Precipitation: Precipitation in eastern Arabia occurs almost entirely as rain. Annual rainfall means range from around 100 mm in the north and northeast of our area to less than 10 mm in the Rub' al-Khali (Fig. 3.2). Dews are not infrequent, especially

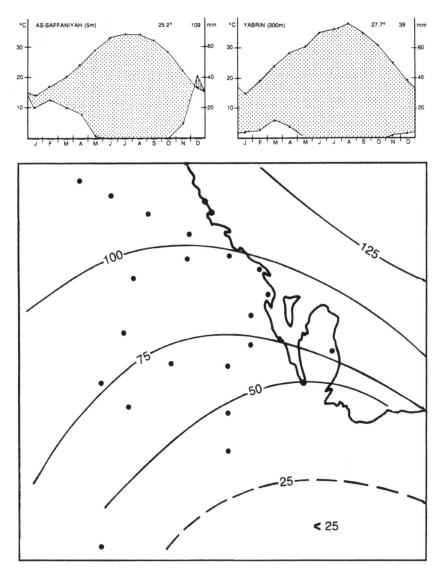

Fig. 3.1. Climate diagrams for a northern coastal station (as-Saffaniyah) and a southern site on the edge of the Rub' al-Khali (Yabrin). The upper line in each diagram follows mean monthly temperature; the lower line charts mean monthly rainfall. Figures at right of station name are (left to right) station elevation above mean sea level, mean annual temperature, and mean annual rainfall. The stippled area represents the period of moisture deficit, while the vertically hatched area (present only in the as-Saffaniyah diagram and then only for a short period in December) indicates the period when moisture balance is positive.

Fig. 3.2. Mean annual rainfall, northeastern Arabia. Isohyets in mm. Curves plotted from station records of the points indicated and from some outside the area of the map. Lengths of records range from 10 to 42 years. Station data from sources cited in Table 3.1

in coastal areas, but they have not been measured or otherwise evaluated. Snow is unknown, although there have been very rare minor falls in the Najd uplands just outside the Eastern Province, such as the one that occurred at Riyadh and al-Kharj on 3 January 1973. Brief intense hail storms sometimes accompany spring squalls.

For plant life the reliability, or lack thereof, of rains from year to year ranks in consequence with its amounts. Total annual rainfall at Dhahran over 39 years has ranged widely between extremes of 5 mm and 277 mm (Arabian American Oil Company n.d.). This extreme variability of rains in deserts is well known, and one measure of it is 'relative interannual variability,' calculated by finding the average of year-to-year rainfall differences over a long period and expressing this as a percent of the mean (Wallén in Hills 1966, 38). This value for the great majority of Eastern Province stations ranges between 70 and 90 percent, showing that the variation in rainfall from year-to-year, on average, approaches the value of the mean itself. In some years it greatly exceeds it. Annual and monthly rainfall records with relative variabilities for stations in eastern Saudi Arabia are summarized in Table 3.1.

Table 3.1 **RAINFALL DATA, SELECTED EASTERN PROVINCE STATIONS**

Mean Annual Rainfall, mm

Station	Mean	Absolute Maximum	Absolute Minimum	(Years)	Percent Variability
as-Saffaniyah	109	206	30	(11)	81
al-Qaysumah	108	348	8	(23)	56
an-Nu'ayriyah	108	300	23	(15)	77
ash-Shumlul	107	283	33	(11)	76
Ras Tanura	94	297	9	(27)	79
Abqaiq	92	181	7	(30)	89
27-08N, 49-12E	86	247	10	(14)	85
as-Sarrar	80	230	43	(14)	71
Dhahran	77	277	5	(42)	89
Khurays	63	158	12	(14)	85
al-Hufuf	59	146	14	(14)	73
Yabrin	39	104	9	(14)	71

Mean Monthly Rainfall, mm

Station	Jan	Feb	Mar	Apr	May	Jun	Jul	Aug	Sep	Oct	Nov	Dec
as-Saffaniyah	20	25	20	16	1	0	0	0	0	0	9	41
al-Qaysumah	23	15	15	20	8	0	0	0	0	4	10	19
an-Nu'ayriyah	20	17	21	24	2	0	0	0	0	0	12	31
ash-Shumlul	20	15	18	22	3	1	0	0	0	3	11	17
Ras Tanura	24	15	11	14	3	0	0	0	0	5	5	21
Abqaiq	21	19	13	10	3	0	0	0	0	0	3	18
27-08N, 49-12E	19	8	18	20	10	1	0	0	0	3	5	12
as-Sarrar	7	10	16	17	4	0	1	0	0	5	6	15
Dhahran	17	12	12	10	2	0	0	0	0	0	9	17
Khurays	6	3	18	26	0	0	1	0	0	0	0	10
al-Hufuf	10	2	19	25	1	0	0	0	0	0	3	6
Yabrin	4	5	12	8	1	0	0	0	0	0	2	3

Note: Monthly means do not add to annual means, having been derived from different data. 'Percent Variability' refers to 'relative inter-annual variability', as defined in the text.

Data from Arabian American Oil Company, 'Meteorologic and Oceanographic Data Book' (n.d.) and Saudi Arabian Ministry of Agriculture and Water Resources, 'Hydrological Publication' (various issues).

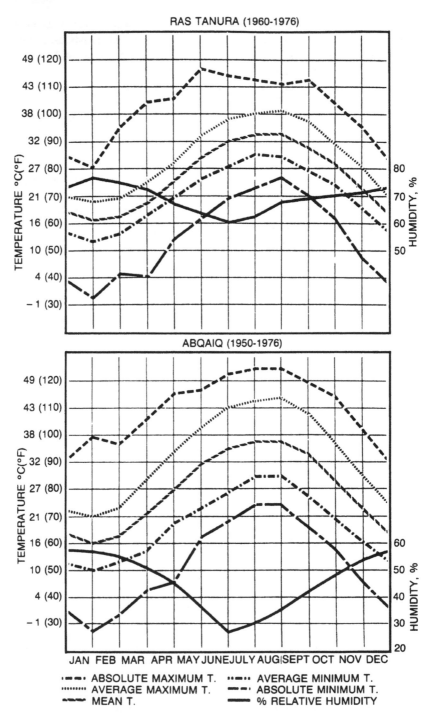

Fig. 3.3. Monthly averages of temperature and relative humidity, Ras Tanura (1960 – 1976) and Abqaiq (1950 – 1976). From Arabian American Oil Company (n.d.)

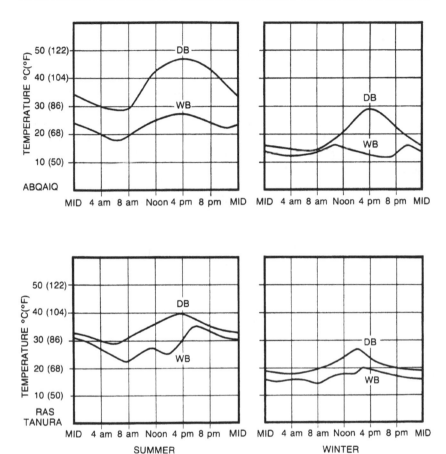

Fig. 3.4. Typical diurnal cycles of dry bulb (DB) and wet bulb (WB) average maximum temperatures for the summer season (May – September) and winter (November – March). Record period: 1951 – 1955. From Arabian American Oil Company (n.d.)

Atmospheric Humidity: As emphasized by Cloudsley-Thompson and Chadwick (1964), relative humidity figures alone bear less relation to transpiration and thus plant water use than parameters such as vapor pressure deficit. Relative humidity data is graphed with temperature in Fig. 3.3, however, for comparisons and as a basis for calculating other indicators. There is a strongly declining relative humidity gradient from the Gulf coast into the hinterland.

Temperature: Temperatures measured by standard means range from an absolute maximum of 52° C (Abqaiq in July) to approximately – 3° C inland in January. Frosts have been reported inland as far south as the central Rub' al-Khali. Points near the coast experience less extreme ranges. Very little data is available for the Rub' al-Khali, but unofficial maximums above 52° C have been reported. The annual temperature

15

graphs for Ras Tanura and Abqaiq (Fig. 3.3) are typical for coastal and inland stations, respectively. Fig. 3.4 shows typical diurnal cycles, summer and winter, of dry and wet bulb temperatures at these stations.

Evaporation: Evaporation rates, because of their close relationship to evapotranspiration and the availability of surface moisture for plant growth, are of particular interest to the field biologist. Most of the data for our area is in the form of open-surface evaporation measurements by the American Class-A Pan method. Some of these are summarized in Table 3.2, which shows rates ranging from 35 up to 100 times the local mean annual rainfall. Direct measurements are not available for our probably most extreme conditions, in the central and southern Rub' al-Khali. The very high rates measured on the western borders of this region at al-Aflaj and as-Sulayyil, however, suggest the values to be expected.

Table 3.2 *Mean Annual Open-Surface Evaporation (Class-A Pan), South-central to Eastern Saudi Arabia*

Station	Evaporation (mm)	Years Record
al-Hufuf	2660	4
al-Qatif	2960	4
al-Kharj	3070	5
Harad	3451	1
al-Aflaj	4130	6
as-Sulayyil	5250	9

Data from Saudi Arabian Ministry of Agriculture and Water Resources, *Hydrological Publication* (various issues).

Winds: The mean wind velocities measured in eastern Arabia are not great by world standards, but wind effects in such an open, loose-soiled desert environment can often prove decisive to plant survival. Winds significantly increase the dessicating power of the already hot dry atmosphere, have a powerful effect in molding topography, particularly in dune sands, and directly affect the root stability of individual plants. The abrasive effects of sand-laden winds on dune plants have probably not been fully appreciated.

There is a diurnal tendency for wind velocities to increase around midday, probably as a result of differential warming. The hottest land masses heat overlying air, causing thermals that temporarily reverse the general pattern of subtropical air subsidence and cause a ground level suction effect on outlying regions (Wallén in Hills 1966).

More spectacular are the results, even far inland, of a strong seasonal pressure pattern over the Gulf. The directional pattern of the basically northeast Trades are strongly distorted beginning about May, as a trough of low pressure moves up the Gulf as an extension of the great seasonal low over the Asian land mass to the east. This leads to the well-known *'shamāl'* winds of early to middle summer which may blow for days, sometimes gusting 30 to 65 km/hr from the north-northwest along the isobars of the Gulf low. From north to south there is a progressive directional shift in this wind from northwest to north and finally northeast — all apparent in the longitudinal dune alignments of the Dahna and the Rub' al-Khali.

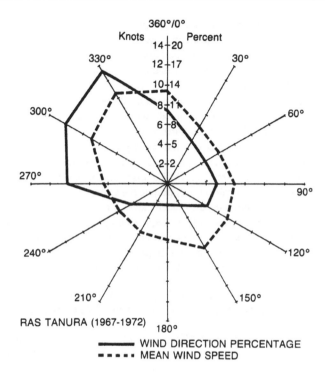

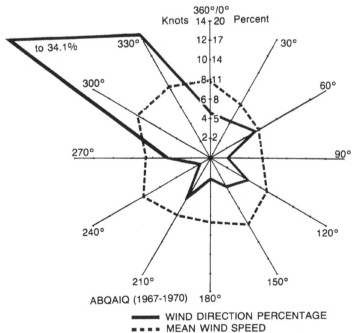

Fig. 3.5. Mean wind speed and direction, Ras Tanura and Abqaiq. From Arabian American Oil Company (n.d.)

The *shamāl* (a term often used somewhat inaccurately by Westerners to refer to the dust storms which may or may not accompany it) is always quite dry. Its coincidence with rising early summer temperatures and longer days probably makes June and July the time of greatest moisture stress for plant life. August, although a hot month, is the calmest, with somewhat rising relative humidity.

Overall, coastal winds are stronger than those inland, averaging 15 – 18 km/hr predominantly from the northwest in fall and winter. Wind speeds over the coast average 14 – 17 km/hr, mainly from the north-northwest during spring and summer. Inland stations show north to northeast winds averaging 9 – 14 km/hr during fall and winter. These means are reduced to 8 – 12 km/hr, mainly from the north-northeast in spring and summer (Arabian American Oil Company n.d.). Mean winter wind speeds thus exceed those of the '*shamāl* season', despite the stronger subjective impression created by those longer-lasting, often dust-laden blows. True sand storms, where sand is raised a meter or more above ground level, require exceptional wind velocities and are rare. The great bulk of sand movement takes place within 50 cm of the surface.

Wind diagrams for Ras Tanura and Abqaiq (Fig. 3.5) show patterns typical for coastal and inland locations in the central parts of the Eastern Province.

4. EASTERN SAUDI ARABIA IN THE PLANT-GEOGRAPHICAL REGIONS OF THE MIDDLE EAST

Floristic Regions of the Middle East: Plant geographers often divide the Middle East into four major floristic regions, each sometimes including several subunits. One long-followed approach, depicted in Map 4.1, divides the area among primary regions known as 'Mediterranean, Irano-Turanian, Saharo-Sindian, and Sudano-Deccanian'. Such schemes imply a certain degree of historical affinity among the plants within each area at the species, genus and sometimes family level, the general supposition being that they share a common geographical area of origin or diversity, often reinforced by present-day patterns of endemism. Considerations of regional climate, topography and vegetation physignomy have also played a significant role in such boundary setting by some authorities. The floristic composition of these regions is seldom neatly exclusive, and assignments are generally based on a preponderance of characteristic species, or at least dominants. The Saharo-Arabian (or Saharo-Sindian) Region perhaps goes farther than the others in including neighboring elements and in having a relatively low proportion of endemics.

The 'Jerusalem School' of plant geography, led first by Alexander Eig and later by Michael Zohary and with its strong emphasis on strict floristic analysis, has strongly influenced many accounts of plant geography in the Middle East. Map 4.1 is based largely on the work of Eig and on Zohary's earlier writings. Zohary later renamed the 'Sudanian-Deccanian Region' the 'Sudanian Region' and extended it up the Arabian Gulf as a narrow coastal strip as far north as Kuwait. The Saharo-Sindian Region, in keeping with his new restriction of the zone to the Arabian Peninsula in the east, became the 'Saharo-Arabian' (Zohary 1973).

The Sudanian — Saharo-Arabian Frontier in Arabia: Eastern Saudi Arabia, historically, has generally been placed entirely in Saharo-Arabian territory or later, in Zohary's maps, given a coastal zone in the Sudanian Region. A major difficulty facing all of these earlier analysts was the extreme paucity of floristic data on which to make such boundary decisions.

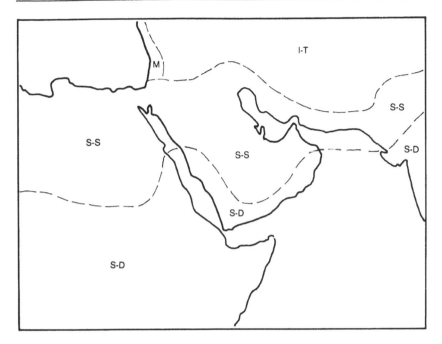

Map 4.1. Conventional floristic regions of the Middle East after Guest (in Townsend and Guest 1966), a scheme based largely on the work of Eig and the earlier accounts of Zohary. M = Mediterranean, I-T = Irano-Turanian, S-S = Saharo-Sindian, S-D = Sudano-Deccanian.

The present author, following Zohary's general approach and terminology but considering newly gained information about the flora of these parts, proposed (1984) a realignment of the frontier between Sudanian and Saharo-Arabian territory in Arabia. This, which would significantly increase the Sudanian area of the Peninsula, is based largely on a consideration of the *Acacia*-dominated plant associations of central Arabia. The Rub' al-Khali, although strongly atypical with its extreme, hyper-arid habitats, is treated as a Sudanian variant. This placement of the boundary, which must be considered provisional pending more detailed analysis of the plant communities concerned, is shown schematically in Map 4.2. In the east the frontier between Saharo-Arabian and Sudanian territory is imprecise, but the line there is marked by scattered occurrences of *Acacia*, along with other associates of southern provenance. The transition is interrupted by the sands of the Rub' al-Khali and perhaps influenced by the persistence, described in the following chapter, of relict Sudanian elements in the Wadi as-Sahba entrant. The frontier is also marked by a strong decline in the abundance and diversity of the annual plants so characteristic of the Saharo-Arabian deserts to the north (chapter 6).

The area of our east Saudi Arabian flora richest in species can thus be assigned to the Saharo-Arabian Floristic Region, while the southern part lies in Sudanian territory — the transition being poorly defined and masked by hyper-arid conditions in the Rub' al-Khali. The following chapter is intended to provide some insights into the historical development of the east Arabian vegetation with particular reference to long-term geological and climatic factors.

19

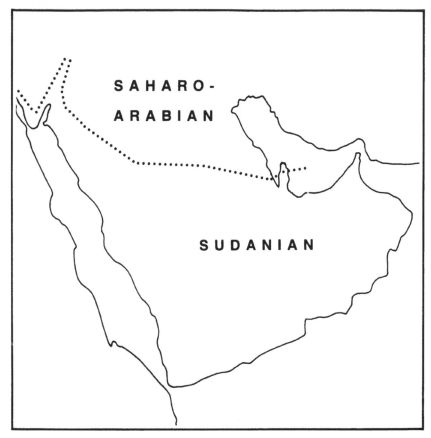

Map 4.2. The Saharo-Arabian — Sudanian floristic frontier in the Arabian Peninula, as proposed schematically by the author (1984). Recent floristic data indicates that much of central Arabia should be assigned to the Sudanian Region *sensu* Zohary.

5. PALEOENVIRONMENTS AND DEVELOPMENT OF THE EAST ARABIAN FLORA

East Arabian Plant Life in the Earlier Geological Eras: Rocks of the Paleozoic and Mesozoic eras lie far below today's land surface in eastern Arabia, but fossil pollen and spores from drill cores provide relatively good evidence of the flora of these early times. Fossil spores and cell sheets in borehole samples from the Ordovician of North Africa have provided perhaps the earliest known evidence to date of vascular land plants (Gray, Massa, and Bout, 1982). Similar material now under study from Arabia appears to contain similar evidence of such extremely early terrestrial flora.

In this, and succeeding, periods through the Mesozoic, eastern Arabia shared in the general sequences of plants — as shown by fossil spore remains — known from many other parts of the world. During those times the continents lay closer together than they do today, and their climates were less diverse. It was this greater global uniformity in environmental conditions that has made palynology such a useful tool in dating and correlating the early Arabian sedimentary rock units — an important procedure in oil exploration.

After this first suggestion of east Arabian vascular plant life in the Ordovician period, some 465 million years ago, the spores of definite land plants appear in lower Silurian deposits. Remains of typical Devonian, Carboniferous, and early Permian plants are found above, following an unconformity.

Recently, for the first time, a determinable assemblage of plant macrofossils has been described from the Paleozoic of central Arabia (El-Khayal, Chaloner and Hill 1980), and this provides evidence of plant formations that certainly extended to the east. In this collection from near 'Unayzah in northern Najd, the most abundant plants are the fern-like genus *Pectopteris*, the early seed plant *Cordaites*, and the arthrophyte *Annularia*. The group, according to the authors, is of late Carboniferous to early Permian age, and its affinities appear to lie clearly with the flora of the northern, Laurasian Province rather than that of the southern proto-continent, Gondwanaland.

Fossil spores reported from the Triassic in eastern Arabia are much like those of contemporary deposits in the rest of the world, while the succeeding Jurassic was for the greater part marine. Middle Cretaceous plant remains are again world-typical, but apparently nearest to Australian taxa. The first flowering plants appeared here in the upper Cretaceous, about 70 million years ago (H. A. McClure, personal communication).

Tertiary Environments and Development of the Modern Flora: Hints of today's east Arabian flora began to appear in Tertiary times, although the fossil record is scanty. Eastern Arabia generally lay above sea level after the withdrawal of the Tethys sea in the early Eocene, although it experienced marine incursions of various, lesser extent during Miocene and Pliocene times. The African-Arabian Shield was as yet undivided by the Red Sea rift, and a paleo-African vegetation presumably extended eastward into the future Arabian Peninsula throughout the middle-late Eocene and Oligocene. Erosion has left scant record of these periods east of the Dahna, but this vegetation certainly persisted in the western highlands as a precursor of today's Sudanian vegetation there. It is only in Miocene and Pliocene deposits that we begin to find direct evidence of Tertiary plant life in eastern Arabia.

Many plant families prominent in today's vegetation, including the Gramineae, Cyperaceae, Compositae, and Caryophyllaceae, have been demonstrated from fossil pollen of Miocene horizons. Families of tropical distribution locally rare or absent today, such as the Meliaceae, Sapotaceae, Combretaceae, Myrtaceae, Ceratopteridaceae, and Palmae, were also represented.

Fossil pollen samples of the eastern Rub' al-Khali in the vicinity of az-Zumul, 150 – 200 m below the present surface, were found to be dominated by the unclassified form genus, *Psilatricolporites* with associates of the Myrtaceae, Palmae and the water fern, *Ceratopteris*. Also present were Jussaceae, Meliaceae/Sapotaceae, and *Echitricolporites*. The age is classed, on the presence of tubiflorate composite pollen, as middle Miocene at oldest. The determinable portion was reported to be 73 percent fern and moss spores, among which *Ceratopteris* was significant, 13 percent grass pollen, and 14 percent pollen of other non-forest taxa (Arabian American Oil Company, unpublished). The environment of deposition would appear to have been fresh water marshland in a humid tropical to subtropical climate. *Ceratopteris* is found today in the freshwater springs of the al-Hasa and Qatif oases.

21

Whybrow and McClure (1980) report fossil mangroves of early to middle Miocene age collected near the edge of an apparent marine incursion inland at Dawmat al-'Awdah and in an upper to late Miocene horizon on the eastern Gulf coast at Jabal Barakah. These, they suggest, indicate a coastal environment of broad alluvial flood plains, with local swamps fed by palm-fringed streams and with mangroves in the intertidal zone. Conditions were apparently dry, but not highly arid, with open savannah grasslands and vegetation-bordered streams that flowed at least seasonally.

Finds of vertebrate fossils also provide good evidence for prevailing vegetation types, and a rich early-middle Miocene assemblage from the edge of the northern Summan at ad-Dabtiyah has been described by Hamilton, Whybrow and McClure (1978). Remains here of turtle, fish, crocodile, rodents, hyrax, rhinoceros, pig, giraffe and bovids provide further evidence for the existence of a late Tertiary tropical savannah vegetation in eastern Arabia, here in a coastal environment.

Inland spreading of the sea during the middle Miocene and a subsequent development of highly arid conditions were responsible for the destruction of this tropical vegetation in zones within about 100 kilometers of the present Gulf coast. The near absence today of Sudanian plant relics in many regions east of the Dahna that were continually emergent can probably be attributed to changes in both climate and edaphic conditions. *Acacia* is now mostly restricted to alluvial soils with fairly fresh water available in its root zone, conditions generally now found only in the larger wadis and basins.

Several of the larger trans-Tuwayq wadis, in late Pliocene or early Pleistocene times, apparently provided avenues for a reintroduction of Sudanian vegetation into the eastern lowlands from its reservoir in the mountains of western Arabia. A highly humid period between 3.5 million and 1.2 million years ago (Hötzl et al. 1978; Hötzl, Krämer and Maurin 1978) saw the great channels of al-Batin in the north, Wadi as-Sahba in the middle, and Wadi ad-Dawasir in the south in energetic flow, carrying huge masses of erosional materials from the western mountains to be spread as cobbles and gravels across the eastern plains. These wadis seem to have provided suitable routes for a limited reinvasion of Sudanian vegetation in the east, and relicts of these intrusions may still be seen: *Acacia gerrardii* and associates such as *Cleome* in al-Batin; and *Acacia ehrenbergiana*, *A. raddiana* and *A. tortilis*, with *Capparis*, *Blepharis*, and *Cleome* in the main channel and extensions of Wadi as-Sahba. The lower reaches of the Wadi ad-Dawasir drainage are now covered by the sands of the Rub' al-Khali, but suggestions of its early flows, too, can perhaps be seen in the unique isolated stand of *Suaeda monoica* at Jawb al-'Asal in the northern Rub' al-Khali. That species is otherwise now unknown in the east but is still found in the mouth of Wadi ad-Dawasir's cut through Tuwayq.

Climate and Topographic Change in the Quaternary: Eastern Arabia, after the wadi-cutting pluvial episode described above, seems to have seen the onset of increasing aridity. The resulting evaporitic sediments must have been favorable for the establishment of the many Chenopodiaceous genera present today, and hotter and dryer conditions exerted strong selective pressures leading to development of the Saharo-Arabian desert flora.

Periods of low sea level in Pleistocene times led to drastic changes in the configuration of the Gulf coastlands and associated littoral vegetation. The most recent of these, between about 20,000 and 15,000 years ago, took the sea more than 100 m below its present level and left much of the Gulf virtually dry land. Such conditions probably brought riverine vegetation down into the Middle Gulf region from lower Mesopotamia. The sea rose rapidly between 12,000 and 8,000 years ago, reaching its present level by about 5,000 B.P. (Vita-Finzi 1978).

Arabia seems to have had an arid or semi-arid climate during Pleistocene and subsequent times, but short periods of increased precipitation had marked effects on local plant life. McClure (1976, 1978) describes clear evidence for two series of freshwater lakes in the central and western Rub' al-Khali that existed, as indicated by radiocarbon dates, about 35,000 – 17,000 years B.P. and 9,000 – 6,000 years B.P. Mammal fossils associated with these sites included *Bos* and *Hippopotamus* and indicated the presence of rich aquatic vegetation and probable surrounding savannah grassland, almost certainly of Sudanian composition. Strong desert conditions appear to have set in at the end of the second lake period and to have persisted across eastern Arabia up to the present.

6. THE VEGETATION: PHYSIOGNOMY AND PLANT COMMUNITIES

Life Form Spectrum of the Vegetation: Several desert-experienced authorities, including Zohary (1962) and Kassas (1966), have emphasized the limited usefulness of Raunkiaer's 'life form' classes as a basis for vegetation description or analysis in arid regions. The difficulty stems in part from the morphological plasticity exhibited by desert plants as an adaptive strategy and from the apparently equal adaptive validity of various resting bud positions in desert perennials. Raunkiaer's system has been applied widely, however, and a life form summary of our flora is perhaps not entirely without value for comparative purposes. It is also of some interest in regard to plant-geographical questions.

A life form analysis for all desert native plants in eastern Saudi Arabia, along with that for Rub' al-Khali species alone, is charted in Fig. 6.1. The high overall proportion of therophytes (annual plants) and chamaephytes (dwarf shrubs) is typical of hot deserts and is also seen in North Africa and similar arid regions. The picture in eastern Arabia varies rather markedly among the various geographical subregions. There is a strong overall decline in the therophyte element from north to south, as demonstrated in Fig. 6.2, and this is paralleled by a similar trend of decreasing total species diversity. This is particularly evident in the South Coastal Lowlands and the Rub' al-Khali, where aridity increases and Sudanian borderlands are entered (chapter 4).

The spectrum of the Rub' al-Khali, marked by the virtual absence of the Saharo-Arabian and Mediterranean-derived annuals so common in the north, is thus quite distinct. Except in its lower count of hemicryptophytes it is very similar to the life form enumeration provided by Shimwell (1971) after Raunkiaer, for Aden, at the southwestern corner of the Peninsula. The Rub' al-Khali flora is unusual also in the very low number of species represented, hardly over 30, compared to the 390-odd desert natives found in Eastern Saudi Arabia as a whole — this despite the relatively immense area covered by the southern sands, over twice that of our other subregions combined.

The Annual Cycle of Plant Growth: Throughout eastern Arabia, as in other hot deserts, summer is the unfavorable season for plant growth, leading to dormancy or the evasion of drought by the production of seeds. The growth cycle for many desert plants thus begins in the autumn or winter, rather than in spring as in temperate regions.

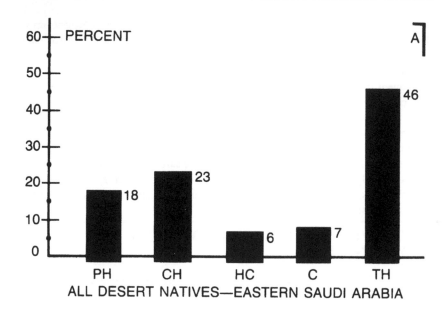

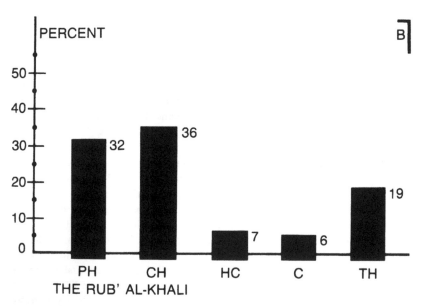

Fig. 6.1. Raunkiaer life form spectra for all desert native species, eastern Saudi Arabia (A), and the flora of the Rub' al-Khali (B). PH – phanerophytes, with perennating buds over 25 cm high; CH – chamaephytes, with buds above ground level and up to 25 cm high; HC – hemicryptophytes with buds at ground level; C – cryptophytes, with perennating organs below ground; TH – therophytes, annual plants. Cryptophytes include marine and fresh water hydrophytes as well as geophytes.

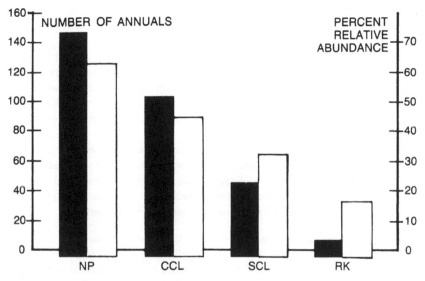

Fig. 6.2. Absolute number (solid bars) and percent relative abundance (open bars) of annual plant species from north to south (left to right) in geographic subregions of eastern Saudi Arabia. NP – Northern Plains; CCL – Central Coastal Lowlands; SCL – South Coastal Lowlands; RK – the Rub' al-Khali. Map 2.2 delimits the geographical areas referred to.

Some perennial plants in our territory show a resumption of active growth as early as September, well before arrival of the first rains. This may be associated with shortening day length, moderating temperatures, or the increase in atmospheric humidity which occurs at this season in coastal areas. The pace quickens, however, with arrival of the first 'Mediterranean depressions' of the season — usually in November or December but occasionally in October. Such rains, if early, may lead to a flush of germinating annuals within a few weeks. If they come in December or January, cold weather may delay germination, or growth much beyond that stage, until the warming of spring. Along with the germination of annuals comes a resumption of new shoot and leaf production by the perennials, many of which have shed leaves or died back completely during the summer drought.

Desert annuals may flower and set seed within a few weeks after germination, completing their life cycle in a fraction of the usual growing season of temperate lands. Some species have apparently evolved chemical germination inhibitors or other moisture detection devices in their seeds to protect them from the danger of germination before soil moisture levels allow growth to maturity (Cloudsley-Thompson and Chadwick 1964). If germination does occur under moisture stress, some plants conserve growth energy by flowering and fruiting in a dwarf stage hardly as large as early seedlings under more normal conditions. In a converse manifestation of such 'form plasticity', growth may be prolonged with the plants achieving unusually great stature if rains are repeated and well spaced. Geographic areas of growth and reproduction of both annuals and perennials may be extremely patchy when rains come from small local storms.

The annuals are often at the peak of maturity in March or early April, although some species are dependable early or late bloomers. Perennials, including some grasses, usually

25

peak somewhat later. By late April or May, depending on local conditions, the annuals wither and are gone for the season; many perennials maintain active growth into May or June before falling again into dormancy.

Some deeply rooted woody plants of Sudanian derivation, such as *Acacia*, *Capparis* and *Ziziphus*, flower and fruit in June or July or even later. A few smaller perennials such as *Zygophyllum* spp. and *Heliotropium bacciferum*, while peaking in late spring or early summer, are able to maintain some degree of active growth through even the hottest part of the dry season. They are in fact rarely seen not bearing at least a few flowers or fruits.

A very clear exception to these patterns is found in the saltbushes of the family Chenopodiaceae. These are in active vegetative growth during the hot season and flower and fruit in October and November. Some of them flower in both spring and autumn. One is tempted to see in their pronounced winter die-back an ancestral adaptation to harsh winter conditions in an Irano-Turanian homeland to the north.

General Characteristics of the Desert Plant Communities: Many of the more important desert plant communities — clearly defined types that cover thousands of square kilometers of eastern Arabia — are characterized by woody dominants of a single species. These are open shrublands with diffuse perennial cover usually totalling well less than 15 percent, the inter-shrub space bare for much of the year but occupied for a few weeks or months by a ground layer of rain-season ephemerals (Plate 2). The total number of species contributing significantly to desert vascular plant biomass production is relatively small. Table 6.1 lists 50 plants assessed subjectively by the author to be most important by virtue of their abundance and contribution to biomass. These in general are also the species of greatest economic importance as grazing resources.

Table 6.1 *Principal Contributors to Biomass*

Dominant Perennials	Lasiurus scindicus	Trigonella stellata
	Dipterygium glaucum	Trigonella hamosa
Haloxylon salicornicum	Anvillea garcinii	Launaea capitata
Rhanterium epapposum	Zilla spinosa	Launaea mucronata
Calligonum comosum		Cutandia memphitica
Calligonum crinitum	**Annuals**	Savignya parviflora
Cornulaca arabica		Arnebia linearifolia
Haloxylon persicum	Plantago boissieri	Brassica tournefortii
Anabasis lachnantha	Plantago ovata	Emex spinosa
Zygophyllum qatarense	Astragalus tribuloides	Horwoodia dicksoniae
Zygophyllum mandavillei	Picris babylonica	Erodium laciniatum
Seidlitzia rosmarinus	Stipa capensis	Senecio glaucus
	Eremobium aegyptiacum	Hippocrepis bicontorta
Abundant Perennials	Schismus barbatus	Lotus halophilus
	Neurada procumbens	Reichardia tingitana
Cyperus conglomeratus	Schimpera arabica	Aaronsohnia factorovskyi
Stipagrostis drarii	Ifloga spicata	Ononis serrata
Stipagrostis plumosa	Astragalus hauarensis	Asphodelus tenuifolius
Moltkiopsis ciliata	Medicago laciniata	Astragalus annularis
Pennisetum divisum		

The dominant shrubs condition the habitat for annual plants, primarily by their soil-binding effects. Annuals sometimes obviously congregate around and under woody plants that catch wind-driven seeds, collect rain, and give wind protection. Much additional field work will be required to determine the degree to which these seasonal annual components are specific to different shrub associations. Certainly many of them are common to clearly different communities, and some evidence suggests they are functioning as 'opportunists' without close community integration. Broad statistical studies will be needed to assay the significance of some apparent differences that do exist in annual layer composition. Much work also remains to be done in determining the edaphic or other factors responsible for the distribution of the dominants themselves. Their boundaries are sometimes correlated with obvious soil, drainage and moisture factors; there remain, however, abrupt transitions without apparent explanation.

Most of these communities appear to be in close equilibrium with their natural environment, and the processes of plant succession, as emphasized in studies of more mesic vegetation, seem to play a minor role on undisturbed sites. Delicately balanced desert communities, however, are manifestly sensitive to interference by man, which is having powerfully deleterious effects in some areas. Clear examples are the destruction of coastal salt marsh habitats in Tarut Bay through artificial land filling, the clearing of shrubland around Dhahran and Dammam, and limestone quarry operations that eradicate vegetation and leave wide areas exposed to wind erosion and in active dust production. The effects of grazing and firewood gathering by Bedouins, except around permanently settled centers, appear to be minor in comparison. The increased mobility of today's Bedouins, who now move flocks and carry water by truck, may well be resulting in a greater dispersion of grazing pressure and less range impairment than in earlier years.

Overall the vegetation is best developed and probably most productive in the boundary region between the Central Coastal Lowlands and the Northern Plains, where favorable soil conditions and higher precipitation occur together.

Important Plant Communities of Eastern Saudi Arabia: Virtually all of the few early published descriptions of plant communities in Saudi Arabia relied heavily on Vesey-Fitzgerald (1957), who summarized some of the broader vegetation groups as observed during anti-locust campaign field surveys. Detailed descriptions have been published more recently of community types in neighboring Iraq (Guest 1966, Thalen 1979) and Kuwait (Halwagy 1986; Halwagy and Halwagy 1974a, 1974b, 1977; Halwagy, Moustafa and Kamel 1982), some of which correlate closely with Saudi Arabian associations and indicate extensions of regional importance.

The following brief descriptions of some of the more significant east Arabian plant communities are based on the author's field experience but not on any formal attempt at community analysis or classification. The use of plant sociological nomenclature is thus not to imply any ordering by specialized methodology such as that of the Braun-Blanquet school. Quantitative data are derived from small samples and are of reconnaissance accuracy.

Haloxyletum salicornici (*Rimth* Saltbush Shrubland): This open shrubland dominated by the *rimth* saltbush, *Haloxylon salicornicum*, probably covers more land than any other community in northeastern Arabia. It ranges from Iraq in the northeast down into the northern edge of the Rub' al-Khali. Its distribution appears to be conditioned by drainage conditions; in the central coastal lowlands it is generally found on sands in wide topographic lows with ground water not far below the surface. It is best developed on deeper sand, where its stature increases and spacing is closer, but it also occurs on gravel plains and even extensive rock and silt surfaces of the Summan, where it is stunted and wide-spaced. In such habitats it is also mixed with or grades into other saltbush dominants such as *Anabasis lachnantha* and *Agathophora alopecuroides*. As pointed out by Halwagy, Moustafa and Kamel (1982), *Haloxylon* should probably not be classed as a clear halophyte; it is not generally found on the highly saline *sabkhah* margins frequented by *Seidlitzia* and *Suaeda*. It often does, however, dominate the regional community surrounding such salt-rich lows at only marginally higher elevations.

Rimth is generally considered to be of somewhat limited value for livestock grazing, and this has been attributed to its salt content although the present writer is not aware of any mineral content measurement data. Its high biomass production in summer and fall, however, makes it useful where plentiful water is available for stock. In the rainy season, annuals of this community are an important grazing resource. *Rimth* provides good firewood, much better than *Rhanterium*, and it is often carried some distance by Bedouins while camping in *'arfaj* country.

In February 1970 the writer studied a stand of Haloxyletum 18 kilometers southwest of as-Saffaniyah with the aim of assembling some quantitative data on shrub density, cover, and on associated annuals. The results are tabulated in Table 6.2, where cover, as in all following data, refers to projected crown area. This near-coastal site may be somewhat atypical in its slight admixture of *Panicum* and *Lycium* from a neighboring Panicetum. The ground layer was very evenly spaced.

Table 6.2 *Survey Data — Haloxyletum salicornici*

18 km SW as-Saffaniyah; 27°53.4'N, 48°37.8'E

Shrub Layer: Sampled by random pairs method
Total shrub density: 772/Ha Total shrub cover: 8.6%

Species	Rel. Dens.	% Total Shrub cover
Haloxylon salicornicum	.90	95
Panicum turgidum	.08	3
Lycium shawii	.02	2

Annual Layer: Sampled by semi-random quadrats (0.1 m^2) along a line transect

Species	Freq.	Density/m^2	Species	Freq.	Density/m^2
Plantago boissieri	.92	145	*Ononis serrata*	.12	2
Schismus barbatus	.84	65	*Ogastemma pusillum*	.08	4
Astragalus spp.*	.64	14	*Picris babylonica*	.08	2
Ifloga spicata	.60	50	*Rostraria pumila*	.08	1
Lotus halophilus	.44	8	*Hippocrepis bicontorta*	.08	1
Paronychia arabica	.32	6	*Atractylis carduus*	.04	0.4
Medicago laciniata	.32	5	*Launaea mucronata*	.04	0.4
Launaea capitata	.28	4	*Euphorbia granulata*	.04	0.4
Crucianella membranacea	.28	4	*Stipa capensis*	.04	0.4
Cutandia memphitica	.16	2			

Astragalus species were too young to be differentiated but appeared to include *A. annularis, A. tribuloides,* and *A. hauarensis.*

Rhanterietum epapposi (*'Arfaj* **Shrubland,** Plate 203): This open shrubland dominated by *Rhanterium epapposum*, or *'arfaj*, is best developed in the northern plains region and part of the northern Summan centered around Qaryat al-'Ulya. It favors better drained soil on higher ground, where bedrock is not far below the surface. This very important northern community, the 'Rhanterium Steppe' of Vesey-Fitzgerald (1957), is also found in Iraq and Kuwait and reaches south to about the 23rd Parallel, where it may still be seen along the eastern edge of the Dahna sands. The community is of great importance as a winter, spring and early summer grazing resource. Survey data for a typical stand 17 kilometers northeast of Qaryat al-'Ulya are given in Table 6.3.

Table 6.3 *Survey Data — Rhanterietum epapposi*

17 km NE Qaryat al-'Ulya; 27°36.7'N, 47°49.2'E

Shrub Layer: Sampled by random pairs method
Total shrub density: 3145/Ha **Total shrub cover:** 16.3%
Composition: *Rhanterium epapposum* (pure stand)

Annual Layer: Sampled by 100 semi-random, 0.25 m² quadrats along a line transect

Species	Freq.	Density/m²	Species	Freq.	Density/m²
Plantago boissieri	.73	22	*Launaea mucronata*	.05	0.2
Picris babylonica	.56	7	*Polycarpaea repens*	.04	0.2
Schismus barbatus	.49	12	*Horwoodia dicksoniae*	.03	0.2
Neurada procumbens	.48	4	*Gastrocotyle hispida*	.03	0.2
Erodium laciniatum	.16	1	*Emex spinosa*	.03	0.1
Asphodelus tenuifolius	.15	1	*Lotus halophilus*	.03	0.1
Medicago laciniata	.13	1	*Silene arabica*	.02	0.1
Plantago ciliata	.12	1	*Anisosciadium lanatum*	.02	0.1
Rostraria pumila	.10	2	*Ifloga spicata*	.02	0.1
Astragalus spp.*	.10	1	*Reseda arabica*	.02	0.1
Trigonella stellata	.08	2	*Stipa capensis*	.02	0.1
Paronychia arabica	.07	0.5	*Loeflingia hispanica*	.01	0.1
Schimpera arabica	.07	0.3	*Hypecoum pendulum*	.01	0.04
Plantago ovata	.05	1	*Leontodon laciniatus?*	.01	0.04

*Young *Astragalus* spp. not differentiated; include *A. tribuloides, A. annularis,* et al.

The greater annual species diversity of this stand, as compared to the Haloxyletum above, is evident. Both carry *Plantago boissieri* as the most frequent and most abundant annual, but *Picris babylonica* and *Neurada procumbens* are important in this example of the Rhanterietum while *Ifloga spicata* and *Lotus halophilus* are characteristic of the Haloxyletum. Single stand data for these communities are of course insufficient for generalizations about fidelity and differential species.

Panicetum turgidi (*Thumām* Grass-shrubland): This community, led by the perennial tussock grass, *Panicum turgidum,* or *thumām,* is widely distributed in the central coastal lowlands and is a characteristic component of Vesey-Fitzgerald's (1957) 'coastal white sand association'. It is also found in limited areas of Kuwait. This is a very important grazing association, and *Panicum* shrublets often are seen cropped by livestock, though rarely to the point of destruction. *Panicum* is practically useless as fuel. The Panicetum appears to require a relatively well-drained sand substratum, although it is often seen on elevated ground near saline terrain. Woody associates include *Lycium shawii,*

29

Leptadenia pyrotechnica and, most frequently, *Calligonum comosum.* In some areas *Calligonum* becomes codominant with *Panicum,* forming a distinctive subassociation. This is best developed in the district of al-Habl, immediately southeast of al-Fadili. It also occurs in some parts nearer the coast.

The data in Table 6.4 are from an extensive, nearly pure stand 25 kilometers south-southwest of al-Fadili, well west of the area where *Calligonum* is a very significant component. The very limited (10-point, 40-distance measurements) sample was by the quarters method.

Table 6.4 *Survey Data (Shrub Layer Only) — Panicetum turgidi*

25 km SSW al-Fadili; 26°46.0'N, 49°05.7'E

Total shrub density: 1515/Ha Total shrub cover: 3.5% **Components:**

Species	Rel. Dens.	% Total Cover
Panicum turgidum	.95	83
Calligonum comosum	.05	17

A stand sampled near Dhahran was somewhat more diverse, with the following relative densities of perennials: *Panicum turgidum* .78, *Cyperus conglomeratus* .16, *Pennisetum divisum* .04, and *Fagonia indica* .02.

Yet another variant of this community, sampled near al-Ju'aymah, was richer in shrubby associates and was notable by the presence of large conspicuous shrubs of *Leptadenia pyrotechnica.* Data for this site, from a small (30-pairs) random pairs sample, is given in Table 6.5.

Table 6.5 *Survey Data (Perennials) — Panicetum turgidi*

Near al-Ju'aymah; 26°47.1'N, 49°54.5'E

Total shrub density: 6250/Ha Total shrub cover: 5% **Components:**

Species	Rel. Dens.	% Total Cover	Species	Rel. Dens.	% Total Cover
Panicum turgidum	.58	65	*Salsola baryosma*	.02	9
Cyperus conglomeratus	.20	4	*Pennisetum divisum*	.02	1
Zygophyllum qatarense	.16	12	*Leptadenia pyrotechnica*	<.01	4
Calligonum comosum	.02	5			

Other perennial associates here, of negligible cover importance, were: *Heliotropium bacciferum, Moltkiopsis ciliata, Polycarpaea repens, Monsonia nivea, Rhanterium epapposum, Stipagrostis plumosa, Lycium shawii, Convolvulus cephalopodus,* and *Centropodia forsskalii.* Standing remains of the previous season's annual plants suggested that the most abundant annual species may have been *Plantago boissieri,* ranging in density between an estimated 25 to 40 individuals per square meter. Also frequent were *Launaea mucronata* and *Arnebia hispidissima,* while one specimen of *Allium sphaerocephalum* was seen. *Silene villosa* had been noted in this stand on other occasions.

This site was on gently rolling sand, about 2-3 m deep, on a *sabkhah* sub-base. At edges of *sabkhah* exposures a halophyte zone included *Cressa cretica, Sporobolus ioclados,* and *Zygophyllum qatarense.* Wild or feral date palms were widely scattered over the entire area except in the *sabkhah* itself.

Calligonetum comosi-Artemisietum monospermae (*'Abal-'Ādhir* Sand Shrubland): This is a distinctive community found on inland sands of the central-northern peninsula, including parts of the Great Nafud and most of the Dahna (Plate 6). Its diagnostic species is *Artemisia monosperma*, which is hardly found outside it. It is characterized by fairly wide-spaced but well developed shrubs of *Calligonum comosum* and *Artemisia monosperma*, with tussocks of the perennial grass, *Stipagrostis drarii*, the sedge *Cyperus conglomeratus*, and in the south, *Limeum arabicum*. *Scrophularia hypericifolia* is an important constituent in parts of the northern Dahna sands of our area, and this species appears to be confined to this community in eastern Arabia. A wide variety of annuals may also be found in this sand association, including the ubiquitous *Plantago boissieri* as well as *Plantago psammophila*, *Eremobium aegyptiacum*, *Cutandia memphitica*, *Linaria tenuis*, and *Anthemis scrobicularis*. The range of this community in the Dahna appears to be limited in the south approximately at 23 degrees north latitude, where it is replaced by the Cornulacetum arabicae (see below) of the Rub' al-Khali.

Ephedretum alatae (Ephedra Shrubland): Essentially pure stands of *Ephedra alata*, seldom of great areal extent, may occasionally be found in scattered locations from the south coastal lowlands to the north. The community may be associated with gypsaceous soils.

Achilletum fragrantissimae-Artemisietum sieberi (Achillea-Artemisia Silt Basin Association): This is a northern community type distinctive by its highly aromatic dominants, *Achillea fragrantissima* and *Artemisia sieberi*. It is much more common in the Hajarah and al-Widyan regions outside the Eastern Province to the northwest, but patches of it may be encountered on heavy silts and clay basins of our Northern Plains and Summan.

Haloxyletum persici (*Ghaḍā* Shrubland, Plate 4): This community, having as dominant the large shrub or semi-tree, *Haloxylon persicum*, or *ghaḍā*, is found in some deeper sand habitats of the southwestern central coastal lowlands, the south coastal lowlands, and the northern Rub' al-Khali. It is the largest of the saltbushes, often well exceeding man-height, and is prized as a desert firewood. Its characteristic sand terrain is less stable than that of *Haloxylon salicornicum* and is notable for its high hummocks and wind shadow drifts. This forms the terrain known to southern tribesmen as *qawz* (pl. *qīzān*), a topographic form quite specific to this distinctive plant community and probably conditioned by it.

    Ghaḍā shrubland often forms islands of small or relatively large extent in *rimth* country (Haloxyletum salicornici). In the Rub' al-Khali, it may border on *ḥādh* shrubland (Cornulacetum arabicae) — another saltbush association. *Ghaḍā* is normally found more widely spaced than the smaller saltbushes, but it contributes much greater cover per individual. Annual associates are relatively few in number in the more mobile sand terrain of this community. *Plantago boissieri*, *Eremobium aegyptiacum* and, at least in northern parts of its range, *Silene villosa* are characteristic.

Cornulacetum arabicae (*Ḥādh* Saltbush Shrubland, Plate 7): This community, like its dominant species, is endemic to the Rub' al-Khali and is one of that hyper-arid region's most important vegetation types. The leading shrub is *Cornulaca arabica*, or

31

ḥādh, spaced 3 to 20 meters or more depending on local conditions. *Cornulaca* may occur as virtually pure shrub stands covering thousands of square kilometers or may have as associates *Calligonum crinitum* (particularly on higher and better drained sands), *Tribulus arabicus,* or *Limeum arabicum. Cyperus conglomeratus* usually occurs with it. The community is limited on the north around the 23rd Parallel; it appears to extend south to near the limit of the sands.

Calligonetum criniti (Rub' al-Khali '*Abal* Shrublands): This is another wide-spaced shrub community of the Rub' al-Khali, where *Calligonum crinitum* subsp. *arabicum* takes the place of the *Calligonum comosum* of the northern sands. The dominant shrub is accompanied almost everywhere by *Cyperus conglomeratus* and often by *Stipagrostis drarii* and *Limeum arabicum.* In the far east, on the great massifs of the 'Sand Mountains', it is associated with *Tribulus arabicus,* which develops luxuriantly after the rare rains. The edges of sand masses there, around the inter-dune salt flats, are fringed with *Zygophyllum mandavillei.*

Succulent Halophyte Associations: A series of distinctive, often sharply zoned halophyte associations, mostly led by succulent chenopods, are found in and around coastal salt marshes and sometimes around inland *sabkhahs.* They range in composition from the mangroves *(Avicennia marina)* of the grey muds in protected Gulf bays through zones led by species such as *Bienertia cycloptera, Halocnemum strobilaceum, Suaeda vermiculata,* and *Seidlitzia rosmarinus.* Halwagy's (1977) description of this vegetation in Kuwait provides a very useful summary with much direct correspondence with units in eastern Saudi Arabia.

Shrubless Community Types: The existence of some wide tracts virtually without woody plants — apart from those on hyper-arid sand sheets, rock, and terrain with other obvious limits — pose an interpretive problem. Perhaps the most notable examples are the Qar'ah 'bald-lands' found in the northeastern parts of the Northern Plains. All woody plants are absent here, and vegetation is almost entirely restricted to a flush of annuals after winter rains. *Stipa capensis* often takes spring dominance here as the 'Stipa Steppe' of Vesey-Fitzgerald (1957), and common associates are *Plantago* spp., *Helianthemum* spp., and a variety of other annuals. Some authorities, such as Guest (Townsend and Guest 1966), believe such conditions represent the last degradation stages, resulting from overgrazing or other disturbance, of other communities. This may well be the case, but further study, particularly of soil conditions, must be undertaken before ruling out the possibility that it is a natural community type.

'Micro-communities': Highly specialized associations, often related to distinct terrain forms, may be found as 'islands' within some of the broad community types. A good example are the '*Ziziphus* basins' scattered through the sparse Haloxyletum of the Summan. Here, in rounded clay-floored basins of various size, often associated with rain pools, are well-developed stands of *Ziziphus nummularia* sheltering associates such as *Ephedra foliata* and *Sisymbrium erysimoides.* Silt-loving herbs like *Althaea ludwigii* and *Notoceras bicorne* occupy the open ground along with a variety of other annuals. The biological importance of these rich spot-associations is often far out of proportion to the limited ground area they occupy, and they are concentration centers for birds and other wildlife as well as the nomads' flocks.

Marine Angiosperm Communities: A marine community type dominated by flowering plants is found on the seabottom of near-shore Gulf waters in protected bays and to a lesser extent along the open coast. These are the so-called 'seagrass beds', important as a habitat for large numbers of marine organisms and characterized by a high rate of primary biomass production.

Four to six species of marine angiosperms are known to be present in Arabian Gulf coastal waters. One of these, *Ruppia maritima*, is generally found in very shallow — almost intertidal — beach waters. Others, such as *Halodule uninervis, Halophila stipulacea,* and *Halophila ovalis*, are found in deeper waters in more or less dense beds of considerable ecological and economic importance. These have been described in some detail by Basson, Burchard, Hardy and Price (1981), and the following summary is based largely on their account:

Halodule and *Halophila* are found on soft bottoms at depths of from 1 to about 20 m depending largely on local light conditions, the dense rhizomes often forming widespread, sediment-trapping mats. All of the species may be found together, but *Halodule* or *Halophila stipulacea* are usually dominant, sometimes apparently alternately by season. It is estimated that in some favorable areas, such as Tarut Bay, bottom coverage of these beds may be 60 – 70 percent. Spring and summer are the main growth periods, succeeding an apparent die-back and loss of leaves during the winter.

Some 500 species of marine animals have been identified from samples taken within this community type, which plays an important role in the life cycle of commercially valuable organisms including the pearl oyster and shrimp. Larger animals associated with it include the green sea turtle and the dugong, both of which are known to feed largely on the angiosperms themselves.

Basson, Burchard, Hardy and Price (1981) estimated that Tarut Bay alone had about 175 square kilometers of such angiosperm beds producing annually some 45,000 metric tons dry weight of biomass (considering leaves only) or an energy input to the ecosystem of 1.4×10^{11} kcal. They evaluate the community as 'one of the key biotopes in the western Arabian Gulf'.

PART II — THE FLORA

7. EXPLANATORY NOTES

The notes below summarize the general approach and conventions adopted in the systematic presentation of the flora.

Scope: Plants enumerated include all recorded non-cultivated vascular species occurring within the geographical area defined in chapter 2, essentially the Eastern Province of Saudi Arabia. A few species known from closely neighbouring areas — mostly weeds or ruderals — are included even though unconfirmed by specimens from within our limits. All such cases are so indicated in the text.

Sequence and Arrangement: The families of flowering plants are presented in the sequence listed by G. L. Stebbins (1974), which is based largely on the classification system of Cronquist. Although the arrangement of families in a manual-flora of this size is not of great practical importance, Stebbins' sequence reflects recent research findings and is thus chosen over more widely used earlier systems such as those of Bentham and Hooker, or Engler. It is also advantageous in being the basic scheme for two very useful family-level references: P. H. Davis and J. Cullen's *The Identification of Flowering Plant Families*, 2nd ed. (1979) and V. H. Heywood's (ed.) *Flowering Plants of the World* (1978). The definitions and scopes of the families are adopted as followed by these two works, which tend to maintain segregate families in their original larger units. The genera, while not strictly following any one system, are in traditional sequences grouping together supposed allies.

Nomenclature and Synonymy: References to the original publications of plant names were not considered necessary for the purposes of this manual. Full citations for the great majority of our species are available in several floras for neighbouring geographical regions. The provision of synonyms is selective, with the objective of including any names under which a plant is likely to be encountered in floras or other literature published since about 1950.

Descriptions: The aim in preparing species descriptions was to provide, without full and formal treatment, enough information to confirm identifications from the keys. Descriptions and dimensions are based for the greater part on the author's specimens. Reference was also made to standard published works, particularly the then-completed parts of the *Flora of Iraq* (Townsend and Guest 1966 – 1985) and *Flora Palaestina* (Zohary and Feinbrun-Dothan 1966 – 1986), for data on dimensional ranges and some anatomical points. A few descriptions are based largely or wholly on such published material and are then attributed to source. In general, more information is provided for plants known to be seldom or incompletely described in other floras.

The user should keep in mind that many desert plants, as adaptive strategy, may exhibit great variation in stature and some organ dimensions under stress of extreme local soil and moisture conditions. An attempt has been made in descriptions to consider this range when known, but exceptionally dwarfed or outsize plants may be encountered occasionally without such warning. Technical terms used in the descriptions are defined in the Glossary of Botanical Terms.

35

Habitat and Frequency: Habitat notes are generally self-explanatory. Notes on frequency and abundance are in subjective terms, based on the author's field experience. Unless otherwise qualified they refer to the overall Eastern Province area; a species listed as 'rare' or 'infrequent' may thus in some cases be rather common or abundant very locally.

Distribution Records: This is a list of specimens collected by the author within the area of coverage with some additional records, when published, by other collectors. All specimens are the author's unless otherwise noted, and numbers without herbarium designators are held at Dhahran.

The specimen records are arranged in the eight major subregions described in chapter 2. Inferences about full geographic range and abundance should be drawn from these lists with caution; in some cases overall range and frequency are not fairly indicated by the numbers of specimens cited.

Place names within the systematic text, except for a very few well known ones with conventional spellings, are presented in a simplified form of the transliteration system followed by the (U.S.) Board on Geographic Names (BGN) and the (British) Permanent Committee on Geographical Names (PCGN). Diacritical marks are omitted, with a simple apostrophe representing Arabic *'ayn,* and the definite article is hyphenated and not capitalized. The names otherwise correspond directly with the spellings used on the best generally available maps for the area (see Bibliographic Notes). Geographical coordinates in north latitude (N) and east longitude (E) are given with degrees separated from minutes by a dash. Brief descriptions and coordinates of all place names are provided in the Gazetteer, and spellings with full BGN/PCGN diacritical marks are given there for each toponym.

Vernacular Names: Unless otherwise indicated, all vernacular names were collected by the author from Bedouin oral sources. The tribal affiliations of informants are provided in parentheses after each name. The abbreviation 'gen.' indicates a name known to be in wide general use by several or more tribes. The absence of this designator, however, should not be assumed to indicate that a name is used only by the tribe or tribes listed. Where the meaning of a name is obvious or obtained from informants, it is provided in loose translation, generally in a form that parallels English plant names. Some names are of obscure derivation, and dictionary speculation is generally better avoided.

Arabic plant names are transliterated in the BGN/PCGN system, including full diacritical marks but with the definite article hyphenated. Names cited from other sources are converted to this system and thus may differ in spelling from their original forms.

Names often vary in pronunciation among tribes or different geographical areas. The spelling adopted by us in most cases is a 'classicized' form, rather than a purely phoenetic one, with the aim of indicating the standard around which the colloquial forms vary. Names in such form are also directly comparable with plant names in classical Arabic botanical works and other literature. Established classical short vowelings are often adopted by us in cases where the colloquial forms are variable or unclear. The grass *'thumām' (Panicum turgidum),* for example, was regularly voweled with a 'u' by the classical Arab botanical lexicographers, although it often sounds more like *'thmām'* or *'thamām'* in current colloquial dialects. Names are maintained as heard, however, whenever they clearly differ from classical cognates in grammatical form.

Readers not familiar with Peninsular colloquial dialects may note the following commonly encountered consonantal shifts from the standard forms given in our transliterations:

'q' (*qāf*) — generally to hard 'g'; in some contexts to 'j' or 'dz'.
'k' (*kāf*) — in some contexts to 'ch'.
'ḍ' (*ḍād*) — to ḍh (emphatic fricative).
'j' (*jīm*) — in the far south-east, to 'y'.

The majority of our plant names were collected among tribes having their home territory, or *dīrah*, in the Eastern Province. In some cases, however, they were obtained from folk recently moved to the east from homelands elsewhere in the Kingdom and whose useage may not always be locally typical. Table 7.1 provides, for reference, a list of tribes frequently cited with general indications of their principal geographic ranges.

Table 7.1 *Tribes Cited as Plant Name Sources*

Tribes of the Eastern Province	
az-Zafir	Northern Plains
Mutayr	Northern Summan, Dahna, and Plains
al-'Awazim	Central Coastal Lowlands (central to north-east)
Bani Khalid	Central Coastal Lowlands (central-east)
al-'Ujman	Central Coastal Lowlands (west), N. Summan
Bani Hajir	Central Coastal Lowlands (central-south)
Al Murrah	North and Central Rub' al-Khali, South Coastal Lowlands
al-Manasir	Northeastern Rub' al-Khali
Al Rashid	Rub' al-Khali (far southeast)
Non-Eastern Province Tribes	
ash-Shararat	Far northwestern Saudi Arabia
'Anazah (incl. Rawalah)	Northern and northwestern Saudi Arabia
Shammar	Northern Najd and adjacent areas
Qahtan	Southern Najd to northwestern 'Asir
ad-Dawasir	Southern Najd, Wadi ad-Dawasir

Vernacular plant names are listed in two index-glossaries, one by Roman alphabet and the other in Arabic script.

Statistical Summary — Numbers of Taxa: The total number of vascular, non-cultivated plants recorded from the 605,000-square-kilometer area of this flora is approximately 565 species in 322 genera. Of these, about 392 species in 236 genera, or 69 percent of all plants, are natives of the desert or other undisturbed habitats. Plants occurring virtually only as segetals, ruderals or on other disturbed sites account for the remaining 173 species and 86 genera. Overall, 73 families are represented.

Table 7.2 provides the numbers of species and genera in the eight overall most important families, which by themselves account for over 60 percent of the total number of species. Two relatively specialized families, the Gramineae and Compositae, are the leading contributors, the former predominating in the total records but the latter slightly more numerous when considering only the desert natives.

Table 7.2 *Numbers of Genera and Species in the Most Important Families, Flora of Eastern Saudi Arabia*

Family	All Non-Cultivated		Desert Natives	
	Genera	Species	Genera	Species
Gramineae	52	91	31	47
Compositae	43	66	33	50
Leguminosae	22	50	16	34
Cruciferae	30	46	22	30
Chenopodiaceae	20	42	17	35
Caryophyllaceae	15	22	13	18
Zygophyllaceae	6	15	5	14
Boraginaceae	8	14	8	13

The flora's endemic elements are few and tend to be concentrated in the more arid southern reaches. The Rub' al-Khali and its immediately adjoining areas have as endemics the dominant subshrubs *Cornulaca arabica* and *Zygophyllum mandavillei*, as well as the bushy herb, *Tribulus arabicus*. *Salsola arabica* is so far known from only a very restricted area in our southern Summan and at al-Aflaj to the southwest, while a probably undescribed *Echinops* is found in the coastal lowlands reaching inland to the Dahna. Eastern Saudi Arabia shares also in the range of a few more widely distributed endemics of Arabia and immediately adjoining deserts, such as *Horwoodia dicksoniae* and *Anthemis scrobicularis*.

Note on Maps: Delimited areas on maps in this book, including the distribution maps, indicate floristic or topographical regions, or the geographical area covered by fieldwork. These limits in a few areas, such as the Kingdom's borderlands in the north with Kuwait, may coincide in part with political boundaries. In general, however, they do not. The lines should not, therefore, be taken to have any significance with regard to international boundaries.

8. SYSTEMATIC ACCOUNT OF THE FLORA

KEY TO THE FAMILIES OF VASCULAR PLANTS

1. Plants without flowers, reproducing by spores
   2. Submerged aquatic ferns .... 3. *Parkeriaceae*
   2. Terrestrial ferns
      3. Sori in a spike; fern of sand desert .... 1. *Ophioglossaceae*
      3. Sori on lower surfaces of leaves; fern of wet habitats .... 2. *Adiantaceae*
1. Plants with flowers, reproducing by seeds
   4. Flowers in heads with common involucre

5. Stamens 5, adnate to inner surface of a tube .... 59. *Compositae*
5. Stamens 4, free .... 58. *Dipsacaceae*
4. Flowers not in heads with common involucre
  6. Flowers with single floral envelope, or with two envelopes of
    similar color and consistency, or without envelope (includes all
    submerged or floating aquatics)
    7. Floral envelope of 6 similar segments; stamens 6, 3, 2, or 1 ... *GROUP I*
    7. Flowers without all these characteristics (includes all submerged
      or floating aquatics) .... *GROUP II*
  6. Flowers with 2 floral envelopes of different color and consistency,
    the inner (corolla) usually larger and colored, the outer (calyx)
    smaller and green
    8. Petals free from each other to their bases
      9. Corolla papilionaceous .... 30. *Leguminosae*
      9. Corolla not papilionaceous
        10. Stamens 12 or more .... *GROUP III*
        10. Stamens fewer than 12 .... *GROUP IV*
    8. Petals coalescent, at least near their bases .... *GROUP V*

● GROUP I

1. Stamens 6
  2. Dioeceous trees with numerous flowers; leaves pinnate, 1 m or
    more long .... 70. *Palmae*
  2. Herbs or shrubby plants with perfect flowers
    3. Perianth segments brownish or black, coriaceous or membranous;
      stiff, pungent, erect shrubby plants of damp saline
      habitat .... 67. *Juncaceae*
    3. Perianth segments white or colored, of normal petal consistency;
      herbs .... 71. *Liliaceae*
1. Stamens 3, 2, or 1
  4. Flowers regular; blue, violet, or rarely white .... 72. *Iridaceae*
  4. Flowers irregular; pink or reddish .... 73. *Orchidaceae*

● GROUP II

1. Aquatic plants, submerged or floating
  2. Ovary inferior
    3. Fresh water plants; leaves finely pinnatisect .... 31. *Haloragaceae*
    3. Marine plants; leaves entire .... 60. *Hydrocharitaceae*
  2. Ovary superior
    4. Leaves divided .... 5. *Ceratophyllaceae*
    4. Leaves entire or denticulate
      5. Flowers bisexual, stalked or spicate
        6. Perianth of 4 segments; fruiting carpels remaining
          sessile .... 62. *Potamogetonaceae*
        6. Perianth 0; beaked fruiting carpels stalked .... 63. *Ruppiaceae*
      5. Flowers unisexual, sessile or shortly peduncled or cymose
        7. Marine plants .... 65. *Cymodoceaceae*

7. Plants of fresh or brackish inland waters
    8. Carpel 1 .... 61. *Najadaceae*
    8. Carpels 5 .... 64. *Zannichelliaceae*
1. Terrestrial plants, or with only base in water
    9. Trees or large shrubs of oases, with linear-lanceolate leaves and
    flowers in catkins .... 23. *Salicaceae*
    9. Plants without all above characteristics
        10. Leaves elongated, flat, ribbon-like or grass-like; perianth replaced
        by hairs, bristles, or membranous bracts (glumes); stamens 3 or 2
           11. Flowers surrounded by hairs, in 2 dense, coaxial spikes
           (the upper staminate, the lower pistillate) forming compact
           cylinders; cat-tail plants, 1 – 2 m high, of pool edges 69. *Typhaceae*
           11. Flowers not as above
               12. Flowers enveloped by glumes; leaves sheathing
               at base: grasses and sedges
                  13. Leaves with longitudinally split sheaths .... 66. *Gramineae*
                  13. Sheaths closed, entirely surrounding the
                  stem .... 68. *Cyperaceae*
               12. Flowers not enveloped by glumes; leaves not
               sheathing at base; coriaceous herb .... 12. *Chenopodiaceae*
        10. Plants without all these characteristics
           14. Shrubs or shrubby climbers
               15. Dioecious plants with flowers in bracted cone-like
               structures .... 4. *Ephedraceae*
               15. Plants monoecious; flowers not cone-like
                  16. Leaves virtually absent, or modified to
                  teretish, sometimes succulent structures
                    17. Flowers with petaloid perianth; stamens
                    10 – 16 .... 14. *Polygonaceae*
                    17. Flowers without petaloid perianth;
                    stamens 5 or less .... 12. *Chenopodiaceae*
                  16. Leaves present, with flattened blade
                    18. Fruits exteriorly 3-lobed or divided; male and female
                    flowers grouped together .... 36. *Euphorbiaceae*
                  18. Fruits not 3-lobed; flowers perfect
                      19. Leaves under 1.5 cm long or, if longer, then
                      fruits enclosed in 2 leaf-like perianth
                      valves .... 12. *Chenopodiaceae*
                    19. Some leaves longer than 1.5 cm; fruits not
                    enclosed in perianth valves .... 13. *Amaranthaceae*
           14. Herbs
               20. Club-shaped fleshy parasite without chlorophyll; flowers in
               a dark reddish spadix .... 34. *Cynomoriaceae*
               20. Plant not as above
                  21. Fruit 3-lobed or 3-divided exteriorly,
                  3-seeded .... 36. *Euphorbiaceae*
                  21. Fruit not 3-lobed; seeds 1 or many

22. Perianth segments 6

    23. Flowers regular; blue or violet, rarely white .... 72. *Iridaceae*

    23. Flowers irregular, pink-red .... 73. *Orchidaceae*

22. Perianth segments not 6

    24. Leaves with petioles modified at base into a sheath
    surrounding the stem .... 14. *Polygonaceae*

    24. Stems without sheaths

        25. Leaves 3-foliolate .... 39. *Zygophyllaceae*

        25. Leaves not 3-foliolate

            26. Flowers in spikes .... 13. *Amaranthaceae*

            26. Flowers not in spikes

                27. Fruit 1-seeded

                    28. Ovary superior .... 12. *Chenopodiaceae*

                    28. Ovary inferior .... 33. *Santalaceae*

                27. Fruit many-seeded

                    29. Petals 4, in 2 dissimilar sets, yellow .. 8. *Fumariaceae*

                    29. Petals not 4, or petals absent

                        30. Fruit 1-loculed .... 10. *Caryophyllaceae*

                        30. Fruit multi-loculed............... 9. *Aizoaceae*

● GROUP III

1. Leaves opposite

    2. Petals numerous; succulent herb .... 9. *Aizoaceae*

    2. Petals 5; non-succulent herbs or shrublets

        3. Carpels 3; sepals 5, of which 2 are much reduced in size .. 19. *Cistaceae*

        3. Carpels 5, each prolonged into a long beak; sepals 5,
        equal .... 40. *Geraniaceae*

1. Leaves alternate

    4. Petals 4

        5. Sepals 4, persistent .... 24. *Capparaceae*

        5. Sepals 2, falling as flower opens; then often apparently 0 . 7. *Papaveraceae*

    4. Petals 5, rarely 6 or 2

        6. Petals with divided borders .... 26. *Resedaceae*

        6. Petals entire

            7. Stamens connate in a column .... 17. *Malvaceae*

            7. Stamens free

                8. Trees with small numerous flowers in dense spherical
                heads; leaves pinnately compound with numerous
                leaflets .,................................. 30. *Leguminosae*

                8. Shrubs or herbs without above characteristics

                    9. Carpels free; fruit an achene .... 6. *Ranunculaceae*

                    9. Carpels adnate or immersed; fruit a capsule

                        10. Leaves reduced, scale-like .... 20. *Tamaricaceae*

                        10. Leaves not scale-like .... 39. *Zygophyllaceae*

● GROUP IV

1. Ovary inferior
    2. Flowers umbelled; stamens 5; styles 5 . . . . 43. *Umbelliferae*
    2. Flowers solitary; stamens 10; styles 10 . . . . 29. *Rosaceae*
1. Ovary superior
    3. Trees or shrubs with scale-like leaves; seeds pappose . . . . 20. *Tamaricaceae*
    3. Plants not as above
        4. Petals 4
            5. Petals in distinctly dissimilar sets; sepals 2, caducous . . 8. *Fumariaceae*
            5. Petals all alike or nearly so; sepals 3 or 4, persistent
                6. Stem leaves opposite
                    7. Glabrous dwarf herb with minute flowers . . . . 28. *Crassulaceae*
                    7. Woody shrub, 2 – 5 m high . . . . 32. *Lythraceae*
                6. Leaves alternate
                    8. Placentation axile; fruits a capsule with 5 valves or with
                    4 apically acute lobes
                        9. Petals fringed; capsules 4-lobed; aromatic weed . 38. *Rutaceae*
                        9. Petals entire; capsules not lobed; plants not
                        aromatic . . . . 16.   *Tiliaceae*
                    8. Placentation parietal; fruit 2-valved, or nut-like, or a
                    berry
                        10. Fruit divided longitudinally by a septum . . . 25. *Cruciferae*
                        10. Fruit without a septum . . . . 24. *Capparaceae*
        4. Petals not 4, sometimes 0
            11. Herbs with flowers having 5 stamens, 5 staminodes, and
            5 indehiscent carpels prolonged into a beak when
            mature . . . . 40. *Geraniaceae*
            11. Plants not as above
                12. Leaves pinnately compound; trees or shrubs . . . 30. *Leguminosae*
                12. Leaves not pinnately compound; herbs or shrubs
                    13. Leaves opposite
                        14. Flowers sessile . . . . 21. *Frankeniaceae*
                        14. Flowers peduncled, sometimes shortly
                            15. Leaves compound; or leaves simple with spiny
                            stipules . . . . 39. *Zygophyllaceae*
                            15. Leaves simple without spiny stipules
                                16. Carpels 3 – 4, distinct . . . . 28. *Crassulaceae*
                                16. Carpels connate or 1
                                    17. Ovary 1-celled
                                        18. Sepals or calyx lobes
                                        4 – 5 . . . . 10. *Caryophyllaceae*
                                        18. Sepals 2, unequal . . 11. *Portulacaceae*
                                    17. Ovary 2 – 5 celled
                                        19. Ovary 2-celled . . . . 9. *Aizoceae*
                                        19. Ovary 4 – 5 celled . . 39. *Zygophyllaceae*
                    13. Leaves alternate or basal
                        20. Leaves digitately trifoliolate . . . . 41. *Oxalidaceae*
                        20. Leaves simple, not digitate

21. Calyx a long tube .... 32. *Lythraceae*
21. Calyx not tubiform
   22. Herbaceous, or slightly woody plants without spines
      23. Flowers irregular
         24. Stamens 8; erect perennial .... 42. *Polygalaceae*
         24. Stamens 5; low annual .... 18. *Violaceae*
      23. Flowers regular
         25. Stamens 10; leaves with pellucid dots, strong
            smelling .... 38. *Rutaceae*
         25. Stamens 1–5; leaves not pellucid-dotted or
            strong smelling
            26. Petals 2; leaves linear .... 26. *Resedaceae*
            26. Petals 5; leaves oval .... 36. *Euphorbiaceae*
   22. Large woody shrubs or trees with spines .... 37. *Rhamnaceae*

● GROUP V

1. Shrubs of marine tidal mud flats with leathery, opposite
leaves .... 51. *Verbenaceae*
1. Terrestrial plants
   2. Ovary inferior
      3. Leaves alternate; fruit spherical, gourd-like, over 3 cm in diameter
         when mature .... 22. *Cuburbitaceae*
      3. Leaves opposite or verticillate; fruit a capsule under 5 mm in
         diameter .... 57. *Rubiaceae*
   2. Ovary superior
      4. Stamens or staminodes inserted opposite petals
         5. Style 1 .... 27. *Primulaceae*
         5. Styles 5; flowers surrounded by membranous
            bracts .... 15. *Plumbaginaceae*
      4. Stamens alternating with petals; no staminodes
         6. Flowers in dense spikes, 4-merous, with membranous bracts and
            calyx; leaves frequently a basal rosette .... 53. *Plantaginaceae*
         6. Flowers not as above
            7. Leaves opposite, verticillate, or densely 4-ranked
               8. Corolla distinctly irregular
                  9. Leaves unarmed; corolla 2-lipped; stem usually 4-angled;
                     ovary exteriorly 4-lobed .... 52. *Labiatae*
                  9. Leaves and bracts spiny-margined .... 56. *Acanthaceae*
               8. Corolla regular, or nearly regular, not 2-lipped
                 10. Fruit a follicle, more than 3 cm long; seeds with silky
                   hairs
                   11. Anthers connate and adnate to the compound
                     stigma; styles 2 .... 46. *Asclepiadaceae*
                   11. Anthers free; style 1 .... 45. *Apocynaceae*
                 10. Fruit a capsule or achene, not exceeding the calyx, less
                 than 3 cm long
                 12. Trees or shrubs

PTERIDOPHYTA

Plants with a distinct alternation of free-living generations, the sporophyte better developed and with vascular tissue. Reproduction is by spores, which grow into gametophytes that are often inconspicuous and thalloid in form. The Pteridophytes form a group traditionally including the ferns and the so-called fern allies, both of which are most common in humid environments. In northeastern Arabia they are represented by three species of ferns, each in a different family.

1. OPHIOGLOSSACEAE

Herbs with leaflike fronds at base sheathing a stalked spike of sporangia. Sporangia sunken in the spike, in two marginal rows, opening by transverse fissures.

● Ophioglossum L.

Sporangia two-ranked along the margins of a linear spike overtopping the leaflike sterile fronds.

Ophioglossum polyphyllum A. Braun
O. aitchisonii (C. B. Clarke) d'Almeida

Annual, glabrous, dwarf stemless fern, generally less than 8 cm high. Sterile fronds usually 1 – 3, leaflike, lanceolate to lanceolate-ovate, acutish, entire, partly folded longitudinally, up to c. 5 cm long, 2 cm broad, with narrow bases clasping the stalk

base of the fertile frond-spike; fertile fronds solitary or several, linear, greenish, stalked from base of the sterile fronds, with a sporangiferous, linear-oblong, longitudinally grooved, acute spike often about as long as its supporting stalk. Sporangia in 2 marginal rows with transverse sutures which open as fissures releasing the spores. *Plate 8.*
Habitat: Drifted or stable sands, usually not far from the coast and often over limestone. Overall rather rare.
Central Coastal Lowlands: 97 (BM), Dhahran; 399, 1453 (BM), al-Midra ash-Shimali; 3218 (BM), 35 km NE an-Nu'ayriyah.
● Dickson reported that children sometimes eat the plant (Burtt and Lewis 1949, 279).

2. ADIANTACEAE

Ferns with creeping or suberect rhizomes and frond segments often obdeltoid in outline, with stalks polished-blackish. Sporangia in mostly marginal sori.

● Adiantum L.

Ferns with fronds usually 1 – 5-pinnate and glabrous. Sori borne below the frond segments, beneath their reflexed margins.

Adiantum capillus-veneris L.
Terrestrial fern with creeping rhizome, the rhizome with brown scales; stipes dark brown to blackish, shining. Fronds up to c. 30 cm long, more or less ovate in outline, compound with widely spaced, alternate pinnae and pinnules; pinnules cuneate at base, usually 5 – 15 mm broad, green and delicate on short stalks, palmately veined, with cut margins. Sori on distal pinnule margins below the reflexed edges of the lobes.
Habitat: Wet, shaded spots in the oases, usually near water level along irrigation channels or at springs. Occasional.
South Coastal Lowlands: 358, 1 km S ad-Dalwah, al-Hasa Oasis; Dickson 539(K), between al-Hufuf and Jabal al-Qarah, al-Hasa Oasis (Burtt and Lewis 1949, 279). Cheesman (1926, 416) also reported it from al-Hasa Oasis. It is certainly also to be found in the Qatif Oasis area.
● One additional terrestrial fern, *Cheilanthes catanensis* (Cos.) Fuchs of the family Sinopteridaceae, may occur in our area in the northern or southern Summan. The author has collected it in shaded limestone crevices on steep wadi walls near Riyadh. It is a highly drought resistant dwarf, recognizable by its hairy or scaly fronds and sessile frond pinnae.

3. PARKERIACEAE

Aquatic ferns with short rhizome. Fronds dimorphic, the fertile ones more divided, with narrower pinnules than the sterile ones. Sporangia sessile along veins on the lower surface of the fertile pinnules.

● Ceratopteris Brongn.

Description as for the family.

Ceratopteris thalictroides (L.) Brongn.
Submerged, bright translucent-green aquatic fern with succulent stipes. Fronds, at least on mature plants, of two distinct types: the sterile ones 10 – 25 cm or more long,

45

pinnate to bipinnate with ovate to triangular segments; the fertile ones bi- to tripinnate, up to c. 50 cm long, with much narrower, linear segments, the ultimate pinnules only c. 2 mm broad, with margins involute below. Sporangia scattered along veins on the lower surface of the fertile frond segments. *Plate 9.*

Habitat: Springs and fresher irrigation channels of the larger oases. Many, if not nearly all, of the al-Hasa Oasis habitats of this aquatic, which contributed so much to the beauty of the clear springs and stream channels there, were destroyed between 1968 and 1971 during construction of the modern irrigation and drainage system. Occasional. **South Coastal Lowlands:** 362, 1 km S Bani Ma'n, al-Hasa Oasis; Umm al-Khursan spring, al-Hasa Oasis (Cheesman 1926, 416); al-Qatif Oasis (sight records by the present author). Fossil spores of *Ceratopteris* have been reported in the Rub' al-Khali from drill core material of late Tertiary age (Aramco, unpublished).

SPERMATOPHYTA

Vascular plants with alternate generations, the sporophyte best developed and usually with some form of flowers. The gametophyte is much reduced and closely associated with the sporophyte, the male developing from pollen grains. These are the seed plants, which are divided into two subordinate groups, the gymnosperms and the angiosperms.

GYMNOSPERMAE

Plants usually with conelike reproductive structures, bearing seeds developing from ovules not enclosed in an ovary. Their wood, except in the Gnetales, is without true vessels. Only two species, of the genus *Ephedra*, are present in the native flora of eastern Saudi Arabia.

4. EPHEDRACEAE

Shrubs or climbers, usually dioecious, with leaves opposite or verticillate, simple, linear, or often apparently entirely absent, reduced to small scales joined with a sheath. Male cones with small naked flowers in the axils of imbricate bracts; filaments coalescent in a column. Female cones 1 – 3-ovuled, the bracts becoming scarious or developing into berrylike fruit structures.

● Ephedra L.

Mostly dioecious shrubs or climbers, often leafless, or with reduced or short-lived opposite or verticillate leaves. Nodes with short, cupulelike sheaths. Female spikes 2-flowered, the bracts or sheaths developing into a dry, conelike, or fleshy, berrylike, structure. Staminate flowers in conelike spikes with imbricate bracts.

Several species of *Ephedra* have long been important as sources of the alkaloid ephedrine, a vaso-constricting drug formerly much used in the treatment of asthma and as a general respiratory decongestant. The alkaloid is often synthesized now, and the natural drug has become less important commercially. Tests for alkaloids have not, apparently, been conducted to date on east Arabian *Ephedras*. Paris and Dillemann (1960) report that one of our species, *E. alata*, was found in North Africa to contain *d*-pseudo-ephedrine. *E. foliata* in India was found to produce none. There is not, apparently, much evidence that our *Ephedras* have special reputations as folk remedies.

Key to the Species of Ephedra

1. Bracts of female cones free to their bases, when mature with scarious margins; male cones sessile in dense axillary clusters; anther column divided at apex, anthers thus stalked individually; stiff erect shrublets .... 1. *E. alata*
1. Bracts of female cones connate for about half their length, fleshy; male cones 1 – 3 at tips of slender branchlets, thus appearing peduncled; anthers sessile at apex of undivided anther column; climbing perennial .... 2. *E. foliata*

1. **Ephedra alata** Decne.

Stiff, yellow-green, densely branched dioecious shrublet, 40 – 100 cm tall and often wider than high. Twigs striate, often whorled and appearing leafless due to the reduction of leaves to scarious cupules or scales less than 5 mm long. Cones sessile, clustered in the axils or at branch tips, the staminate ones with the stamens somewhat exserted, the anthers 5 – 6, stipitate from a common column; the pistillate ones c. 1 cm long, composed of several pairs of broadly scarious-margined bracts. *Plate 10.*

Habitat: Light-colored, gritty calcareous or sometimes gypsaceous soil. Occasional, sometimes forming pure stands of limited extent.

Central Coastal Lowlands: 158, 15 km SW 'Ayn Dar; Dickson 525(K), between Abqaiq and al-Hasa (Burtt and Lewis 1949, 280).

Northern Summan: 2836(BM), 26 – 21N, 47 – 25E; 2853(BM), 26 – 10N, 48 – 17E; 8728, near Mashdhubah, 27 – 20N, 45 – 03E.

● Numerous sight records in other areas.

Vernacular Names: ʿALANDÁ ('thick, strong', gen.), ʿADÁM (Rawalah).

2. **Ephedra foliata** Boiss. ex C. A. Meyer

E. ciliata C. A. Meyer
E. peduncularis Boiss. et Hausskn.

Dioecious, or rarely monoecious, shrub, often climbing and straggly, to 2 m or more high when climbing on large shrubs, sometimes unsupported with shoots from a woody base. Leaves variable in size, from virtually absent to up to 2 cm long in earlier growth stages. Staminate cones solitary or several at the tips of fine branches, with the anthers usually slightly exserted; anthers 3 – 4, sessile on a single column. Pistillate cones sessile or peduncled at branchlet tips, with 2 of the bracts becoming fleshy, forming a berrylike fruit at maturity.

Less common than the preceding species and readily distinguishable from it even when sterile by its straggly, climbing habit.

Habitat: Very often associated with *Ziziphus nummularia*, rooting at its base and climbing up into its branches; also found in steep rocky ravines of the Summan. Rare to occasional.

Northern Plains: 579(K), 15 km SSW Hafar al-Batin; 3083(BM), al-Batin, 28 – 01N, 45 – 29E; 3094(BM), 3104(BM), 3105(BM), al-Batin, 28 – 00N, 45 – 28E; 8743, ravine to al-Batin, 28 – 50N, 46 – 24E.

Northern Summan: 4041, al-Batin, 27 – 55N, 45 – 22E.

ANGIOSPERMAE

True flowering plants, with ovules enclosed in an ovary and with woody tissue including vessels. The group is traditionally composed of two classes, the dicotyledons and the monocotyledons.

Dicotyledoneae

Plants with two seed leaves (cotyledons) when embryonic and with vascular bundles arranged circularly in the stem, net-veined leaves, and floral parts usually in fours, fives, or their multiples.

5. CERATOPHYLLACEAE

Submerged aquatic herbs with finely divided, whorled leaves. Flowers unisexual, solitary, with perianth of 10 – 15 segments. Staminate flowers with 10 – 20 stamens; pistillate flowers with 1, 1-celled ovary. Fruit a nut.

● Ceratophyllum L.

Monoecious aquatic herbs with whorled leaves dissected into fine segments. Flowers solitary in the axils, minute, the staminate ones with 8 – 20 anthers. Fruit a small nut, usually spined.

Ceratophyllum demersum L.

Submerged perennial herb growing in strands to 1 m or more long. Leaves whorled, 10 – 20 mm long, once or twice divided dichotomously into fine, linear, toothed segments. Flowers minute, solitary in the axils, the staminate ones with about 12 tepals and 10 – 16 stamens. Fruit narrowly ovoid, to 5 mm long, with 2 spinules at base and a spine-like style at apex, the spinules as long as or longer than the fruit body.
Habitat: Fresh-water springs and water channels in the oases. Sometimes growing covered with filamentous algae. Occasional.
South Coastal Lowlands: 2895(BM), 'Ayn al-Khadud, al-Hasa Oasis.

6. RANUNCULACEAE

Annuals or perennials, usually herbaceous; leaves mostly alternate, simple or compound, often lobed. Flowers bisexual, mostly actinomorphic, hypogynous; petals and sepals several to numerous; stamens numerous. Carpels 1 to numerous; fruit an achene or a follicle.

● Ranunculus L.

Annual or perennial herbs with leaves alternate or spirally inserted, usually palmately lobed or divided. Flowers with sepals (or sepaloid bracts) 3 – 5; petals 5; stamens numerous. Carpels numerous, developing into a head of apiculate or beaked, usually compressed, achenes.

Key to the Species of Ranunculus
1. Fruiting peduncle much longer than its subtending leaf; lower leaves ovate in outline, divided into 3 main lobes which are in turn double cleft nearly to the middle leaving all segments less than 20 mm wide .... 1. *R. cornutus*

1. Fruiting peduncle equal to or shorter than its subtending leaf; lower leaves suborbicular to reniform in outline, divided into 3 main lobes which are in turn crenate or lobed to middle with entire portion below these lobes more than 20 mm wide .... 2. *R. muricatus*

1. **Ranunculus cornutus** DC.

Erect, glabrescent or sparsely hairy annual to about 35 cm high. Lower leaves with petioles exceeding the blades; blades ovate in outline, divided into 3 main lobes, these in turn divided into toothed or lobed sublinear segments. Flowers c. 15 – 20 mm in diameter, with glossy yellow petals 6 – 10 mm long, 4 – 7 mm broad; peduncles mostly 2 – 5 cm long. Achenes compressed, beaked, 4 – 5 mm long, with faces tuberculate, surrounded by a flat, smoothish margin.

Habitat: Not a native of eastern Arabia and not a desert plant. So far recorded only in moist artificial habitats such as nursery plots, where it is apparently introduced occasionally with cultivated plants. Rare.

Central Coastal Lowlands: 3844(BM), Sayhat, al-Qatif Oasis, a weed in nursery plots. The author's duplicate of this one specimen, with immature fruits, has some features characteristic of *R. marginatus* Urv.

2. **Ranunculus muricatus** L.

Branched, ascending annual herb to about 35 cm high, glabrous or with sparse scattered hairs. Basal leaves long-petioled, suborbicular to reniform in outline, often slightly wider than long (ours up to c. 5×7 cm), with 3 main lobes, each of which has a toothed or short-lobed distal margin. Flowers c. 10 – 18 mm in diameter, with glossy yellow petals in ours c. 5 – 8 mm long, 3 – 4 mm wide. Peduncles mostly 1 – 2 cm in flower, elongating in fruit to c. 10 cm. Achenes compressed, 6 – 7 mm long including the beak, the flat faces with subspinous tubercles, surrounded by a narrow green margin. *Plate 11.*

Habitat: Like the preceding species an introduced weed to be expected only around and in garden plots, and then usually on shaded ground. Rarely seen, but has been observed spreading abundantly in a shaded fallow field in the al-Hasa Oasis.

Central Coastal Lowlands: 3843(BM), Sayhat, al-Qatif Oasis; a weed in a plant nursery.

South Coastal Lowlands: 8198, 8232, near al-Qurayn, al-Hasa Oasis.

7. PAPAVERACEAE

Annuals or perennials, mostly herbaceous, sometimes with white or colored latex. Leaves alternate, usually lobed or more or less dissected. Flowers solitary or in cymose, racemose or paniculate inflorescences, bisexual, actinomorphic or zygomorphic, hypogynous or perigynous. Sepals usually 2, caducous; petals 4(– 6) or 8 – 12. Stamens 4 to numerous, free. Ovary unilocular or nearly so; fruit a capsule, usually many-seeded, opening by valves or pores.

Key to the Genera of Papaveraceae

1. Capsule linear, hairy, opening by valves along its full length; petals purple-violet .... 1. *Roemeria*
1. Capsule (in ours) ovoid, glabrous, opening by pores near the apex; petals red-scarlet .... 2. *Papaver*

● 1. **Roemeria** Medik.

Annual herbs with dissected leaves. Flowers solitary, terminal, with 2 caducous sepals and 4 violet to red petals. Stamens numerous. Ovary 3 – 4 carpellate; fruit an oblong-linear, many-seeded capsule opening by valves from the apex.

Roemeria hybrida (L.) DC.
R. dodecandra (Forssk.) Stapf
Erect or ascending annual, 10 – 30 cm high, often with stiff whitish erect hairs. Leaves 3 – 5 cm long, deeply pinnatisect into linear lobes. Petals 0.8 – 2 cm long, deep violet-purple, fragile and soon falling. Capsule 2 – 4 cm long, 2 – 3 mm wide, often with stiff, erect, c. 1.5 mm-long whitish hairs, or almost hispid. Sepals caducous and seldom seen except in bud. *Plate 12.*

Habitat: Shallow silty sand in basins of the north. Occasional.

Northern Plains: 1501(BM), 30 km ESE al-Qaysumah; 1546(BM), 1552(BM), 36 km NE Hafar al-Batin; 1641(BM), 1677A(BM), 2 km S Kuwait border at 47 – 14E; 1683(BM), al-Hamatiyat, 28 – 50N, 47 – 30E; 4008, ad-Dibdibah, 28 – 10N, 45 – 55E; 7571, 22 km SW Hafar al-Batin; 1309(BM), 5 km NW Jabal Dab'; 8717, 20 km, SW ar-Ruq'i wells, 28 – 51N, 46 – 25E.

Northern Summan: 3349, 5 km WSW an-Nazim; 626(BM), Wabrah; 2826, 30 km N ash-Shumlul; 1423(BM), 27 km W Qaryat al-'Ulya; 1352(BM), 1368(BM), 42 km W Qaryat al-'Ulya; 8428, edge 'Irq al-Jathum, 25 – 54N, 48 – 36E.

Vernacular Names: ḤASSĀR (az-Zafir), BAKHATRĪ (Dickson, 1955).

● 2. **Papaver** L.

Annual or perennial herbs with lobed to dissected leaves. Flowers solitary, racemed, or panicled, with sepals usually 2, petals 4. Stamens numerous. Ovary multicarpellate, mostly 1-locular, with stigmas sessile in radiating lines in a disc at apex of the ovary; fruit a many-seeded capsule opening beneath the apical disc by pores.

Papaver rhoeas L.
Erect annual, more or less pubescent to hispid, 20 – 50 cm high. Leaves pinnatisect with lobes serrate or pinnately divided. Flowers showy, solitary on erect, leafless, hairy to setose peduncles; buds ovoid to ellipsoid, sparsely pubescent-setose, 15 – 25 mm long, 11 – 12 mm wide; petals red-scarlet, with or without a black spot near base, 20 – 35 mm long, somewhat broader than long. Disc with 9 – 13 stigmatic rays which overlap near its margin. Capsule ovoid to semi-globose, 10 – 18 mm long, surmounted by the disc.

Habitat: Seen by the author only once, in a remote desert location but on an abandoned Bedouin camp site with a variety of Mediterranean weeds that were probably brought in with sacked grain stored there. Rare.

Central Coastal Lowlands: 7944, 8094, N Batn al-Faruq, 25 – 43N, 48 – 53E, at abandoned Bedouin camp.

8. **FUMARIACEAE**

Annual or perennial herbs with leaves mostly alternate and pinnatisect. Flowers racemed or spicate, bisexual, zygomorphic, with 2 caducous sepals and 4 petals in 2 dissimilar sets. Stamens 6 in 2 groups; ovary unilocular with style 1 and stigmas 2 – 8; fruit a 2-valved capsule or an indehiscent nutlet.

Key to the Genera of Fumariaceae
1. Flowers yellow; fruit an elongated, segmented, podlike loment; desert plants .... 1. *Hypecoum*
1. Flowers white or pinkish; fruit a 1-seeded nutlet; garden or farm weed .... 2. *Fumaria*

● 1. Hypecoum L.

Glabrous, glaucous, annual herbs with finely dissected, mostly rosetted leaves. Sepals 2, caducous. Petals 4, the outer pair entire or 3-lobed, the inner pair deeply tripartite, with the lateral lobes entire and the central lobe fimbriate. Stamens 4. Fruit a linear loment, finally breaking into 1-seeded joints.

Key to the Species of Hypecoum
1. Central lobe of the trifid inner petals as broad as or narrower than the whole petal below the dividing point; inner petals minutely flecked dark purple-black; fruits subpendulous to pendulous .... 2. *H. pendulum*
1. Central lobe of the trifid inner petals broader than the whole petal below the dividing point; inner petals not flecked; fruits mostly ascending to erect .... 1. *H. geslinii*

Many authors separate these two very similar species in keys by the fruit habit alone: pendulous vs. erect. This is often not a decisive feature, however, particularly when the fruits are immature, and determinations generally require close, magnified examination of fresh flowers or the dissection of dried specimens. The dark flecking on the inner petals of *H. pendulum* appears to be a generally reliable feature. The pigment seems to be water-soluble, however, and it often disappears completely in hot water dissections. The proportional width of the inner petals' central lobe is usually decisive. The two species have different habitat preferences, *H. geslinii* usually being found on sand and *H. pendulum* on silty ground. In transitional zones they may grow side by side.

1. Hypecoum geslinii Coss. et Kral.
Ascending or decumbent glabrous annual, 5 – 25 cm high, with several to many main stem branches ascending, or more or less decumbent from the base outside the leaf rosette, the stems often branched once or twice above. Leaves mostly basal, rosulate, lanceolate-oblong in outline, usually 3 – 10 cm long, 0.5 – 2.5 cm wide, short-petiolate, several-pinnatisect into linear to filiform lobes; some smaller leaves present at the upper stem branch points. Flowers 5 – 7 mm long, yellow; petals 4, the outer 2 oblong and entire, the inner pair 3-lobed from about the middle with the central lobe distally broadened and fringed-ciliate, longer than the lateral 2 and usually wider than the whole petal below the trifurcation. Fruits cylindrical-linear, tapering at the apex, sometimes somewhat angular, transversely jointed, 2 – 3.5 cm long, 1.5 – 2 mm wide, often ascending from a reflexed pedicel. *Plate 13.*
Habitat: Sand. Occasional to locally frequent.

Northern Plains: 583, 15 km SSW Hafar al-Batin; 8778, 18 km WSW ar-Ruq'i Post, 28 – 59N, 46 – 31E.

Northern Summan: 3885a(BM), 5 km SW Umm 'Ushar; 685, 20 km WSW Khubayra; 8430, edge of 'Irq al-Jathum, 25 – 54N, 48 – 36E; 8550, 10 km W Nita'.

Northern Dahna: 4027(BM), ad-Dahna, 27 – 35N, 44 – 51E; 3872(BM), 15 km WSW Umm 'Ushar.

Central Coastal Lowlands: 1442(BM), 2 km E Jabal Ghuraymil; 1796(BM), 5 km NE al-Khursaniyah; 7522, Ras Tanaqib.

Vernacular Names: UMM ATH-THURAYB (Suhul).

2. Hypecoum pendulum L.

Ascending or decumbent glabrous annual, 5 – 30 cm high, with several stems ascending or decumbent from the base, usually arising outside the leaf rosette, often branched once or twice above. Leaves mostly basal, rosulate, lanceolate-oblong in outline, usually 3 – 10 cm long, 0.5 – 2.5 cm wide, short-petiolate. Flowers c. 5 – 6 mm long, yellow. Petals 4, the outer 2 oblong-lanceolate and entire, the inner pair flecked dark purple, 3-lobed from about the middle, with the central lobe distally broadened and fringed-ciliate, longer than the lateral 2 and usually narrower than the width of the whole petal below the trifurcation. Fruits cylindrical-linear, straight or curved, 2 – 4 cm long, 1.5 – 3 mm wide, transversely jointed, usually pendulous from a deflexed pedicel.

Habitat: Silty basins. Occasional to locally frequent.

Northern Plains: 1535(BM), 36 km NE Hafar al-Batin; 558(BM), 676, Khabari Wadha; 1282(BM), 15 km ENE Qaryat al-'Ulya; 4007(BM), ad-Dibdibah, 28 – 10N, 45 – 55E; 8713, 20 km SW ar-Ruq'i wells, 28 – 51N, 46 – 25E.

Northern Summan: 2800, 15 km ESE ash-Shumlul; 728, 33 km SW al-Qar'ah wells; 1373(BM), 42 km W Qaryat al-'Ulya; 2789, 41 km SE ash-Shumlul; 3219(BM), 74 km S Qaryat al-'Ulya; 3237(BM), Dahl al-Furayy; 3348(BM), 5 km WSW an-Nazim; 3832, ash-Shayyit, 27 – 29N, 47 – 22E; 3885(BM), 5 km SW Umm 'Ushar; 8429, edge of 'Irq al-Jathum, 25 – 54N, 48 – 36E.

Northern Dahna: 4021, near ath-Thumami, 27 – 38N, 44 – 56E.

Central Coastal Lowlands: 1193(BM), Jabal an-Nu'ayriyah; 3215(BM), 35 km NE an-Nu'ayriyah.

Southern Summan: 2078, Jaww ad-Dukhan, 24 – 38N, 44 – 56E.

Vernacular Names: UMM ATH-THURAYB (Suhul).

● 2. Fumaria L.

Glabrous annual herbs with alternate, multi-pinnatisect leaves. Flowers bracteate, racemed, irregular, pink or white, with upper petal spurred at base, the lateral ones connate at apex. Stamens 6, joined in 2 groups; fruit an indehiscent nutlet with 1 or 2 ovules.

Fumaria parviflora Lam.

Glabrous, often branched, annual, 10 – 50 cm high with angular stems. Leaves several-pinnatisect with linear lobes up to c. 6 mm long, 1 mm wide. Racemes dense in flower, elongating in fruit, with bracts c. equalling the 1.5 mm-long pedicels. Flowers c. 4 – 6 mm long with very small dentate sepals and corolla white to pink with tips of the lateral petals blotched purplish. Fruit c. 2 mm long, keeled and somewhat compressed. (Description largely after Townsend in *Flora of Iraq* 4, pt. 2).

Habitat: A weed of gardens or farms. Probably occasional.

● No certain records from the strict confines of the Eastern Province, but collected by the author as a farm weed in 'Unayzah, north-central Arabia (3955,BM) and reported by S. A. Chaudhary (personal communication) and Chaudhary, Parker and Kasasian (1981) as a fairly common weed of Najd farms not far from the limits of our area.

9. AIZOACEAE

Herbs or small shrubs with leaves usually succulent, opposite or alternate. Flowers bisexual, actinomorphic, with calyx 5-lobed and petals absent or numerous (or replaced by petaloid staminodes); stamens 3 to numerous. Ovary superior or inferior; fruit a few- to many-seeded capsule.

Key to the Genera of Aizoaceae
1. Petals (actually petaloid staminodes) numerous, linear .. 3. *Mesembryanthemum*
1. Petals definite or absent
   2. Flowers with petals; leaves less than 8 mm long; desert
      perennials .... 1. *Limeum*
   2. Flowers apetalous (but sepals sometimes colored); leaves more than
      8 mm long; weeds or desert annuals
      3. Perianth green, whitish, or yellowish within; capsules opening by
         apical slits; desert annuals .... 2. *Aizoon*
      3. Perianth pink or reddish within; capsule circumscissile; weeds
         4. Style 1 .... 5. *Trianthema*
         4. Styles 2 – 5
            5. Flowers mostly solitary; seeds numerous .... 4. *Sesuvium*
            5. Flowers clustered; seeds 4 per capsule .... 6. *Zaleya*

● 1. Limeum L.

Herbs or small shrublets, often glandular, with leaves subopposite and 1 of each pair often strongly reduced. Flowers in axillary cymes; sepals 5, nearly free; petals 5, whitish, or sometimes absent. Fruit a schizocarp breaking into 2 hard, hemispherical mericarps.

A genus sometimes segregated with some others in a separate, closely related family, the Molluginaceae. The group is characterized by leaves not or hardly succulent, calyx segments nearly free, and stamens definite, in 2 whorls.

Key to the Species of Limeum
1. Seeds reticulately wrinkled, dull, grey-tan .... 1. L. *arabicum*
1. Seeds smooth, brown .... 2. L. *humile*

1. Limeum arabicum Friedr.

Intricately branched tangled shrublet, to c. 0.5 m high, densely covered with minute knobbed glands and often viscid with adherent sand. Leaves opposite or subopposite with 1 of each pair reduced in size, 3 – 4.5(5) mm long, 2 – 4.5 mm wide, ovate to orbicular, on petioles about 1 mm long or subsessile. Flowers solitary in the axils,

4–5 mm long, on pedicels 1–1.5 mm long; sepals green, white-margined; petals white, clawed, about equal to the sepals in length, obscurely toothed at apex; stamens 7, dilated and ciliate at base; style deeply bifid. Fruit a schizocarp splitting at maturity into 2 hard, hemispherical mericarps, grey-tan and distinctly reticulate-wrinkled on the outer rounded surface. *Map 8.1.*

Habitat: Deep sands of the southern Jafurah and the Rub' al-Khali, where it is an important constituent of several plant communities. Locally common.

South Coastal Lowlands: 3798, the Jafurah, 25–03N, 50–03E.

Southern Dahna: 8351, 51 km W Harad.

Rub' al-Khali: 468(BM,K), 22–27N, 51–04E; 1019(BM), 22–45N, 50–20E; 7003(BM), 3 km NE Shalfa; 7081, 2 km NW Camp S–3 in 22–11N, 54–19E; 7677, eastern Sanam, 21–59N, 51–13E.

Vernacular Names: BIRKAN (Al Rashid, Al Murrah).

2. **Limeum humile** Forssk.

L. indicum Stocks

Divaricately branched glandular perennial herb or semi-shrub, to 20 cm high, often with adherent sand. Leaves obovate or broadly elliptical to orbicular, 3–7 mm long, 2–6 mm wide, sometimes mucronate at apex, on 1–3 mm-long petioles. Flowers in our specimens mostly solitary in the axils, c. 3 mm long with ovate sepals green with white hyaline margins; petals white, slightly shorter than the calyx. Fruit a schizocarp splitting at maturity into 2 hard hemispherical mericarps which are brown and smoothish on their rounded surfaces. *Map 8.1.*

Plant less glandular than *L. arabicum* (above) and readily distinguished from it by the surface texture of the seed. Seedlings and young plants have the larger leaves; summer leaves are smaller.

Habitat: Unlike the above, usually found in silt flats on limestone or gravel plains. Rare.

Southern Summan: 2927(BM), Jaww ad-Dukhan, 40 km SSW al-'Udayliyah.

South Coastal Lowlands: 7703, 11 km S Wadi as-Sahba in 23–52N, 49–11E.

● 2. **Aizoon** L.

Annual or perennial herbs or small shrubs, often papillose, with opposite or alternate, more or less succulent, entire leaves. Flowers axillary, with a 5-fid, sepaloid perianth with short tube; petals absent. Stamens numerous, in groups. Ovary superior; stigmas 3–5. Capsule 3–5-loculate, opening loculicidally or by a star-shaped apical slit.

Key to the Species of Aizoon

1. Leaves spathulate, petiolate; flowers less than 10 mm long with
   perianth segments shorter than their tube .... 1. *A. canariense*
1. Leaves lanceolate, sessile; flowers more than 10 mm long with
   perianth segments longer than their tube .... 2. *A. hispanicum*

1. **Aizoon canariense** L.

Annual procumbent, pubescent-papillose herb with stems branching radially from the base, rather stiff, often zig-zag, to c. 15 cm long. Leaves alternate, spathulate to oblong-obovate, obtusish, mostly 1–2 cm long, 0.5–1 cm wide, long-tapering at base to an indistinct petiole. Flowers sessile, apetalous, greenish outside but yellowish within;

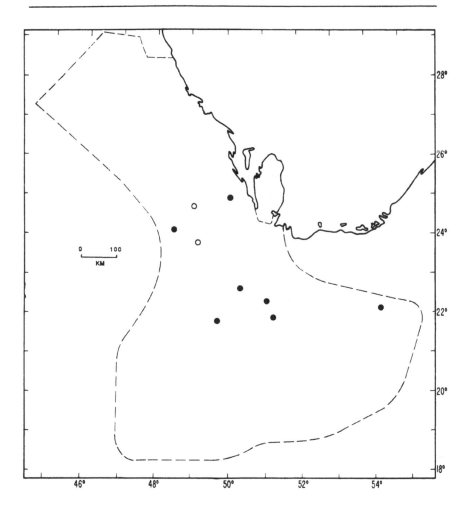

Map 8.1. Distribution as indicated by collection sites of *Limeum arabicum* (●) and *Limeum humile* (○) in the coverage area of this flora (dashed line), eastern Saudi Arabia. Each symbol represents one collection record.

perianth 3 – 5 mm long with triangular, acute lobes, whitish hyaline-margined, with crisped hairs at back; stigmas 5. Capsule flattish, star-shaped at apex, sometimes with 5 ascending lobules from the angles. *Plate 14.*

Habitat: Coastal, sometimes inland, sands; rarely as a weed on farms. Frequent.

Northern Plains: 1740, Jabal al-'Amudah.

Northern Summan: 3165(BM), 11 km SSW Hanidh.

Central Coastal Lowlands: 190, 2 km N Dhahran Airport; 3745, 24 km W as-Saffaniyah; 400, al-Midra ash-Shimali; 1208(BM), Jabal an-Nu'ayriyah; 1751(BM), Abu Hadriyah; 2735, ar-Ruqayyiqah, al-Hasa Oasis; 3299, 2 km W Sabkhat Umm al-Jimal; 3742, Ras az-Zawr; 7517, Ras Tanaqib; 7736, 2 km S Ras Tanaqib, 27 – 49N, 48 – 53E.

Vernacular Names: DU'A' (Shammar, Rawalah).

2. Aizoon hispanicum L.

Ascending, papillose-succulent, branched annual, 5 – 15 cm high. Leaves opposite or subopposite, oblong-lanceolate to linear, sessile or subsessile, to c. 3 cm long, 0.8 cm wide, with narrow but obtusish apex. Flowers apetalous, greenish outside, whitish within, short-pediceled, solitary in the axils. Perianth 11 – 16 mm long, with lobes lanceolate, much longer than the tube. Capsule somewhat flattened at apex. *Plate 15.*

Habitat: Shallow sand or gritty silt. Usually found further inland than the preceding. Occasional.

Northern Plains: 1504(BM), 30 km ESE al-Qaysumah; 1528(BM), 36 km NE Hafar al-Batin; 4052, ad-Dibdibah, 28 – 04N, 46 – 05E; 1305(BM), 8 km WNW Qaryat al-'Ulya; 8763, trib. to al-Batin, 28 – 49N, 46 – 23E.

Northern Summan: 1370(BM), 42 km W Qaryat al-'Ulya.

Central Coastal Lowlands: 76a(BM), Thaj; 356(BM), near al-Fudul, al-Hasa Oasis; 401, al-Midra ash-Shimali.

Vernacular Names: MULLAYḤ, 'saltwort' (Al Murrah, Shammar, Qahtan).

● 3. Mesembryanthemum L.

Annual or perennial herbs, sometimes shrubs, often succulent or papillose. Leaves usually opposite or subopposite. Flowers mostly axillary, solitary or cymose, the calyx with 4 – 5, sometimes unequal lobes, the petals (petaloid staminodes) numerous, linear. Stamens numerous. Ovary half-inferior; stigmas 4 – 5; capsule fleshy or indurate, 5-loculate, opening by an apical star-shaped slit.

Key to the Species of Mesembryanthemum

1. Leaves 5 – 15 mm thick; petals longer than the calyx . . . . 1. *M. forsskalei*
1. Leaves 2 – 4 mm thick; petals not or hardly longer than the
   calyx . . . . 2. *M. nodiflorum*

1. Mesembryanthemum forsskalei Hochst.

Opophytum forskahlii (Hochst.) N. E. Br.

Annual papillose herb, very succulent, erect or ascending with stems 10 – 25 cm high. Leaves conical, subterete, decurrent above, up to c. 5 cm long, 1.5 cm thick. Flowers axillary, on pedicels shorter than the calyx. Calyx with unequal conical lobes. Petals white to cream, yellowish at base, exceeding the calyx at anthesis. Capsule 12 – 15 mm long. *Plate 16.*

Habitat: (In our area) coastal sands. Rare.

Central Coastal Lowlands: 3764A, Dhahran, along fences of stables boundary. Good (1955) reports it as common in Bahrain.

Vernacular Names, Uses: SAMḤ (gen. north). Seeds from this plant are collected by Bedouins in northern Arabia and ground into a flour to make bread. The plant is rare in eastern Arabia and not so used here. In the north, Shammar tribesmen use the terms *sabīb* for the seed, *ka'bar* for the capsules. The best sort of *samḥ* is called *ḥurr* (Shammar); an inferior one, which is grown around Baq'a and Kahfah in northern Arabia, is *ḥamr wāqif* (Shammar). Musil (1927, 122 – 124 and 1928a, 16 – 17) describes how the Bedouins collect the capsules, dry them and thresh them with sticks. He mentions another variety called *du'ā'* (1927, 464), a name recorded by the present writer for *Aizoon canariense.*

A long-keeping food is made in al-Jawf from dates and roasted ground *samh* seeds. The flavor is reportedly chocolate-like (Musil 1928b, 6).

2. **Mesembryanthemum nodiflorum** L.

Cryophytum nodiflorum (L.) L. Bol.

Annual, low ascending or decumbent succulent herb, branching from the base with stems 5 – 20 cm long, papillose. Leaves teretish, linear, succulent-papillose, opposite or alternate or often both, to c. 3 cm long, 2 – 4 mm wide, more or less ciliate at base. Flowers axillary, subsessile. Calyx 6 – 10 mm long with teretish, succulent, often very unequal, lobes. Petals often 20 – 30, narrowly linear or subfiliform, obtuse, white to cream, yellowish near base, usually shorter than the longer calyx lobes but often overtopping the calyx at anthesis. Capsule c. 5 – 8 mm long, somewhat pentagonal-pyramidal at apex. Seeds triangular-ovoid, somewhat compressed, c. 1 mm long, brown. *Plate 17.*

● This dwarf succulent may form dense colonies, with near continuous cover, in coastal regions in spring. The foliage often takes on an overall reddish color, particularly under drying conditions.

Habitat: Coastal sands and *sabkhah* margins; found on many of the Gulf islands. Locally frequent.

Central Coastal Lowlands: 1829, Dhahran; 2893(BM), Jurayd Island, Arabian Gulf; 7033, Jana Island, Arabian Gulf; 3781, 1 km SW Ras al-Ghar; 7523, Ras Tanaqib. Sight records on Gulf islands other than the above.

● 4. **Sesuvium** L.

Prostrate or repent herbs with subalternate to opposite leaves. Flowers axillary with sepaloid perianth, the lobes green dorsally, colored within, with a subapical mucro; petals absent; stamens 5-numerous. Capsule circumscissile.

Sesuvium verrucosum Raf.

Glabrous, much-branched succulent herb to c. 30 cm or more high. Leaves mostly opposite, narrowly to broadly spathulate, 1 – 4 cm long. Flowers mostly solitary in the axils, subsessile; sepals ovate-lanceolate, short-horned near apex, pink-mauve inside, greenish at back. Capsules opening by a conical lid; seeds numerous, black.

Habitat: A weed of waste places, abandoned fields.

Central or South Coastal Lowlands: Not seen by the author in our area but a common weed on Bahrain (Phillips 1988, Cornes 1989) and likely to become established in eastern Saudi Arabia. On Bahrain it is commonly seen on waste ground and is tolerant of poorly drained, saline soil. Another species, *Sesuvium sesuvioides* (Fenzl) Verdc., has been reported as a weed in central Saudi Arabia by Chaudhary and Zawawi (1983) and by Chaudhary, Parker and Kasasian (1981).

● 5. **Trianthema** L.

Prostrate herbs or shrublets with unequal, opposite, entire leaves. Flowers axillary; perianth lobes 5, each with a subapical mucro; petals absent; stamens 5 – 10. Ovary superior, 1 – 2-loculate with 1 – 2 stigmas. Capsule bilocular, operculate.

Trianthema portulacastrum L.

Usually prostrate weedy herb, glabrescent but with lines of whitish hairs on the young parts. Leaves obovate, opposite with 1 of each pair distinctly smaller; blades 1.5 – 3.5(4) cm long, 1 – 3.5 cm wide, tapering below to a petiole about ½ to ¾ the length of the blade, the petiole winged and sheathing at base. Flowers solitary, sessile, axillary, c. 4 – 5 mm long, sheathed by the expanded basal wings of the petioles. Perianth lobes with a recurved mucro inserted dorsally just below the apex. Stamens about 15, with anthers pink; styles 1; capsule immersed in the petiolar sheaths; seeds usually c. 6 – 7, reniform, c. 2 mm broad, black and rather coarsely wrinkled.

Habitat: A weed of gardens, farms. Rare.

Central Coastal Lowlands: 1926, Dhahran, at stables farm; 8336, Dhahran.

● 6. Zaleya Burm. f.

Weedy herbs with opposite, succulent leaves. Flowers clustered, without corolla; fruit a 2-loculate, 4-seeded capsule with a 2-valved lid.

Zaleya pentandra (L.) Jeffr.
Trianthema pentandra L.

Prostrate or decumbent perennial herb with young parts somewhat succulent-papillose. Leaves opposite, oblong to elliptical, obtusish, mostly 1 – 3.5 cm long, 0.5 – 2 cm wide, more or less papillose, with petioles shorter than the blades and broadened to scarious wings sheathing at the base. Flowers clustered 3 – 10 in the axils, sessile or subsessile, c. 3 mm long. Corolla absent; perianth lobes 5, each with a subterminal dorsal hornlet. Stamens 5; ovary 2-loculed; styles 2. Capsule red, dehiscing near the middle. Seeds 4, black, wrinkled, 2 falling with the 2-chambered lid and 2 remaining in the base.

Habitat: A weed of cultivated or disturbed ground. Probably rare to occasional.

Central or South Coastal Lowlands: Listed as a common weed, at least in central Saudi Arabia (Chaudhary and Zawawi 1983). The author has not seen specimens from the Eastern Province, but the species is likely to occur as an occasional weed.

10. CARYOPHYLLACEAE

Annual or perennial herbs, rarely shrublets, with leaves opposite or whorled, entire, stipulate or exstipulate. Flowers usually bisexual, hypogynous or perigynous, actinomorphic, in cymose inflorescences. Sepals usually 5, connate or free; petals mostly 5, sometimes absent; stamens 2 – 10. Ovary superior, 1-locular; stigmas 2 – 5. Fruit a capsule, nut or berry.

An important family in the desert flora of northeastern Arabia, with 15 genera, most of which are herbaceous annuals. As treated here it includes the group with scarious stipules sometimes segregated as a separate family, the Illecebraceae.

Key to the Genera of Caryophyllaceae

1. Sepals connate for at least half their length; exstipulate plants
    2. Calyx with 5 green ribs maturing into prominent wings .... 3. *Vaccaria*
    2. Calyx without ribs or wings
        3. Calyx campanulate, 5-nerved; petals reddish-striped
           or lined .... 2. *Gypsophila*

3. Calyx tubular, 10- or 30-nerved; petals not reddish-lined .... 1. *Silene*
1. Sepals free, or connate for less than half their length; plants, except
Stellaria, stipulate (stipules inconspicuous in *Herniaria*)
  4. Fruits indehiscent, 1-seeded
    5. Shrublet, definitely woody above .... 11. *Gymnocarpos*
    5. Herbs, annual or perennial
      6. Flowers in 3's, crowded in short dichasia on a broadened,
        flat, green, leaf-like peduncle .... 15. *Pteranthus*
      6. Flowers not as above
        7. Inflorescence hardened, globose, spiny .... 14. *Sclerocephalus*
        7. Inflorescence not hardened and not spiny
          8. Inflorescence scarious, with conspicuous silvery
            bracts .... 12. *Paronychia*
          8. Inflorescence green, not scarious; stipules
            inconspicuous .... 13. *Herniaria*
  4. Fruits dehiscent, of multi-seeded capsules
    9. Leaves subulate, or narrowly linear with revolute margins
      10. Stipules adnate to leaves for half their length; calyx lobes
        with a subulate appendage on either side .... 10. *Loeflingia*
      10. Stipules and sepals not as above
        11. Petals pink .... 6. *Spergularia*
        11. Petals white (in *Polycarpaea* inconspicuous in silvery
          bracts)
          12. Stamens 5; prostrate tomentose perennial usually
            woody at base .... 8. *Polycarpaea*
          12. Stamens 10 (or 6 – 8); nearly glabrous ascending
            annual .... 5. *Spergula*
    9. Leaves flat, not revolute margined
      13. Plants exstipulate; leaves ovate, acute .... 4. *Stellaria*
      13. Plants stipulate, leaves elliptical to linear-lanceolate
        14. Bracts and stipules wholly scarious, sepals
          keeled .... 7. *Polycarpon*
        14. Bracts and stipules with herbaceous middles; sepals not
          keeled .... 9. *Robbairea*

● 1. **Silene** L.

Annual or perennial herbs with opposite, entire, exstipulate leaves. Flowers generally bisexual, solitary or in cymose inflorescences. Calyx tubular, 5-toothed at apex, with 10 (or, less commonly, 20, 30 or 60) longitudinal nerves. Petals 5, or rarely 0, often somewhat showy, long-clawed, mostly white or shades of pink to purple. Stamens 10, unequal. Ovary 3 – 5-celled at base, 1-celled above; styles 3. Capsule stipitate on a carpophore, or sessile, dehiscing by 6 – 10 teeth. Seeds numerous, reniform or ear-shaped.

Two of our species — *S. villosa* and *S. arabica* — are frequent to common. The others are rarer and most likely to be encountered in the far north of the Province. *S. villosa* is psammophilous, extending south as far as Yabrin and probably into the northern Rub' al-Khali. *S. arabica* is more often seen on silty soils and has a more northerly distribution but extends into some sand areas and into the southern Summan. Living flowers of *Silene* are best observed early or late in the day; the petals often roll in at midday except in overcast weather.

Key to the Species of Silene

1. Calyx 30-nerved .... 4. *S. conoidea*
1. Calyx 10-nerved
    2. Pedicels pubescent, shorter than the flowers
        3. Seeds with 2 undulate wings at back; plants appressed-pubescent
            4. Petals shorter than the calyx, thus scarcely visible, or absent
               altogether .... 1. *S. apetala*
            4. Petals well developed and clearly exserted .... 3. *S. colorata*
        3. Seeds without undulate wings (although sometimes strongly grooved on back); plants glandular-pubescent
            5. Calyx 12 – 22 mm long; petals divided less than ¾ way to base of limb; some leaves 4 mm or more wide .... 6. *S. villosa*
            5. Calyx 10 – 11 mm long; petals parted more than ¾ way to base of limb; leaves less than 4 mm wide .... 2. *S. arabica*
    2. Pedicels glabrous, equalling or often exceeding the flowers ... 5. *S. linearis*

1. Silene apetala Willd.

Erect, appressed-pubescent annual, 5 – 35 cm high, sometimes flowering in dwarf condition. Leaves 10 – 25 mm long, 1 – 3(5) mm wide, linear to linear-oblanceolate, acute. Flowers on pedicels of variable length. Calyx 6 – 9 mm long, with recurved-spreading, acute, ciliate teeth; petals reduced and included, or absent. Seeds ear-shaped, c. 1.2 mm long, dark brown, with dorsal groove bordered by undulate, transversely fine-striate wings.

Quickly indentifiable by its apetalous or near-apetalous state.
Habitat: Silty basins or wadis. Rather uncommon.
Northern Plains: 4042(BM), tributary to al-Batin, 27 – 56N, 45 – 23E; 8712, 20 km SW ar-Ruq'i wells; 8761, tributary to al-Batin, 28 – 49N, 46 – 23E.
Northern Summan: 749, 33 km SW al-Qar'ah wells.

2. Silene arabica Boiss.

Erect, glandular-pubescent, often moderately viscid annual, 10 – 30 cm high; stem several-branched from near base or simple. Leaves linear-lanceolate, 10 – 30 mm long, 2 – 3 mm wide. Flowers in 1-sided racemes on pedicels mostly shorter than or equalling the flower. Calyx 9 – 11 mm long, club-shaped in fruit. Petals white or faintly pink, sometimes with pinkish veins beneath, the limb parted more than ¾ the way to its base into 2 linear lobes. Capsule ovoid-oblong, c. as long as the carpophore. Seeds ear-shaped, c. 0.6 mm long, dark brown to black, with a dorsal groove bordered by obtuse, finely transversely striate ridges; lateral faces excavated, ear-shaped. *Plate 18.*
Habitat: Prefers silty soil but sometimes also found in pure sand. Frequent to common; the most common *Silene* of the north, also extending to the middle coast and through the Summan to south of Wadi as-Sahba.
Northern Plains: 566(BM), 678, Khabari Wadha; 873, 9 km S 'Ayn al-'Abd; 1503(BM), 30 km ESE al-Qaysumah; 1300(BM), 15 km ENE Qaryat al-'Ulya; 1323(BM), 8 km W

Jabal Dab'; 1644(BM), 2 km S Kuwait border in 47–14E; 1712(BM), ash-Shaqq, 28–30N, 47–42E; 2716, 3 km NW Jabal Dab'; 8752, tributary to al-Batin, 28–50N, 46–24E.

Northern Summan: 1401(BM), 42 km W Qaryat al-'Ulya; 715, 33 km SW al-Qar'ah wells; 3876, 15 km WSW Umm 'Ushar; 8441, edge of 'Iraq al-Jathum, 25–54N, 48–36E.

Northern Dahna: 4022, 27–38N, 44–56E; 7530, 40 km NE Umm al-Jamajim.

Central Coastal Lowlands: 7513, Ras Tanaqib; 371, Jabal Sha'ban, al-Hasa; Dickson 531(K), SW Abqaiq (Burtt and Lewis 1952, 343); 1415(BM), 7 km NW Abu Hadriyah; 1229, Jabal an-Nu'ayriyah; 1779(BM), Abu Hadriyah; 1809(BM), 5 km NE al-Khursaniyah; 3275, 3279, 8 km W Qatif-Qaisumah Pipeline KP–144; 3356, Jabal al-Ahass; 3720, 8 km N Abu Hadriyah; 3740, Ras az-Zawr; 3778, 3 km E Ras al-Ghar; 4056, az-Zulayfayn, 26–51N, 49–51E; 7467, 7 km SW Qannur, Darb al-Kunhuri; 7719, Ras az-Zawr.

Southern Summan: 8005, 8010, 24–00N, 49–00E; 8053, 24–06N, 48–48E.

3. **Silene colorata** Poir. var. **oliveriana** (Otth) Muschl.

S. oliveriana Otth in DC.

Erect annual, 10–30 cm high, appressed-pubescent, not glandular, often unbranched or sometimes branched from near base. Leaves oblanceolate to linear-oblanceolate, narrower above, 10–35 mm long, 1–7 mm wide, acute. Inflorescence a 1-sided raceme with pedicels shorter than the calyx. Calyx 12–15 mm long. Corolla pinkish. Seeds ear-shaped, with a deep dorsal groove bordered by undulate wings.

Habitat: Silt-sand. Infrequent to rare.

Northern Plains: 593(BM), 16 km ENE Qulban Ibn Busayyis; 4044(BM), tributary to al-Batin, 27–56N, 45–23E.

4. **Silene conoidea** L.

Erect, rather stout annual, 10–50 cm high, few or several-branched with stems glandular-pubescent. Leaves lanceolate-linear, acute, 30–80 mm long, 4–10 mm wide, finely puberulent. Flowers on pedicels mostly shorter than the calyx. Calyx 20–30 mm long, 30-nerved, more or less cylindrical when young, broadening and ovoid in fruit, with finely acuminate teeth up to ⅓ its length. Corolla a strong pink.

Habitat: Primarily a weed, with desert occurrences around former Bedouin encampments where seeds may be associated with sacked grain. Rare.

Northern Plains: 7525, 27 km WSW al-Qaysumah.

Central Coastal Lowlands: 7951, N Batn al-Faruq, 25–43N, 48–53E.

5. **Silene linearis** Decne.

Annual, 10–30(50) cm high with stems glabrescent or sparingly fine-pubescent near the nodes, with or without narrow viscous bands on upper parts of the internodes and pedicels. Leaves narrowly elliptical-linear, tapering to an acute apex, 20–50(65) mm long, 1.5–4(5) mm wide, glabrescent above, more or less pubescent below between the somewhat revolute margins, the lower margins ciliate. Inflorescence dichotomously much branched, panicle-like, the pedicels equalling or exceeding, and up to 2–2.5 times as long as, the flowers, often sharply reflexed near the flower. Calyx 12–15 mm long, 2.5–4 mm wide, whitish and somewhat pubescent between the nerves, with acute, acuminate teeth. Corolla white to pale pinkish, the petals pale yellowish-green at back, with lamina usually divided to the middle or sometimes only ⅓ the way, with linear lobes. Capsule c. 8 mm long, equalling the carpophore or slightly longer. Seeds ear-shaped, c. 0.8 mm long, shallowly grooved dorsally, the lateral faces radially fine-striate. *Plate 19.*

● This species differs rather distinctly in habit from our other *Silenes* by its relatively broad, glabrous, subpaniculate inflorescence and longish, fine pedicels. Our eastern specimens do not always have the distinctive brown viscid bands on the internodes that are generally present in the species.

Habitat: Silty ravine or wadi bottoms. So far collected only on the extreme northeastern edge of our territory.

Northern Plains: 1645(BM), 2 km S Kuwait border in 47 – 14E; 8770, 49 km NE Hafar al-Batin; 8791, 30 km SW ar-Ruq'i Post; 8719, 20 km SW ar-Ruq' wells; 8740, tributary to al-Batin, 28 – 50N, 46 – 24E; 8760, tributary to al-Batin, 28 – 49N, 46 – 23E.

6. Silene villosa Forssk.

Glandular-pubescent ascending, branched annual, 10 – 30 cm high, viscid and often with adherent sand. Leaves oblong-oblanceolate, 20 – 40 mm long, 3 – 5 mm wide. Flowers in cymes, with pedicels mostly shorter than the calyx. Calyx (10)12 – 22(25) mm long, in fruit contracted below and thus club-shaped. Corolla showy, white to very pale pink, the petal limb 2-parted or divided up to c. ½ the way to base, the lobes usually obtuse. Seeds ear-shaped, dorsally grooved, brown, finely reticulate-tuberculate, c. 0.6 mm long. *Plate 20.*

● This species varies significantly, and apparently geographically, in stature and corolla form. Specimens from the northern Dahna appear to be smaller in all dimensions than those from the central coast. Populations in the middle coastal lowlands have the largest flowers, with particularly broad petals.

Habitat: Sand. Frequent to common.

Northern Plains: 538(BM), 8 km SE Qaryat al-'Ulya.

Northern Summan: 2841, 32 km ESE ash-Shumlul; 3146(BM), 9 km SSW Hanidh.

Northern Dahna: 7531, 40 km NE Umm al-Jamajim; 4036, 27 – 35N, 44 – 51E; 3249, S edge 'Irq Jaham; 2318, 22 km ENE Rumah; 8568, E edge 'Urayq al-Khufaysah, 26 – 24N, 47 – 12E.

Central Coastal Lowlands: 629(BM), 10 km SSW Nita; 1795(BM), 5 km E al-Khursaniyah; Dickson 488(K), Ras Tanura (Burtt and Lewis 1952, 347); Dickson 584(K), W al-Jubayl (*ibid.*); 366, 1818(BM), Dhahran; 1209, Jabal an-Nu'ayriyah; 2724, 23 km N Jabal Fazran; 2835, 26 – 21N, 47 – 25E; 3278, 8 km W Qatif-Qaisumah Pipeline KP – 144; 3355, Jabal al-Ahass; 3783, Ras al-Ghar; 7466, 7 km SW Qannur, Darb al-Kunhuri.

South Coastal Lowlands: Dickson 529(K), 531(K), SW Abqaiq (*ibid.*); 2114, 3 km SW al-Jawamir, Yabrin; 2272, 13 km S 'Ayn Dar GOSP – 4; 8023, Yabrin, 32 – 11N, 48 – 53E; 1432(BM), 2 km E Jabal Ghuraymil.

Rub' al-Khali: Likely to occur in the northern parts.

Vernacular Names: TURBAH (classically TARIBAH; 'earth-weed', no doubt derived from Arabic *tarib, turb,* 'earth, dust', referring to the adherent sand usually clothing this viscid plant; gen., Al Murrah), BALLAH ('wet-weed', referring to the viscid surface of the plant, Shammar, Rawalah), QAHWĪYAN (a name more usually applied to *Anthemis melampodina,* another annual with rather showy white flowers, Qahtan).

● 2. Gypsophila L.

Annual or perennial herbs or shrubs with opposite, exstipulate leaves. Flowers in open, branched cymes, bisexual or unisexual, small. Calyx more or less campanulate below, 5-toothed above. Petals 5, white or pink, tapering below to a claw. Stamens 10. Ovary 1-celled; capsule dehiscing by valves.

Gypsophila antari Post et Beauv.

Glabrous, glaucous, branched annual, to c. 0.5 m high, with fine upper branches. Leaves at base spathulate, to c. 5 cm long, 1 cm wide, the higher ones becoming narrower, oblong-lanceolate to narrowly linear in the upper stems. Inflorescence a rich but open capillary panicle. Flowers c. 5 mm long, terminating filiform pedicels mostly several times as long as the flower and reflexed near the calyx; calyx campanulate; petals c. 1.5 times as long as the calyx, white with red-purple lines. Capsule c. as long as the calyx.
Habitat: Shallow sand or silty sand, often over a limestone substrate. Occasional to locally frequent.
Northern Plains: 562(K,BM), Khabari Wadha; 994, 12 km N Jabal al-'Amudah; 1526(BM), 30 km ESE al-Qaysumah; 1651(BM), 2 km S Kuwait border in 47 – 14E; 1476(BM), 50 km WNW an-Nu'ayriyah.
Northern Summan: 213(BM), 25 km S Qaryat al-'Ulya.
Central Coastal Lowlands: 424, Na'lat Shadqam; 1758(BM), Abu Hadriyah; 3725, 8 km N Abu Hadriyah; 8640, 4 km SW Thaj.
Vernacular Names: SALIH (Rawalah).

● 3. Vaccaria Medik.

Annual herbs with opposite, exstipulate leaves. Flowers bisexual, in loose paniculate inflorescences. Calyx herbaceous, strongly keeled or winged longitudinally, 5-toothed. Petals 5, clawed, pink to purple. Stamens 10. Ovary 1-celled, 2-styled; fruit a capsule opening by 4 teeth.

Vaccaria hispanica (Mill.) Rauschert
V. pyramidata Medik.
Saponaria vaccaria L.
Erect glabrous annual to 60 cm high. Leaves lanceolate, sessile, 4 – 10 cm long, 0.5 – 4 cm wide, subcordate at base. Flowers in a paniculate inflorescence. Calyx ovoid, pale, with 5 prominent green ribs, 10 – 14(17) mm long, 8 – 10 mm wide, short-toothed. Petals pink. Capsule enclosed in the calyx, ovoid or subglobular, 4-valved. Seeds globose, finely tuberculate, rust-brown to black.
Habitat: An uncommon weed of croplands; very rarely found in desert where grain has been spilled.
Central Coastal Lowlands: 8084, N Batn al-Faruq, 25 – 43N, 48 – 53E, Bedouin camp ground.

● 4. Stellaria L.

Annual or perennial, often tender herbs with opposite, exstipulate leaves. Flowers in cymose inflorescences or solitary. Sepals 5, free. Petals 5 – 4, usually bifid. Stamens usually 10. Ovary 1-celled; fruit a 6-valved capsule with numerous seeds.

Stellaria media (L.) Vill.

Ascending or decumbent, somewhat weak-stemmed annual to c. 25 cm high, glabrous or pubescent. Leaves opposite, ovate, acute, long-petioled below to sessile above, 0.5 – 2.5(3) cm long, to c. 1.5 cm wide. Flowers in terminal or axillary cymes, on pedicels equalling or well exceeding the calyx. Calyx 4 – 5 mm long. Petals white, bifid, somewhat shorter than the sepals; anthers red-violet. Capsule exceeding the calyx, c. 6 mm long, with (5)6 valves. Seeds somewhat compressed, tuberculate, brown, c. 1 mm in diameter.

Habitat: A weed of cultivated areas, usually on shaded ground. Rare except very locally but sometimes abundant in individual fields.

Central Coastal Lowlands: 3841, Sayhat; 8269, Dhahran.

South Coastal Lowlands: 8196, near al-Qurayn, al-Hasa Oasis.

● 5. Spergula L.

Slender annual herbs with leaves whorled, narrowly linear to filiform. Stipules scarious, small. Flowers in cymose, mostly terminal inflorescences. Sepals 5, free, herbaceous with scarious margins. Petals 5, white. Stamens (5 –)10. Ovary 1-celled; fruit a many-seeded capsule. Seeds keeled or winged.

Spergula fallax (Lowe) E.H.L. Krause

Glabrous dwarf annual, 5 – 15(20) cm high, with stems branching from the base, ascending or decumbent. Leaves 10 – 40 mm long, whorled, filiform; stipules ovate-triangular, white-scarious. Flowers mostly on the upper stems, on pedicels shorter than, or considerably longer than the calyx. Sepals ovate, green-herbaceous, narrowly white-margined, 3 – 4 mm long. Petals white, 2 – 4 mm long. Stamens (8 –)10. *Plate 21.*

● In habit rather resembles *Spergularia diandra*, with which it may be found growing, but readily separable by its white corolla and higher stamen number.

Habitat: Silt basins; also on higher rocky ground. Infrequent except in Summan terrain.

Northern Summan: 3244, 5 km N Umm al-Hawshat; 2796, 26 km SE ash-Shumlul; 2815, 12 km ESE ash-Shumlul; 3225, 74 km S Qaryat al-'Ulya on Riyadh track; 8438, 'Irq al-Jathum, 25 – 54N, 48 – 36E; 8605, Wadi Mibhil, 26 – 40N, 47 – 34E.

Central Coastal Lowlands: 7040, Karan Island, Arabian Gulf.

Southern Summan: 2039, 25 – 00N, 48 – 30E; 2095, 8 km N Harad; 2193, 24 – 34N, 48 – 45E.

● 6. Spergularia (Pers.) J. et C. Presl

Annual or perennial herbs with opposite, linear to filiform leaves with scarious stipules. Flowers generally bisexual, in cymose, mostly terminal, inflorescences. Sepals 5, free, scarious-margined. Petals 5, pink or white. Stamens 2 – 5 – 10. Ovary 1-celled; fruit a valved capsule. Seeds winged or wingless.

Key to the Species of Spergularia

1. Capsule 2 – 3 mm long; stamens 2 – 3; desert plant .... 1. *S. diandra*
1. Capsule 4 – 6 mm long; stamens 4 – 8(10); weed of cultivation
   or waste ground .... 2. *S. marina*

1. **Spergularia diandra** (Guss.) Heldr. et Sart.

Ascending dwarf annual, glabrescent, or glandular-puberulent above, with several to numerous stems branching from the base, 5 – 15(20) cm long. Leaves opposite or rarely whorled, mostly 15 – 30 mm long, 0.7 – 1 mm wide. Stipules triangular-ovate, white-scarious. Flowers in cymose, mostly terminal inflorescences, on pedicels up to c. 3 times as long as the flower. Sepals oblong, scarious-margined, 2 – 3 mm long, sometimes sparsely glandular. Petals oblong, pink, slightly shorter than the calyx. Stamens 2(3). Capsule ovoid, c. as long as the calyx, opening apically by 3 valves. Seeds numerous, sub-triangular, somewhat compressed, obtusely keeled, c. 0.4 mm long, dark brown, nearly smooth. *Plate 22.*

Habitat: Silt or sand, often over limestone. Frequent, and unlike the following species, found in undisturbed desert habitats.

Northern Plains: 1308(BM), 8 km WNW Qaryat al-'Ulya; 1493(BM), 30 km ESE al-Qaysumah; 1530(BM), 36 km NE Hafar al-Batin; 1569(BM), al-Batin/al-'Awja junction; 1657(BM), 2 km S Kuwait border in 47 – 14E; 7979, *sabkhah* at Khawr al-Khafji.

Northern Summan: 592, 16 km ENE Qulban Ibn Busayyis; 1348(BM), 1390(BM), 42 km W Qaryat al-'Ulya; 3242, 5 km N Umm al-Hawshat; 3339, 5 km WSW an-Nazim; 8433, edge 'Irq al-Jathum, 25 – 54N, 48 – 36E.

Central Coastal Lowlands: 1825(BM), Dhahran; 1187(BM), Jabal an-Nu'ayriyah; 3256, 8 km W Qatif-Qaisumah Pipeline KP – 144; 3325, 12 km S Nita'; 3737, Ras az-Zawr; 4065, az-Zulayfayn, 26 – 51N, 49 – 51E.

Southern Summan: 2083, 24 – 27N, 49 – 05E; 2040, 25 – 00N, 48 – 30E; 2094, 8 km N Harad; 2176, 24 – 25N, 48 – 41E; 2245, 26 km ENE al-Hunayy.

2. **Spergularia marina** (L.) Griseb.

S. salina J. et C. Presl

Ascending or decumbent annual 5 – 25 cm high, rarely with base somewhat woody, glabrescent below, glandular-pubescent above, with numerous stems branching from the base. Leaves opposite or whorled, 20 – 50 mm long, filiform, somewhat succulent; stipules white-scarious, connate at base. Flowers mostly in terminal cymes, on pedicels usually about equalling the flower but sometimes longer. Calyx 3 – 5 mm long. Petals pink, somewhat shorter than the calyx. Stamens 5 – 10. Capsule slightly exceeding the calyx, 3-valved. Seeds numerous, triangular-ear-shaped, obtusely keeled, c. 0.5 – 0.8 mm long, often with minute knobbed tubercles, nearly all unwinged but usually with a few (often only c. 2 – 3) circularly winged seeds present near base of the capsule, the wings erose-margined.

● None of the author's specimens appear to be attributable with any certainty to another *Spergularia* sometimes listed as a weed in Arabia and neighboring countries, the closely related *S. rubra* (L.) J. et C. Presl. This is usually described as differing by its hardly succulent leaves, longer and nearly free stipules and somewhat shorter capsules with seeds always entirely wingless. The few plants in our collections with seeds definitely all wingless – and it is easy to overlook the few winged seeds usually present near the capsule base – do not have enough other correlating characters to justify a separation. It seems preferable, at least provisionally, to treat our plants as a single, rather variable species composed of various forms perhaps originating as weed introductions from different geographical areas.

Yet another species, *S. media* (L.) C. Presl, which has all seeds winged, may be found to occur as a weed in the Eastern Province. No confirmed records, however, are known to the author.

Habitat: Weed of farmlands and lawn edges.

Central Coastal Lowlands: 1452(BM), 3919, 8245, Dhahran; 2962(BM), 3763, Dhahran stables farm; 8158, near al-Jishsh, al-Qatif Oasis; 8260, 8261, 5 km S al-Qatif town.

South Coastal Lowlands: 8204, near al-Mutayrifi, al-Hasa; C. Parker No. 4(BM), farm near old airport, al-Hufuf; 8298, King Faysal University grounds, al-Hufuf.

● 7. Polycarpon L.

Small annual or perennial herbs with leaves opposite or whorled, scarious-stipulate. Flowers in dichasial cymes; calyx with 5 keeled lobes, hooded at apex, scarious-margined; petals 5; stamens 3 – 5. Ovary 1-celled; fruit a 3-valved capsule with several to numerous seeds.

Polycarpon tetraphyllum (L.) L.

Glabrous, prostrate to ascending, rather dense and somewhat succulent, dichotomously branched dwarf annual with stems usually only 2 – 5(10) cm long. Leaves (in ours) 3 – 6(10) mm long, 1 – 2(3) mm wide, linear-spathulate or narrowly oblong, tapering to a narrow base. Stipules conspicuous, wholly white-scarious, acuminate. Flowers c. 2 mm long, numerous, on pedicels c. equalling or somewhat shorter than the flowers. Calyx lobes oblong, scarious-margined, nearly twice as long as the petals, mucronate or (as in ours) not or hardly so. Capsule enclosed in the calyx, with valves twisted after dehiscence. Seeds pale, smooth, c. 0.3 – 0.4 mm long.

● Miss Dorothy Hillcoat has pointed out (personal communication) that our plants show some characteristics of *P. succulentum* (Del.) J. Gay. King and Kay (1984) place all Arabian material in *tetraphyllum* and question the distinctness of *succulentum* as a species.

Habitat: Sands, usually not far from the coast. Infrequent.

Central Coastal Lowlands: 1232, Jabal an-Nu'ayriyah; 1761, Abu Hadriyah; 1793, 5 km NE al-Khursaniyah; 1983, ad-Dawsariyah; 7717, Ras az-Zawr; 7720, 2 km S Ras Tanaqib; 7746, 5 km N Ras Tanaqib.

● 8. Polycarpaea Lam.

Annual or perennial herbs or shrubs with scarious-stipulate, opposite or whorled leaves. Flowers in paniculate, mostly terminal, cymes. Calyx with 5 scarious-margined lobes; petals 5; stamens 5; 5 staminodes sometimes present. Ovary 1-celled; fruit a 3-valved capsule.

Polycarpaea repens (Forssk.) Aschers. et Schweinf.

Prostrate trailing perennial, close woolly-tomentose to glabrescent, with stems woody at base. Horizontal stems to 25 cm or considerably longer under favorable conditions. Leaves opposite or whorled, 3 – 10 mm long, 1 – 2 mm wide, lanceolate-linear, mucronate, with revolute margins. Stipules scarious. Flowers numerous, mostly near ends of branches, c. 2 – 2.5 mm long, the sepals scarious-margined, sometimes weakly keeled above, mucronate.

● A highly drought-resistant plant found in the most arid sand habitats. It dies back to a short woody base during the dry season but rapidly sends out new herbaceous shoots in the cool part of the year, particularly after winter rains.

Habitat: Sand. Widespread and frequent; found in virtually all sandy areas including parts of the Rub' al-Khali.

Northern Plains: 1169, Jabal al-Ba'al; 877, 9 km S 'Ayn al-'Abd; 883, Jabal al-'Amudah; 929, Khafji-as-Saffaniyah road junction.

Northern Summan: 3156(BM), 9 km SSW Hanidh; 7470, Mishash Ibn Jum'ah.

Northern Dahna: 7542, 8 km NE Umm al-Jamajim; 796, 15 km SE Hawmat an-Niqyan.

Central Coastal Lowlands: 138, 1084, Dhahran; 1440(BM), 2 km E Jabal Ghuraymil; 1977, ad-Dawsariyah; 2103, 49 km S Harad; Dickson 462(K), 10 km N al-Jubayl (Burtt and Lewis 1952, 338); 5(BM), Sabkhat an-Nu'ayriyah.

Rub' al-Khali: sight records.

Vernacular Names and Uses: MAKR (Al Murrah), LA'LA'AH (Al Rashid), RUQAYYIQAH ('thin-weed', Shammar). According to Al Murrah Bedouins, this plant is traditionally used to treat sarcoptic mange of camels, either as ash or crushed leaves or in combination with yellow arsenic.

● 9. **Robbairea** Boiss.

Glabrous, prostrate or ascending annual or perennial herbs with leaves opposite or whorled. Stipules and bracts with herbaceous middles, scarious-margined. Flowers in a loose inflorescence. Calyx with 5 scarious-margined lobes; petals 5, clawed; stamens 5. Ovary 1-celled, with trifid style; capsule 3-valved.

Robbairea delileana Milne-Redhead

Glabrous, prostrate to ascending annual, apparently sometimes perennating, with branching stems 4 – 20 cm long. Leaves mostly basal or on lower parts of stem, whorled or opposite, elliptical-spathulate, tapering to base, 3 – 10(15) mm long, 1 – 3(4) mm wide. Stipules and bracts with herbaceous middles. Flowers in a rather loose, paniculate inflorescence, pedicellate; calyx c. 2 – 3 mm long, scarious-margined; petals c. equalling or slightly exceeding the calyx. Capsule shorter than the calyx.

Habitat: Sand or silty ground. Infrequent.

Northern Plains: 1479(BM), 50 km WNW an-Nu'ayriyah.

Southern Summan: 2150, 23 – 15N, 48 – 33E; 2232, 15 km WSW al-Hunayy; 2244, 26 km ENE al-Hunayy.

● 10. **Loeflingia** L.

Annual dwarf herbs with leaves opposite or whorled, subulate, with stipules adnate to the blade. Flowers minute, sessile, solitary or clustered. Calyx lobes 5, keeled, rigid, scarious-margined with a ciliate appendage at each side. Petals 3 – 5, minute, or absent. Capsule 3-valved, with several to numerous seeds.

Loeflingia hispanica L.

Glandular-pubescent, prostrate to ascending, dwarf annual, 5 – 10 cm high, branched from the base. Leaves subulate, 3 – 6 mm long; stipules filiform, connate with the leaf blades and forming lateral appendages. Flowers sessile, minute, axillary; calyx

c. 2 – 4 mm long. Capsule ovoid or subtrigonous, c. 2 mm long; seeds subovoid to ear-shaped, shining, minutely punctuate, c. 0.5 mm long, brown-grey.

Habitat: Sand or silty sand. Frequent.

Northern Plains: 1676(BM), 2 km S Kuwait border in 47 – 14E.

Northern Summan: 1388(BM), 42 km W Qaryat al-'Ulya; 1478(BM), 50 km WNW an-Nu'ayriyah.

Northern Dahna: 4023, 27 – 38N, 44 – 56E.

Central Coastal Lowlands: 1179(BM), Jabal an-Nu'ayriyah; 1418(BM), 7 km NW Abu Hadriyah; 1764(BM), Abu Hadriyah; 3727, 8 km N Abu Hadriyah; 3739, Ras az-Zawr; 7519, Ras Tanaqib; 1730(BM), Jabal al-'Amudah; 3276, 8 km W Qatif-Qaisumah Pipeline KP-144; 1427(BM), 2 km E Jabal Ghuraymil; 7722, 2 km S Ras Tanaqib.

● 11. Gymnocarpos Forssk.

Woody shrublet with opposite or whorled leaves. Flowers clustered in short-peduncled cymes. Calyx 5-lobed with an urn-shaped base; petals 5, minute, setaceous. Stamens 5. Ovary 1-locular; fruit an indehiscent utricle, 1-seeded.

Gymnocarpos decandrum Forssk.

Intricately branched stiff shrublet, 0.1 – 0.5 m high, with grey-white bark. Leaves opposite or whorled, teretish, linear, somewhat succulent, mostly 4 – 10 mm long, 1 – 2 mm wide. Flowers in short-peduncled, pubescent, bracteate clusters; calyx c. 5 – 7 mm long, with linear-oblong, hyaline-margined lobes connate at base, hooded at apex with a subapical mucro. Petals filiform, resembling the filaments, shorter than the calyx.

Habitat: Rocky, well-drained, usually elevated ground such as the limestone walls of wadis. Infrequent.

Northern Plains: 578(K), 15 km SSW Hafar al-Batin; 3096(BM), al-Batin, 28 – 00N, 45 – 28E.

Northern Summan: 2867, 26 – 01N, 48 – 29E; 3903, 27 – 41N, 45 – 35E; 8727, 3 km SE Mashdhubah, 27 – 19N, 45 – 04E.

● 12. Paronychia Mill.

Annual or perennial herbs with leaves mostly opposite, with conspicuous, silvery-scarious stipules and bracts. Flowers small, often immersed in scarious bracts. Calyx lobes mostly 5, nearly free, sometimes hooded or mucronate. Petals 5, filiform. Stamens 5 or 3. Ovary 1-celled; fruit a 1-seeded utricle.

Paronychia arabica (L.) DC.

Prostrate annual, reportedly sometimes perennnating, with stems puberulent, sometimes tinged reddish, to c. 30 cm long. Leaves sessile, narrowly oblanceolate or elliptical to linear, mucronate, 4 – 13 mm long, 1 – 2 mm wide, glabrescent. Stipules silvery-white, conspicuous. Flowers c. 1.5 mm long, axillary, sessile or short-pedicellate, hidden in bracts. *Plate 23.*

Generally easily recognizable by its silvery-white bracts and stipules.

Habitat: Sands and silts. Frequent.

Northern Plains: 3069(BM), 3075(BM), ad-Dibdibah, 28 – 45N, 47 – 01E; 1482(BM),

50 km WNW an-Nu'ayriyah; 3098(BM), 3112(BM), al-Batin, 28 – 00N, 45 – 28E; Dickson 603(K), Jabal al-'Amudah; 1268(BM), 15 km ENE Qaryat al-'Ulya; 563(BM), Khabari Wadha; 1531(BM), 36 km NE Hafar al-Batin; 876, 9 km S 'Ayn al-'Abd; 1155, Jabal al-Ba'al.

Northern Summan: 712, 33 km SW al-Qar'ah wells; 764, 69 km SW al-Qar'ah wells; 2813, 12 km ESE ash-Shumlul; 3222, 74 km S Qaryat al-'Ulya; 8624, Dahl Abu Harmalah, 26 – 29N, 47 – 36E.

Central Coastal Lowlands: 1778(BM), Abu Hadriyah; 65, Thaj; 413, 7426(BM), Dhahran; 1197, Jabal an-Nu'ayriyah; 1462(BM), 2 km E al-Ajam; 2715, 18 km SW as-Saffaniyah; 3217, an-Nu'ayriyah-as-Saffaniyah road junction; 3266, 8 km W Qatif-Qaisumah Pipeline KP – 144; 3303, 2 km W Sabkhat Umm al-Jimal; 7518, Ras Tanaqib; 2120, 3 km SW al-Jawamir (Yabrin area); 7042, Karan Island, Arabian Gulf; 7740, 5 km NW Ras Tanaqib.

Southern Summan: 2213, 26 km SSE al-Hunayy.

Vernacular Names: SHUHAYBA' ('silver-wort', Shammar), BUWAYDA' ('white-wort', Musil 1927, 594), SHADQ AL-JAMAL ('camel-jaw', Musil 1928b, 362; 1927, 623d).

● 13. Herniaria L.

Annual or perennial herbs or dwarf shrublets, often prostrate, with leaves opposite, sometimes alternate above, with small stipules. Flowers bisexual or unisexual, in axillary clusters. Calyx lobes 4 – 5; petals 5, filiform, short; stamens 2 – 5. Fruit a utricle.

Key to the Species of Herniaria
1. Perennial; sepals 4, unequal, fleshy .... 1. *H. hemistemon*
1. Annual; sepals 5, equal or subequal .... 2. *H. hirsuta*

1. Herniaria hemistemon J. Gay
Dwarf, often prostrate, finely pubescent perennial, 3 – 10 cm high, herbaceous from a weakly woody base. Leaves greyish green, 2 – 6 mm long, opposite (or appearing alternate through reduction of one of the pair), elliptical-oblong, ciliate, sessile. Stipules red-brown. Flowers minute, in axillary clusters. Seeds globular-ovoid, slightly compressed, c. 0.5 mm long, smooth and shining-brown. *Plate 24.*

Habitat: Most commonly in rocky places, in crevices on limestone or thin sand cover on limestone. Locally frequent but not widespread.

Northern Plains: 882, Jabal al-'Amudah.
Northern Summan: 815, 61 km NE Jarrarah.
Central Coastal Lowlands: 652, Dhahran; 1981, ad-Dawsariyah; 3211, Jalmudah.

2. Herniaria hirsuta L.
Prostrate dwarf annual, hirsute, with stems 4 – 20 cm long. Leaves sessile, mostly alternate, 2 – 7 mm long, elliptical-oblanceolate, tapering to base, ciliate. Stipules white, inconspicuous. Flowers sessile in axillary clusters, minute; calyx lobes 1 – 2 mm long, with white erect hairs; stamens 2 – 5.

Habitat: On sand or silty ground. More widespread and common than the above.

Northern Plains: 1479(BM), 30 km ESE al-Qaysumah; 1538(BM), 36 km NE Hafar al-Batin; 1650(BM), 2 km S Kuwait border in 47 – 14E; 7566, al-Faw al-Janubi, 2 km N al-Majma'ah highway.

Northern Summan: 1363(BM), 12 km W Qaryat al-'Ulya; 2814, 12 km ESE ash-Shumlul.
Central Coastal Lowlands: 3773(BM), Ras az-Zawr; 7741, 5 km NW Ras Tanaqib.
Southern Summan: 2082, al-Ghawar, 24 – 27N, 49 – 05E.

● 14. Sclerocephalus Boiss.

Annual herb with leaves opposite, terete-linear; stipules scarious. Flowers in dense, spherical, peduncled heads falling as a unit at maturity. Calyx lobes 5, hooded, spinescent. Petals 0; stamens 2 – 5. Ovary 1-celled; fruit a membranous utricle, 1-seeded.

Sclerocephalus arabicus Boiss.

Procumbent or ascending branched, glabrous annual with stems rather rigid, 2 – 10(15) cm long. Leaves 4 – 15 mm long, c. 1 mm wide, terete-linear, mucronate; stipules scarious. Flowers in dense, short-peduncled spherical heads, indurate and spiny, burr-like in fruit, 0.7 – 1 cm in diameter. Calyx c. 4 mm long, woolly below. *Plate 25.*

Easily identified by its spinescent heads.

Habitat: Shallow sands, occasionally silts, often in rocky areas. Frequent.

Northern Plains: 1159(BM), Jabal al-Ba'al; 1280(BM), 15 km ENE Qaryat al-'Ulya.
Northern Summan: 2775, 25 – 50N, 48 – 00E.
Central Coastal Lowlands: 174, 443, Dhahran; Dickson 579(K), W Abu Hadriyah (Burtt and Lewis 1952, 342); 3216, an-Nu'ayriyah-as-Saffaniyah road junction; 1251, Jabal an-Nu'ayriyah; 3272, 8 km W Qatif-Qaisumah Pipeline KP – 144.
Southern Summan: 2050, Jaww ad-Dukhan, 24 – 48N, 49 – 05E; 2184, 24 – 34N, 48 – 45E; 2034, 44 km E Khurays; 7510, Khashm az-Zaynah.
Vernacular Names: ḤARĀS ('thorn-wort', Qahtan, Bani Hajir), ḌURAYSAH ('tooth-wort', recorded as 'Al Thraisa' by Dickson 1955, 83).

● 15. Pteranthus Forssk.

Annual herbs with leaves opposite or whorled, small-stipulate. Flowers in dichasia with 3 flowers on a common peduncle that becomes dilated, leaf-like. Fertile flowers with 4 oblong-linear calyx lobes, hooded and spiny-tipped. Petals absent; stamens 4. Ovary 1-celled; fruit a membranous utricle.

Pteranthus dichotomus Forssk.

Puberulent annual, 5 – 25 cm high, branched, usually ascending. Leaves 8 – 20 mm long, 1 – 1.5 mm wide, linear. Inflorescence a 3-flowered dichasium on a common peduncle which becomes greatly broadened and leaf-like in fruit, the fruiting inflorescence becoming somewhat spinescent. Calyx 3 – 4 mm long; utricle c. 2 mm. *Plate 26.*

Readily recognizable by its very distinctive inflorescence structure.

Habitat: Usually on rocky, somewhat elevated ground. Locally frequent.

Northern Plains: 1628(BM), 3 km ESE al-Batin/al-'Awja junction; 1713(BM), 11 km NW Abraq al-Kabrit.
Northern Summan: 1350(BM), 42 km W Qaryat al-'Ulya; 3351, 3 km SE an-Nazim; 3869, 15 km WSW Umm 'Ushar; 3899, 27 – 41N, 45 – 19E; 8602, 17 km ESE Jabal Burmah, 26 – 41N, 47 – 45E.

Central Coastal Lowlands: 1227, Jabal an-Nu'ayriyah; 3286, 8 km W Qatif-Qaisumah Pipeline KP-144.

11. PORTULACACEAE

Annual or perennial herbs or shrubs with alternate or opposite, often succulent, stipulate leaves. Flowers bisexual, actinomorphic. Sepals often 2. Petals mostly 4 – 6. Stamens 3 to numerous. Ovary superior or partly inferior, 1-celled; fruit a capsule dehiscing by a lid.

● Portulaca L.

Annual or perennial herbs with leaves alternate or subopposite, stipulate, usually succulent. Flowers solitary or clustered. Sepals 2, connate at base, partly adherent to the ovary. Petals 4 – 6, yellow or red. Stamens 6 to numerous. Ovary with central placentation; capsule membranous, opening by a lid.

Key to the Species of Portulaca
1. Petals yellow or greenish; leaves obovate-cuneate .... 1. *P. oleracea*
1. Petals red; leaves linear to narrowly lanceolate .... 2. *P. pilosa*

1. Portulaca oleracea L.
Glabrous, succulent, branching herb, 10 – 30 cm high, ascending to procumbent, rarely erect. Leaves opposite or alternate, obovate-cuneate, obtuse, sometimes weakly emarginate, succulent, tapering at base, sessile or indistinctly petiolate, 1 – 2(4) cm long, 0.5 – 1(1.5) cm wide. Flowers mostly terminal or subterminal. Sepals 2, unequal, obtuse. Petals yellow, deciduous. Stamens 6 – 15. Capsule 6 – 8(10) mm long, ovoid, opening by a lid dehiscing less than ½ way above the base. Seeds reniform, nearly black, 0.5 – 0.8(1.2) mm long, finely tuberculate.
Habitat: A weed of gardens and walk edges. Common.
Central Coastal Lowlands: 7763, 8337, Dhahran. No. 8337 was an unusually erect and stout-stemmed plant, growing with a few others like it among much more numerous representatives of the usual procumbent, smaller-leaved form. It had capsules to c. 10 mm long and seeds 1 – 1.2 mm across; it perhaps represents a cultivated form.
South Coastal Lowlands: 7785, adh-Dhulayqiyah, al-Hasa.
Vernacular Names and Uses: BARBĪR (al-Hasa and al-Qatif farmers). The plant is sometimes cultivated and sold as a salad vegetable known by various names including BAQL.

2. Portulaca pilosa L.
Ascending, perennating succulent herb, 5 – 15 cm high, mostly glabrous but with conspicuous 3 – 6 mm-long white hairs in the axils; plant often turning reddish overall in sunny sites. Leaves alternate, spirally inserted, lanceolate-linear, acute, 10 – 15 mm long, 2 – 3 mm wide, subsessile, sometimes with a vermiform-reticulate pattern visible beneath the epidermis. Flowers 5 – 6 mm long, subtended by white hairs and with bright rose-red petals somewhat exceeding the paired sepals. Stamens more than 10. Capsule ovoid, c. 4 mm long, circumscissile near the base. Seeds reniform, dark brown to black, finely muricate, 0.5 – 0.7 mm long.
Habitat: Appeared and spread in Dhahran in the early 1970s and now a common weed

around lawns and walks. Usually occupies drier spots thant *P. oleracea*, near which it may be seen.

Central Coastal Lowlands: 3790(BM), 7764, Dhahran.

12. CHENOPODIACEAE

Annual or perennial herbs or shrubs with leaves alternate or opposite, exstipulate, often succulent, or reduced to scales. Flowers bisexual or unisexual, actinomorphic, with a uniseriate, sepaloid perianth of 3 – 5 lobes. Stamens 2 – 5, sometimes alternating with staminodes. Ovary usually superior, 1-celled, 1-ovulate, with styles 2 – 3. Fruit an achene, utricle or nut, with embryo circular or spiral; seed vertical or horizontal.

An important family in eastern Arabia with 42 species, including 2 endemics, in 20 genera. The largest genus is *Salsola*, with 8 species.

The family is maintained here with its traditional, broad limits although in the author's view there is much to be said for recent moves to divide it into three families: Chenopodiaceae *sensu stricto*, Salsolaceae (comprising in general the genera with winged or otherwise appendaged perianth), and Salicorniaceae (genera with flowers sunk in cavities in often joint-like stems). Scott (1977) keys these taxa and provides useful background data with a detailed review of the Salicorniaceae.

For practical purposes our chenopods tend to fall into three general groups, the first comprising less modified herbaceous weeds of wide Mediterranean or worldwide distribution, a succulent group found generally in salt marshes, and a larger one composed of the inland desert saltbushes. The latter are strongly adapted to dry and saline conditions, and some are dominant in important and widespread plant communities. The saltbushes differ from most other plants of eastern Arabia in their seasonal development, making active growth in summer and fruiting in the fall. Many species flower in September-October and have mature fruits in late October to middle November. Species in some genera, such as *Salsola* and *Halothamnus*, also have spring or summer fruiting seasons.

The saltbushes are difficult to key adequately because of their strong tendency to morphological convergence, often across generic and even family section lines. Seed position, whether vertical or horizontal, is often used in keys to the Chenopodiaceae at the generic level. The seed contains an embryo which is curved or spiral in one plane, and seed position is defined by the relationship of this plane to the axis of fruit insertion, whether normal to it and thus horizontal, or parallel to it and thus vertical. It is determined most easily in dissections of fresh, somewhat immature, ovules.

The presence of enlarged, transverse, membranous wings on the fruiting perianth lobes of several genera is an important diagnostic feature. These wings develop, however, only with fruit maturation and thus will be still absent or only a fraction of full size in specimens with young flowers.

Key to the Genera of Chenopodiaceae

1. Leaves (even when small or very narrow) flat and of 'normal' aspect, not or only moderately succulent.
    2. Fruit perianth with conspicuous radiating, membranous, wing-like appendages . . . . 15. *Salsola*
    2. Fruit perianth unwinged
        3. Leaves hairy

    4. Fruits with radiating spinelets, sometimes immersed in fleece;
       plants 10–50 cm high .... 4. *Bassia*

    4. Fruits unarmed, exposed; coarse herbs 0.5 to over 1 m
       high .... 5. *Kochia*

  3. Leaves glabrous or mealy

    5. Leaves linear, grasslike, tapering to apex, with parallel
       veins .... 6. *Agriophyllum*

    5. Leaves widened into a blade, veins not parallel

      6. Fruits enclosed in two flat, leaflike bracts .... 3. *Atriplex*

      6. Fruits without enclosing leaflike bracts

        7. Fruits separate, with herbaceous perianth .... 2. *Chenopodium*

        7. Fruits connate in pairs or 3's, falling as a unit; perianth
          indurated .... 1. *Beta*

1. Leaves completely absent or modified to non-'leaflike' structures

  8. Leaves or modified leaves virtually absent; plants with stems
   cylindrical, appearing composed of articulated segments

    9. Fruit perianth winged; desert shrubs

      10. Seed vertical; 2 dense bundles of fleece inside the perianth;
        flat-topped shrublet rarely exceeding 50 cm high ..... 18. *Anabasis*

      10. Seed horizontal; dense fleece bundles absent within perianth;
        rounded shrubs usually 60–300 cm high .... 17. *Haloxylon*

    9. Fruit perianth wingless; succulent seashore plants

      11. Flower triads immersed in a single, entire pit; many-stemmed,
        shrubby decumbent perennial .... 9. *Arthrocnemum*

      11. Each flower of triad in a separate septum-walled pit; annual
        with a single or few erect stems or branches .... 10. *Salicornia*

  8. Modified leaves present but reduced to budlike, scalelike, linear,
   succulent or prickly structures

    12. Leaves perfoliate, very succulent, subglobular, often turning
       red .... 7. *Halopeplis*

    12. Leaves not perfoliate, rarely red

      13. Leaves on lower branches modified to rounded, fleshy, budlike
        knots; upper stems appearing as crowded joints when in flower,
        often yellow with exserted stamens; low-growing procumbent
        shrublets of coastal salt marshes .... 8. *Halocnemum*

      13. Leaves not as above; plants ascending to erect, of desert,
        coastal, or *sabkhah* habitats

        14. Leaves modified to clasping, hardened, pointed, often
          prickly, scales; desert shrublet or herb .... 20. *Cornulaca*

        14. Leaves not clasping and not prickly

          15. Leaves and branches opposite

            16. Leaves linear, subterete, tapering at base, those
             on sterile branches at least 1 cm long, whitish;
             shrubs usually 60 cm or more high ..... 14. *Seidlitzia*

            16. Leaves club-shaped, green, usually 1 cm or less
             long, sometimes terminating in a deciduous bristle;
             shrublets 10–40 cm high .... 18. *Anabasis*

15. Leaves and branches alternate, or (in *Salsola*) leaves sometimes apparently absent.
    17. Fruiting perianth with membranous wings
        18. Leaves tipped with a fine spinule 1 – 4 mm long; seed vertical .... 19. *Agathophora*
        18. Leaves unarmed or absent; seed horizontal
            19. Fruiting perianth base indurate, bony, distinctly 5-pitted beneath, *or:* fruit 12 mm or more in diameter (including wings) .... 16. *Halothamnus*
            19. Fruiting perianth base not bony, not distinctly 5-pitted; fruits (across wings) under 12 mm in diameter ...... 15. *Salsola*
    17. Fruiting perianth not winged
        20. Densely villous plants; fruits with 5 radiating fine spines .. 4. *Bassia*
        20. Plants glabrous, or fleecy only in the axils; fruits spineless
            21. Stiff, hardly succulent shrublet with distant, somewhat hardened, triquetous, hornlike leaves .... 13. *Traganum*
            21. Shrubs or herbs with soft, succulent, terete or somewhat flattened leaves
                22. Fruits orbicular with fleshy, nearly entire wing; herb of intertidal mud or wet saltmarsh habitats .... 12. *Bienertia*
                22. Fruits not orbicular or winged; shrubs or shrubby herbs .... 11. *Suaeda*

● 1. Beta L.

Annual or perennial herbs with alternate leaves. Flowers bisexual, bracteate, usually clustered and more or less connate at base, axillary or in branching paniculate inflorescences. Perianth 5-lobed, growing in fruit. Stamens 5. Ovary with 2 – 3 stigmas. Seed horizontal.

Beta vulgaris L.

Erect or decumbent annual or perennial herb, stout, 20 – 100 cm high. Basal leaves oblong-ovate, obtuse, to 10 cm long; cauline leaves lanceolate to elliptical, acute. Flowers in loose spikes, 2 – 3 connate back-to-back in sessile groups. Perianth lobes growing to as long or longer than diameter of the fruit, incurved.

Habitat: Known only as a weed of farm fields or ruderal sites. Rare to occasional.
Central Coastal Lowlands: 8095, N Batn al-Faruq, 25 – 43N, 48 – 53E, Bedouin camp ground.
South Coastal Lowlands: 2740, farm near ar-Ruqayyiqah, al-Hasa.

● 2. Chenopodium L.

Annual or biennial herbs, sometimes subfrutescent, with leaves alternate, usually petiolate and more or less lobed or toothed. Flowers sessile, ebracteate, mostly bisexual, in spicate or paniculate clusters. Perianth lobes free or connate. Stamens (2 –)5. Stigmas 2(– 5). Fruit a depressed-globose, membranous utricle. Seeds mostly horizontal.

    A weedy genus of which two species, *C. murale* and *C. glaucum,* are well established and abundant in eastern Saudi Arabia. *C. glaucum* is a very troublesome farm weed;

both it and *C. murale* are frequently seen on waste ground around town and camp edges. The other two species are cosmopolitan weeds relatively rare in the Eastern Province but which may be encountered on disturbed ground and around cultivation, particularly near grain fields.

Key to the Species of Chenopodium

1. Leaves with obtusely toothed margins; plant usually procumbent, decumbent, or low-spreading .... 2. *C. glaucum*
1. Leaves with margins acutely toothed or entire; plants erect
   2. Younger leaves green below as above; seeds dull with dorso-ventrally acute margin .... 3.. *C. murale*
   2. Younger leaves grey-white below; seeds shining, with obtuse margin
      3. Nearly all leaves at least twice as long as broad, with teeth, if present, more or less regular .... 1. *C. album*
      3. Some leaves nearly as broad as long, with teeth mostly irregular .... 4. *C. opulifolium*

1. Chenopodium album L.

Erect glabrous annual, green or greyish mealy, to about 60 cm high. Leaves very variable in outline but usually at least twice as long as wide, generally lanceolate or deltoid, toothed more or less regularly, sometimes with a pronounced lobe near base on each side, the upper ones often lanceolate and nearly entire, greyish below at least when young, mostly 1 – 5 cm long, 0.3 – 3 cm wide; petioles c. half the length of the blade to c. equalling it. Flowers in dense terminal and axile clusters. Tepals keeled, greenish and often white-mealy. Seeds round, compressed but with rounded, obtuse margins, glossy black after falling from the caducous, dullish-membranous sheath, c. 2 mm in diameter.

Habitat: A farm and garden weed, rather infrequent in our area except sometimes around grain fields. Found in the desert once on a disturbed campsite where grain had been stored.

Central Coastal Lowlands: 7954, 7955, N Batn al-Faruq, 25 – 43N, 48 – 53E, on abandoned Bedouin camp site with Mediterranean weeds.

South Coastal Lowlands: 8213, Aramco farm near al-Qarn, al-Hasa; 8187, 8217, near ash-Shuqayq, al-Hasa; 8305, al-Hufuf, King Faysal University.

2. Chenopodium glaucum L.

Procumbent, decumbent or (rarely) erect annual, often red-flushed. Leaves narrowly oblong, petioled, fleshy, with a few obtuse teeth or sinuate, white mealy beneath, 12 – 35(45) mm long, 3 – 18(22) mm wide. Flowers grouped in somewhat distant clusters in lax axillary and terminal spikes. Seeds 1 – 1.2 mm in diameter, dullish.

Habitat: Weed of gardens, farms, waste ground. Frequent.

Central Coastal Lowlands: 2960(BM), 3761, Dhahran stables farm; C. Parker 119(BM), Qatif experimental station; 7809, Dhahran; 8157, near al-Jishsh, al-Qatif Oasis.

South Coastal Lowlands: 8214, Aramco farm, near al-Qarn, al-Hasa; 2733(BM), ar-Ruqayyiqah, al-Hasa; C. Parker 18(BM), al-Hasa experimental farm.

Vernacular Names and Uses: GHUBAYRA' ('dust-weed', referring to the grey-glaucous, somewhat mealy leaves, Qatifi farmers). A very troublesome and persistent weed in farm plots, often producing seedlings in great numbers.

3. Chenopodium murale L.

Erect green annual, to 40 cm high. Leaves 1.5 – 8 cm long, 1 – 4(6) cm wide, broadly triangular with coarsely and acutely toothed to incised margins, about equally green on both faces. Flowers in terminal and axillary clusters. Seeds suborbicular, compressed, c. 1.8 mm in diameter, dull due to persistence of the membranous sheath, the margins compressed dorso-ventrally and thus acute.

Habitat: Weed of garden, walk edges, waste places. Occasionally seen in desert, especially around old camp sites. Common.

Northern Plains: 1224(BM), Jabal an-Nu'ayriyah; 7848, Qaisumah Pump Station.

Northern Summan: 3164(BM), 11 km SSW Hanidh.

Central Coastal Lowlands: 128, Dhahran; Dickson 729(K), 'Ayn Najm, al-Hasa (Burtt and Lewis 1954, 380); 1072, 3762, Dhahran stables farm; 7076, al-'Arabiyah Island; 7044, Karan Island; 7036, Jana Island.

South Coastal Lowlands: 8186, 8216, near ash-Shuqayq, al-Hasa; C. Parker 21(BM), farm near al-Hufuf airport.

Vernacular Names: 'UWAYJIMĀN (Qahtan).

4. Chenopodium opulifolium Schrad.

Glabrous ascending annual, sometimes stout and up to 80 cm high. Leaves rather distant, ovate to deltoid, 1 – 5 cm long, 0.5 – 4 cm wide, (in ours) acutely serrate often with 5 – 8 teeth on each side, greyish beneath especially when young; upper leaves narrower, nearly entire; petioles rather fine, mostly at least half as long as the blades. Flowers in somewhat distant clusters in spicate-paniculate terminal and upper-axile inflorescences. Tepals green and keeled medially, with white margins.

The assignment of the records below to this species must be considered provisional. Although they appear to be closer to *opulifolium* than *album*, they could conceivably be one of the various atypical forms of the latter.

Habitat: A weed of farm or garden. Rare.

Central Coastal Lowlands: 7774, 7824(BM), 1 km S al-Qatif town in shaded garden; 7872, Dhahran.

● 3. Atriplex L.

Annual or perennial, monoecious or dioecious herbs or shrubs with leaves usually alternate and often more or less deltoid. Flowers unisexual, in axillary or terminal clusters. Staminate flowers with perianth 5-lobed and with 5(3) stamens. Pistillate flowers usually without normal perianth but enclosed in 2 leaf-like bracts or valves which often enlarge in fruit. Ovary ovoid, 2-styled; fruit a utricle; seed vertical.

Key to the Species of Atriplex

1. Annual; leaves (2.5)3 – 5(6) cm long, when fresh silvery with
   crystalline papillae .... 1. *A. dimorphostegia*
1. Shrubby perennial, leaves 0.3 – 2.5 cm long, all mealy without
   crystalline papillae .... 2. *A. leucoclada*

1. Atriplex dimorphostegia Kar. et Kir.

Whitish-canescent branched annual, 10 – 25 cm high, with stems prostrate to ascending. Leaves 2.5 – 6 cm long, alternate, entire to obscurely sinuate-dentate, deltoid or rhomboid to ovate-elliptical, usually tapering toward the base but sometimes subcordate, petioled or nearly sessile, silvery with shining crystalline papillae when fresh. Flowers in both axillary clusters and short terminal spikes. Fruiting valves broadly cordate, obtuse, reticulately nerved, c. 0.8 cm broad, obscurely denticulate. *Plate 27.*

Habitat: Inland sands, usually in *rimth* shrubland. Rare to locally occasional.

Northern Summan: 622(BM), al-Lihabah.

Central Coastal Lowlands: 3324, 12 km S Nita'; 3352, 10 km SSW as-Sarrar; 1447(BM), 2 km E Jabal Ghuraymil.

South Coastal Lowlands: 7835, 8 km SE al-Hufuf; 8301, al-Hufuf.

Vernacular Names: RUGHAYLAH ('little *rughl*', diminutive form of the name commonly applied to the following species; northern tribes, Musil 1928b, 359; 1927, 619; 1928a, 705).

2. Atriplex leucoclada Boiss.

Ascending shrub, 20 – 80 cm high, pale greyish or yellowish mealy-canescent. Leaves 0.3 – 2.5 cm long, deltoid, sinuate-dentate, rarely entire. Flowers both axillary and terminal, often densely clustered at the nodes. Fruit valves incised-dentate, tubercled, c. 4.5 mm long and broad. A polymorphic species, with leaf size and shape varying seasonally and by habitat, the summer and autumn leaves being smaller, often less than 5 mm long. *Plate 28.*

Habitat: Coastal zones and inland saltbush country. Occasional.

Northern Plains: 240(BM), 7451, Qaisumah Pump Station; 863, 9 km S 'Ayn al-'Abd; 3194, al-Batin, 28 – 00N, 45 – 27E; 3200, al-Batin, 5 km NE Dhabhah; 598, al-Batin, 16 km ENE Qulban Ibn Busayyis; 8810, 18 km WSW ar-Ruq'i Post.

Northern Summan: 803, 44 km SW Jarrarah; 2953(LE), 3128, 6 km SE ad-Dabtiyah.

Central Coastal Lowlands: 3733, Ras al-Ghar; 1107, 1109, 10 km W Abqaiq; 2926(LE), Za'l Island; 7520, Ras Tanaqib.

South Coastal Lowlands: 7837, 10 km SE al-Hufuf; 3814, 7 km SE al-Hufuf.

Vernacular Names: RUGHL (gen.).

● A sterile specimen (1565, BM) of an *Atriplex* from the Northern Plains, at the junction of al-Batin with Wadi al-'Awja, was reported as possibly attributable to *A. tatarica* L.

● 4. Bassia All.

Annual or perennial herbs or shrubs with alternate, sessile, entire leaves. Flowers bisexual or unisexual, axillary or terminal. Perianth urn-shaped, 5-lobed, sometimes prickly or tubercled in fruit. Stamens usually 5. Styles with 2 – 3 filiform lobes. Utricles included in the perianth; seed horizontal.

Key to the Species of Bassia

1. Fruit very cottony, entirely hidden in dense fleece; leaves
   oblong or linear-oblong, 2 – 3(4) mm wide .... 1. *B. eriophora*
1. Fruit fleecy, but spines protruding from hairs and
   visible; leaves narrowly linear, 1 – 2 mm wide .... 2. *B. muricata*

1. Bassia eriophora (Schrad.) Aschers.

Ascending, cottony-villous annual, 5 – 20 cm high, the fruiting stems appearing as white columns of cotton with the leaves partly hidden in the fleece. Leaves narrowly oblong to elliptical, 8 – 14 mm long, 2 – 3(4) mm wide. Flowers clustered, hidden in dense fleece. Fruiting perianth clothed in extremely dense white, cottony wool, with the perianth lobes and spines thus invisible without dissection, falling as dispersal units resembling small balls of cotton. *Plate 29.*

Habitat: Usually on somewhat silty soil on limestone or gravel plains. Occasional to locally frequent; sometimes seen on disturbed ground.

Northern Plains: 1618(BM), 3 km ESE al-Batin-al-'Awja junction.

Northern Summan: 610, 18 km NE Umm 'Ushar; 786, 100 km SW al-Qar'ah wells.

Southern Summan: 8344, al-Ghawar, 69 km N Harad.

Northern Dahna: 795, 15 km SE Hawmat an-Niqyan.

South Coastal Lowlands: 8293, al-Hufuf.

Vernacular Names: QUṬAYNAH ('cotton-weed', gen., Rawalah), QUṬṬAYNAH, ḤUMAYḌAT AL-ARNAB ('rabbit's salt-bush', Qahtan), ṬIRF (Shammar), ḤUMAYḌ ('little salt-bush', Al Rashid).

2. Bassia muricata (L.) Aschers.

Erect villous annual, 8 – 50 cm high, often somewhat frutescent. Leaves greyish-green, narrowly linear, 5 – 15 mm long, 1 – 2 mm wide, villous. Flowers solitary or clustered in the axils, fleecy. Fruit perianth indurate, villous below, the lobes prolonged in yellow, spreading, tapering spinules c. 2 – 4 mm long, exserted from the fleece of the perianth.

Habitat: Occurs in desert but more common as a weed of waste and disturbed ground. Frequent.

Northern Plains: 845, 9 km S 'Ayn al-'Abd; 1233(BM), Jabal an-Nu'ayriyah.

Northern Summan: 3131(BM), 6 km SE ad-Dabtiyah; 3147(BM), 9 km SSW Hanidh.

Central Coastal Lowlands: 229, Dhahran; 1859, Abqaiq.

South Coastal Lowlands: 1435(BM), 2 km E Jabal Ghuraymil; 2127, 4 km SW al-Jawamir (Yabrin area); 8021, Yabrin area, 23 – 11N, 48 – 53E.

Vernacular Names: QUṬAYNAH ('cotton-weed', Al Murrah), URAYNIBAH ('little hare-weed', Rawalah), SHUHAYBĀ' ('grey-weed', Shammar), LAYBID ('felt-weed', Qahtan), DHINNABĀN ('tail-weed', Bani Hajir).

● 5. Kochia Roth

Herbs or shrubs, often pubescent, with alternate, linear to lanceolate, sessile leaves. Flowers bisexual, or polygamous, sessile in clusters. Perianth 5-lobed, winged or unwinged dorsally. Stamens 5. Style bifid. Fruit a utricle included in the indurating perianth. Seeds mostly horizontal.

Kochia indica Wight

Stout, erect, much-branched, shrubby annual herb, 40 – 120 cm high. Upper branches glabrescent, whitish to yellowish. Leaves 5 – 15 mm long, 2 – 4 mm wide, linear-lanceolate, soft-hairy. Flowers in leafy axillary spikes which are loose below, tight above. Perianth lobes usually minutely and obscurely winged.

Habitat: A weed of roadsides and waste ground in agricultural areas. Rare to occasional.

South Coastal Lowlands: 3935, near 'Ayn al-Khadud, al-Hasa; 3939, 2 km ENE al-Hufuf; also seen in al-Qatif Oasis.

● 6. **Agriophyllum** Bieb.

Coriaceous annual herbs with alternate leaves. Flowers bisexual, solitary in the axils of densely clustered bracts. Perianth lobes 1 – 5, membranous. Stamens 1 – 5. Ovary compressed with style bilobed, persistent; fruit a compressed utricle with a wing at each side of the apex. Seed vertical.

Agriophyllum minus Fisch. et C. A. Meyer
Coriaceous, rather stiff annual, 8 – 30 cm high, green, usually entirely glabrous but sometimes, especially in the younger parts, with branched hairs; stems finely striate, often branched from near the base. Leaves alternate, 1 – 4(5) cm long, 1 – 5 mm wide, linear-lanceolate, tapering to a fine point with parallel veins and thus having the appearance of grass leaves. Flowers sessile in the axils of densely clustered, long-triangular, pungent, spinescent bracts. Perianth lobes whitish-membranous, included in the bracts. Stamens mostly c. 5. Ovary transversely compressed with a bilobed style. Utricle c. 3 mm long, compressed, with a persistent style and 2 long-triangular spreading wings at apex, somewhat exserted from the bracts. Seed oblong, brown, c. 1.5 mm long.

An interesting occurrence of this trans-Caspian Asian plant, also found in the deserts of Iran, in Qatar, and reportedly in the coastal United Arab Emirates.

The author is indebted to Mr. Ian Hedge, of Edinburgh, for calling attention to the description of a new species of *Agriophyllum* from eastern Saudi Arabia by Dr. A. El-Gazzar of King Faisal University: *A. montasiri* El-Gazzar (1988). El-Gazzar's specimens doubtless represent the same taxon as our material. The present writer, however, after some deliberation has decided to maintain provisionally here our original choice of name pending further study. El-Gazzar appears to split his new species from *A. minus* primarily if not entirely on the glabrous state of our plant (El-Gazzar 1988). At least one early authority, however (Boissier, *Fl. Orient.* 4:928 – 29, 1879), has described *A. minus* as either 'beset with sparse branched hairs' or 'glabrous'. Additionally, although nearly all of our specimens are glabrous, one (No. 8102) is quite densely furnished with branched hairs in young parts and sparsely so in older leaves. If further study proves our plant to be distinct at the species level on other morphological grounds, the new name should of course be adopted.

Habitat: Sand cover over coastal *sabkhah*, open flats among *Tamarix* shrubs; also further inland on sand; appearing sometimes as an invader on disturbed ground. Occasional, with range possibly increasing.

Central Coastal Lowlands: 634(K), near Safwa; 1455(BM), 2 km NE al-Ajam; 7391, 5 km S Safwa; 7914, 3 km S Jabal Ghuraymil, 25 – 46N, 49 – 32E; 8102, 5 km NW 'Uray'irah.

South Coastal Lowlands: 7791, 7796, adh-Dhulayqiyah, al-Hasa; 8139, 2 km N al-Muhtaraqah, al-Hasa.

● 7. **Halopeplis** Bge.

Annual herbs or shrubs with alternate, very succulent, clasping leaves. Flowers bisexual, grouped by threes in spike-like inflorences. Perianth 3-lobed. Stamens 1 – 2. Ovary compressed, with 2 stigmas. Fruit a utricle included in the perianth.

Halopeplis perfoliata (Forssk.) Aschers. et Schweinf.

Erect, glabrous, succulent shrublet, 20 – 40 cm high. Leaves very succulent, subglobular or pyriform, perfoliate, giving the stems a swollen, jointed appearance. Flowers in dense terminal spikes. Some or all of the plant, particularly the inflorescence, often is strongly red-colored in the succulent parts.

Habitat: Salt marshes at the coast or in *sabkhahs* near the coast.

Central Coastal Lowlands: 2925(LE, BM), Tarut Island; frequent around Dammam and Ras Tanura (*sabkhahs* near Rahimah).

South Coastal Lowlands: 1936(LE), 21 km SSE al-'Uqayr; 3811, Salwah.

Vernacular Names: KHURRAYZ (gen., Bani Hajir), a name derived from a word meaning 'glass beads', referring to the appearance of the stems strung with bead-like, perfoliate and often red-colored leaves.

● 8. Halocnemum M. B.

Spreading succulent salt marsh shrub with jointed stems and leaves opposite, reduced to connate scales. Flowers bisexual, mostly clustered by threes in the axils, forming dense spikes. Perianth lobes 3, free. Stamen 1. Stigmas 2. Fruit a utricle with vertical seed.

Halocnemum strobilaceum (Pall.) M. B.

Low, straggling shrub, 15 – 40 cm high, ascending at base but often at length procumbent with stems spreading on or near the ground. Leaf rudiments nearer base forming opposite, decussate, subspherical bud-like structures. Flowering branches nearer extremities apparently leafless, cylindrical, composed of articulated joints, branching or sometimes one-sided on the ground. Flowers immersed in the nodes, with the single stamens exserted, often leaving the whole stem yellow with anthers and pollen. *Plate 30.*

Habitat: Always in damp salt marsh, usually at the seashore but sometimes on hummocks or at margins of inland *sabkhahs*.

Northern Plains: Dickson 610(K), Dickson 716(K), N Ras al-Mish'ab (Burtt and Lewis 1954, 381).

Central Coastal Lowlands: 462, Dammam; 318(BM), 1103, Tarut Island; 9(BM), Sabkhat an-Nu'ayriyah.

South Coastal Lowlands: 3809, Salwah.

Vernacular Names and Uses: THULLAYTH (gen., Bani Hajir), 'UJAYRIMĀN (Bani Hajir). H. Dickson (1951, 318) notes a herdsmen's belief that overfeeding on this plant is a cause of lung disease in camels.

● 9. Arthrocnemum Moq.

Branching, succulent, salt-marsh shrubs with jointed, apparently leafless stems. Leaves reduced to minute connate scales forming a short cupule at each joint. Flowers bisexual, concealed by the bracts, immersed at equal height in groups of 3 in a common cavity in the spike-like inflorescence. Stamen 1. Seed with endosperm and crustaceous testa, granular or smooth.

Arthrocnemum macrostachyum (Moric.) Moris et Delponte
Arthrocnemum glaucum (Del.) Ung.-Sternb.
Glabrous, succulent shrub, 20–50 cm high with ascending to decumbent branches sometimes rooting in contact with the ground. Leaves virtually absent, reduced to minute, opposite, connate lobes 1 mm or less long, forming a cupule at each joint. Flowers minute, in terminal spikes in opposed groups of 3 at the nodes, partially exserted within connate bracts from an undivided cavity, the central flower of each triad slightly larger. The dispersal unit is the group of connate bracts of each flower, somewhat quadrangular, wedge-shaped at base, with the seed loosely held below. Seeds (in ours) ovoid, somewhat compressed, c. 1.5 mm long, containing endosperm; testa dark brown to black and somewhat shining, with a hard, muriculate surface.

● The seed characters of our plant clearly place it in this genus as critically defined by Scott (1977). Our plants have been seen to root from the nodes, however, a character Scott does not consider typical of *Arthrocnemum* as he delimits it.

Habitat: Protected seacoast and inlets, in intertidal mud or just above high tide line. May occur rarely around edges of inland *sabkhahs*. Locally frequent.

Central Coastal Lowlands: 460(K), Dammam; 1104(LE), 8682, west shore of Tarut Island; 1119, as-Saffaniyah.

Vernacular Names: SHU' (Bani Hajir).

● 10. **Salicornia** L.

Annual herbs with succulent, jointed, apparently leafless branches. Flowers bisexual, grouped by threes at the nodes and immersed in individual cup-like cavities arranged as a triangle, forming spikes. Perianth 3–4-angled with lobes connate. Stamens 1–2. Stigmas 2. Fruit an achene or utricle with seed vertical; seed without endosperm, with membranous hairy testa.

Salicornia europaea L.
S. herbacea (L.) L.
Glabrous, glaucous, succulent annual, 10–30 cm high, often with a single strict main stem with a few erect side branches, rarely rounded and bushy with many branches. Leaves virtually absent, reduced to opposite connate scales forming a minute cupule at stem joints. Flowers in terminal and terminal-lateral spikes, deeply immersed in triangular groups of 3 in the stem, each of the 3 in a pit separated by a septum from that of its neighbors. Seeds c. 1.2–1.5 mm long, greyish to pale brown, with a soft hairy surface.

Habitat: Wet coastal salt marshes and intertidal mud in protected bays. Occasional to locally common.

Northern Plains: 7972, bay at Ras al-Mish'ab; 1118, as-Saffaniyah.

Central Coastal Lowlands: 316(K), 1106, 8680, 8681, Tarut Island.

● 11. **Suaeda** Forssk. ex Scop.

Annual or perennial herbs, or shrubs, with succulent, mostly alternate leaves. Flowers usually bisexual, solitary or clustered in the axils, forming leafy spikes. Perianth 5-lobed. Stamens 5. Ovary free from the perianth or adnate to it, with 2–5 stigmas. Fruit a utricle with seed horizontal or vertical.

The author is indebted to Prof. H. Freitag of the University of Kassel for comments on some of our specimens. Some of our names must be considered provisional pending a badly needed regional revision of the genus.

Key to the Species of Suaeda

1. Stem leaves flattened above and below; large shrub or small tree, 1.5 – 3 m high . . . . 3. *S. monoica*
1. Stem leaves terete, club-shaped, or flattened above and rounded below; shrubs or herbs under 1.5 m high
  2. Ovary adnate to the perianth, thus appearing inferior; shrubby herb, usually of waste or disturbed ground . . . . 1. *S. aegyptiaca*
  2. Ovary free from the perianth
    3. Herb (sometimes semi-woody below) of intertidal seashore muds . . . . 2. *S. maritima*
    3. Shrubs, not of intertidal mud (but sometimes of coastal zones)
      4. Lower leaves oblong or obovate, flat above and rounded below, short petioled . . . . 4. *S. vermiculata*
      4. Lower leaves club-shaped, terete or only slightly flattened, sessile or subsessile . . . . 5. *S. fruticosa*

1. Suaeda aegyptiaca (Hasselq.) Zoh.

Schanginia aegyptiaca (Hasselq.) Aellen
Schanginia baccata (Forssk.) Moq.-Tand.

Shrubby, densely leafy, soft-succulent herb, glabrous, glaucous or somewhat mealy, up to about 60 cm high with stems erect and decumbent. Leaves teretish, linear-cylindrical, or sometimes somewhat flattened above, 10 – 20(30) mm long, 1 – 3(5) mm wide. Flowers clustered in leafy spikes. Fruiting perianth about 3 mm long, top-shaped with lobes becoming inflated, spongy, green, sometimes ripening to purple or black; ovary immersed in and adnate to the perianth.

Habitat: A weed of waste farm land and roadsides. Often on saline ground. Frequent.

Central Coastal Lowlands: 237(BM), Dammam; 1965, *sabkhah* W Abqaiq. Common around al-Qatif and al-Hasa Oases.

Vernacular Names: SUWWĀD ('black-bush', Philby 1922), HAṬLAS (Bani Hajir).

2. Suaeda maritima (L.) Dum.

Ascending, glabrous, blue-green annual, 10 – 40(80) cm high, sometimes branched. Leaves succulent, linear-semiterete, sessile, 10 – 20 mm long, 2 – 2.5 mm wide, shorter above. Flowers often reddish, clustered 3 – 6 in the axils near stem apices. *Plate 31.*

Habitat: Beach salt marshes, often in intertidal mud above the mangrove zone. May form dense stands locally but infrequent overall. Known only from the seashore in protected bays.

Central Coastal Lowlands: 1105(BM), 7809-A(LE), Tarut Island, W side.

3. Suaeda monoica Forssk. ex J.F. Gmel.

Shrub or small tree, up to 3 m high, with densely leafy, sometimes drooping, branches. Leaves linear, succulent, mostly 15 – 20 mm long, about 2 mm wide, flattened on both

1 Hills and escarpment marking the eastern edge of the northern Summan.

2 Spring annuals after good rains on the northern plains. *Diplotaxis acris* (pink flowers) and *Picris babylonica* (yellow) predominate here, along with scattered tussocks of *Stipa capensis*.

3 Sand ripples in the coastal lowlands, with *Calligonum comosum* (1.2 m high). Such terrain is typical of the sands that merge gradually in the south with the heavier dunes of the Jafurah and the Rub' al-Khali.

4 A *qawz*, or *Haloxylon persicum* shrub belt, in the south coastal lowlands (shrubs 1.3 – 2 m high). Larger areas of similar terrain are dominated by another saltbush, *Haloxylon salicornicum*.

5 Knolls and basins of the northern Summan. Spring annuals cover the silt-floored bottoms while the limestone hillocks bear only scattered, still-dormant shrublets of the *rimth* saltbush, *Haloxylon salicornicum*.

6 Typical dune terrain in the northern Dahna. *Calligonum comosum* and *Artemisia monosperma* are codominant in the characteristic plant community of these red sands.

Massive dunes rising 200 m
above saline flats in the
southeastern Rub' al-Khali.
Calligonum crinitum shrubs
may be found on the higher
sand bodies; scattered
Cornulaca arabica ascend the
flanks.

8 (Above left) *Ophioglossum polyphyllum* x 1 1/2

9 (Above right) *Ceratopteris thalictroides* (growing submerged in a slow-flowing irrigation channel)

10 *Ephedra alata* (staminate plant) x 1

11 *Ranunculus muricatus*
x 1 1/2

12 *Roemeria hybrida* x 1/5

13 *Hypecoum geslinii* x 2/3

14 *Aizoon canariense* x 1/2

15 *Aizoon hispanicum*
x 1 1/5

18 *Silene arabica*   x 1

17 *Mesembryanthemum nodiflorum*   x 1

16 *Mesembryanthemum forsskalei*   x 1 1/2

19 *Silene linearis* x 1 1/2

20 *Silene villosa* x 2/5

21 *Spergula fallax* x 1

22 *Spergularia diandra* x 1

24 *Herniaria hemistemon*
x 1 2/5

23 *Paronychia arabica* x 1

25 *Sclerocephalus
arabicus* x 1

26 *Pteranthus dichotomus* x 1/2

27 *Atriplex dimorphostegia* x 1 1/4

28 *Atriplex leucoclada* x 1/5

29 *Bassia eriophora* x 2/5

31 *Suaeda maritima* x 2/3

30 *Halocnemum strobilaceum* x 3/4

32 *Suaeda monoica* 3.5 m high

33 *Suaeda fruticosa* (island form) x 1

34 *Suaeda vermiculata* x 1/5

35 *Bienertia cycloptera* x 1

36 *Traganum nudatum* x 1

37 *Seidlitzia rosmarinus* x 1/4

38 *Salsola arabica* x 1/3

40 *Salsola cyclophylla* x 1

39 *Salsola arabica* x 2/3

41 *Salsola vermiculata* x 2

43 *Halothamnus iraqensis*
x 1/6

44 *Halothamnus iraqensis* x 1

42 *Salsola drummondii* x 1

45 *Haloxylon salicornicum* x 1/6

46 *Haloxylon salicornicum* x 1

48 *Anabasis setifera* x 1 1/3

47 *Anabasis lachnantha* x 1 1/2

49 *Agathophora alopecuroides* x 1 1/3

50 *Cornulaca arabica* x 1/4

52 *Aerva javanica* x 1/6

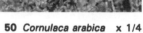

51 *Cornulaca monacantha*
x 1 1/2

53 *Rumex dentatus* x 1/10

54 *Rumex pictus* x 1/4

55 *Rumex vesicarius* x 1 2/3

sides, approximate. Flowers axillary in loose, leafy spikes. Fruit perianth 1 – 2 mm long, reddish when ripening. *Plate 32.*
Habitat: Sand hummocks near *sabkhah.* Rare.
Rub' al-Khali: 1016(BM, LE), Jawb al-'Asal, 22 – 55N, 49 – 59E.
Vernacular Names: ʿAṢAL (Al Murrah, gen.).
● This plant is known in our area only from a single remarkable site, a low, *sabkhah-*floored basin in the northern Rub' al-Khali. It is a Sudanian species, and the isolated stand here most probably is a relict of a wider population that extended northeast from the mouth of Wadi ad-Dawasir in earlier, wetter times. The nearest present population is in the mouth of that wadi, near as-Sulayyil, about 600 km from our site.

4. Suaeda vermiculata Forssk.
Shrub, 30 – 80 cm high, much branched with rather remote leaves, mostly glabrous but with youngest branches more or less pubescent. Leaves glabrous, blue-green, glaucous, oblong to obovate, flattened above, short petioled, 4 – 15 mm long, 2 – 6 mm wide. Flowers axillary in often loose, leafy terminal spikes. *Plate 34.*
Habitat: Saline ground at the coast; coastal or inland *sabkhahs.* Frequent.
Northern Plains: Dickson 717(K), 1.5 km N Ras al-Mish'ab (Burtt and Lewis 1954, 387).
Northern Summan: 3135(BM), 6 km SE ad-Dabtiyah.
Central Coastal Lowlands: 7048, Abu 'Ali Island; 317, 2922(LE, BM), Tarut Island; 12, Sabkhat an-Nu'ayriyah; 1116(BM, LE), as-Saffaniyah.
South Coastal Lowlands: 3808, Salwah; 2124, 4 km SW al-Jawamir (Yabrin area); 1941, 28 km S al-'Uqayr; 1950(LE), Nibak; 3802, 2924(LE, BM), 10 km NW Salwah; 7504, as-Samamik Island.
Vernacular Names: SUWWAD ('black-bush', gen.), ṬAḤMAʾ ('thick-bush', Rawalah, Shammar).

5. Suaeda fruticosa Forssk. ex J. F. Gmel.
Shrub, glabrous or nearly so, to about 100 cm high with stems often much branched. Leaves terete or slightly flattened above, linear or club-shaped, subsessile, 5 – 20(25) mm long, 1 – 2 mm wide, straight or bent. Flowers clustered 3 – 5 in the axils, forming leafy spikes. Stigmas 3 – 5, dark, filiform or subulate. Seeds c. 1 mm long, compressed-ovoid with a nearly black, shining, smooth surface. *Plate 33.*
Habitat: Apparently found both on saline roadsides and waste areas in oases and on offshore Gulf islands (see notes below). Infrequent except locally.
Central Coastal Lowlands: 2923(BM), Tarut Island, NW az-Zawr village; 7031, 1093, 2894, Jurayd Island; 7035, 7036, 7502(LE), Kurayn Island; also seen on Jana Island.
South Coastal Lowlands: 8674, 8678, 8679, 3 km NE al-Hufuf, al-Hasa Oasis; 8667, 6 km S al-'Uyun, al-Hasa Oasis.
Vernacular Names: No specific names recorded, but the general name for *Suaeda,* SUWWAD, may be expected.
● Our specimens from the Gulf islands differ rather remarkably in habit from the al-Hasa Oasis plants and merit further study. The oasis form is more typical of the species as usually described, having leaves longer (to c. 20 – 25 mm) and somewhat distant, and not being very densely leafy in the inflorescence. The island bushes are shorter, denser overall and densely leafy even into the inflorescnce, with leaves shorter

(generally under 12 mm). It had been suggested that this form was assignable to *S. vera* Forssk., but Prof. H. Freitag (personal communication) reports that our plant is incompatible with the type of *S. vera,* which has distinctive flattened, peltate stigmas. In his view, it lies within the range of variation of *S. fruticosa* as observed by him through much of the Middle East. Our two forms indeed do not appear to differ significantly in features of the flower, fruit or seeds. The different facies of the island plants may be an environmental effect of its maritime habitats.

● 12. **Bienertia** Bge.

Succulent, annual salt-marsh herb with alternate, linear, semi-terete, deciduous leaves. Flowers bisexual or polygamous, in terminal racemes. Perianth 5-lobed, in fruit surrounded by a circular, spongy wing. Stamens 5. Stigmas 2 – 3. Fruit a utricle adnate to the perianth. Seed horizontal.

Bienertia cycloptera Bge. ex Boiss.

Erect, glaucous, very succulent annual, 20 – 50 cm high. Leaves sessile, succulent, linear to oblong, obtuse, 1 – 3(3.5) cm long, 5 – 10(13) mm wide, deciduous. Flowers single or clustered at the nodes, in dense or loose racemes. Perianth adherent to the ovary. Fruit fleshy-orbicular, berry-like, surrounded by a fleshy circular wing 6 – 7 mm in diameter when dry, somewhat broader when fresh. *Plate 35.*

Coastal and inland forms of this species appear to differ in habit. Beach plants at Tarut Island were erect and less branched, with narrower leaves and fruits inserted singly in open racemes. Inland around 'Uray'irah, where the plant is common on disturbed saline ground, it grows in densely branched rounded forms with broader leaves and fruits clustered in dense racemes.

Habitat: Salt marshes, at the seashore or at inland *sabkhahs.* Infrequent except locally.

Central Coastal Lowlands: 466(K), 238, 461, 471, Dammam; 319(BM), Tarut Island; 11(BM), Sabkhat an-Nu'ayriyah; 1117, as-Saffaniyah; 1964, *sabkhah* W Abqaiq; 7810-A, Tarut Island, W side; 8815, 3 km S 'Uray'irah.

South Coastal Lowlands: 8140, 2 km N al-Muhtaraqah, al-Hasa.

Vernacular Names: HARṬALLAS (Al Murrah), HAṬALLAS (Bani Hajir), HARṬABÍL (northern), HURṬUMĀN (Shammar), ṬARṬAY' (Shammar), GHIḌRAF (Bani Hajir).

● 13. **Traganum** Del.

Shrubs with alternate, reduced, somewhat succulent triquetrous leaves. Flowers bisexual, axillary, 2-bracted. Perianth 5-lobed, furnished in fruit with 2 hornlike, obtuse, protuberances. Stamens 5. Stigmas 2. Fruit a depressed utricle included in the indurated perianth, with horizontal seed.

Traganum nudatum Del.

Diffuse shrub, 20 – 60 cm high with glabrous, whitish, rather virgate branches. Leaves triangular-lanceolate, subtriquetrous, sessile, somewhat fleshy, distant, up to 8 mm long, 4 mm wide, often decurved, shorter above, usually with 2 smaller, sometimes fleshy, rounded bracts at each side of base. Flowers sessile in the densely short-woolly axils. *Plate 36.*

Habitat: Shallow sand near *sabkhahs* or on higher, rocky ground, usually as scattered individuals. Occasional.

Northern Summan: 2957(BM), 6 km SE ad-Dabtiyah; 2846, 26 – 18N, 47 – 56E.

Central Coastal Lowlands: 6(BM), 1113, Sabkhat an-Nu'ayriyah; 222(BM), near 'Irj.

South Coastal Lowlands: 324(BM), Jaww ash-Shanayin, 25 – 50N, 49 – 18E; 1968, hill above Abqaiq salt pits; 2930(BM), 2935(BM), 24 – 09N, 48 – 37E; 2943(BM), 24 – 11N, 48 – 39E.

Vernacular Names: ḐUMRĀN ('thin-bush', gen.). Herdsmen often speak of grazing camels' special fondness for this shrub.

● 14. Seidlitzia Bge.

Glabrous shrubs or annual herbs with opposite, terete, succulent leaves. Flowers bisexual, solitary or clustered in the axils, 2-bracted. Perianth 5- or 2-lobed, furnished in fruit with transverse, scarious wings. Stamens 5; staminodes 5, thickened, glandular-ciliate. Stigmas 2. Fruit a depressed utricle with horizontal seed.

Seidlitzia rosmarinus Ehrenb. ex Bge.

Rounded glabrous shrub up to about 80 cm high, often raised on hummocks. Branches mostly opposite but sometimes alternate, often white-glossy. Leaves opposite, terete-succulent, club-shaped and obtuse, tapering toward base, up to about 18 mm long, 2 – 2.5 mm thick; floral leaves shorter, broader, 4 – 8 mm long. Flowers clustered 2 – 3 in the axils. Fruiting perianth about 10 mm in diameter (including wings); wings unequal, somewhat broader than long, inserted about ¼ below apex of the perianth lobes. Seed horizontal, black, compressed, coin-shaped. *Plate 37.*

Habitat: Usually on hummocks in or bordering coastal or inland *sabkhahs;* rarely on elevated rocky ground. Frequent.

Northern Summan: 2956(BM), 6 km SE ad-Dabtiyah.

Central Coastal Lowlands: 4(BM), 1114, Sabkhat an-Nu'ayriyah; Dickson 715, as-Saffaniyah (Burtt and Lewis 1954, 386); 8668, 8685, 8704, Dhahran, hill S of al-Midra ash-Shimali.

South Coastal Lowlands: 321(BM), Jaww ash-Shanayin, 25 – 47N, 49 – 22E; 1125, 10 km S Abu Shidad; 1953(LE), Jawb al-'Abd; 1962(LE), 11 km S Jal al-Wutayd; 1966, 7605, *sabkhah* W Abqaiq.

Rub' al-Khali: 7058, edge Sabkhat Matti, 23 – 16N, 51 – 43E; 7057, al-Mahakik, 22 – 42N, 51 – 43E; 7606, 5 km W 'Amad 2 well.

Vernacular Names and Uses: SHINAN, USHNĀN (gen.), DUWWAYD ('worm-bush', probably referring to the terete, succulent, leaves, Shararat, northern 'Anazah). Both Musil (1928a, 134) and Dickson (1955, 86) report the use by Bedouins of dried pounded *Seidlitzia* leaves as a soap substitute.

● 15. Salsola L.

Shrubs or annual herbs with leaves alternate or opposite, often reduced and more or less succulent. Flowers mostly bisexual, sessile in the axils, 2-bracteate, forming loose or dense spikes. Perianth 5-lobed, membranous or indurated at base, the fruiting lobes each furnished dorsally with a transverse, scarious wing. Stamens 5(4); staminodes usually 0. Stigmas 2. Fruit usually a utricle with seed horizontal.

Salsola includes important constituents of the desert saltbush vegetation, although it rarely provides dominants as do *Haloxylon* and some other Chenopodiaceous genera. The size and other characters of the leaf are variable in *Salsola,* with several species producing spring foliage quite different, usually longer and softer, than that of summer and fall. The dimensions of the mature fruiting perianth are generally constant and provide a character useful in many cases for confirming species identifications. *S. baryosma* is the common weedy species and is very often found on disturbed ground. The others, except rarely *S. volkensii* and *S. jordanicola,* are desert plants.

Key to the Species of Salsola

1. Leaves glabrous, succulent, terete, club-shaped or linear
    2. Leaves linear; fruiting perianth 3.5 – 5 mm wide ...... 7. *S. schweinfurthii*
    2. Leaves club-shaped or obpyriform; perianth
       5 – 7 mm wide .... 6. *S. drummondii*
1. Leaves pubescent or glandular, not succulent and terete; or leaves absent
    3. Perianth segments above the wings appressed, flattened-connivent, forming a disc concave toward the center; each segment with a raised flat medial band .... 4. *S. jordanicola*
    3. Perianth segments above the wings conical-connivent, ascending without interruption to an apex; segments without raised flat band.
        4. Perianth segments above the wings pubescent, at least near apex
            5. Annual; both older and young stems with erect hairs . 8. *S. volkensii*
            5. Shrublets with woody base; adult stems woody without erect hairs
                6. Plants having both: fruits more than 4.5 mm in diameter including wings, and some pubescent lanceolate-subulate leaves 2 mm or more long .... 5. *S. vermiculata*
                6. Plants not with above characters together
                    7. Fruit bracts swollen, strongly hooded, as high as or higher than the wings. Fruit perianth somewhat obscure even when mature .... 1. *S. arabica*
                    7. Fruit bracts not or faintly hooded, seldom reaching the wings; perianth conspicuous when mature 2. *S. cyclophylla*
        4. Perianth segments above the wings glabrous
            8. Fruit bracts appressed-hairy .... 5. *S. vermiculata*
            8. Fruit bracts glabrous, sometimes mealy .... 3. *S. baryosma*

1. Salsola arabica Botsch.

Shrublet, 20 – 60 cm high, rounded and many-branched, often grey-tan canescent above. Leaves sessile, lanceolate to broad triangular, somewhat triquetrous, 2 – 5 mm long, 1.5 – 2 mm wide, pubescent when young, at length glabrescent, sometimes absent. Softer but hardly larger pubescent leaves are produced in spring. Flowers in very dense congested lateral spikes 5 – 15 mm long, 4 – 5 mm wide, crowded with swollen, hooded, pubescent bracts. Fruiting perianth including wings 3 – 3.5 mm in diameter, somewhat obscure and the spikes sometimes appearing sterile without close examination. Wings proportionally smaller than in other local species. The plant has been found flowering in late March as well as in autumn. *Plates 38, 39. Map 8.2.*

Habitat: Shallow sand or silty soil on rocky ground along wadi banks. Rare.
Southern Summan: 2934, S of Wadi as-Sahba, 24–09N, 48–37E; 2945 (holotype, LE), 2941, tributary in N slope of Wadi as-Sahba, 24–11N, 48–39E; 7014(BM), near Khashm az-Zaynah, Wadi as-Sahba, 24–12N, 48–33E; 8068, N slope of Wadi as-Sahba, 24–12N, 48–38E; 8689, Wadi behind Khashm az-Zaynah, 24–12N, 48–36E; 8692, N slope of Wadi as-Sahba, 24–12N, 48–32E; 8694, N slope of Wadi as-Sahba, 24–12N, 48–34E; 8700, S slope of Wadi as-Sahba, 24–10N, 48–36E.

● The range of this recently described endemic, as far as known, is the smallest of any species in our territory, where it is limited to the banks of small runnels and ravines tributary to Wadi as-Sahbah around Khashm az-Zaynah. Its first record outside this 4 by 9 km discovery area was reported by Chaudhary (1986c) from a specimen collected near al-Aflaj by Dr Abdallah El-Sheikh. This extension in range to the southwest suggests it may be a southern Najd species with other occurrences to be expected along the rocky banks of wadis draining eastward from the back of the Tuwayq Escarpment, perhaps extending to central Arabia west of Tuwayq. In the author's experience, the plant is regularly associated with other saltbushes in mixed stands including *Salsola cyclophylla*, *Salsola schweinfurthii*, and *Traganum nudatum*. It appears to be allied to *Salsola cyclophylla* in some respects, and if it is not just a relict species, a possible hybridization or polyploid relationship with that relative might be worth investigating.

2. Salsola cyclophylla Bak.

Stiff shrublet, 10–50 cm high, sometimes dwarfed but always strongly woody at base, intricately branched. Leaves suborbicular, crowded into bud-like knots. Flowers in very dense short lateral spikes, 5–15 mm long, 3–7 mm in diameter, which are often rather distant on terminal branches. Fruiting perianth 3–5 mm in diameter, including wings. *Plate 40.*

● As suggested by its morphology, this plant is extremely drought resistant and may be found in the driest imaginable spots on elevated rocky ground. Northern and southern specimens differ in inflorescence dimensions, the far northern ones having broader spikes, 5–7 mm in diameter, and the broader perianth. This species is most probably the *'Salsola tetrandra'* noted by Vesey-Fitzgerald (1957, 796) from the southeastern edge of the Eastern Province.
Habitat: Usually in rocky, hilly country with sand cover thin if present. Frequent.
Northern Plains: 3191(BM), 3192, al-Batin, 28–00N, 45–27E; 3201(BM), al-Batin, 28–02N, 45–32E; 7853, al-Batin, 28–05N, 45–35E.
Central Coastal Lowlands: 1, 16(BM), 3210, 640(BM), Dhahran.
Southern Summan: 1121, 40 km W Khurais-al-'Udayliyah junction; 2932(BM), 24–09N, 48–37E; 2942(BM), 24–11N, 48–39E; 3828, 25–18N, 48–59E.
South Coastal Lowlands: 327(BM), Jaww ash-Shanayin; 1131, al-'Udayliyah; 1937(LE), 21 km SSE al-'Uqayr; 1948, 33 km WSW Salwah; 1959(LE), 40 km WNW Salwah; 1970, hill above Abqaiq salt pits; 7063, Dawhat Duwayhin, 24–25N, 51–19E.
Rub' al-Khali: 8402, 8403, 8404, near Camp G–3, 20–25N, 55–08E; 8407, 8408, Camp G–3, 20–26N, 55–07E.
Vernacular Names: ‘ARAD ('hard-bush', gen.), ḤAMḌ AL-ARNAB ('hare's salt-bush', Bani Khalid).

3. Salsola baryosma (Roem. et Schult.) Dandy

Ascending shrub 30–60 cm high, glabrescent or mealy, or soft pubescent when young.

Leaves varying seasonally, those in spring linear-subtriquetrous, hairy, about 3–10 mm long, often on reddish new stems; those of late summer and autumn suborbicular and minute, glabrescent, sometimes imbricated in dense groups. Flowers often in dense lateral spikes that are 10–25(150) mm long, 5–10 mm in diameter, the spikes sometimes in a rather large, branching, paniculate inflorescence. Fruiting perianth 4–6 mm in diameter including the wings. The fresh plant is more or less fetid when crushed.

We follow this long-used name rather than the recently revived *S. imbricata* Forssk. because of the uncertainties regarding the plant to which Forsskal applied his name, as pointed out by Prof. H. Freitag (personal communication 1989).

Habitat: More frequent as a weed of waste ground in inhabited areas than as a desert shrub, but sometimes occurring in the desert in silty or rocky terrain. Frequent.

Northern Plains: 3118(BM), 3095(BM), al-Batin, 28–01N, 45–29E; 3201(BM), al-Batin, 28–02N, 45–32E.

Northern Summan: 2954(BM), 6 km SE ad-Dabtiyah.

Central Coastal Lowlands: 15(LE), Dhahran airport; 458(LE), Dhahran stables farm; 470(LE), 1930(LE), Dhahran; 1115(LE), Sabkhat an-Nu'ayriyah; 639(BM), Ras al-Ghar; probable record from Jurayd Island.

South Coastal Lowlands: 3817(BM), 513(LE), 8 km N al-Mutayrifi, al-Hasa; 1896(BM), 1 km NE al-Mutayrifi, al-Hasa; 2908(BM), al-'Arfaj farm, al-Hasa; 1940(LE), 1942(LE), 28 km S al'Uqayr.

Vernacular Names: KHURRAYṬ (Bani Khalid, gen.) KHARĪṬ (Mutayr).

4. Salsola jordanicola Eig

Ascending annual, sometimes frutescent, 15–30 cm high and much branched from the base. Stems sometimes pubescent when young but mature fruiting branches mealy or scurfy without hairs. Leaves linear or oblong-linear, 10–20 mm long, 1.5–2 mm wide, dilated at the base, but soon deciduous and the plant often leafless when in fruit. Flowers solitary in the axils, distant or sometimes somewhat crowded, spread over the stems. Fruit bracts persistent, forming 3–4 mm-wide shallow cupules in the axils after fall of fruit. Fruiting perianth 9–11 mm in diameter including wings, with mature perianth segments (especially after drying) depressed-connivent above the wings. Perianth above wings scurfy-mealy, sometimes appearing scurfy-pubescent, each lobe with a slightly raised, scurfy medial band.

Habitat: Shallow silty soil over limestone. Rare.

Northern Plains: 8812, 2 km E ar-Ruq'i wells.

Northern Summan: 508(K,LE), 15 km N Jibal al-Harmaliyat.

Southern Summan: 3829, 25–18N, 48–59E.

Central Coastal Lowlands: 8817, 3 km N Judah.

South Coastal Lowlands: 512(BM), 8 km N al-Mutayrifi, al-Hasa; 7839, 5 km SE al-Hufuf; 7776, 7803, 7 km SE al-Hufuf.

Vernacular Names: QAḌQAḌ ('Anazah).

5. Salsola vermiculata L. s.l.

Salsola mandavillei Botsch.

Shrublet, 15–60 cm high, much branched with stems glabrous below, pubescent toward extremities. Leaves 2.5–10 mm long, 0.4–0.7 mm wide, long-triangular to linear-subulate, pubescent, longer in spring. Spring flowers axillary, forming somewhat loose spikes up to c. 1.5 – 12 cm long on upper branches; autumn flowers often in shorter, more congested spikes, 1 – 3(6) cm long. Fruiting perianth 5 – 8(9) mm wide (including wings), with the connivent lobes above the wings pubescent in spring flowers but in autumn often glabrous or hairy only near the apex. The wings may be straw-yellow or pinkish. *Plate 41.*

This species has a spring, as well as a fall, flowering phase.

Habitat: Rocky slopes with little sand cover or silty bottoms of basins and the larger wadis in the north. Relatively rare in the Eastern Province.

Northern Plains: 3093(BM), 3119 (holotype of *S. mandavillei* Botsch., LE), al-Batin, 28 – 01N, 45 – 29E; 3190(BM), 3195(BM), al-Batin, 28 – 00N, 45 – 27E; 3202(BM), al-Batin, 28 – 02N, 45 – 32E. Sterile specimens probably attributable to this species are: 1583(BM), al Batin, Wadi al-'Awja junction; 754, 33 km SW al-Qar'ah wells; and 3907(BM), 27 – 41N, 45 – 48E.

● This plant is the well-known *rūth* saltbush of northern Arabia; the Eastern Province records represent the southeastern edge of its range.

Botschantzev, in his revision (1975b) of *Salsola* subsection *Vermiculatae*, described our plant as a new species, *S. mandavillei*, a segregate of the confused complex to which the names *S. villosa* Del. ex Roem. et Schult., *S. vermiculata* L. subsp. *villosa* (Del.) Eig, and *S. vermiculata* var. *villosa* Moq. have been variously applied. It would seem preferable here to follow the early name, in a broad sense, pending resolution of some apparent remaining taxonomic questions in the group.

Vernacular Names and Uses: RŪTH (Mutayr, gen. N); reported by Bedouins to be a favorite grazing plant of the camel, and often used for firewood. Many formerly good stands in wadis northwest of our territory have been much depleted by the pressure of these uses. Its importance to the northern tribes is suggested by a folk anecdote that relates how a Rawalah tribesman, upon death, asked his heavenly judge whether *rūth* was to be found in paradise. When told that it was not, he replied that in that case, he would simply prefer to go elsewhere.

6. **Salsola drummondii** Ulbrich

Ascending glabrous shrublet, 30 – 60 cm high, with youngest branches whitish, becoming brown and knotty below. Leaves usually alternate, sometimes opposite, sessile, terete-succulent, short-club-shaped or obovoid, often obpyriform, mostly 3 – 7 mm long. Bracts succulent, concave above beneath the perianth. Flowers in lateral and terminal spikes (1)2 – 5(9) cm long, 0.5 – 0.8 cm wide, the fruiting perianth 5 – 7 mm in diameter including wings. Perianth lobes above the wings membranous or chartaceous, less than 1 mm long, not connivent and leaving the ovary exposed; wings inserted near apex of the perianth lobes, imbricate, suborbicular, equal or somewhat unequal, sometimes involute at base, flushed rose or straw-colored. *Plate 42, Map 8.2.*

Habitat: Saline ground around near-coastal *sabkhahs* and depressions; rarely on rocky, elevated ground. Infrequent.

Central Coastal Lowlands: 8669, 8670, 8683, 8684, Dhahran, hill south of al-Midra ash-Shimali.

South Coastal Lowlands: 1939(LE), 10 km N al-'Uqayr; 1951(LE), Nibak; 3810(BM), Salwah; 3812, as-Sikak; 7062, near Batn at-Tarfa, 23 – 55N, 51 – 33E; 7064, Dawhat Duwayhin, 24 – 25N, 51 – 19E; 7065, Jaww Butayhin, 24 – 09N, 51 – 33E; 7503, az-Zakhnuniyah Island.

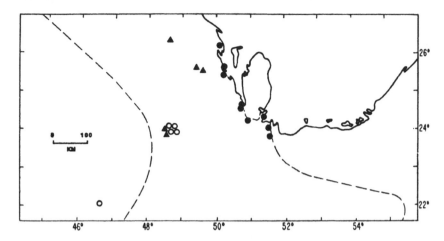

Map 8.2. Distribution as indicated by collection sites of *Salsola drummondii* (●), *Salsola schweinfurthii* (▲), and *Salsola arabica* (○). The record for *S. arabica* outside the area of this flora (dashed lines) was reported by Chaudhary (1986c). The specimen was collected at al-Aflaj.

● In our territory the distribution of this species, originally described from northwestern India, is coastal and southern. Our records apparently represent the westerly limit of its total range. The author is indebted to Prof. H. Freitag of the University of Kassel for the identification of this plant, which has been confused with *S. schweinfurthii* (below). It differs from the latter by its larger fruiting perianth and short-clavate to obpyriform (not cylindrical) leaves.

7. Salsola schweinfurthii Solms-Laub.

Ascending glabrous shrublet, 15 – 50 cm high with branches opposite or alternate, the younger ones smooth, whitish. Leaves opposite, subopposite or alternate, sometimes clustered, sessile with short-woolly axils, terete, succulent, straight or sometimes curved, glaucous, grey- to bluish-green, sometimes tinged reddish, 5 – 12(14) mm long, 2.5 – 3.5 mm wide when fresh, rounded at apex with a minute bluntish mucro. Flowers tightly grouped 3 – 5 in the upper axils, forming dense continuous or interrupted terminal spikes up to c. 16 cm long, 0.4 – 0.7 cm broad. Bracts succulent, more or less ovoid, obtuse, rounded and hardly keeled at back, concave and narrowly white-hyaline-margined beneath the flowering perianth. Fruiting perianth 3.5 – 4.5 mm wide (including wings); lobes ovate, c. 1.7 mm long, white-hyaline-margined, mostly obtuse and somewhat erose at apex; wings often unequal, 1 – 1.5 mm long, more or less circular, inserted near the middle of the perianth lobe. Staminodes hardly developing, very short, broader than long; ovary ovoid, narrowing at apex, with stigmas lanceolate-triangular, flattened, darkening with age. Seed horizontal, short-conical at base, flat above. (Description based wholly on our specimens). *Map 8.2.*

Habitat: Rocky limestone slopes or rocky ravines. In mixed saltbush communities; infrequent to rare.

Northern Summan: 3172(LE,BM), 15 km SSW Hanidh; 8835, 18 km S Hanidh.
Southern Summan: 8686, 8687, N slope of Wadi as-Sahba near Khashm az-Zaynah; 8690, wadi behind Khashm az-Zaynah; 2931(BM), 24 – 09N, 48 – 37E.

South Coastal Lowlands: 7594, Jaww ash-Shanayin, 25 – 45N, 49 – 22E; 7832, Burq Aba ad-Dalasis, 25 – 43N, 49 – 31E.

Our specimens were determined by comparison with type-locality material through the kind assistance of Prof. H. Freitag.

8. Salsola volkensii Aschers. et Schweinf.

Erect or ascending, somewhat fetid annual, 10 – 40 cm high, covered with erect white hairs on both young and older stems; blue-green when fresh, drying to a yellow-brown. Leaves alternate, sessile, linear, 3 – 6 mm long, 2 mm wide, deciduous, the floral ones shorter; summer leaves smaller, often suborbicular and crowded. Flowers solitary, in loose or dense spikes. Fruiting perianth 7 – 10 mm in diameter including the wings, the lobes above the wings white-margined and beset medially with white hairs. Aspect of a small *S. baryosma* but differing by its larger and pubescent fruiting perianth and by the erect pubescence of its stems.

Habitat: Recorded mainly in our extreme north and on more or less disturbed ground, where it occupies the same ruderal habitats favored by *S. baryosma* in more southerly parts of the Province. In 1988 it was collected as far south as Judah, at a roadside, and may be increasing in range. Rare except very locally.

Northern Plains: 314(BM), 7452, 7845, Tapline Pump Station grounds, Qaisumah, 28 – 19N, 46 – 07E; 8809, ar-Ruq'i wells, 29 – 00N, 46 – 32E.

Central Coastal Lowlands: 8818, 3 km N Judah.

Vernacular Names: KHIḌRĀF (north).

● 16. **Halothamnus** Jaub. et Spach
Aellenia Ulbr.

Annual or perennial herbs or shrublets with alternate leaves mostly reduced to more or less triquetrous or semiterete scales, or rarely subulate. Flowers bisexual, mostly solitary and axillary, 2-bracted. Perianth 5-lobed, in fruit indurate with the base hardened and bearing 5 pits around the point of insertion, each lobe furnished dorsally with a transverse, scarious, wing. Stamens 5; staminodes 0. Ovary depressed-globose with a bifid style. Fruit a utricle enclosed in the indurate perianth; seed horizontal. Closely related to *Salsola* and distinguished from it by the perianth with broadened, 5-pitted base.

Key to the Species of Halothamnus

1. Fruiting perianth (including wings) 4 – 8 mm in diameter; plant finely glandular .... 1. *H. bottae*
1. Fruiting perianth (including wings) 12 – 15 mm in diameter; plant not glandular .... 2. *H. iraqensis*

1. **Halothamnus bottae** Jaub. et Spach
Salsola bottae (Jaub. et Spach) Boiss.

Ascending, many-branched shrublet, sometimes dwarfed, 10 – 30 cm high, finely glandular all over, blue-green greyish when growing, dead stems yellow-brown. Leaves reduced to triangular-triquetrous, subclasping and sometimes decurrent rudiments, 0.5 – 1.5 mm long. Flowers usually solitary in the axils, distant, over much of the upper lateral branchlets. Fruits, including wings, 4 – 8 mm in diameter with perianth segments connivent above the wings in a narrow cone that is higher than its basal diameter. Plants have been collected flowering and fruiting in May and through mid-summer as well as in the autumn. *Map 8.3.*

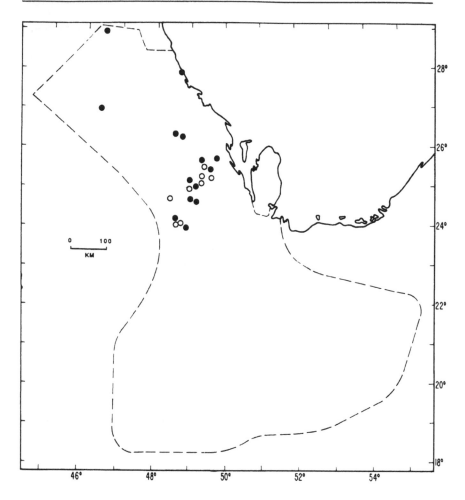

Map 8.3. Distribution as indicated by collection sites of *Halothamnus iraqensis* (●) and *Halothamnus bottae* (○) in the coverage area of this flora (dashed line), eastern Saudi Arabia. Each symbol represents one collection record.

Habitat: Rocky limestone hills with very thin sand cover. Infrequent except rather locally.

South Coastal Lowlands: 1052(BM), 24 km N al-'Udayliyah; 1132(LE), al-'Udayliyah; 7073, N al-'Udayliyah, 25 – 15N, 49 – 20E; 7843, 8317, 5 km SE al-Hufuf; 8324, 18 km WNW al-Hufuf; 8331, 25 km NW al-Hufuf.

Southern Summan: 8382, 39 km WSW al-Hunayy on Darb Mazalij; 8385, Darb Mazalij, 24 km ENE al-Hunayy; 8391, Darb Mazalij, 25 – 02N, 49 – 00E; 8695, Wadi as-Sahba near Khashm az-Zaynah.

Vernacular Names: ḤAMḌ AL-ARNAB (Al Murrah), ṬIḤYAN (Al Rashid; cf. 'tahyin' of Thesiger 1948, 3).

2. **Halothamnus iraqensis** Botsch.

Salsola subaphylla C. A. Meyer quoad nom.

Aellenia subaphylla (C. A. Meyer) Botsch. ex Aellen quoad nom.

Ascending, much-branched shrublet, glabrous or glabrescent, 20 – 50 cm high. Spring season leaves fine-linear, 3 – 15(20) mm long, 1 – 2 mm broad, terete or somewhat flattened ventrally near their bases; summer and autumn leaves reduced to triangular rudiments, 1 – 1.5 mm long. Flowers in spikes on lateral shoots, the spikes often rather loose but sometimes crowded with the wings of adjacent perianths overlapping. The rather showy fruiting perianth of this species, with its wings, is the broadest of the wing-fruited saltbushes in eastern Arabia, measuring (11)12 – 15 mm in total diameter. Perianth wings striate, translucent, yellowish or tinged rosy-pink. *Plates 43, 44; Map 8.3.*

Habitat: In saltbush country in shallow sand in hilly, somewhat rocky terrain and usually as scattered, single individuals. Occasional.

Northern Plains: 8808, ar-Ruq'i wells; 29 – 00N, 46 – 32E.

Northern Summan: 785, 69 km SW al-Qar'ah wells; 2952, 6 km SE ad-Dabtiyah; 224, Hanidh road near 'Irj.

Central Coastal Lowlands: 510(K), 1969, hill near Abqaiq salt pits; 7(BM), as-Saffaniyah; 326, 7595, Jaww ash-Shanayin; 1124, 10 km S Abu Shidad; 1129, al-'Udayliyah; 3816, 8 km N al-Mutayrifi.

Southern Summan: 3827(BM), 25 – 18N, 48 – 59E; 2929, Jaww ad-Dukhan, 55 km SSW al-'Udayliyah; 1867, 24 – 42N, 49 – 08E; 1123, 23 km W Khurais-al-'Udayliyah junction; 2936, 24 – 11N, 48 – 37E; 8701, Wadi as-Sahba near railroad km 340.

Vernacular Names: QADQAD (Al Murrah, 'Ujman, ad-Dawasir).

● 17. **Haloxylon** Bge.

Shrubs or small trees with terete, jointed stems, appearing leafless. Leaves opposite, reduced to minute, connate scales forming cupules at the nodes. Flowers bisexual, axillary, in dense lateral spikes. Perianth lobes 5, in fruit furnished dorsally with scarious, translucent, transverse wings. Stamens 5; staminodes 5. Ovary with 2 – 5 stigmas. Fruit a depressed utricle with horizontal seed.

Key to the Species of Haloxylon

1. Flowering shoots (the final lateral spikes) 2 – 3 mm in diameter and
   flowering nodes usually more than 5 per spike, crowded; fruiting
   perianth, including wings, 5 – 7 mm in diameter; shrublet usually
   0.6 – 1 m high . . . . 1. *H. salicornicum*
1. Flowering shoots 1 – 1.5 mm in diameter with flowering nodes
   usually 1 – 4 per shoot, distant; fruiting perianth, including wings,
   c. 8 mm in diameter; large shrub 1.5 – 3 m high with thick
   woody base . . . . 2. *H. persicum*

1. **Haloxylon salicornicum** (Moq.) Bge.

Hammada salicornica (Moq.) Iljin

Hammada elegans (Bge.) Botsch.

Diffuse, rounded, many-branched shrub, (30)60 – 100 cm high, often raised on hummocks. Leaves apparently absent, reduced to minute connate scales forming cupules at articulations of the naked cylindrical stems. Flowers in dense, 2 – 8 cm or more-long lateral and terminal spikes on upper branchlets. Fruiting perianth with 5 membranous yellow or pink wings inserted ½ to ⅔ from base of the lobes, 5–6 mm in diameter, including the wings. Fruiting perianth segments connivent above the wings, enclosing the ovary. *Plates 45, 46.*

● This widespread and ecologically important shrub is sometimes placed in the genus *Hammada* Iljin and assigned the names we give above as synonyms. In view of some doubts about the distinctiveness of Iljin's genus, as well as that of the form given the epithet *elegans,* we prefer to follow the earlier, broader nomenclature. This is the dominant species of a very important plant community covering thousands of square kilometers of parts of our area from its northern borders to the northern fringes of the Rub' al-Khali. Thalen (1979) provides a detailed review of its ecology in Iraq.

Habitat: Shallow to deep sands or silts, mostly in lower, less well-drained terrain. Also in gravel plains and in rocky country, where it is usually stunted. May also be found outside its usual community habitat on roadsides, where it sometimes becomes established on disturbed ground. Common.

Northern Plains: 3199, al-Batin, 5 km NE Dhabhah; 3206, al-Batin, 28 km SW Hafar al-Batin.

Northern Summan: 2950(BM), Jaww Ghanim.

Central Coastal Lowlands: 3767, 70 km NW Qatif Junction; Dickson 524(K), S Abqaiq (Burtt and Lewis 1954, 383); Dickson 580B(K), Tapline road W Jubail (Burtt and Lewis 1954, 383).

South Coastal Lowlands: 325(BM), 7598, Jaww ash-Shanayin, 25–45N, 49–22E; 1943(LE), 28 km S al-'Uqayr; 1949(LE), 33 km WSW Salwah; 1961(LE), 40 km WNW Salwah; 7070, al-Jafurah, 24–15N, 49–50E.

Southern Summan: 2938(BM), 24–09N, 48–31E; 8697, 23–53N, 48–50E.

Rub' al-Khali: 1954, 1 km W Bunayyan; 7027, 23–02N, 49–41E; 7507, 13 km SE Niqa Fardan, 21–53N, 48–34E.

Vernacular Names and Uses: RIMTH (gen.). An important camel grazing plant during summer and fall when annuals and grasses are not available. It may provide salts beneficial to the camel, although long and excessive saltbush grazing is recognized to be physiologically damaging. The milk of camels feeding on the plant is sometimes somewhat salty-tasting. It also provides good firewood.

2. Haloxylon persicum Bge.

H. ammodendron (C. A. Meyer) Bge.

Large shrub or small tree, 1.5–3(4) m high with thick woody base, erect, sometimes with drooping terminal shoots. Leaves apparently absent, the stems appearing naked-cylindrical, jointed. Flowers in short lateral spikes on thin, 1–1.5 mm-diameter lateral shoots or terminal branchlets. Fruiting perianth with spreading membranous yellowish wings inserted near tips of the lobes, c. 8 mm in diameter including the wings. Perianth segments above wings less than 1 mm long, shorter than the ovary, not connivent and leaving the ovary exposed.

The stem and leaf structure of this plant is very similar to that of *Haloxylon salicornicum* and *Anabasis lachnantha.* This is a much larger shrub, however, and it is easily recognized by habit apart from other features of the key.

Habitat: Forming large hummocks in deep drift sand.

South Coastal Lowlands: 322(BM), 7599, Jaww ash-Shanayin, 25 – 46N, 49 – 22E; Dickson 527, SW 'Ayn Jadidah, N al-Hufuf (Burtt and Lewis 1954, 382).

Rub' al-Khali: 7026, 'Irq al-Ghanam, 21 – 53N, 49 – 40E.

● This is an important shrub of sand areas south of about 26°30' N, where it leads a distinctive saltbush community forming hummocky terrain called *qawz* (pl. *qīzān*). Well developed stands are found around 'Ayn Dar, between 'Ayn Dar and 'Uray'irah, and to the north around 'Irj. It also forms shrubland 'islands' in the sands of the northern Rub' al-Khali. The hummocks around the shrubs are often tenanted by jerds (*Meriones*) and other rodents. This shrub is an Irano-Turanian species apparently originating in Central Asia, where it is an important component of the desert 'saxaul' forests. Zohary (1940) discussed its identity and distribution in the Middle East.

Vernacular Names and Uses: GHAḌA (gen.). For the Bedouins and desert villagers, *ghaḍā* provides highly-valued firewood that burns long and clear. This use has led to some destruction of the shrub around permanent settlements. It is also a grazing plant for camels.

● 18. Anabasis L.

Shrubs or perennial herbs with jointed stems and leaves opposite, terete-succulent, or reduced to minute scales. Flowers axillary, bisexual, 2-bracteate. Perianth with 5 free lobes, of which all, or 3, are winged in fruit. Stamens 5, alternating with 5 staminode-like lobes. Utricle membranous; seed vertical.

Key to the Species of Anabasis

1. Leaves apparently absent; shrubs with cylindrical branches and shoots appearing jointed .... 1. *A. lachnantha*
1. Leaves succulent, opposite, subterete club-shaped .... 2. *A. setifera*

1. Anabasis lachnantha Aellen et Rech. f.

Shrublet, 20 – 60 cm high, with branches and shoots woody below. Leaves virtually absent, reduced to opposite triangular scales or lobes 1 mm long, forming cupules at the joints. Flowers axillary, in spikes near ends of the terminal shoots. Fruit perianth furnished outside with 5 spreading, membranous, yellow to pink wings, which are often crowded and thus compressed transversely to the shoot axis; fruiting perianth 5 – 7 mm in diameter, including the wings. Ovary higher than wide, somewhat flattened laterally, with vertical embryo and seed; bundles of tangled fleece present at each side of the ovary inside the perianth. *Plate 47.*

In overall appearance this plant closely resembles *Haloxylon salicornicum*, near which it may be found growing. With experience the two may be separated by habit even when sterile, *Anabasis* being lower, with somewhat thicker shoots and often of flat-topped shape.

Habitat: Usually in shallow gritty to silty soil over limestone; seldom on sand of any depth. Occasional to locally frequent, perhaps more common than supposed because of confusion with *Haloxylon*.

Northern Summan: 783, 69 km SW al-Qar'ah wells; 2951(BM), Jaww Ghanim; Dickson 397(K), the Summan (Burtt and Lewis 1954, 378).

Central Coastal Lowlands: 320(BM), 7600, Jaww ash-Shanayin, 25 – 44N, 49 – 22E; 1127, 10 km S Abu Shidad.

Vernacular Names: ʿUJRUM (gen.), ʿUJAYRIMĀN (Al Murrah), ḤURḌ ('alkali, potash-bush', probably referring to use of the plant's ashes as a source of potash or lye, Shararat), GHASLAH ('wash-bush', possibly a reference to use of the plant in washing or a source of lye in soap-making, Shammar).

2. Anabasis setifera Moq.

Glabrous succulent shrublet, 10 – 30 cm high with erect jointed stems. Leaves opposite, club-shaped, obtuse, 3 – 9 mm long, 4 – 5 mm thick, furrowed above, sometimes ending in a deciduous bristle. Flowers clustered in the upper axils. Fruiting perianth with 5 wings, which are often compressed laterally due to crowding of the fruits. Ovary and seed erect, vertical. The terminal leaf bristle or weak spine usually described as a character of this species is not always seen in our specimens. *Plate 48.*

Habitat: Gritty soil or shallow sand, often on rocky ground. Occasional to locally frequent.

Northern Summan: 784, 69 km SW al-Qar'ah wells.

Central Coastal Lowlands: 65, Dhahran; Dickson 481(K), Jubail (Burtt and Lewis 1954, 378).

Southern Summan: 2928(BM), Jaww ad-Dukhan, 55 km SSW al-'Udayliyah; 2940(BM), 24 – 11N, 48 – 39E; 3830, 25 – 18N, 48 – 59E.

South Coastal Lowlands: 1140, 15 km N al-Muhassan junction; 1952, Khashm al-'Abd; 1958(LE), 11 km S Jal al-Wutayd.

Vernacular Names: SHAʿRĀN (gen.); SHAʿR (Rawalah, Shararat). Both of these names appear to be derived from Arabic *shaʿr*, *'hair'*, but the association is not clear.

● 19. Agathophora (Fenzl) Bge.

Shrubs or annual herbs with alternate, subterete, succulent, often bristle-tipped leaves. Flowers axillary in groups with both bisexual and unisexual flowers, the outer ones 2 – 3-bracteolate, the inner ones ebracteate. Perianth lobes 5, the outer 2 – 3(5) winged or gibbous dorsally. Stamens 5 (or rarely 2 – 3); staminodes 4 – 5. Ovary with 2 stigmas; seed vertical or horizontal.

Agathophora alopecuroides (Del.) Fenzl ex Bge.

Halogeton alopecuroides (Del.) Moq.-Tand.
Agathophora iraqensis Botsch.

Rather stiff glabrous shrublet with whitish bark, 15 – 40 cm high. Leaves scattered or clustered, rather distant, teretish, club-shaped or sometimes nearly globose, those of spring season 5 – 12 mm long, terete, succulent, tipped with a conical spinule 1 – 2 mm long; autumn leaves mostly 2 – 5(7) mm long, with a fine, straight or bent, needle-like spinule 2 – 4 mm long. Flowers in bracteate clusters in the woolly axils, forming dense or somewhat loose cylindrical spikes 7 – 12 cm long, 10 – 12 mm broad. Fruiting perianth usually 3(2 – 4) winged. Seed vertical. *Plate 49.*

Habitat: Usually in barren limestone country, rocky wadis, or gravel plains. Rare to occasional.

Northern Plains: 3077(LE,BM), al-Musannah, 28 – 43N, 46 – 47E; 8733, 8811, al-Batin ravine near ar-Ruq'i wells, 29 – 01N, 46 – 32E.

Northern Summan: 788(BM), 142 km SW al-Qar'ah; 223(BM), near 'Irj; 3913, 27 – 41N, 46 – 02E.

● 20. Cornulaca Del.

Shrubs or annual herbs with linear or subulate, spinescent leaves woolly in the axils. Flowers axillary, solitary or clustered, bracteate. Perianth lobes 5, indurating, with one to several becoming spinescent or prickly in fruit. Stamens 5; staminodes 5. Style with 2 filiform lobes. Fruit a nut, often spined or horned at apex and subtended by hairs at the base.

Key to the Species of Cornulaca

1. Dwarf annual, 5 – 10 cm high; ribbon-like hairs at axils equalling or exceeding the leaves .... 2. *C. aucheri*
1. Shrubs or tough shrubby herbs 15 – 100 cm high; axils sometimes woolly but no hairs approaching leaf-length.
   2. Mature fruit with 2 minute horns, 0.5 – 1 mm-long, at apex, all immersed in subtending hairs; shrub of deep Rub' al-Khali sands .... 1. *C. arabica*
   2. Mature fruit with 1 or 2, 4 – 8 mm-long, spines at apex clearly exserted above the subtending hairs; shrubs or shrubby herbs not of deep sands
      3. Shrub; flowers clustered 2 – 5 in the upper axils; fruits with 1 or 2 spines .... 4. *C. monacantha*
      3. Shrubby herb; flowers clustered 2 – 10 in both lower and upper axils; all fruits with only 1 spine .... 3. *C. leucacantha*
   For couplet 3, above, see description of *C. monacantha* for further differential characters.

1. Cornulaca arabica Botsch.

Rounded, grey-green, prickly, somewhat tangled many-branched shrub, 30 – 80 cm high, glabrous but with dense short wool in the axils. Leaves clasping, approximate or remote, 1.5 – 3(5) mm long, broadly to narrowly triangular, pungent, ending in a spine. Flowers solitary or few in axils of terminal branchlets. Stamens exserted in anthesis. Fruit subpyramidal, 3.5 – 4.5 mm long, 2 – 2.5 mm wide, with 1 side flattened-concave, the other convex-rounded; apex with 2 nearly equal hornlets or spinules 0.5 – 1 mm long. Fruit subtended at base and entirely or nearly hidden by dense straight, erect, simple white hairs, which grow with fruit maturation up to c. 5 mm long and fall with the fruit. *Plate 50, Map 8.4.*

● The type (Mandaville 467) is fairly representative in morphology. It is, however, a young flowering specimen collected in mid-October without fruits, which were listed by Botschantzev as *ignotus*. The fruit is in fact highly diagnostic, differing from that of other regional species in being nearly spineless.

Our records suggest that this Rub' al-Khali endemic flowers in autumn beginning around September but often does not develop mature fruits until January – February. Flowering in this southern, highly arid environment might also be somewhat irregular, depending on local moisture conditions. Small windrows of the hair-tufted fruits may be seen in *ḥādh* country of the northeastern Rub' al-Khali in early spring; the hairs are certainly an aid to dissemination by the wind. The miniscule nuts may well be important as food for local rodents and insects in such habitats, where vegetable matter of any kind is at a premium.

The species is very important in the vegetation of the Rub' al-Khali and is dominant in a community type with a broad patchy distribution throughout the southern sands. Specimens of it may have been collected as early as 1932 by H. St. J. B. Philby during his exploratory Rub' al-Khali crossing. The list of plants in his book, *The Empty Quarter*, includes a '*Cornulaca monacantha*' called 'hadh'.

Habitat: Deep sands of the Rub' al-Khali, where it is locally frequent to common.

Rub' al-Khali: 467 (holotype, K), al-Hadidah meteorite craters; 1018(BM), 21 – 45N, 50 – 35E; 1934(LE), Zumul camp, 22 – 20N, 54 – 57E; 4002, 21 – 46N, 53 – 34E; 7019, 22 – 17N, 53 – 22E; 7059, 23 – 16N, 51 – 43E; 7082, 22 – 11N, 54 – 19E; 7086, 7603, 22 – 10N, 54 – 21E; 7487, ar-Rumaylah, 21 – 40N, 47 – 42E; 7505, ash-Shuqqan, 21 – 50N, 48 – 38E; 7618, S Hibakah, 20 – 35N, 50 – 20E; 7637, W Qa'amiyat, 18 – 28N, 47 – 07E; 8400, Camp G – 3, 20 – 25N, 55 – 08E. This endemic is so far known only from the Rub' al-Khali proper. It may extend into the southern Jafurah as far north as about 24 – 30N.

Vernacular Names and Uses: ḤĀDH, ḤADHDH (gen. south). An important camel grazing plant of the Rub' al-Khali, *ḥādh* has a specialized vernacular terminology denoting its developmental stages. When in young flower it is *wāris*, 'yellowing', or *muwarris*, 'yellow-tinting' (both referring to the yellow, exserted anthers), adjectives also applied to other saltbushes (Al Murrah, Al Rashid). The term *jādir*, 'flowering, sprouting', (Al Murrah, Al Rashid) refers to the stage of developing fruits, when it is said to be somewhat better for camels. After the seeds have fallen it is called *muraykhī*, probably referring to some likeness to *markh*, the *Leptadenia* shrub (Al Rashid). Al Rashid refer to seedlings and young plants as *jarw*, literally 'whelps', a term said not to be used in this way for other shrub species.

2. Cornulaca aucheri Moq.

Dwarf annual, 5 – 10 cm high, often branched from the base. Leaves 8 – 10 mm long, 0.5 – 1 mm wide, pungent, ending in a fine spine, nearly or fully exceeded by white, ribbon-like hairs originating in the axils. Flowers grouped 2 – 3 in the upper axils. Fruits with 0 or 2 spines. *Map 8.5.*

Habitat: Sand-silt, often over limestone. Rare.

Northern Summan: 2790(BM), 41km SE ash-Shumlul; 7393, Jibal al-Harmaliyat, 26 – 10N, 48 – 10E; 7931, 25 – 32N, 48 – 46E.

Northern Dahna: 8564, near 'Urayq al-Huwaymil, 25 – 25N, 47 – 37E.

● A sterile specimen (7676) matching *C. aucheri* rather well was collected in the Rub' al-Khali in eastern Sanam, 21 – 59N, 51 – 10E, where *C. arabica* occurs sparsely. It is suspected of being a seedling of that shrub, particularly in view of the fact that the habitat is atypical for *C. aucheri*.

3. Cornulaca leucacantha Charif et Aellen

Coarse, very prickly annual or perennial herb, 10 – 30 cm high, erect to ascending, rarely procumbent, often branched from the base. Leaves 4 – 6(9) mm long, partially clasping, triangular-subtriquetrous, very spiny. Flowers clustered 2 – 10 in both upper and lower axils. Fruits with a single yellowish to whitish spine c. 5 mm long. *Map 8.5.*

Habitat: Shallow sand or silt, often in rocky limestone country, or on gravel plains. Occasional to frequent.

Central Coastal Lowlands: 1931(LE), 7602, 7604, Dhahran.

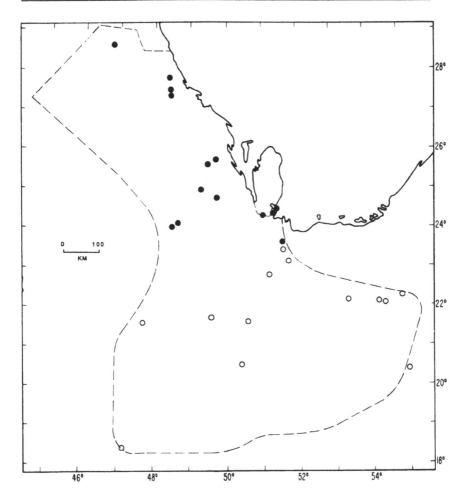

Map 8.4. Distribution as indicated by collection sites of *Cornulaca monacantha* (●) and *Cornulaca arabica* (○) in the coverage area of this flora (dashed line), eastern Saudi Arabia. Each symbol represents one collection record.

Southern Summan: 1866(BM), Jaww al-Hawiyah, 24–42N, 49–08E; 2503(BM), 26 km E Khurays.
South Coastal Lowlands: 1135(BM), 1139(BM), al-'Udayliyah; 464(K), al-Mustannah; 469(BM), 23–39N, 49–49E; 1947, 27 km W Salwah; 1023(BM), 23–36N, 49–34E; 1960(LE), 40 km WNW Salwah; 7597, Jaww ash-Shanayin, 25–45N, 49–22E; 1895(BM), 1 km NE al-Mutayrifi.
Vernacular Names: SILLAJ (Al Murrah, gen.), SULLAYJ (Qahtan).

4. Cornulaca monacantha Del.

Glabrous prickly shrublet, 10–40 cm high, much branched. Leaves clasping, triquetrous, 3–10 mm long, spiny-tipped, short-woolly in the axils. Flowers clustered

99

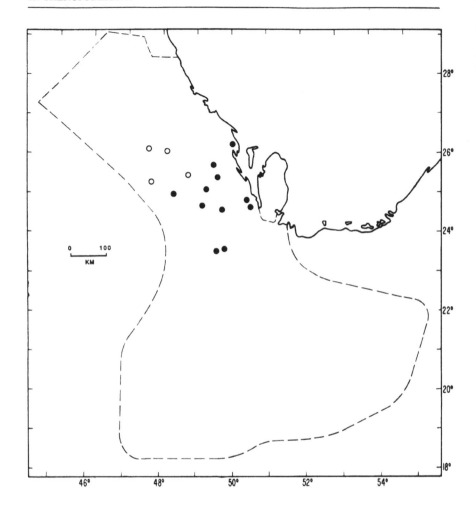

Map 8.5. Distribution as indicated by collection sites of *Cornulaca leucacantha* (●) and *Cornulaca aucheri* (○) in the coverage area of this flora (dashed line), eastern Saudi Arabia. Each symbol represents one collection record.

2–5 in the upper axils of the terminal branchlets. Fruits subtended by hairs, with 1 or 2, 4–6 mm-long clearly exserted spines. *Plate 51, Map 8.4.*
● This species is rather variable in habit, leaf, and fruit form. Some specimens from around al-'Udayliyah have been particularly confusing, seemingly intermediate between *C. monacantha* and *C. leucacantha*. Following, to supplement diagnostic features in the key, are some additional if more subjective differences between the two species:

C. monacantha	*C. leucacantha*
Definitely woody at base	Hard but seldom truly woody
Leaves on intermediate stems often more than half clasping	These leaves less than half clasping
Branched complexly in several directions	Often with several ascending or erect parallel main branches from the base
Smaller old stems tending to crack at the axils, appearing jointed.	Such stems seldom appearing jointed.

Habitat: Shallow soils often on rocky ground. Not found in deep sand.

Northern Plains: 642(BM), 20 km N an-Nu'ayriyah; 838(BM), 30 km SW al-Mish'ab; 3076(LE, BM), al Musannah, 28–43N, 46–47E; 1013(BM), 2 km NE an-Nu'ayriyah.

Southern Summan: 2946(BM), 24–11N, 48–39E.

Central Coastal Lowlands: 3831, 1932(LE), Dhahran.

South Coastal Lowlands: 465(K,BM), al-Mustannah; 1136(BM), 1134, al-'Udayliyah; 2939(LE), 24–09N, 48–31E; 7061, al-Majann, 23–42N, 51–34E; 7067, Sabkhat al-Mashakhil, 24–29N, 51–01E; 7596, Jaww ash-Shanayin, 25–45N, 49–22E; 1971, 1972, 1973, hill above Abqaiq salt pits; 7456, al-'Udayd, 24–29N, 51–20E; 7458, al-'Udayd, 24–37N, 51–26E; 7461, al-'Udayd, 24–36N, 51–24E.

Rub' al-Khali: 7060, 23–31N, 51–32E.

Vernacular Names: ḤUWAYDHĀN, ḤUWAYDHDHĀN (Al Murrah), ḤADH (Shammar), 'ARĀD (sic, usually applied to *Salsola cyclophylla*, 'Awazim). Northern tribes appear to use the name *ḥādh* for this species, while southern tribesmen who know the Rub' al-Khali use *ḥuwaydhān*, to differentiate it from their *ḥādh*, *C. arabica*.

13. AMARANTHACEAE

Herbs or shrubs with alternate or opposite, simple exstipulate leaves. Flowers bisexual, or unisexual with the plants monoecious or dioecious, 3-bracteate, mostly in axillary or terminal spicate cymes. Perianth 3–5-lobed, the segments more or less scarious. Stamens 3–5. Ovary superior, 1-celled and generally 1-ovuled. Fruit a utricle or capsule.

Key to the Genera of Amaranthaceae

1. Shrubby grey-tomentose perennial; flowers in dense spikes, immersed in very dense white wool .... 2. *Aerva*
1. Glabrous or glabrescent annuals; flowers not woolly
   2. Each fertile flower subtended by 2 sterile, bractlike ones; style distinct, longer than fruit; stigmas 2; fruits crustaceous .... 3. *Digera*
   2. Fertile flowers without sterile companions; style absent; stigmas 3; fruit membranous around seed .... 1. *Amaranthus*

● 1. Amaranthus L.

Annual herbs with alternate, mostly simple and petiolate leaves. Flowers usually unisexual, 2–3-bracteolate, clustered in axillary or terminal, spicate or paniculate inflorescences. Perianth 3–5-lobed, the lobes free and usually scarious. Stamens 3–5. Ovary with 2–3 stigmas, 1-ovuled. Fruit a 1-seeded utricle.

Key to the Species of Amaranthus

1. Tepals 5; bracteoles subulate, longer than the tepals .... 3. *A. hybridus*
1. Tepals 3; bracteoles shorter than the tepals
   2. Leaves with distinct apical notch about 1/10 as deep as blade
     length .... 1. *A. blitum*
   2. Leaves unnotched, or with obsolete notch much less than 1/10 as
     deep as blade length
      3. Flowers in axillary clusters; leaf blades elliptical, mostly 1 – 3(4) cm
        long, 0.5 – 1.5(1.8) cm wide .... 2. *A. graecizans*
      3. Flowers mainly in a terminal, spicate inflorescence; leaf blades
        ovate to lanceolate, 3 – 8 cm long, 1.5 – 5.5 cm wide .... 4. *A. viridis*

The common weed Amaranths of our area are *A. graecizans* and *A. viridis*. *A. hybridus* and *A. blitum* are rarer.

1. Amaranthus blitum L.

A. ascendens Lois.
A. lividus L. var. *ascendens* (Lois.) Thell.

Erect or decumbent herb, often reddish. Leaves rhomboid, obtuse, 2 – 4 cm long, 1 – 3 cm wide, with a blunt apical notch about 1/10 as deep as blade length, usually with a fine mucro in the notch; petioles subequalling the blades. Flowers in dense terminal or axillary spikes; bracts acute. Tepals narrowly white-margined, equalled or exceeded by the 1.5 mm-long, somewhat laterally compressed utricle. Seed round, compressed, smooth-glossy, nearly black, c. 1.5 mm in diameter.

Habitat: A weed of gardens. Infrequent.

Central Coastal Lowlands. 7822, 8156, Dhahran.

2. Amaranthus graecizans L.

Glabrescent annual, 10 – 50 cm high, ascending or prostrate. Leaves elliptical, alternate, 1 – 3(4) cm long, 0.5 – 1.5(2) cm wide, obtuse at apex, tapering to a short petiole, prominently white-veined beneath. Flowers in axillary clusters distributed over most of the plant. Utricle laterally compressed, circumscissile, somewhat exceeding the white-margined tepals. Seed round, compressed, smooth-glossy, dark brown, c. 1.5 mm in diameter.

Habitat: A weed of gardens, farms, waste ground. Common.

Central Coastal Lowlands: 1919(BM), al-Jarudiyah, Qatif Oasis; 2961(BM), Dhahran, stables farm.

South Coastal Lowlands: 1901(BM), ash-Shuqayq, al-Hasa Oasis; 1892, al-Mutayrifi experimental farm, al-Hasa Oasis; 2907(BM), al-'Arfaj farm, al-Hasa; 3938, 2 km ENE al-Hufuf.

3. Amaranthus hybridus L.

A. chlorostachys Willd.

Glabrescent erect annual, 20 – 150 cm high; sometimes very stout. Leaves alternate, ovate, acute or obtuse, up to 12 cm long, 8 cm wide on longish petioles about ⅓ to ½ length of the blade. Flowers 2 – 2.5 mm long, with acuminate, mucronate or subspinescent bracteoles longer than the perianth, in a very dense terminal inflorescence composed of a longer terminal spike with smaller lateral ones, or of upwards-shortening lateral spikes.

Habitat: A weed of waste ground and gardens. Infrequent.

Central Coastal Lowlands: 3785(BM), 7078(BM), 7820, Dhahran.

South Coastal Lowlands: 8663, 2 km E al-Hufuf, al-Hasa Oasis.

4. Amaranthus viridis L.

A. gracilis Desf.

Glabrous erect annual, 20 – 60 cm high. Leaves alternate, ovate to lanceolate, obtuse and often obscurely retuse at apex, 3 – 8 cm long, 1.5 – 6 cm wide, with lighter colored veins beneath, on petioles ½-⅔ the length of the blade. Inflorescence terminal, on one or more upper stems, composed of a long spike with several shorter lateral ones below; spikes somewhat loose, ripening brownish. Tepals with a green, medial, longitudinal stripe, white-margined, somewhat exceeded by the laterally compressed, wrinkled, 1.5 mm-long utricle. Seed round, compressed, shining, nearly black, minutely punctate, c. 1 – 1.2 mm in diameter.

Habitat: A weed of gardens, farms, waste ground. Common.

Central Coastal Lowlands: 632(K), Dhahran, stables farm; 1918(BM), al-Jarudiyah, Qatif Oasis; 7812, al-Khubar; 7819, 7871, Dhahran; 7826, 7827, 1 km S Qatif town; 8159, near al-Jishsh, al-Qatif Oasis.

South Coastal Lowlands: 1880(BM), Abqaiq; 2897(BM), 2901, 2910(BM), al-'Arfaj farm, al-Hasa; C. Parker 23(BM), farm near al-Hufuf Airport; 8294, al-Hufuf.

● 2. Aerva Forssk.

Tomentose-woolly herbs or shrubs with alternate, entire leaves. Flowers mostly bisexual, 3-bracteate, in spicate or paniculate inflorescences. Perianth segments 5, free, all or some of them densely white-fleecy. Stamens 5; staminodes 5, tooth-like. Stigmas 2. Utricle with a single vertical seed.

Aerva javanica (Burm. f.) Spreng.

A. persica (Burm. f.) Merrill

A. tomentosa Forssk.

Grey shrub, tomentose with dense, stellate hairs, 30 – 70 cm high with erect stems branching from near the base. Leaves elliptical-oblanceolate, obtusish, grey-tomentose below, tapering to a short petiole, 1 – 5 cm long, 0.5 – 1 cm wide. Flowers extremely white-woolly, in leafless, mostly terminal, spikes 1.5 – 5 cm long, 1 – 1.5 cm wide. Perianth c. 3 mm long, the lobes lanceolate, scarious, densely white-fleecy. *Plate 52.*

Habitat: Usually on rocky ground with shallow sand-silt. Rare.

Central Coastal Lowlands: 173(BM), 436, Dhahran; 364, Jabal Umm ar-Rus, Dhahran; 648, Qatif-Safwa road junction.

Vernacular Names, Uses: RĀ' (gen.), TUWWAYM ('pearly-bush'?, Bani Hajir). The densely woolly parts of this shrub's inflorescence were used by Bedouins in earlier times for stuffing saddle pads and cushions.

● 3. Digera Forssk.

Annual herbs with alternate, petiolate, simple leaves. Flowers 3-bracteate, grouped in threes with the central one of each group bisexual and fertile, the lateral two rudimentary and sterile. Perianth with 4 – 5 unequal lobes. Stamens 5. Ovary superior, with 2-stigmas, 1-ovuled. Fruit a hard utricle.

Digera muricata (L.) Mart.
D. alternifolia (L.) Aschers.
Ascending glabrescent annual, 20 – 50 cm high. Leaves alternate, ovate to lanceolate, acute at apex, up to about 5 cm long, 3 cm wide; petioles ½ to ⅔ length of the blade, with sparse erect, crisped hairs. Flowers subsessile in slender erect spikes up to 15 cm long, the peduncles axillary, exceeding the leaves.

Habitat: A weed of gardens, farms.

Central Coastal Lowlands: 1924(BM), al-Jarudiyah, al-Qatif Oasis. Rare.

14. POLYGONACEAE

Herbs or shrubs with stems often swollen at the nodes and leaves mostly alternate, the stipules connate forming a short sheath (ochrea). Flowers bisexual, or unisexual and the plant monoecious or dioecious, actinomorphic, solitary or clustered in cymes forming various larger inflorescence units. Perianth with 3 – 6 segments, sepaloid or petaloid, sometimes modified in fruit. Stamens usually 6 – 9. Ovary superior, 1-celled, 1-ovuled, with 2 – 4 styles. Fruit an angular or flattened achene.

Key to the Genera of Polygonaceae
1. Shrubs, usually leafless; fruit with branching bristles or with indurated denticulate wings .... 3. *Calligonum*
1. Herbs, with leaves; fruit without bristles, if winged not indurate
  2. Outer 3 fruit perianth segments terminating in rigid spines 1 – 3 mm long .... 1. *Emex*
  2. Perianth without rigid spines
    3. Fruit perianth plain, without appendages .... 4. *Polygonum*
    3. Fruit perianth developing into wings or valves, entire or toothed at margins .... 2. *Rumex*

● 1. Emex Neck. ex Campd.

Annual herbs with alternate entire leaves; ochreae membranous, truncate, cleft. Flowers unisexual (or sometimes bisexual), clustered or racemed at the axils, the plant monoecious or polygamous. Staminate flowers with herbaceous, 3 – 6-lobed perianth and 4 – 6 stamens. Pistillate flowers with perianth of 6 herbaceous lobes growing and indurating in fruit, the outer 3 at length spinescent. Ovary with 3 styles. Achene triquetrous.

Emex spinosa (L.) Campd.
Glabrous annual, 5 – 20 cm high and sometimes tinged reddish, single-stemmed with a basal rosette of leaves or with decumbent branches from the base. Taproot thickened, carrot-like, whitish and fleshy. Leaves ovate to oblong, truncate at base and rounded at apex, 1 – 6 cm long, 0.5 – 3 cm wide, on petioles about equalling the blades. Flowers clustered in the axils. Perianth growing in fruit, the outer 3 segments indurating, medially keeled and transversely pitted, with recurved apical spinules 1 – 3 mm long.

Habitat: Shallow stabilized sands or on silty soil; rarely in deeper sand. Occasionally a weed on sandy farmland. Frequent to common.

Northern Plains: 668, Khabari Wadha; 1680(BM), 2 km S Kuwait border in 47 – 14E; 1292(BM), 15 km ENE Qaryat al-'Ulya; 1216, Jabal an-Nu'ayriyah.

Northern Summan: 747, 33 km SW Qar'ah wells; 776, 69 km SW Qar'ah wells; 1386(BM), 42 km W Qaryat al-'Ulya.

Central Coastal Lowlands: 904, Mulayjah.

Southern Summan: 2228, 15 km WSW al-Hunayy; 2057, Jaww ad-Dukhan, 24 – 48N, 49 – 05E.

South Coastal Lowlands: 2734, ar-Ruqayyiqah, al-Hasa.

Vernacular Names and Uses: ḤAMBIZĀN (gen., Shammar), ḤIMBĀZAH (Shammar), ḤUMBAYZ (Rawalah), 'AMBAṢIṢ (Al Murrah). All Bedouins know the edible qualities of this plant's sweet, carrot-like taproot. Vesey-Fitzgerald (1957, 791) notes that the petioles are also plucked and eaten raw.

● 2. Rumex L.

Annual or perennial herbs with tubular ochreae and alternate leaves. Flowers bisexual, or unisexual and the plant monoecious, polygamous or dioecious, whorled in racemose or paniculate inflorescences. Perianth herbaceous, of 6 segments, of which the inner 3 are often much growing in fruit to form conspicuous, often colored, winged valves enclosing the achene. Stamens 6. Ovary with 3 styles, 1-ovuled, trigonous. Achene trigonous.

Key to the Species of Rumex
1. Leaves pinnately cleft or parted .... 2. *R. pictus*
1. Leaves entire, wavy or denticulate at margins
    2. Wings of fruiting valves entire; desert plant .... 3. *R. vesicarius*
    2. Wings of fruiting valves toothed; weeds of wet cultivated
       areas .... 1. *R. dentatus*

1. Rumex dentatus L.

Annual or biennial erect branching herb, 15 – 70(100) cm high. Basal leaves up to about 15(20) cm long, 3(4) cm wide, oblong-lanceolate, with petioles as long as or shorter than the blade. Inflorescence terminal, spicate, the flowers clustered in dense remote whorls subtended by small leaves. Fruiting valves ovate-lanceolate, green, c. 4 – 5 mm long, reticulately nerved, with an elongated, swollen tubercle at base of the face and with 3 – 5 subulate teeth at each margin. *Plate 53.*

Chaudhary and Zawawi (1983) refer to *R. pulcher* L. as an occasional weed in central and eastern Saudi Arabia. None of our specimens, however, seem to be attributable with any certainty to that closely related species.

Habitat: An oasis weed of gardens, farms. Usually on wet ground or even standing in the shallow water of rice fields. Locally frequent.

South Coastal Lowlands: 3298, near 'Ayn al-Khadud, al-Hasa; C. Parker 16(BM), al-Hasa farm; 8201, near 'Ayn Umm Sab', al-Hasa Oasis; 8229, 8230, 8231, near al-Qurayn, al-Hasa Oasis.

2. **Rumex pictus** Forssk.

R. lacerus Balbis

Glabrous, often reddish and somewhat brittle-succulent annual, 10 – 25 cm high with many ascending and decumbent branches from the base. Leaves 1.5 – 4 cm long, to 2 cm wide, pinnately parted, usually with 2 pairs of main side lobes and one terminal lobe, on petioles often about as long as the blades. Flowers clustered in the upper nodes, the inflorescence appearing spicate or narrowly racemose. Fruit perianth 5 – 9 mm broad (including the wings), pink or turning yellow, the valves rounded, sometimes broader than long, furnished medially with a smooth, elongated wart. *Plate 54.*

Habitat: Stable sands. Occasional to locally frequent in the northern parts of our area.

Northern Plains: 1249(BM), Jabal an-Nu'ayriyah.

Northern Dahna: 4030, 27 – 35N, 44 – 51E; 7538, 8 km NE Umm al-Jamajim.

Central Coastal Lowlands: 3283, 10 km W Abu Hadriyah; 3321, 12 km S Nita'; 3780, 3 km E Ras al-Ghar; 7516, Ras Tanaqib; 7716, Ras az-Zawr, 27 – 29N, 49 – 16E.

Vernacular Names and Uses: ḤAMBAṢIṢ (Dickson 1955, 81), ḤAMṢIṢ (Musil 1927, 95). Carter (1917, 180), Musil (1927, 95) and Dickson (1955, 81) report that the leaves are eaten raw by Bedouins.

3. **Rumex vesicarius** L.

Glabrous, slightly succulent annual, 10 – 30 cm high, branching from the base. Leaves ovate to deltoid, truncate or subcordate at base, to c. 6 cm long, 4 cm wide, obtuse or sometimes weakly acute, on petioles mostly equalling or exceeding the blade. Flowers clustered in the upper axils, forming dense terminal racemes. Fruiting perianth growing, showy, greenish-yellow when young, becoming bright pink to reddish; the valves entire, up to c. 2 cm broad, the wings with radiating, branching-reticulate, red nerves. *Plate 55.*

Habitat: Usually on shallow sand in rocky, hilly terrain; also sometimes in the shelter of shrubs on open ground. Occasional to frequent.

Northern Plains: 1204, Jabal an-Nu'ayriyah.

Northern Summan: 2761, 25 – 49N, 48 – 01E.

Central Coastal Lowlands: 397, al-Midra ash-Shimali.

Southern Summan: 2197, 24 – 34N, 48 – 45E.

Vernacular Names and Uses: ḤUMMAYḌ ('sour-wort', gen., Shammar), ḤAMMAḌ ('sour-wort', Rawalah), ḤAMBAḌ (Musil 1927, 602); the name ḤAMBAṢIṢ may also be used for this species. The use of this plant's raw leaves as a salad vegetable by Bedouins is well known. According to a Bani Hajir tribesman, the plant is sometimes added during the preparation of *iqṭ* (dried milk shards) to increase its acidity. Carter (1917, 181) reports it is also eaten cooked with meat.

● 3. **Calligonum** L.

Shrubs with rigid main branches, flexible young shoots, and minute, soon deciduous leaves; ochreae short, scarious. Flowers bisexual, short-pedicellate, clustered in the axils. Perianth with 5 nearly equal lobes. Stamens 10 – 18. Ovary 4-angled, with 4 styles. Achene indurate-woody, much exceeding the perianth, furnished with wings or dense rows of branching bristles.

Key to the Species of Calligonum

1. Fruits covered with bristles
    2. Bristles originating on 4 pairs of longitudinal wings, each pair
       on one of the quadrangular fruit sections; pre-fruiting flowers
       with pedicels longer than the perianth .... 1. *C. comosum*
    2. Bristles (although sometimes broadened at base) not originating
       on defined wings but directly from the quadrangular fruit
       sections; pedicels mostly shorter than the perianth; Rub'
       al-Khali shrub .... 2. *C. crinitum*
1. Fruits without bristles but with 4, weakly spiraled, sinuate to
   denticulate, longitudinal wings .... 3. *C. tetrapterum*

1. **Calligonum comosum** L'Hér.

C. polygonoides L. subsp. *comosum* (L'Hér.) Sosk.

Ascending shrub, 40 – 120 cm high, with whitish woody older branches, swollen
and knotty at the nodes, and younger flexible green shoots 1 – 2 mm in diameter.
Leaves minute, soon deciduous and usually absent. Flowers on pedicels about
equalling or somewhat exceeding the perianth lobes. Perianth lobes oblong, obtuse,
c. 2.8 – 3 mm long, white-pink or white-greenish, with darker medial part. Fruits
red or greenish yellow, 1 – 1.5 cm long, 0.5 – 1 cm broad, covered in branching
stiff bristles arising from 4 pairs of short longitudinal wings. *Plates 56, 57, 58,
Map 8.6.*

● This shrub occurs in two color forms, with plants of both types often apparently
randomly distributed in the same stand but with flowers and fruits on each individual
plant uniform: one with red fruits and reddish flowers (Plate 57), the other with yellow
or greenish yellow fruits and green-white flowers (Plate 58). This is apparently a non-
adaptive genetic feature affecting the red pigment trait, but its distribution has not
been studied quantitatively.

Habitat: Almost always in the deeper sands, where it is a dominant constituent of several
communities. Frequent to common in its habitat.

Northern Summan: 2854(BM), 2856(BM), 26 – 09N, 48 – 18E; 2873(BM), 5 km NE
al-Jawwiyah wells; 2857(BM), 26 – 03N, 48 – 25E.

Northern Dahna: 4026, 27 – 35N, 44 – 51E; 8726, 27 – 24N, 44 – 56E.

Central Coastal Lowlands: 68, Thaj; 392, Dhahran; 1454, 2 km NE al-Ajam; 1786,
13 km E Abu Hadriyah; 2725(BM), 23 km N Jabal Fazran; 4069, 4072, 26 – 51N,
49 – 51E; 7089, 7090, 7091, near al-Ajam.

Southern Dahna: 2139(BM), E edge of the Dahna in 23 – 05N; 7660, 24 – 09N,
48 – 17E.

Southern Summan: 2207(BM), 26 km SSE al-Hunayy.

South Coastal Lowlands: 1450, 2 km E Jabal Ghuraymil.

Vernacular Names and Uses: 'ABAL (gen. south), ARŢÁ (gen. north); NATHARAH, the
red fruits (Al Murrah). Both Musil (1927, 154) and Dickson (1955, 29) report the use
by Bedouins of the leaves (or young shoots) in the tanning of hides. The shrub is well
known throughout Arabia for its excellent, long- and clear-burning firewood, and it
is often cut and gathered for this product. Women in the Thursday market at al-Hufuf
sell fine twigs of the plant, whole or already pounded to a fine powder, which are said
to be added to milk as a flavoring or a tonic.

2. **Calligonum crinitum** Boiss. subsp. **arabicum** (Sosk.) Sosk.
   C. arabicum Sosk.

Ascending shrub, 0.5 – 2 m high, with rigid, grey-whitish main branches, swollen and knotty at the nodes; young shoots fine, flexible, 1 – 2 mm in diameter, the ochreae nearly entire and forming short scarious cupules at the nodes. Leaves virtually absent. Flowers solitary or 2 – 3 at the nodes of the younger branches, on pedicels mostly shorter than the perianth. Perianth lobes oblong, obtuse, c. 3 mm long, white-pink or white-greenish with darker medial stripe. Fruits red or yellow-green, 1 – 1.5 cm long, 1 – 2 cm broad (including bristles), clothed in simple or branched spreading bristles up to 10 mm long arising from backs of the quadrangular fruit lobes (as seen in cross-section). Bristles not or hardly joined into longitudinal wings at base. Like the preceding, occurs in both red and yellow fruit color forms. *Plate 59. Map 8.6.*

Habitat: Deep sands in the Rub' al-Khali and adjacent sands in the south. Frequent and often dominant within its range.

Rub' al-Khali: 4000(BM), Shaybah Camp; 4001(BM), 21 – 46N, 53 – 34E; 4004(BM), 21 – 38N, 53 – 21E; 7052, Rawakib well site; 7083, 7084, 22 – 15N, 54 – 16E; 7389, 7390, 21 – 09N, 54 – 52E; 7613, at-Tara'iz, 21 – 00N, 50 – 27E; 7617, southern al-Hibakah, 20 – 35N, 50 – 20E; 7620, 7621, 7622, Hadh Bani Zaynan, 20 – 17N, 49 – 45E; 7624, southern ash-Shuwaykilah, 19 – 57N, 49 – 01E; 7625, southwestern ash-Shuwaykilah, 19 – 48N, 48 – 46E; 7633, 'Uruq al-Awarik, 19 – 14N, 47 – 37E; 7634, eastern al-Awarik, 19 – 00N, 47 – 32E; 7638, SW al-Qa'amiyat, 18 – 15N, 46 – 57E; 7679, as-Sanam, 21 – 45N, 50 – 53E; 7680, W as-Sanam, 21 – 53N, 50 – 26E.

Vernacular Names and Uses: 'ABAL (Al Murrah, Al Rashid, gen. south). Provides excellent firewood and, like the preceding, reportedly used in tanning hides.

3. **Calligonum tetrapterum** Jaub. et Spach

Shrub, 30 – 60 cm high with stiff, knotty, grey-white main branches. Leaves much reduced and usually appearing absent. Flowers grouped 1 – 3 at nodes of the younger branches, on pedicels about equalling or shorter than the perianth. Perianth lobes pink-white with darker medial portions. Fruits red, usually less than 1 cm long, with 4 slightly spiralled, denticulate, indurate wings but no bristles. Yellow flower and fruit forms of this species have not been seen by the author but might be found to occur. *Plate 60, Map 8.6.*

Habitat: The single recorded Eastern Province site, as well as the collection point of the author's specimens from central Arabia east of Riyadh, were in very shallow sand on stony ground. It thus appears to differ significantly in habitat preference from the other, deep sand-loving, *Calligonums* of Arabia. Apparently rare in the Eastern Province.

Northern Summan: 2844, 2845, 26 – 18N, 47 – 56E.

Vernacular Names: None recorded but most probably called 'ABAL or ARṬÁ as are other *Calligonums*.

● 4. **Polygonum** L.

Annual or perennial herbs, or shrubs, with membranous ochreae and alternate, simple leaves. Flowers bisexual, solitary or clustered in spicate or paniculate inflorescences. Perianth mostly 5-lobed, sepaloid or petaloid. Stamens 8. Ovary superior, 1-ovuled; style or stigmas 2 – 3. Achenes triquetrous or lenticular.

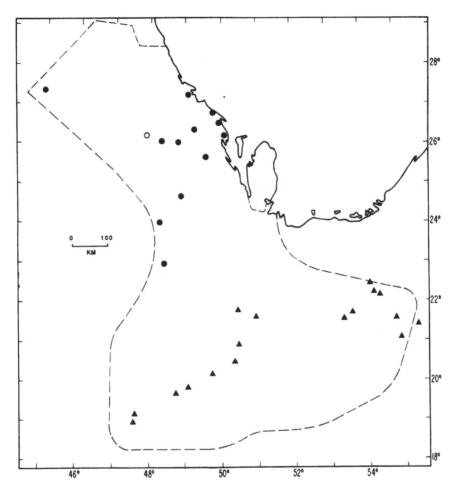

Map 8.6. Distribution as indicated by collection sites of *Calligonum comosum* (●), *Calligonum tetrapterum* (○) and *Calligonum crinitum* subsp. *arabicum* (▲) in the coverage area of this flora (dashed line), eastern Saudi Arabia. Each symbol represents one collection record.

Polygonum argyrocoleum Steud. ex Kunze

Glabrous, rather straggly, branched annual, 20 – 50 cm high, sometimes few-leaved and nearly leafless above, the stems finely striate with ochreae membranous, longitudinally nerved, lacerate, divided partly or entirely to base. Leaves oblanceolate to lanceolate-linear, up to 4 cm long, 0.8 cm wide but usually smaller, tapering below to a short petiole, deciduous. Floral leaves much smaller or absent. Internodes much shortened terminally, leading to appearance of a terminal, spicate inflorescence. Flowers 2 – 5(8) together at the nodes on pedicels (partly concealed by the ochreoles) as long or longer than the perianth. Perianth segments somewhat hooded, greenish medially, white-margined below, rose-red-margined above. Seeds c. 2 mm long, ovoid-3-angled, shining, smooth brown, not at all or only most obscurely punctate.

109

A weed rather variable in leaf form and in habit, more tender and leafy in shady sites, sometimes nearly leafless and almost woody below in exposed habitats. Further study of Eastern Province specimens is required to determine their relationships with plants described elsewhere as *P. patulum* M. B., *P. bellardii* All., and possibly *P. equisetiforme* Sibth. et Sm.

Habitat: Found only as a weed of gardens or farms, or sometimes as a ruderal. Occasional.

Northern Plains: 7849, Qaisumah Pump Station.

South Coastal Lowlands: 2731(BM), ar-Ruqayyiqah, al-Hasa; 8185, near ash-Shuqayq, al-Hasa Oasis; 8212, Aramco farm near al-Qarn, al-Hasa Oasis; 8296, al-Hufuf; 7709(E), roadside at Harad Agricultural Project; 7784, 7788, adh-Dhulayqiyah, al-Hasa.

15. PLUMBAGINACEAE

Annual or perennial herbs, or shrubs, with leaves alternate or spirally inserted, exstipulate, often rosulate. Flowers bisexual, actinomorphic, hypogynous, in cymose or racemose inflorescences. Calyx usually gamosepalous, tubular or funnel-shaped, with the limb 5 – 10-lobed or dentate, plicate, often scarious, colored and showy. Corolla lobes 5, free or connate. Stamens 5. Ovary superior, 1-loculed, 1-ovuled, with styles or stigmas 5. Fruit a utricle enclosed in the calyx.

● **Limonium** Mill.
  Statice L.

Annual or perennial herbs, or shrubs, with leaves mostly rosulate. Flowers in panicled cymes or spikes. Calyx funnel-shaped, 5-veined, persistent, with a scarious, conspicuous, sometimes showy limb. Corolla inconspicuous, divided nearly or entirely to base, deciduous. Stamens 5. Styles 5.

Key to the Species of Limonium
1. Leaves pinnately lobed; stems winged .... 3. *L. thouinii*
1. Leaves entire; stems not winged
    2. Leaves oblanceolate-spathulate, 4 – 8 mm wide .... 1. *L. axillare*
    2. Leaves linear-spathulate, 1 – 3 mm wide .... 2. *L. carnosum*

1. **Limonium axillare** (Forssk.) O. Kuntze
    Statice axillaris Forssk.
Ascending shrublet, 10 – 50 cm high. Leaves grey-green, minutely punctate and white-dotted with excreted salts, sessile, clasping at base, fascicled or closely alternate, oblanceolate to oblanceolate-spathulate with rounded to subacute apices, long tapering to base, mostly 4 – 8 mm wide and up to 4.5 cm long. Flowers in dense spikelets on the terminal branches of a paniculate-spicate, alternately branching inflorescence up to about 20 cm long. Bracts reddish, white margined. Calyx c. 4 mm long, plicate, pubescent on the tube below, with shallowly lobed, white limb. Corolla purple but soon deciduous leaving the white calyx. *Plate 61.*

Habitat: Edges of coastal salt marshes and *sabkhahs* in coastal regions. Frequent.

Central Coastal Lowlands: 119, Ras Tanura; 226, Abu Ma'n; 7093, 1 km E Abu Ma'n.

South Coastal Lowlands: 3807(BM), Salwah.

Vernacular Names: QAṬAF (Bani Hajir).

2. **Limonium carnosum** (Boiss.) O. Kuntze

Statice carnosa Boiss.

Ascending to suberect perennial, 20–40 cm high, woody at base. Leaves densely fascicled, greyish green, linear-spathulate to long-cuneate, mostly 0.5–2 cm long, 1–3 mm wide, somewhat fleshy, finely punctate and covered with fine whitish crystals of excreted salt. Flowers in spicate panicles at extremities of scapes, tightly grouped within ovate-oblong, white-margined bracts; calyx c. 3 mm long, more or less plicate, green or reddish and hairy at angles of the tube, with shallowly lobed whitish, somewhat erose limb. Corolla deciduous, in ours white to very pale pink, c. 4.5 mm long with somewhat retuse lobes. Stamens c. equalling the corolla or slightly exserted.

Habitat: In, or at margins of, coastal salt marshes. Apparently rather frequent locally.

Central Coastal Lowlands: 7080, Dammam, near base of port causeway; 8657, Tarut Island, between Darin and Tarut; 8659, Tarut Island, east of causeway.

The author is indebted to Mrs. Sheila Collenette for pointing out the presence of this species in our territory. It is the dominant *Limonium* on Tarut Island and occurs at other coastal points.

Vernacular Names: QAṬAF (Bani Hajir).

3. **Limonium thouinii** (Viv.) O. Kuntze

Annual, glabrous but finely dotted with minute white excretions, 10–25 cm high, with several ascending stems from a basal rosette of leaves. Upper stems with 2–3 mm-wide veined wings. Leaves oblanceolate to spathulate, pinnately lobed, runcinate, usually with 3–4 pairs of rounded lobes, finely ciliate, to 8 cm long, 2 cm wide, long tapering at base to a short petiole. Flowers in terminal helicoid cymes with the inflorescence branches below the flowers expanded into a cuneate, leaf-like wing with branching nerves, up to c. 1 cm wide, asymmetrical at the apex. Upper bracts ending in recurved, subulate teeth. Calyx limb papery, c. 2 mm long, divided ⅓ to ½ to top of the tube, with 5 lobes alternating with 5 bristles. Corolla cream-yellow but soon deciduous, leaving the pale blue to white persistent calyces. *Plate 62.*

Habitat: Silty soil on the inland northern plains. Apparently restricted to the far northwestern corner of the Province although reported on the coast in southern Kuwait (Dickson 1955, 60). Frequent very locally.

Northern Plains: 3879, 15 km WSW Umm 'Ushar; 3886, 5 km ESE Umm 'Ushar; 3888, 14 km ESE Umm 'Ushar; 3898, 27–41N, 45–19E; 3904, 27–41N, 45–35E; 3905, 27–41N, 45–48E; 7527, 62 km E Umm 'Ushar.

Vernacular Names: KITA'AH (Mutayr), SIBSAB (Dickson 1955, 60).

16. TILIACEAE

Herbs, shrubs or trees, often stellate-pubescent, with leaves usually alternate. Flowers bisexual, or rarely unisexual and the plant monoecious, actinomorphic, in cymose inflorescences. Sepals (3–)5. Petals (4)5, mostly free. Stamens (4) to numerous. Ovary usually superior, multicarpellate, with 2 to many cells; style 1. Fruit a capsule, or berry- or nut-like.

● Corchorus L.

Herbs or shrubs with alternate serrate leaves. Flowers bisexual, grouped 1 to several on bracteate axillary peduncles. Sepals 5(4), caducous. Petals 5(4), free, yellow. Stamens numerous, or rarely 4 – 8. Ovary with 2 – 6 cells and several ovules in each. Fruit a capsule dehiscing by 2 – 5 valves.

Key to the Species of Corchorus
1. Flowers 4-merous; leaves less than 2 cm long; prostrate perennial
   plant .... 1. *C. depressus*
1. Flowers 5-merous; leaves 3 – 10 cm long; annual erect plant .... 2. *C. olitorius*

1. **Corchorus depressus** (L.) Stocks
   C. antichorus Raeusch.

Prostrate perennial, matlike and very close to the ground. Leaves alternate, broadly elliptic to ovate, 0.5 – 2 cm long, petioled, with prominent nerves and crenate margins. Flowers yellow, about 7 mm in diameter. Sepals 4; petals 4; stamens 7 – 8. Capsule 0.8 – 1.5 cm long, cylindrical, opening by 4 valves. *Plate 63.*

Habitat: Silty basins in desert, or as a weed in cultivated areas. Rare; the only *Corchorus* likely to be found outside the oases.

Southern Coastal Lowlands: Apparently not yet recorded from the Eastern Province but collected by the author in Qatar, where it is reported common by Batanouny (1981). It is very likely to be found on desert silt basins in the far east of the Province near the Qatar border or immediately east of the Qatar Peninsula in the coastal strip near the UAE border. Also reported from Bahrain.

2. **Corchorus olitorius** L.

Erect, glabrescent annual, 30 – 150 cm or more high. Leaves alternate, serrate, acute, lanceolate-elliptical with 2 tails near base of blade, 4 – 10 cm long, 1 – 4 cm wide, on petioles about half as long as blade. Flowers at the nodes on peduncles shorter than the subtending petiole; sepals 5; petals 5, yellow; stamens numerous. Capsule glabrous, 5-valved, 5 – 10 cm long, 0.5 cm broad, beaked.

Habitat: Nearly always near or in farm plots, where it is cultivated and sometimes escapes. Rare.

Central Coastal Lowlands: 7768, 7830, 1 km S Qatif town.

Uses: Cultivated as a pot herb. Grown in some tropical countries as a source of jute fiber, but not known to be so used locally.

● Two other species of *Corchorus*, well known as tropical weeds, may be found to occur sporadically in our area although records are so far lacking. In many features they resemble *C. olitorius* but may be distinguished by the characters indicated: *C. trilocularis* L. (reported by Chaudhary and Zawawi, 1983, as a weed in central Arabia) — capsule 3-valved, scabrous; stem hairy; leaf apices obtusish. *C. tridens* L. (also reported, *ibid.*, as a rarer weed in central Arabia) — capsules usually less than 3 – 4 cm long, with 3 ascending-spreading teeth or horns at apex (cf. apical teeth in the others lacking, or minute and close-connivent).

17. MALVACEAE

Herbs, shrubs or trees, usually stellate-pubescent, with alternate, stipulate, often dentate or lobed leaves. Flowers usually bisexual, actinomorphic, solitary and axillary or in racemose or paniculate inflorescences, sometimes subtended by an 'epicalyx' of bracteoles free from, or adnate to, the calyx, appearing as a second or outer calyx. Sepals usually 5, connate below. Petals 5. Stamens generally numerous, connate below in a column. Ovary superior, multicarpellate, with style several- to many-lobed at apex. Fruit a loculicidal capsule or schizocarp, breaking into several to numerous mericarps. Seeds often pubescent.

Key to the Genera of Malvaceae
1. Flowers yellow or orange, without epicalyx; shrubby perennials with dense, velvety indumentum .... 3. *Abutilon*
1. Flowers white to purplish or pinkish, with epicalyx of 3 – 10 bracteoles; annuals
    2. Bracteoles of epicalyx 7 – 10; upper leaves palmately parted more than ½ way to base .... 2. *Althaea*
    2. Bracteoles 3; upper leaves entire or lobed less than ½ way to base .... 1. *Malva*

● 1. Malva L.

Annual or perennial herbs with leaves generally palmately lobed. Flowers bisexual, solitary or clustered in the axils. Epicalyx of 3 free bracteoles inserted closely beneath the calyx. Sepals 5, connate below, often growing in fruit. Petals free, emarginate or deeply notched, purplish, pinkish, or white. Stamens numerous above a column. Ovary multicarpellate. Fruit a schizocarp with several to numerous, 1-seeded mericarps surrounding a torus.

Key to the Species of Malva
1. Petals twice as long as the sepals .... 1. *M. neglecta*
1. Petals less than 1½ times as long as the sepals
    2. Calyx broadly flattened at base, with lobes broader than long; fruit carpels sharply angled at margins .... 2. *M. parviflora*
    2. Calyx ascending, not broadly flattened, with lobes longer than broad; fruit carpels somewhat rounded at margins ... 3. *M. verticillata*
    M. parviflora is the common weed *Malva* of eastern Arabia. The other two species are rare.

1. Malva neglecta Wallr.
   M. rotundifolia auct. non L.
Variably pubescent annual with procumbent or decumbent stems, 10 – 40 cm high. Leaves orbicular in outline, 1 – 5 cm broad, shallowly lobed, on petioles often longer than the blades. Flowers axillary, solitary or clustered. Calyx longer than the epicalyx, hairy. Petals twice as long as the calyx, c. 10 mm long, white or pink, sometimes veined darker, bearded at base. Fruit mericarps smooth, not or hardly reticulate or wrinkled at back, with rounded margins and pubescent lateral faces.
Habitat: A weed of gardens, farms. Rare.
Central or South Coastal Lowlands, or Northern Summan: Vesey-Fitzgerald (1957, 791) reported dense stands of this plant at Rawdat Khuraym, a much-frequented spring camping ground for Riyadh residents on the western edge of the Eastern Province.

The species is also reported as a central Arabian weed by Chaudhary and Zawawi (1983). Records are apparently still lacking for its extension into the Eastern Province, but it probably will be found to occur as an occasional weed or ruderal.

2. Malva parviflora L.

Ascending annual, 4 – 40 cm high, sparsely to moderately stellate-pubescent, or the leaves sometimes glabrescent. Leaves orbicular or reniform in outline, crenate-dentate, 2 – 8 cm broad, nearly entire or palmately lobed less than half way to base, on petioles mostly 2 – 6 cm long. Flowers few or densely clustered, subsessile or on pedicels up to c. 3 mm long. Calyx broadly flattened at base, with the lobes erect or spreading, broader than long. Petals slightly longer than the calyx, white to pink or bluish. Carpels c. 11, in fruit with wavy radiating ridges at sides, sharp-angled and obscurely dentate at dorso-lateral margins, reticulate-wrinkled at back, often short-pubescent. On drying ground the plant may flower and produce mature fruit in dwarf form, less than 5 cm high.

Habitat: A weed of cultivated areas and waste and disturbed ground. Sometimes in the desert, particularly on old camp sites. Very common.

Northern Plains: 852, 9 km S 'Ayn al-'Abd; 7962, 3 km S as-Saffaniyah; 7980, al-Khafji town.

Northern Summan: 3163(BM), 11 km SSW Hanidh; 1341(BM), 42 km W Qaryat al-'Ulya.

Central Coastal Lowlands: 132, 1035, 7423(BM), Dhahran; 1068, Dhahran stables farm.

South Coastal Lowlands: Dickson 730, 'Ayn Najm, al-Hasa (Burtt and Lewis 1954, 391); 2737, ar-Ruqayyiqah, al-Hasa; C. Parker 22(BM), farm near al-Hufuf.

Vernacular Names and Uses: KHUBBAYZ ('baker-weed', gen.), ṬUBBAQ (said by Bani Hajir to be a synonym of the preceding). Carter (1917, 196) and Dickson (1955, 65) note that the leaves of this plant were often eaten as a vegetable by Persians in Kuwait.

3. Malva verticillata L.

Erect annual or perennial, 30 – 70 cm high. Leaves broadly cordate to suborbicular, 10 – 15 cm in diameter, rather deeply 5-lobed, on petioles 10 – 15 cm long. Flowers subsessile in dense clusters; calyx lobes to c. 7 – 8 mm long in fruit. Carpels 10 – 12, pitted and wrinkled dorsally.

● This description (after Rechinger 1964, 428) indicates a plant larger overall than *M. parviflora*. Burtt and Lewis (1954, 392 – 393) note that it differs from *M. parviflora* var. *microcarpa* in its more erect habit and in having carpels more rounded at the edges and less reticulate-wrinkled. These same authors *(ibid.)* also discuss the somewhat questionable status of the species and list specimens from islands in the northern and central Arabian Gulf. Seeds of specimens from al-Farisiyah Island in mid-Gulf were grown at Kew.

Habitat: To be expected as an occasional weed of farms and disturbed ground. Will probably also be found on Gulf islands. Rare.

Central or South Coastal Lowlands: Not seen by the author but listed as a weed of central Arabia by Chaudhary and Zawawi (1983). Recorded from al-Farisiyah Island (Iran) and thus to be expected on neighboring Saudi Arabian islands of the central Gulf, such as al-'Arabiyah.

● 2. Althaea L.

Annual or perennial herbs with leaves often more or less palmately lobed. Flowers bisexual, axillary, solitary or racemed. Epicalyx of 6 – 12 bracteoles. Calyx campanulate, 5-lobed. Petals 5. Stamens numerous above the tube. Ovary with numerous 1-ovuled carpels. Fruit a schizocarp of 8 – 25 mericarps surrounding a short central torus.

Althaea ludwigii L.

Annual, variably stellate-pubescent, usually with several decumbent stems from the base, 10 – 50 cm long. Leaves 1 – 4 cm across, broadly ovate to orbicular and sometimes broader than long, the upper ones palmately parted nearly to base with 3 – 5 main lobes, each with shorter secondary lobes; petioles 1 – 10 cm long, shorter above. Flowers axillary, c. 7 – 10 mm long, subtended by an epicalyx of linear-lanceolate bracteoles. Petals whitish-blue or whitish-pink, somewhat exceeding the villous calyx; peduncles 3 – 10 mm long. Fruits enclosed in the enlarging calyx; carpels faintly ridged. Seeds c. 2.5 mm long. *Plate 64.*

Habitat: Silt-clay basins on limestone. Infrequent except in the Summan.
Northern Plains: 4015, 27 – 56N, 45 – 35E.
Northern Summan: 3890, 14 km ESE Umm 'Ushar; 3238, Dahl al-Furayy; 2882, Rawdat Musay'id, 26 – 02N, 48 – 43E; 2811, 12 km ESE ash-Shumlul; 804, 44 km SW Jarrarah; 745, 33 km SW al-Qar'ah wells.
Southern Summan: 2249, 26 km ENE al-Hunayy; 2086, al-Ghawar, 24 – 27N, 49 – 05E.
South Coastal Lowlands: 7834, 8 km SE al-Hufuf.

● 3. Abutilon Mill.

Annual or perennial herbs or shrubs, usually stellate-tomentose, with alternate, dentate leaves. Flowers generally solitary in the axils, without epicalyx. Calyx more or less campanulate, 5-lobed or dentate. Petals 5, free, usually yellow or orangish. Stamens numerous from a column. Ovary with several to numerous carpels. Fruit a schizocarp, with mericarps 1 to several-seeded, often opening longitudinally at backs.

Abutilon pannosum (Forst. f.) Schlecht.

Grey- or yellow-green shrublet, 15 – 100 cm high, with dense velvety indumentum of stellate hairs. Leaves broadly ovate or cordate, 4 – 12 cm long, acute to shortly acuminate, on petioles ¼ to ¾ the length of the blade. Flowers yellow, c. 4 cm in diameter, sometimes with purple center. Fruit a schizocarp 1 – 2 cm in diameter, centrally depressed, consisting of a ring of 20 – 30 carpels rounded at back and apex. Seeds brown, stellate-pubescent.

Habitat: Silt-floored basins; perhaps most likely to be found on somewhat disturbed ground. Rare.
South Coastal Lowlands: Not yet recorded from the strict limits of our area but collected by the author (4128, BM) in south-central Qatar within 30 km of Saudi Arabian territory. It is likely to be found in the extreme southeast near the Qatar border or east of the Qatar Peninsula near the UAE boundary.
● Another *Abutilon* occasionally found in Qatar, *A. fruticosum* Guill. et Perr. (Batanouny, 1981), may also be found to occur, rarely, in the same Saudi Arabian border zones. It is a shrubby plant similar to *A. pannosum* but differing in having only 8 – 11 carpels in the fruit and the carpels truncate and angled at apex rather than rounded.

18. VIOLACEAE

Annual or perennial herbs, or shrubs, with alternate, stipulate leaves. Flowers bisexual, or sometimes unisexual and the plant polygamous, actinomorphic or zygomorphic, axillary, solitary or in cymose or racemose inflorescences. Sepals 5. Petals 5, generally unequal, with 1 sometimes enlarged and spurred. Stamens 5, connivent around the ovary, with 2 sometimes spurred basally. Ovary superior, several-carpellate, 1-loculate. Fruit usually a 3-valved capsule, rarely berry- or nut-like.

● Viola L.

Perennial or annual herbs, rarely somewhat frutescent, with alternate leaves. Flowers axillary, mostly solitary, zygomorphic, sometimes cleistogamous. Sepals subequal, with basal appendage. Petals unequal, with the lowest enlarged, spurred or saccate at base. Stamens with anthers nearly sessile, the 2 lowermost with spur at base entering the petal spur. Ovary 3-carpellate, 1-loculate. Fruit a 3-valved capsule. Seeds ovoid or globose.

A genus centered in north temperate regions, represented in our area by one weed species, collected only once. The nearest native violet population in Arabia is probably that of the drought-tolerant *Viola cinerea* Boiss., in the rocky foothills and mountains of Oman.

Viola arvensis Murr.
V. tricolor L.
Ascending, branching annual, 5 – 20 cm high, glabrescent or somewhat pubescent, particularly at lower margins of the stipules. Stipules enlarged, leaflike, to c. 2 cm long, mostly tripartite with an enlarged, oblanceolate central lobe and two smaller lateral lobes from near the base, sometimes finely hispid-ciliate at lower margins. Leaves ovate, 0.5 – 2 cm long, obscurely crenate, on petioles 2 – 5 cm long. Flowers c. 10 mm long, on pedicels 2 – 5 cm long, the pedicel with 2 minute bracteoles fixed just below insertion of the flower. Corolla about equalling the acute sepals, cream-whitish, yellow within, with some violet spots. Capsule c. 5 mm long; seeds ovoid, c. 1.5 mm long, smooth, light brown.

The above description is based primarily on our single collection; larger vegetative dimensions are reported for the plant in other areas.
Habitat: Collected only once, as an obviously introduced weed in remote desert at an abandoned Bedouin campsite where other locally unusual Mediterranean weeds were growing in spring. Most probably introduced with sacked grain. Rare.
Central Coastal Lowlands: 7952, N Batn al-Faruq, 25 – 43N, 48 – 53E.

19. CISTACEAE

Herbs or shrubs, often stellate-pubescent, with opposite or alternate leaves. Flowers bisexual, actinomorphic, solitary or in cymose inflorescences. Sepals 3 – 5, sometimes unequal. Petals (3)5. Stamens numerous. Ovary superior, 3 – 10-carpellate, 1-locular or incompletely several-locular, with simple style and 1 – 5 stigmas. Fruit a capsule.

● Helianthemum Mill.

Annual or perennial herbs, or small shrublets, with leaves opposite, subopposite, or alternate. Flowers in cymose, sometimes compound, inflorescences, bracteate. Sepals 5, free, the outer 2 often smaller and bractlike. Petals 5, generally yellow, sometimes pink, purple or white. Stamens numerous. Ovary 1- or 3-celled; style simple, with capitate stigma. Capsule ovoid to globose, 3-valved.

Four species of *Helianthemum* are known from eastern Arabia, and our records extend the known southern ranges of two of them. *H. ledifolium* and *H. salicifolium* apparently were not previously known south of Kuwait (Burtt and Lewis 1949, 305, 307).

Several writers including Carter (1917, 197) and Dickson (1955, 49) have remarked on the habitat association — well known among the Bedouins of eastern Arabia — of *Helianthemum* spp. and the desert truffles, *Tirmania* and *Terfezia*. Certainly one does not find truffles wherever *Helianthemum* occurs, but the converse does hold true: where truffles are found, one or more species of *Helianthemum* are almost always present. The plants are thus an indication of favorable conditions for establishment of the fungus, which may or may not be present with it.

The application of vernacular names to our four species appears to be rather capricious, with several names, particularly *raqrūq*, certainly used for more than one. All the name variants appear to be developments on the root themes of *raqrūq*, *jurrayd*, *suwayqah*, and *arqā'*.

Key to the Species of Helianthemum
1. Dwarf shrublets
   2. Flowers sessile or subsessile .... 3. *H. lippii*
   2. Flowers on pedicels 2 – 4 mm long .... 1. *H. kahiricum*
1. Annual herbs
   3. Pedicels longer than the sepals; capsules glabrous .... 4. *H. salicifolium*
   3. Pedicels shorter than the sepals; capsules ciliate at
      apical angles .... 2. *H. ledifolium*

1. Helianthemum kahiricum Del.
Many-branched dwarf shrublet, 10 – 30 cm high, grey-green with stellate pubescence. Leaves 3 – 12 mm long, 1 – 3 mm wide, elliptical-lanceolate, revolute margined, subsessile or on petioles 1 – 3 mm long. Flowers racemed on upper stems, rarely opening, on pedicels half as long to as long as the calyx. Inner sepals 4 – 6 mm long, lanceolate-ovate, pubescent, with the outer 2 shorter and narrower, bract-like. Petals yellow. Capsule ovoid with ascending-spreading hairs above, c. 4 mm long. Seeds angular, compressed, brown, 1 – 1.5 mm long.

Resembles and sometimes found with *H. lippii* (below) but readily recognized by its pedicellate flowers.

Habitat: Shallow sand on rocky ground. Occasional.
Northern Plains: 1222(BM), Jabal an-Nu'ayriyah.
Northern Summan: 8600, 26 – 41N, 47 – 45E.
Central Coastal Lowlands: 1039, 7754, Dhahran; Dickson 483, Dhahran (Burtt and Lewis 1949, 303).
Vernacular Names: RAQRŪQ (gen., Bani Hajir), UMM AS-SUWAYQAH (Al Murrah), HASHMAH (Dickson in Burtt and Lewis 1949, 304).

2. Helianthemum ledifolium (L.) Mill.
Erect, single-stemmed or branched pubescent annual, 5 – 20 cm high. Leaves elliptical-lanceolate, up to 15 mm long, 7 mm broad, subsessile or on 1 – 2 mm-long petioles.

Flowers solitary in upper axils, distant or crowded, 5–8 mm long, on rather thick pedicels 1–4 mm long. Inner sepals ovate-lanceolate, acute, pubescent, (in ours) c. 5–8 mm long, the outer 2 smaller, bract-like. Capsule ovoid, glabrous but ciliate along the upper margins of the valves, in ours c. 7 mm long. Seeds angular, smoothish, light brown, 0.8–1 mm long.

Habitat: Silty soils in the northern plains. Frequent in its range.

Northern Plains: 1714(BM), 11 km NW Abraq al-Kabrit; 1307(BM), 8 km WNW Qaryat al-'Ulya; 1328(BM), 8 km W Jabal Dab'; 1642(BM), 2 km S Kuwait border in 47–14E; 4054, ad-Dibdibah, 28–06N, 46–25E; 7568, al-Faw al-Janubi; 7574, Qatif-Qaisumah Pipeline KP–377.

Northern Summan: 744(BM), 33 km SW al-Qar'ah wells; 1378, 1392(BM), 42 km W Qaryat al-'Ulya; 1425(BM), 27 km W Qaryat al-'Ulya; 2817, 12 km ESE ash-Shumlul.

Central Coastal Lowlands: 1417(BM), 7 km NW Abu Hadriyah; 3260, 8 km W Qatif-Qaisumah Pipeline KP–144; 3316, 5 km S Nita'.

Vernacular Names: RAQRŪQ (gen.), JURRAYD (az-Zafir).

3. Helianthemum lippii (L.) Dum.-Cours.

Ascending, stellate-pubescent dwarf shrublet with branches often white-glossy beneath the hairs. Leaves elliptical-lanceolate, narrowly oblong or linear, revolute-margined, 4–15(18) mm long, 2–5(8) mm wide, on 1–2 mm-long petioles. Flowers sessile or subsessile in 5–10-flowered terminal, often 1-sided spikes. Inner sepals ovate, prominently 3(–4)-nerved, pubescent, 3–4 mm long; the outer 2 much reduced, narrower and bractlike. Capsule ovoid, hairy.

The size and form of the leaves in this species (as in *H. kahiricum*) are variable. Spring leaves under wetter conditions are longer, broader and more flat-margined. Summer leaves may be much reduced, narrower, more pubescent and tightly revolute.

Habitat: Shallow sands over limestone. Frequent.

Northern Plains 1278(BM), 15 km ENE Qaryat al-'Ulya; 1212(BM), Jabal an-Nu'ayriyah; 1311(BM), 8 km W Jabal Dab'; 1511(BM), 30 km ESE Qaisumah; 1627(BM), 3 km ESE al-Batin/al-'Awja junction; 3071(BM), ad-Dibdibah, 28–45N, 47–01E; 3109(BM), al-Batin, 28–00N, 45–28E.

Northern Summan: 690, 20 km WSW Khubayra; 717, 33 km SW al-Qar'ah wells; 758, 69 km SW al-Qar'ah wells; 787, 142 km SW al-Qar'ah wells; 2744, 25–18N, 48–20E; 2768, 25–49N, 48–01E; 2851, 26–18N, 47–56E; 7486, Batn Sumlul, 26–26N, 48–15E; 3143(BM), 9 km SSW Hanidh.

Central Coastal Lowlands: 1038, 419, Dhahran.

Southern Summan: 2227, 15 km WSW al-Hunayy.

South Coastal Lowlands: 2110, 49 km S Harad; Dickson 560(K), between Abqaiq and al-Hufuf (Burtt and Lewis 1949, 305).

Vernacular Names: RAQRŪQ (gen.), UMM AS-SUWAYQAH (Al Murrah), LIRQAH (sic, arqa'?, Bani Hajir), ARQA' (as *arja*, Musil 1927, 598), JURRAYD ('Amarat; Musil 1927, 598).

4. Helianthemum salicifolium (L.) Mill.

Ascending pubescent annual, 5–25 cm high, often branching from near base with one erect stem and 2 or more ascending or decumbent lateral branches. Leaves ovate to lanceolate, to 20 mm long, 5 mm wide, acute or obtuse, tapering at base to a short petiole. Flowers in terminal racemes, distant, on horizontally spreading, distally reflexed pedicels 8 – 10 mm long. Inner sepals ovate, 4 – 5(6) mm long, pubescent, nerved, the outer 2 reduced, bractlike. Capsule globose-ovoid, 4 – 5 mm long, glabrous. Seeds yellow-brown, c. 1 mm long, minutely tuberculate at the angles.

Habitat: Silty soils on the northern plains. Occasional to frequent.

Northern Plains: 654, Khabari Wadha; 1306(BM), 8 km WNW Qaryat al-'Ulya; 1327(BM), 8 km W Jabal Dab'; 1522(BM), 30 km ENE Qaisumah; 1643(BM), 2 km S Kuwait border in 47 – 14E; 1701(BM), ash-Shaqq, 28 – 30N, 47 – 42E; 4053, ad-Dibdibah, 28 – 06N, 46 – 25E; 7573, Qatif-Qaisumah Pipeline KP – 377.

Northern Summan: 1335(BM), 42 km W Qaryat al-'Ulya; 1424(BM), 27 km W Qaryat al-'Ulya.

Central Coastal Lowlands: 3309, 5 km S Nita'.

Vernacular Names: RAQRŪQ (gen.), JURRAYD (az-Zafir, also 'Amarat, Musil 1927, 598), SUWAYQAH (Musil 1927, 625; 1928b, 364), ARQA' (as *arja*, Musil 1927, 598).

20. TAMARICACEAE

Shrubs or trees with alternate, often reduced and scale-like leaves. Flowers mostly bisexual, actinomorphic, solitary or in racemose to paniculate inflorescences. Sepals and petals 4 – 5(6), free or nearly free to the base. Stamens 4 – 15, inserted on a disc. Ovary superior, 1-locular, 3 – 4 carpellate. Fruit a capsule; seeds hairy at least at apex.

Dr Bernard Baum, of the Canadian Biosystematics Research Institute, Ottawa and recent (1978) world monographer of *Tamarix*, has greatly assisted our attack on this difficult genus by identifying duplicates of specimens. His nomenclature is followed, and descriptions and keys are based on the present author's study of duplicates of these named specimens.

The assignment of *Tamarix* specimens to species generally requires examination of the minute floral parts under good magnification, preferably by use of a dissecting microscope on fresh material. The chief diagnostic characters are provided by the androecium, including the disc. Baum virtually ignores the gynoecium, although we have found the relatively very large fruit size of *T. pycnocarpa* to be diagnostic by itself in our area.

We follow Baum's separation of *T. arabica* and *T. mannifera* with reservations. His (1978) key character of entire vs. finely toothed sepals does not appear to be definitive in our specimens, and the mode of insertion of one or two stamens seems a rather insecure basis for splitting them. The author is thus inclined toward Zohary's view (1972) that these represent subspecific taxa in a '*T. nilotica*' (or other) series. Controlled breeding experiments and cytotaxonomic techniques may be required to clear up such questions.

A considerably larger body of *Tamarix* specimens has been collected by the author than is represented by those listed below under each species. Listings are here restricted to those determined by Baum and the few others kept in duplicate by the writer and thus available for direct comparison with authenticated material.

Only two vernacular names are used for the seven *Tamarix* species of our area: *athl* (exclusively for *T. aphylla*) and *ṭarfā'*, for the remaining six.

● **Tamarix** L.

Shrubs or trees with reduced, alternate, scalelike leaves often dotted with salt-excreting glands. Flowers bisexual, bracteate, in dense spikelike racemes. Sepals 4 – 5(6), free nearly or entirely to base. Petals 4–5(6), free, white or pinkish. Stamens 4–15, sometimes somewhat unequal, inserted on a disc. Ovary 3–4-carpellate with styles usually 3–4. Capsule loculicidal, pyramidal. Seeds numerous, with an apical pappus.

Key to the Species of Tamarix
1. Stamens 5
   2. Leaves vaginate, fully encircling stem without blade, stems thus appearing articulate; racemes 4–6 cm long; cultivated tree ... 1. *T. aphylla*
   2. Leaves more or less clasping but not fully encircling stem, with at least a short triangular blade; racemes 1.5–3.5(4) cm long; wild shrubs or small trees
      3. Petals persistent; all stamens hypodiscal, i.e. inserted slightly beneath edge of the disc and between its lobes .... 7. *T. ramosissima*
      3. Petals caducous after anthesis; at least 3 stamens peridiscal, i.e. inserted directly on edge of disc between the lobes
         4. All stamens peridiscal .... 5. *T. mannifera*
         4. 1–2 stamens hypodiscal, 3–4 stamens peridiscal ...... 2. *T. arabica*
1. Stamens 9–16
   5. Stamens (9)10, not or hardly broadened toward their bases .... 4. *T. macrocarpa*
   5. Stamens 11–15(16), broadened toward their bases
      6. Broadened basal portions of stamens overlapping; some flowers with 14 stamens; mature fruits 10(12) mm long .... 6. *T. pycnocarpa*
      6. Basal portions of stamens not overlapping; stamens 11–13; fruits not exceeding 7 mm long .... 3. *T. aucheriana*

1. Tamarix aphylla (L.) Karst.
   T. orientalis Forssk.
   T. articulata Vahl

Cultivated tree or tall shrub, sometimes up to 20 m or more high, usually with well developed trunk. Leaves vaginate, wholly encircling the stem without lamina, the stems thus appearing to consist of fine joints. Racemes longer than in other 5-stamened species in our area, usually over 4 cm and up to 6 cm long, 4–5 mm wide. Flowers pink.
Habitat: Cultivated in saline or non-saline soils where water table is within reach of its roots. Sometimes seen at sites of ruins or remote wells but never, apparently, self-propagating in the Eastern Province. Common.
Central Coastal Lowlands: 1091, Dhahran.
Vernacular Names and Uses: ATHL (gen.). A tree of great economic importance in eastern Arabia, not only as an ornamental in towns but as an agricultural windbreak and a sand stabilizer. Its high transpiration rates also assist ground water control when planted in poorly drained waste water runoff areas. Virtual forests of *athl* have long been established on dunes in the vicinity of 'Unayzah and Buraydah in central Arabia. In the east it is planted around edges of farms and is the primary plant used in a large-scale dune stabilization project on the northern edge of al-Hasa Oasis. Its trunks and branches provided roofing beams in traditional house construction.

2. Tamarix arabica Bge.

Large shrub or small tree, c. 1.5 – 3 m high, with brown or red-brown branches. Leaves partially clasping with acute triangular blade. Flowers in dense spiciform racemes 1.5 – 4 cm long, 3 – 4 mm wide. Petals caducous, c. 1.5 mm long, pale pink to white, often pinkish when in bud and turning near white at anthesis. Anthers pink, or sometimes cream-white. Fruits pyramidal, about 3 times as long as broad, tapering to apex with 3 stigmas, up to c. 3 mm long, reddish or greenish. *Plate 65.*

This plant appears to occur in two color forms: the more common with pink flowers (at least when in bud) and with pink stamens, and the other almost totally devoid of red pigmentation and appearing cream-white throughout the flower. See family notes, above, for remarks on the delimitation of this species.

Habitat: Apparently more common in disturbed, ruderal situations than in natural habitats, this is probably the most frequently seen *Tamarix* along roadsides and camp edges, usually on somewhat saline, poorly drained ground. Common.

Northern Plains: 7846(DAO), Qaisumah Pump Station.

Central Coastal Lowlands: 7862(DAO), 7864(DAO), Dhahran.

South Coastal Lowlands: 3942(DAO), 3 km ENE al-Hufuf.

Vernacular Names and Uses: ṬARFA' (gen.). This species has potential for cultivation as an ornamental shrub, being ready-adapted to disturbed and somewhat saline soil and bearing seasonally attractive albeit small flowers. At least one such shrub is already successfully so used in Dhahran, perhaps having been introduced accidentally or in confusion with *T. aphylla*.

3. Tamarix aucheriana (Decne.) Baum

Shrub, 1.5 – 2.5 m high, with brown to purplish branches. Leaves on younger parts strongly clasping, rather short, with acute triangular lamina often incurved at apex, papillose. Flowers often aestival, i.e. on green stems of current year's growth, in spiciform racemes 2 – 5 cm long, 7 – 10 mm wide. Petals somewhat persistent, pink to white, 2.5 – 3 mm long. Stamens (11)12 – 13, somewhat broadened at base (but not overlapping); anthers weakly apiculate; fruits pyramidal, 4 – 6 mm long. Some specimens of what otherwise appears to be this species have been seen to have only 10 stamens.

Habitat: Saline ground at roadsides or *sabkhah* edges. Apparently less frequent than the other *Tamarix* species of the Eastern Province.

Central Coastal Lowlands: 7443(DAO), Hanidh.

Vernacular Names: ṬARFA' (gen.).

4. Tamarix macrocarpa (Ehrenb.) Bge.

Shrub or small tree, 1.5 – 3 m high, with brown to grey-purplish branches. Leaves clasping with short-triangular lamina. Flowers predominantly aestival, with bracts equalling or somewhat exceeding the pedicels, in dense or somewhat open spiciform racemes 2 – 6 cm long, 4 – 6 mm wide. Petals pale pink, 2.5 – 3 mm long; stamens regularly 10 (rarely 9), not or hardly broadened at base, somewhat grouped in pairs; anthers pink, distinctly apiculate; fruits 3 – 7 mm long with 3 distally free stigmas at apex.

Habitat: Saline ground near wells or margins of *sabkhahs*. Frequent.

Northern Summan: 628, 820, near Wabrah wells.

Central Coastal Lowlands: 7448(DAO), al-Badrani junction.

Vernacular Names: ṬARFA' (gen.).

5. **Tamarix mannifera** (Ehrenb.) Bge.

Shrub, 1.5 – 3 m high, with brown to red-brown branches. Leaves clasping, with acute triangular lamina often incurved at apex. Flowers in dense, spiciform racemes 0.7 – 3 cm long, 3 – 5 mm wide. Petals 1.5 – 2 mm long, pink to white, stronger pink in bud; stamens 5, pink, or rarely cream, all inserted peridiscally. See family notes, above, for remarks on status of the species.

Habitat: Saline ground at roadsides or *sabkhah* edges. Frequency of occurrence not well known but apparently less common than *T. arabica.*

Central Coastal Lowlands: 7817A(DAO), al-Khobar; 7856(DAO), 6 km NW al-'Aba, 26 – 46N, 49 – 41E.

Vernacular Names: ṬARFA' (gen.).

6. **Tamarix pycnocarpa** DC.

Shrub or small tree, 1 – 3 m high, with grey-brown to grey-purplish branches. Leaves strongly clasping with short-triangular lamina. Flowers pink-rose, rather showy, in spiciform racemes 2 – 6 cm long, c. 10 mm broad or sometimes up to 20 mm broad in fruit. Petals 5, rarely 6 or 7, up to 4.5 mm long 3 mm wide, more or less emarginate at base when detached; stamens (11)12 – 15, broadened and overlapping basally; anthers not apiculate. Fruits 8 – 11(12) mm long; stigmas sometimes coalescent above forming a common, lobed stigmatic surface.

● Usually immediately identifiable by its large flowers and fruits, this species has a tendency to exhibit supernumerary floral parts. Individual flowers, or sometimes all flowers on a plant, may be found with 4- or 5-branched stigmas, and this condition may be associated with the presence of 1 or 2 extra petals.

Habitat: Saline ground at roadsides, village edges, or margins of *sabkhahs.* Frequent.

Central Coastal Lowlands: 1459(BM), 2 km E al-Ajam; 7092(DAO), W Umm as-Sahik, 26 – 38N, 49 – 53E.

South Coastal Lowlands: 7838, 10 km S al-Hufuf.

Vernacular Names: ṬARFA' (gen.).

7. **Tamarix ramosissima** Ledeb.

Large shrub or small tree with grey-purplish branches, 1.5 – 5 m high, often forming large hummocks on sand terrain. Leaves half-clasping with triangular, acute lamina. Flowers in spiciform racemes 2 – 5 cm long, 3 – 5 mm wide (sometimes to 8 – 10 mm wide in fruit). Petals persistent, white to pink, 1 – 1.75 mm long; stamens 5, all inserted hypodiscally near margin of disc. Fruits 2.5 – 4 mm long. The pedicels in this plant tend to be longer than in other local species, and the bracts often do not reach the middle of the calyx.

Habitat: Sand over saline ground. Occasional to locally frequent.

Central Coastal Lowlands: 531(K, DAO), Safwa, 26 – 40N, 49 – 56E; 1457(DAO, BM), 2 km E al-Ajam; 143(BM), 26 – 10N, 49 – 47E; 650(BM), N edge of Qatif Oasis.

Vernacular Names: ṬARFA' (gen.).

21. FRANKENIACEAE

Annual or perennial herbs or small shrublets with opposite leaves. Flowers usually bisexual, actinomorphic, solitary or in cymose inflorescences. Sepals 4 – 7, connate at least below. Petals 4 – 7, clawed, with a scalelike appendage. Stamens 3 – 6 or numerous. Ovary superior, 2 – 4-carpellate. Fruit a valved capsule.

● Frankenia L.

Annual or perennial herbs or subshrubs with opposite or whorled leaves. Flowers bisexual, axillary or terminal. Calyx tubular, (4 –)5-lobed. Petals (4 –)5, pink. Stamens (3 –)6, in 2 series. Ovary (2)3(4)-carpellate with style 3 – 4-lobed. Capsule 1-locular, 3-valved.

Frankenia pulverulenta L.

Low, subprostrate annual, greyish green or often turning reddish, finely pulverulent with whitish glands or excreted salt crystals. Stems branching at base, 5 – 30 cm long. Leaves opposite or whorled, oblong-oblanceolate, obtuse or retuse, sometimes slightly revolute-margined, 2 – 6 mm long, 1 – 3 mm wide, often with some crisped hairs. Flowers solitary or in loose cymes; calyx tubular, 3 – 4 mm long, longitudinally ribbed; petals pink, exserted from the calyx; stamens 6. Capsule 2 – 2.5 mm long. *Plate 66*.

Habitat: Saline ground, often in disturbed areas but sometimes in desert. Occasional.

Northern Plains: 1741(BM), Jabal al-'Amudah.

Central Coastal Lowlands: 75, Thaj; 912, Mulayjah; 1463(BM), 2 km E al-Ajam; 1827(BM), Dhahran; 3274, 3289, 8 km W Qatif-Qaisumah Pipeline KP – 144.

South Coastal Lowlands: 1898, 1 km NE al-Mutayrifi, al-Hasa; 2738, ar-Ruqayyiqah, al-Hasa; 3804, 10 km NW Salwah; 8299, al-Hufuf; 8311, adh-Dhulayqiyah.

Vernacular Names: ABŪ THURAYB, UMM THURAYB (Qahtan).

22. CUCURBITACEAE

Creeping, vinelike herbs, sometimes shrubby, with tendrils and alternate, often palmately-veined, leaves. Flowers usually unisexual with the plant monoecious or dioecious, epigynous, usually actinomorphic, axillary, solitary or cymose. Calyx and corolla 5-lobed. Stamens (1)3(5). Ovary inferior, 1-locular. Fruit a pepo (gourd), or berry-like.

Key to the Genera of Cucurbitaceae
1. Fruit spherical, smooth .... 1. *Citrullus*
1. Fruit ellipsoid, tubercled or bristly .... 2. *Cucumis*

● 1. Citrullus Schrad. ex Eckl. et Zeyh.

Monoecious annual or perennial vinelike herb with leaves deeply 3 – 5-lobed. Flowers solitary in the axils, yellow. Calyx tube shortly 5-lobed at apex. Corolla 5-lobed. Stamens 3. Ovary ovoid; fruit a globose gourd with numerous, compressed, obovate to elliptical seeds.

Citrullus colocynthis (L.) Schrad.
Colocynthis vulgaris Schrad.
Creeping scabrid perennial with stems tendril-bearing, sometimes over 1 m long. Leaves alternate, ovate- or triangular-cordate in outline, 3 – 10 cm long, 2 – 6 cm wide, petiolate, parted more than half way to base with 3 – 5 main lobes. Flowers solitary, pedicelled,

with yellow corolla c. 2 cm broad. Fruit a globose, smooth gourd, variously striped green and yellow-white, 4 – 10 cm in diameter, turning all yellow and hollowing when ripe and drying. *Plates 67, 68.*

Habitat: Sandy or silty ground, particularly in wadis. Common.

Central Coastal Lowlands: 230, 1085, Dhahran. *Citrullus* may be seen in almost any part of the Eastern Province, most frequently in northern and inland wadis, where the fruits appear to reach larger proportions than in the south.

Vernacular Names and Uses: SHARY (Al Murrah, gen. south), ḤANẒAL (gen. north). This gourd is the well known colocynth of long medicinal use in both Eastern and Western practice. The pulp and seeds of the fruit are strongly laxative and, in excessive dose, poisonous.

● 2. Cucumis L.

Annual or perennial, mostly monoecious, trailing herbs with simple tendrils and palmately-lobed leaves. Flowers solitary or clustered. Calyx tube 5-lobed. Corolla 5-lobed, yellow. Stamens 3. Fruit a many-seeded gourd with smooth or spiny-tuberculate surface.

Cucumis prophetarum L.

Perennial monoecious herb, with long trailing stems. Leaves scabrid, orbicular to cordate in outline, palmately 3 – 5 lobed. Flowers solitary, with yellow corolla. Fruit ellipsoid, 2 – 6 cm long, covered with prominent turbercles or bristles, longitudinally striped green and white, ripening all yellow. Generally resembles *Citrullus* (above) but easily recognizable by its different fruit characters.

Habitat: Inland wadi beds. Rare.

Northern Summan: 3241, Dahl al-Furayy. Only collected once, in the mouth of a limestone water hole, where it may have been introduced from the west. It is more common in Sudanian territory west of the Dahna.

23. SALICACEAE

Trees or shrubs with simple, mostly alternate leaves. Flowers in catkins, unisexual and the plant dioecious. Perianth generally absent, or minute, cup-like. Stamens 2 to numerous. Ovary 1-locular, 2 – 4-valved. Fruit a capsule with numerous hairy seeds.

● Salix L.

Trees or shrubs with leaves alternate, often elongate. Flowers in bracteate catkins with persistent scales. Stamens usually 2. Ovary 2-carpellate; fruit 2-valved.

Salix acmophylla Boiss.

S. persica Boiss.

Large glabrous, dioecious shrub or small tree, 3 – 6 m high, with red-brown twigs. Leaves linear-lanceolate, 8 – 14 cm long, 1 – 3 cm wide, green above and grey-glaucous below, long-tapering to a fine apex, with more or less serrulate margins, glabrous except sometimes when very young. Flowers in cylindrical, 2 – 5 cm-long catkins, the male with 4 – 5 stamens, the female with style less than 0.4 mm long on a 1.5 – 2 mm-long ovary. Ripe capsules 4 – 5 mm long, 1.5 – 2 mm wide (description after Meikle in Townsend and Guest 4(1), 1980, 33).

Our oasis willow probably belongs to this species, but some element of doubt remains due to the absence of mature flowering material in our specimens. Its status, whether truly wild or escaped or remaining from plantings, is uncertain.

Habitat: Along irrigation or drainage ditches in the larger oases. Occasional to rare.

Central Coastal Lowlands: 7404(BM), al-Qatif Oasis. Also seen in al-Hasa Oasis area.

24. CAPPARACEAE (including Cleomaceae)

Herbs or shrubs with alternate leaves; stipules spiny or small and unarmed. Flowers usually bisexual, actinomorphic, solitary, racemed or in compound inflorescences. Sepals 4(– 8). Petals 4, rarely absent or 5 – 8. Stamens 4 to numerous. Ovary superior, 1-several-locular, sessile or borne on a gynophore. Fruit a capsule, or rarely a drupe or berry.

Key to the Genera of Capparaceae
1. Shrubs with hooked stipular spines .... 1. *Capparis*
1. Shrublets or herbs without spines
   2. Shrublet; fruit nutlike, 3 – 8 mm long, surrounded by a
      scarious wing .... 3. *Dipterygium*
   2. Herbs (often perennials); fruit a 2-valved dehiscent capsule
      10 – 55 mm long, not winged .... 2. *Cleome*

A plant of another Capparaceous genus, *Maerua crassifolia* Forssk., has been collected by the author on rocky hills in the southwestern edge of the Rub' al-Khali outside the limits of the Eastern Province. *Maerua*, ranging in stature from a shrub to a large tree, may be found occasionally in the uplands of Sudanian territory west of the Eastern Province boundaries south of al-Kharj. It has apetalous flowers and obovate to cuneate leaves.

● 1. Capparis L.

Shrubs or trees, prostrate, climbing or ascending, with simple entire leaves and stipules often spinescent. Flowers bisexual, more or less zygomorphic, solitary or in various axillary or terminal inflorescences. Sepals 4, often unequal. Petals 4, usually white to pinkish and sometimes showy. Stamens numerous. Ovary generally 1-locular. Fruit an indehiscent or valved berry.

Capparis spinosa L.
Scrambling, branched shrub, 0.2 – 0.6 m high, often over 1 – 3 m broad, with young parts more or less grey-tomentose but becoming glabrescent with age. Leaves alternate or rarely opposite, short petioled, orbicular or broadly ovate, often with an apical spinelet, more or less tomentose when very young, becoming glabrescent, 1 – 4 cm long, 1 – 4 cm wide. Stipules modified to sharp, hooked spines 3 – 5 mm long. Flowers showy, to about 8 cm across; petals (in our No. 8377) white to pale pink, to c. 4 cm long and somewhat retuse. Fruits obovate-ellipsoid, c. 3(3.5) cm long, 2 cm wide, glabrous, dark green with 7 lighter longitudinal stripes, opening by valves from the apex, the valves then strongly reflexed, exposing the red pulp and numerous seeds. Seeds c. 2 – 2.5 mm, globose-reniform, smooth, brown to grey. *Plates 69, 70.*

Habitat: Silty ground over limestone. Rare to occasional.

Northern Plains: 3122(BM), al-Batin, 28–12N, 45–45E; sight record, al-Batin, 18 km NE Umm 'Ushar.

Southern Summan: 1879, 8377, 4.5 km NNE Mishash Jaww Dukhan.

Vernacular Names: SHAFALLAḤ (gen.)

● Two other, Sudanian, species of *Capparis* have been collected by the author near but outside the western boundaries of the Eastern Province: *C. cartilaginea* Decne. and *C. decidua* (Forssk.) Edgew. The former, found at al-Aflaj, differs from *C. spinosa* in having, instead of subequal sepals, the upper sepal much larger than the other 3 and hooded. It also has thicker and more leathery, less orbicular leaves and smaller stipular spines. *C. decidua* occurs at the very edge of the Province, on silt basins immediately outside the Dahna sands. It is a much taller shrub, up to 3–4 m high with ascending spiny branches, leafless when flowering, with pink to red flowers.

● 2. Cleome L.

Annual or perennial herbs or subshrubs with simple or palmately compound leaves. Flowers bisexual, solitary or racemed. Sepals 4. Petals 4, sometimes clawed. Stamens (4)6 to numerous. Ovary linear; fruit a capsule dehiscing by 2 valves.

A tropical and subtropical genus placed by some authors in a segregate family, the Cleomaceae.

Key to the species of Cleome

1. Lower leaves palmately divided into 3 leaflets . . . . 1. *C. amblyocarpa*
1. Leaves all simple, entire . . . . 2. *C. glaucescens*

1. Cleome amblyocarpa Barr. et Murb.

    C. arabica auct. non L.
    C. africana Botsch.

Erect perennial glandular, somewhat fetid herb, 10–40(50) cm high, sometimes subfrutescent at base. Leaves trifoliolate, with elliptical-oblanceolate leaflets 4–25 mm long, 2–5 mm wide, on petioles about as long as the blade. Uppermost leaves sometimes simple, smaller. Flowers pedicelled, with a rather obscure corolla; petals white, broadly veined yellow, with purple-veined tips. Fruit a compressed, glandular, 2-valved capsule, 10–50(55) mm long, 3–10 mm wide, mucronate at apex, on pedicels 3–10(15) mm long. Seeds densely covered with white hairs.

Habitat: Silty, sometimes disturbed, ground, mostly in the larger wadis. Occasional.

Northern Plains: 3117(BM), al-Batin, 28–12N, 45–45E; 8737, al-Batin bottom, 29–05N, 46–34E.

Northern Summan: 611(BM), al-Batin, 2 km ENE Qulban Ibn Busayyis; 3883, al-Batin, 5 km SW Umm 'Ushar.

Southern Summan: 2937(BM), 10 km WSW railroad KP 350; 2165, as-Summan, 24–08N, 48–43E; 3784(BM), 17 km E Khurays; 7501(BM), 7494(BM), Khashm az-Zaynah; 8017, 24–00N, 49–00E.

South Coastal Lowlands: 16 km SSE Harad Station.

Vernacular Names: KHUNNAYZ ('stink-weed', Qahtan), ḌURRAYṬ AN-NA'ĀM ('ostrich fart'; north, Musil 1927, 631; 1928a, 712), 'UFAYNAH ('stench-weed', north, Musil 1927, 597).

2. Cleome glaucescens DC.

Erect, glandular, perennial herb, 15 – 40 cm high, sometimes somewhat woody at base. Lower leaves 3-nerved, obovate, 1 – 3 cm long, 0.5 – 1.5 cm wide, on petioles about equalling the blade, the upper ones narrower, elliptical, on shorter petioles. Flowers axillary in the upper stem, forming loose terminal racemes, on pedicels longer than the flowers. Petals yellowish, richly veined brownish-red, sometimes purplish. Capsules compressed, glandular, 20 – 50 mm long, 4 – 8 mm broad, mucronate, on pedicels ¼ to ⅓ their length. Seeds covered with white hairs when mature. *Plate 71.*

Habitat: Silty wadi bottoms; found both in desert and on disturbed ground. Rare.

Southern Summan: 7907, edge of Na'lat Shadqam, 25 – 43N, 49 – 29E; 3836(BM), roadside at Harad Agricultural Project; 8045, Wadi as-Sahba, 23 – 59N, 49 – 10E; 8703, Wadi as-Sahba, 24 – 11N, 48 – 54E.

● 3. Dipterygium Decne.

Shrublet with small, entire leaves. Flowers bisexual, racemed. Sepals 4. Petals 4. Stamens 6. Fruit nutlike, indehiscent, 1-seeded, surrounded by a scarious wing.

A monotypic genus placed by some authors in the Cruciferae. Hedge, Kjaer and Malver (1980) discuss the history of, and arguments for, its family assignment and point out that phytochemical data strongly reinforce its inclusion in the Capparaceae.

Dipterygium glaucum Decne.

Yellowish-green glabrous, finely scabridulous shrublet, 30 – 80 cm high, with many thin ascending stems, often nearly leafless in the dry season. Leaves alternate, obtuse, oblong, 3 – 12 mm long, 1 – 6 mm wide, tapering to a 1 – 2 mm-long petiole. Flowers yellowish, 3 – 4 mm long, on 1 – 2 mm-long pedicels, solitary, or rather distant in axillary or terminal racemes. Fruits indehiscent, nutlike, slightly compressed, 3 – 8 mm long, 2 – 4 mm wide, obovate with wrinkled excrescences on the lateral faces, surrounded by a scarious wing 1 – 3 mm wide; pedicel 1 – 3 mm long. *Map 8.7.*

An important, although rarely dominant, constituent of several deep sand communities, particularly in the Rub' al-Khali.

Habitat: Deep sands of semi-stabilized dune areas. Frequent.

Central Coastal Lowlands: 236, 2 km W Dammam Port; 457, Dhahran.

South Coastal Lowlands: 195, 5 km E Abu Shidad, 25 – 52N, 49 – 22E; 1945, 28 km S al-'Uqayr, 25 – 24N, 50 – 21E; 1430(BM), 2 km E Jabal Ghuraymil, 25 – 47N, 49 – 36E; 3806, Bu'ayj, 24 – 40N, 50 – 39E; 1025, 23 – 36N, 49 – 34E.

Southern Dahna: 3820, 24 – 26N, 48 – 05E.

Rub' al-Khali: 1933, Zumul Camp, 22 – 20N, 54 – 57E; 1022, 22 – 57N, 50 – 12E; 7020, ST – 38, 22 – 17N, 53 – 22E; 7030, Camp S – 3, 21 – 29N, 55 – 24E; 7614, at-Tara'iz, 21 – 00N, 50 – 27E; 7615, southern Hibakah, 20 – 39N, 50 – 26E; 7619, Hadh Bani Zaynan, 20 – 17N, 49 – 45E; 7623, Shuwaykilah, 20 – 03N, 49 – 12E; 7627, southwestern Shuwaykilah, 19 – 48N, 48 – 46E; 7636, western al-Qa'amiyat, 18 – 28N, 47 – 07E; 7700, western al-Jawb, 23 – 10N, 49 – 35E; 8401, near camp G – 3, 20 – 25N, 55 – 08E; 8405, near camp G – 3, 20 – 26N, 55 – 07E; 8424, near G – 1 airstrip, 20 – 44N, 55 – 09E.

Vernacular Names and Uses: ῾ALQÁ (Al Rashid, gen. S), ῾ALANDÁ (sic, a name usually reserved for *Ephedra alata;* Bani Hajir). Of some importance as a grazing plant in the more remote sand regions.

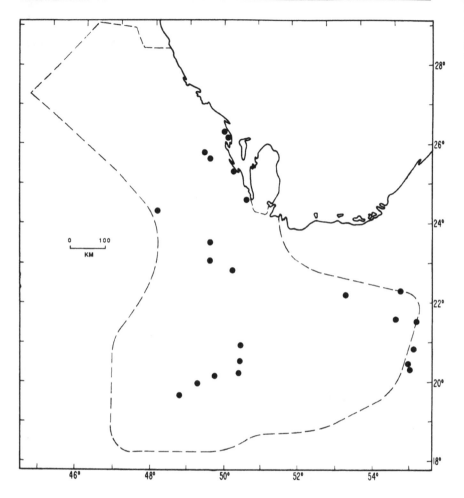

Map 8.7. Distribution as indicated by collection sites of *Dipterygium glaucum* (●) in the coverage area of this flora (dashed line), eastern Saudi Arabia. Each symbol represents one collection record.

25. CRUCIFERAE (Brassicaceae)

Annual or perennial herbs, sometimes suffrutescent, with alternate, rarely opposite, leaves. Flowers bisexual, actinomorphic, hypogynous, racemed. Sepals 4, in 2 series. Petals 4, free and clawed. Stamens 6, tetradynamous with 2 shorter in an outer whorl. Ovary superior, 2-carpellate, usually 2-locular with a false septum. Fruit a capsule dehiscing by 2 valves, called a silique (or siliqua) when at least 3 times as long as broad, a silicle (or silicula) when less than 3 times as long as broad; or, less commonly, indehiscent or a loment breaking into 1-seeded segments.

An important family in the annual flora of northeastern Saudi Arabia, with 46 species in 30 genera, including one genus, *Horwoodia,* endemic to the Arabian Peninsula and

128

bordering desert regions. Many of these plants represent Mediterranean elements in the Saharo-Arabian region, and there is a very strong decline in the diversity of crucifers from north to south in our area as the frontiers of Sudanian vegetation are approached. Members of the family are easily recognized by their 4-petaled, 'cruciform', flowers and mustardy taste.

Key to the Genera of Cruciferae

1. Fruit a silique, i.e. at least 3 times as long as broad (for fruits reaching this proportion because of having oblique, not straight, beaks or flat, tongue-like appendages, see the second lead of this couplet, below)
   - 2. Petals (at least ground color) white or yellow
     - 3. Petals veined, the veins usually dark-contrasting but sometimes nearly concolorous with the petal
       - 4. Silique strongly compressed, oblong-elliptical, not jointed; desert plants with close indumentum .... 21. *Farsetia*
       - 4. Silique more or less linear, terete or slightly compressed; weeds or escapes from cultivation
         - 5. Racemes bracteate; petals mostly 6 – 7 mm long . 7. *Enarthrocarpus*
         - 5. Racemes ebracteate; petals 8 – 25 mm long
           - 6. Fruit dehiscing by 2 valves
             - 7. Fruiting pedicels appressed, c. 2 – 4 mm long ..... 5. *Eruca*
             - 7. Fruiting pedicels spreading, c. 10 mm long ... 4. *Diplotaxis*
           - 6. Fruit breaking into 1-seeded segments .... 6. *Raphanus*
     - 3. Petals veinless
       - 8. Leaves subfiliform .... 25. *Leptaleum*
       - 8. Leaves flat, not filiform
         - 9. Silique with seeds in 2 parallel rows or planes
           - 10. Petals bright yellow .... 4. *Diplotaxis*
           - 10. Petals whitish .... 21. *Farsetia*
         - 9. Silique with seeds in 1 row
           - 11. Silique beaked
             - 12. Silique 10 – 15(25) mm long, appressed; beak swollen, often bent .... 3. *Hirschfeldia*
             - 12. Silique 18 – 50(70) mm long; beak not swollen, not bent
               - 13. Silique with 1 lateral nerve .... 1. *Brassica*
               - 13. Silique with 3 – 7 lateral nerves .... 2. *Sinapis*
           - 11. Silique beakless
             - 14. Petals yellow .... 29. *Sisymbrium*
             - 14. Petals white
               - 15. Silique obliquely 2-horned at apex ..... 23. *Notoceras*
               - 15. Silique not 2-horned
                 - 16. Leaves entire, linear .... 27. *Eremobium*
                 - 16. Leaves dentate or lobed .... 30. *Neotorularia*
   - 2. Petals purple or pink
     - 17. Plant glabrous, glabrescent, or with scattered hairs
       - 18. Leaves dentate; silique not 2-jointed

19. Ovules and seeds in 2 parallel rows in each locule;
pedicels 5 mm or more long .... 4. *Diplotaxis*
19. Ovules and seeds in 1 row in each locule; pedicels
shorter than 5 mm .... 26. *Malcolmia*
18. Leaves lyrate to pinnatifid; silique 2-jointed
20. Distal joint of silique compressed-tetragonal, ensiform
at apex; lower joint 1-seeded .... 9. *Cakile*
20. Distal joint ovoid or cylindrical, beaked
21. Petals 11–14 mm long; fruit 1–3 mm in
diameter .... 8. *Erucaria*
21. Petals 15–25 mm long; fruit 8–15 mm in
diameter .... 6. *Raphanus*
17. Plant densely pubescent, tomentose
22. Silique flat .... 21. *Farsetia*
22. Silique terete
23. Petals strongly wrinkled; silique 2-horned or
(in *Matthiola arabica*) tubercled at apex .... 24. *Matthiola*
23. Petals smooth; silique hornless
24. Pedicels shorter than calyx
25. Petals 8–12 mm long; basal leaves more
than 5 mm wide .... 26. *Malcolmia*
25. Petals shorter than 5 mm; all leaves less
than 5 mm wide .... 27. *Eremobium*
24. Pedicels longer than calyx; dwarf herb .... 28. *Maresia*
1. Fruit a silicle, i.e. less than 3 times as long as broad
26. Silicle conspicuously compressed, or appearing so because
of the presence of broad wings
27. Silicle broadly winged around full or nearly full
circumference, oblong-orbicular, more than 10 mm
long including wings .... 16. *Horwoodia*
27. Silicle not winged or only partially winged; if winged
at all, less than 10 mm long
28. Silicle compressed with wall perpendicular to flat
sides, thus appearing as a line on the face
29. Locules of silicle 1-seeded
30. Silicle cordate, without notch or indentation at
apex; upper leaves clasping .... 15. *Cardaria*
30. Silicle ovate, elliptical, or orbicular or with 2
suborbicular valves, notched, indented or
horned at apex; leaves not clasping
31. Silicle indehiscent, of 2 suborbicular valves,
2 mm or less long; prostrate or decumbent
weed .... 14. *Coronopus*
31. Silicle dehiscent, longer than 2 mm; erect
weeds or prostrate desert herb .... 13. *Lepidium*
29. Locules of silicle many-seeded .... 17. *Capsella*
28. Silicle strongly compressed with wall parallel to flat sides

      32. Leaves entire
         33. Silicle pubescent, at least 10 mm long .... 21. *Farsetia*
         33. Silicle glabrous, less than 10 mm long .... 22. *Alyssum*
      32. Leaves dentate .... 12. *Savignya*
26. Silicle not conspicuously compressed (although with flattish
  appendages in *Schimpera* and *Carrichtera*)
  34. Petals yellow
      35. Fruiting pedicels longer than fruit; fruits minutely
         globose, 1–3 mm long .... 20. *Neslia*
      35. Fruiting pedicels shorter than the fruit; fruits
         (including beak if present) more than 5 mm long
         36. Fruits with oblique, transversely compressed,
            beak-like appendage at apex; petals strong
            yellow .... 19. *Schimpera*
         36. Fruits with 2 diverging horns at apex; petals
            pale yellow .... 23. *Notoceras*
  34. Petals pink, whitish, or cream, sometimes with darker veins
      37. Petals pink; spiny shrub .... 10. *Zilla*
      37. Petals white or cream, with or without darker veins;
         unarmed annuals
         38. Petals cream with dark veins; silicle with stiff erect
            hairs and apical flat, tongue-like appendage .... 11. *Carrichtera*
         38. Petals whitish or cream, unveined; silicle without
            apical tongue-like appendage
            39. Silicle linear-oblong, 2-horned at apex ...... 23. *Notoceras*
            39. Silicle ovoid-globose with rounded, ear-like
               auricles on each side .... 18. *Anastatica*

● 1. Brassica L.

Annual or perennial herbs with simple or pinnatifid leaves rosulate at base, alternate above. Flowers yellow or white with inner 2 sepals sometimes saccate at base. Siliques teretish, linear to oblong, dehiscent by 2 valves, beaked at apex. Seeds in a single row.

Key to the Species of Brassica
1. Upper stem leaves clasping .... 2. *B. rapa*
1. Upper stem leaves petioled or narrowed to a non-amplexicaul base
  2. Beak of silique 4–8 mm long; lower stem leaves glabrous or
    glabrescent at margins .... 1. *B. juncea*
  2. Beak of silique 10–20 mm long; lower stem leaves rough-ciliate
    at margins .... 3. *B. tournefortii*

● *B. tournefortii* is a common desert annual. The others are infrequently seen weeds.

1. Brassica juncea (L.) Czern.
Glabrescent annual, to 1 m high. Lower leaves lyrate-pinnatisect with enlarged terminal lobe, petioled. Stems glabrous. Petals yellow, 7–9 mm long. Fruits 20–50 mm long, 2–3 mm wide, with constricted, almost filiform beak 4–8 mm long.

Habitat: A weed of garden, waste areas, or roadsides; rarely in desert and then on disturbed ground. Occasional.

Central Coastal Lowlands: 389, 3932(BM), Dhahran; 396, Dhahran Airport; 7999, 75 km WNW Qatif junction; 8128, 5 km ESE an-Nu'ayrirah.

2. Brassica rapa L.

B. campestris L.

Erect annual or biennial herb. Lower leaves lyrate-pinnatifid, more or less hispid; upper leaves glabrous, more or less entire, clasping. Flowers yellow. Siliques 40–60 mm long, compressed, with a long tapering beak comprising up to half the length of the fruit.

Habitat: A weed of farms, disturbed ground; rarely found on desert camp sites. Rare.

Central Coastal Lowlands: 3845(BM), Qatif farm; 8091, N Faruq, 25–43N, 48–53E, Bedouin camp ground; 8446, 27–01N, 49–19E, highway edge.

● This is the weed form of the cultivated turnip.

3. Brassica tournefortii Gouan

Ascending, branched annual, 20–75 cm high, with basal rosette of leaves. Lower stem hispid. Basal leaves lyrate-pinnatifid, with 5–15 pairs of dentate, rough, ciliate-hispid-margined lobes. Upper leaves smaller, entire, smoother. Flowers yellow, sometimes tinged violet, apparently very rarely pink (see below), with petals 5–7 mm long. Siliques 20–70 mm long, 2–2.5 mm wide, with beak 10–15(20) mm long. Fruit pedicels 12–20 mm long.

Habitat: A desert annual, common in the northern parts of the Province; also sometimes found as a weed.

Northern Plains: 989(BM), 12 km N Jabal al-'Amudah; 1177(BM), 1252(BM), Jabal an-Nu'ayriyah.

Northern Dahna: 4021, near ath-Thumami, 27–38N, 44–56E.

Northern Summan 1332(BM), 42 km W Qaryat al-'Ulya; 2776, 2782, 25–50N, 48–00E; 3327, 12 km WSW an-Nazim; 3877, 15 km WSW Umm 'Ushar.

Central Coastal Lowlands: 7514, Ras Tanaqib.

● Specimen No. 3877, from the Northern Summan, had flowers that were generally pink-violet overall. The plant otherwise was typical of the species. It is not known whether this was an individual aberration or a form that might be encountered elsewhere. Yellow petals sometimes dry pinkish.

Vernacular Names: KHAFSH (Shammar), ḤURAYSHA' (Dickson in Burtt and Lewis 1949) meaning 'roughweed', ṢUFFAYR ('yellow-weed', al-Hasa farmer), ZAHR (northern, Musil 1928b, 367). The last could be a mistaken recording of the common word for 'flower'; *zahr*, however is sometimes used as a proper specific name, cf. use of the name for *Tribulus* among southern tribes.

● 2. Sinapis L.

Herbs, mostly annual, with leaves lyrate to pinnatifid at base, sometimes nearly entire above. Flowers yellow or white; sepals equal at base. Siliques linear, teretish, beaked, with valves 3–7-nerved. Closely related to *Brassica*, which has the fruit valves with a single, median nerve.

Sinapis arvensis L.

Annual, 20 – 60 cm high, with stiff erect or retrorse hairs. Lower leaves lyrate, petiolate, with more or less dentate lobes. Petals 5 – 10 mm long, yellow. Siliques 18 – 40 mm long, 2 – 3 mm broad, with a conical beak and several prominent longitudinal lateral nerves, on spreading pedicels 3 – 5 mm long. Plants with glabrous fruits and plants with retrorsely hairy fruits may be found together.

Habitat: A weed of gardens or farms, rarely found in the desert around camp sites or roadsides. Infrequent.

Northern Plains: 3746(BM), 24 km W as-Saffaniyah.

Central Coastal Lowlands: 7424(BM), 8249, Dhahran; 8092, 8093, N Faruq, 25 – 43N, 48 – 53E; 8161, al-Jishsh, al-Qatif Oasis; 8129, 7 km WNW Abu Hadriyah, roadside.

South Coastal Lowlands: 8677, 1 km N 'Ayn al-Luwaymi, al-Hasa Oasis.

● 3. Hirschfeldia Moench

Annual or biennial herbs with runcinate-pinnatifid leaves. Flowers yellow or white. Siliques dehiscent, with a somewhat swollen beak. Seeds in one row, ovoid.

Hirschfeldia incana (L.) Lag.-Foss.

Pubescent annual or biennial with erect stems, 20 – 80 cm high. Basal leaves petiolate, rosulate, those above becoming linear. Racemes dense, elongating below with yellow flowers 6 – 9 mm long. Siliques 10 – 15(25) mm long including beak, appressed to stem, the beak 3 – 6 mm long, somewhat swollen and sometimes bent. Seeds ovoid or oblong, reddish-brown.

Habitat: A plant of disturbed ground. Collected only once, as a weed in a new lawn. Rare.

Central Coastal Lowlands: 7406(BM,E), Dhahran. Det. I. Hedge.

● 4. Diplotaxis DC.

Annual or perennial herbs with leaves entire to dentate or pinnatifid. Flowers yellow, white or pink-violet, with inner sepals not or hardly saccate. Silique compressed, linear, dehiscing by 2 valves, with numerous seeds 2-rowed in each locule.

Key to the Species of Diplotaxis

1. Petals purple-pink or bright yellow; desert plants
    2. Petals purple-pink . . . . 1. *D. acris*
    2. Petals bright yellow . . . . 3. *D. harra*
1. Petals white with pale greenish claw; weed of garden or
    farms . . . . 2. *D. erucoides*

1. Diplotaxis acris (Forssk.) Boiss.

Erect annual, rarely perennating, 5 – 50 cm high, sometimes branched near base, glabrescent with some scattered erect hairs on flower and fruit pedicels. Leaves mostly at base, somewhat fleshy, obovate to oblong, dentate, on petioles ⅓ to ½ as long as the blade. Flowers in an often flat-topped terminal inflorescence; petals pink-purple, very rarely white to yellowish, about 12 – 20 mm long. Siliques ascending, 20 – 50 mm long, 2 – 3 mm wide, on 7 – 15 mm-long pedicels. *Plates 72, 73.*

● Haines (1951), describing 'potential annuals' of the Egyptian desert, says that *D. acris* is a 'strict annual', unlike *D. harra* which may perennate. The author, however, has collected *acris* blooming in July in northern Arabia (15 km W Badanah). In the Eastern Province it has been so far recorded only as a winter-spring annual.

Habitat: Silty soil, sometimes in basins but often on rocky ground. Frequent.

Northern Plains: 1605(BM), 3 km ESE al-Batin/al-'Awja junction; 1570(BM), al-Batin/al-'Awja junction; Dickson 552(K), al-Ju'uf (Burtt and Lewis 1949, 286); Dickson 156, Abraq al-Khaliqah (Burtt and Lewis 1949, 286); 3115(BM), al-Batin, 28–00N, 45–28E; 7993, 14 km SSW Abraq al-Kabrit; 8758, ravine to al-Batin, 28–50N, 46–24E; 8759, (white flowered form), same place as preceding.

Northern Summan: 602(BM), al-Batin, 5 km ENE Qulban Ibn Busayyis; 627, Wabrah; 1339(BM), 42 km W Qaryat al-'Ulya; 7545, 65 km ESE by E of Umm al-Jamajim.

Northern Dahna: 607(BM), ath-Thumami.

Vernacular Names: JAHAQ (Shammar).

2. Diplotaxis erucoides (L.) DC.

Ascending annual, 15–50 cm high, rather sparsely pubescent with short recurved or retrorsely appressed hairs. Leaves oblong to oblanceolate in outline, strongly or obscurely pinnatipartite, obtuse or acute, to c. 10(15) cm long, 3(4) cm wide, petioled below, sessile and somewhat clasping above. Petals c. 7–14 mm long, white with pale greenish or purplish claw, nerved, with the nerves somewhat contrasting or quite concolorous with the white field. Siliques mostly 20–40 mm long, 1.5–2 mm broad, on spreading pedicels c. 10 mm long.

Habitat: A Mediterranean weed of garden or farmlands, apparently introduced with seeds of cultivated plants. Rare.

Central Coastal Lowlands: 8248, Dhahran.

3. Diplotaxis harra (Forssk.) Boiss.

Erect, more or less densely pubescent annual, sometimes perennating, 10–50 cm high. Leaves mostly basal, ovate to oblong-oblanceolate, greyish-green pubescent, dentate. Petals yellow, 7–10 mm long. Fruits 20–40 mm long, c. 2 mm broad, erect-spreading, at length somewhat pendulous, on pedicels 5–15 mm long. *Plate 74.*

Habitat: Silty soils, often on rocky, hilly ground. Frequent.

Northern Plains: 1673(BM), 2 km S Kuwait border at 47–14E; 541, 9 km SE Qaryat al-'Ulya; 1313(BM), 8 km W Jabal Dab'; 1243(BM), Jabal an-Nu'ayriyah; 1158, Jabal al-Ba'al.

Northern Summan: 330, 3 km SE an-Nazim; 819, near Wabrah; 1337(BM), 42 km W Qaryat al-'Ulya.

Southern Summan: 2268, 30 km ENE al-Hunayy; 346, 16 km N Harad; 2080, al-Ghawar, 24–30N, 49–05E.

Vernacular Names: KHAFSH (Shammar, Qahtan; also in Musil 1927, 601), KHUNNAYZ (Qahtan), ḤAWDHĀN (Al Murrah).

● 5. Eruca Mill.

Annual or perennial herbs, glabrescent or with simple hairs, with leaves usually pinnatifid. Flowers ebracteate with the inner sepals saccate at base, the petals clawed, yellowish with darker veins. Silique compressed, linear to oblong, with a flattened, seedless beak at apex; seeds 2-rowed in each locule.

Eruca sativa Mill.

Erect, branched puberulent to glabrescent annual, 10 – 50 cm high. Leaves pinnatifid with enlarged terminal lobe; upper leaves entire or serrate. Petals 15 – 20 mm long, cream or yellowish with greenish to violet veins. Siliques 15 – 25 mm long, 3 – 5 mm wide, appressed to stem, with a prominent longitudinal nerve on each side and a compressed beak 5 – 8 mm long; fruiting pedicels 3 – 6 mm long.

Habitat: Known only as a weed in farm fields and occasionally along roadsides.

Northern Plains: 7976, al-Khafji, roadside.

Central Coastal Lowlands: 546, Qatif Experimental Farm; 64, Abu Ma'n; 7998, 18 km WNW Abu Hadriyah, roadside.

Northern Summan: 8427, 16 km SW Judah, roadside.

Vernacular Names: JIRJĪR (Qahtan resident in al-Qatif area).

● 6. Raphanus L.

Annual, biennial, or perennial herbs, often stiff-hairy, with lyrate basal leaves and linear to oblong stem leaves. Flowers pinkish, white or yellow with the inner sepals somewhat saccate at base, the petals often veined. Siliques terete, 2-jointed, with the lower joint narrow and pedicel-like, the upper one indehiscent or breaking into 1-seeded sections.

Key to the Species of Raphanus

1. Fruit constricted between the seeds; taproot non-tuberous;
    petals usually pale yellow or whitish with darker veins .... 1. *R. raphanistrum*
1. Fruit not or hardly constricted; taproot tuberous, fleshy;
    petals usually pink-violet with darker veins .... 2. *R. sativus*

1. **Raphanus raphanistrum** L.
  R. rostratus DC.

Erect, branched, glabrescent to puberulent annual to c. 50 cm high. Lower leaves lyrate-pinnatisect, petioled; stem leaves narrower, dentate to entire. Petals 14 – 18 mm long, often cream or pale yellow (but also known pink-lilac), with darker veins. Fruits terete, 15 – 55 mm long, 3 – 7 mm broad, 2-jointed, the upper constricted between the seeds and breaking into 1-seeded segments.

Habitat: A weed of disturbed ground and waste places. Rare.

Central Coastal Lowlands: 363(BM), Dhahran; a rather doubtful specimen, but the plant has been recorded from Bahrain (I. Hedge, personal communication) and may be expected in our area as a weed).

2. **Raphanus sativus** L.

Annual or biennial herb to c. 80 cm high. Leaves lyrate-pinnatifid. Flowers pink to violet, sometimes white or cream, with purple veins or with veins virtually concolorous with the petals. Fruits terete, 2-jointed, 7 – 10 mm or more wide, inflated with spongy wall, hardly constricted and not breaking into joints.

Habitat: The cultivated radish; known outside cultivation as an escape or weed around farms. Occasional.

South Coastal Lowlands: 2726, ar-Ruqayyiqah, al-Hasa; 8190, near ash-Shuqayq, al-Hasa.

Vernacular Names: FUJL, FUJUL (often pronounced FIJAL).

135

● 7. **Enarthrocarpus** Labill.

Annual, pubescent to hispid herbs with pinnately divided, more or less lyrate, radical leaves. Flowers with yellow, purple-veined petals, bracteate at least in lower part of the inflorescence, in racemes elongating in fruit. Fruit an indehiscent, 2-jointed, terete to weakly compressed silique, the upper joint breaking into 1-seeded segments.

Enarthrocarpus lyratus (Forssk.) DC.

Annual herb with scattered simple hairs, branched from the base with stems ascending, to c. 50 cm long. Radical leaves petioled, lyrate, pinnately divided with dentate lateral lobes. Stem leaves more or less lobed, becoming dentate above. Racemes numerous, bracteate, elongating in fruit to 10 – 35 cm long. Petals 6 – 7 mm long, yellow with purple veins. Siliques short-pedicellate, 15 – 30(40) mm long, c. 2 mm wide, terete or weakly compressed, tapering at apex, straight or curved, appressed or ascending, longitudinally striate and more or less pubescent, with the lower joint about ¼ to ⅓ the length of the silique. Upper joint 4 – 6-seeded, weakly torulose, breaking at maturity into 1-seeded segments.

Habitat: A weed of farm lands. Rare to infrequent.

Central Coastal Lowlands: 8143, 13 km WNW Dammam, farm.

● 8. **Erucaria** Gaertn.

Annual herbs, glabrous or with simple hairs; leaves pinnatisect. Flowers pink or purplish, rarely white, with inner sepals hardly saccate and petals clawed. Siliques longitudinally nerved, consisting of 2 joints: the lower cylindrical, 2-locular, 2 – 8-seeded, the upper ovoid or beaklike, 1 – 4-seeded.

Erucaria hispanica (L.) Druce
E. lineariloba Boiss.

Erect glabrous branched annual to 75 cm high. Leaves succulent, bipinnatisect with linear lobes. Petals 10 – 14 mm long, purple to white, finely veined. Siliques 10 – 17 mm long, spreading or somewhat appressed on 2 – 3 mm-long pedicels, slightly constricted between the seeds; upper segment broader than the lower, 1.5 – 2 mm wide, terminating in a beaklike filiform style 3 – 4 mm long. Aspect of *Cakile arabica* but may be recognized in fruit by its terete upper fruit segment terminating in a filiform style rather than the distinctive flat, broad beak of *Cakile*.

Habitat: Northern Sands. Occasional.

Northern Plains: 4046(BM), al-Batin, 28 – 00N, 45 – 27E; 8732, ravine near ar-Ruq'i wells, 29 – 01N, 46 – 32E; 8774, 18 km WSW ar-Ruq'i Post, 28 – 59N, 46 – 31E.

Northern Summan: 4041(BM), al-Batin, 27 – 52N, 45 – 15E.

Northern Dahna: 4024(BM), near ath-Thumami, 27 – 38N, 44 – 56E; 3878, 15 km WSW Umm 'Ushar; 608(K), ath-Thumami.

Vernacular Names: SALĪḤ (north, Dickson).

● 9. **Cakile** Mill.

Annual herbs with entire or pinnatifid, somewhat succulent leaves. Flowers white, pink, or violet, with inner sepals slightly saccate at base. Siliques more or less tetragonous, 2-jointed, the distal joint flattened, deciduous.

Cakile arabica Vel. et Bornm.

Ascending, branched, somewhat succulent, glabrescent herb to 50 cm high. Lower leaves pinnately divided almost to base into narrowly linear lobes. Upper leaves often absent. Petals purple, 8 – 10 mm long. Siliques spreading, in elongated loose racemes, 2-jointed, 10 – 20 mm long, 2 – 3 mm wide, on pedicels 2 – 4 mm long; lower joint of silique terete, narrowing somewhat toward the pedicel, 3 – 5 mm long, furnished inside the articulation with 2 horns which fit into internal recesses of the upper joint. Distal joint compressed-tetragonal, tapering gradually toward a flattened, obtusish apex, 2 – 3 times as long as the lower joint. *Plate 75.*

Habitat: Stable sands; also found once as a weed in a new lawn. Occasional to locally frequent.

Northern Plains: 1166(BM), Jabal al-Ba'al; 1184(BM), 1235(BM), Jabal an-Nu'ayriyah; 1733(BM), Jabal al-'Amudah; 7971, Ras al-Mish'ab.

Northern Dahna: 4038, 27 – 35N, 44 – 51E; 7533, 40 km NE Umm al-Jamajim.

Northern Summan: 8432, edge 'Irq al-Jathum, 25 – 54N, 48 – 36E.

Central Coastal Lowlands: 3734, Ras az-Zawr; 1810(BM), 33 km SE Abu Hadriyah; 7521, Ras Tanaqib; 7738, 2 km S Ras Tanaqib, 27 – 49N, 48 – 53E; 7047, Ras as-Saffaniyah; 3142(BM), 3158(BM), 9 km SSW Hanidh; 3173(BM), 13 km SSW Hanidh.

South Coastal Lowlands: 1449(BM), 2 km E Jabal Ghuraymil; 8308, al-Hufuf, as a weed.

Vernacular Names: ZAMLŪQ (Bani Hajir), SALIH (north, Musil 1927, 606; Dickson in Burtt and Lewis 1949).

● 10. Zilla Forssk.

Glabrous, spinescent, rigid shrublet with leaves somewhat fleshy, few and deciduous. Flowers pink-violet to white, rather showy; calyx bisaccate at base. Fruit an ovoid to globose woody silicle, tapering at apex to a conical beak. Seeds 2, solitary in each of the 2 locules.

Zilla spinosa Prantl

Intricate, rounded, glabrous and nearly leafless spiny-branched shrublet, up to 75 cm high. Basal leaves on young green plants rosette-forming, often pinnately lobed; upper leaves on fresh green growth oblong-linear, entire, fleshy. Older growth nearly leafless with stems hardening to tapering spines. Petals pink or violet, sometimes nearly white, with darker veins, rather showy, 15 – 18 mm long. Fruit ovoid-globose with obscure obtuse ribs, becoming bony, 8 – 10 mm in diameter with a broad-conical 3 – 4 mm-long beak at apex, on a stout pedicel 2 – 3 mm long. *Plate 76.*

Habitat: Wadis and silt basins. Frequent.

Northern Plains: Frequently seen in main course and tributaries of al-Batin.

Northern Summan: 695, 1 km N al-Qar'ah wells.

Southern Summan: 2270, 30 km ENE al-Hunayy.

Vernacular Names: SHUBRUM (Al Murrah, Bani Hajir, 'Ujman, Qahtan), SHIBRIQ (Shararat), SILLA' (Harb, 'Utaybah, Mutayr, Hutaym). *'Shibriq'* is probably derived from the verb *shabraqa,* 'to tear to pieces, tatters', a reference to the very spiny habit of this plant. *'Shubrum'* may be a dialectal variant of the same word.

● 11. Carrichtera DC.

Annual herb with pinnatisect leaves. Inner sepals not or hardly saccate; petals white or cream with purple veins. Silicle ovoid with a flattened, tonguelike appendage.

Carrichtera annua (L.) DC.
C. vellae DC.

Ascending annual, 5 – 30 cm high, hispid with stiff, straight, white hairs. Leaves about 2 – 4 cm long, 1 – 1.5 cm wide, pinnately divided into narrow, linear lobes. Stems densely hispid below, becoming glabrescent in upper flowering parts. Flowers in elongated terminal racemes with petals cream to yellowish, veined purple or brownish, c. 5 – 8 mm long. Fruits 6 – 8 mm long on deflexed pedicels 2 – 3 mm long; silicle consisting of a main ovoid body, beset with erect white hairs, with a glabrous, flat, tonguelike distal appendage equalling or exceeding it in length. *Plates 77, 78.*

Habitat: Silty soils. Occasional.

Northern Plains: 1547(BM), 36 km NE Hafar al-Batin; 1576(BM), al-Batin/al-'Awja; 8754, tributary to al-Batin, 28 – 50N, 46 – 24E.

Northern Summan: 604(BM), 5 km ENE Qulban Ibn Busayyis; 7551, 65 km ESE by E of Umm 'Ushar; 4138, W as-Sarrar, 25 – 58N, 48 – 18E.

Vernacular Names: KHUSHSHAYNAH (north, Musil 1927, 603), 'rough-weed'; NAFKH (Dickson 1955); KHUZĀMĀ (Philby 1922; questionable — this name is usually applied to *Horwoodia*).

● 12. Savignya DC.

Annual herb with simple, somewhat fleshy leaves. Flowers pinkish or violet to white; inner sepals slightly saccate at base. Silicle strongly compressed, elliptical-oblong, stipitate, dehiscent, with seeds in 2 rows. Seeds flat, surrounded by a scarious wing.

Savignya parviflora (Del.) Webb

Ascending, glabrous, branched annual, 10 – 40 cm high. Lower leaves obovate-oblong, obtuse, sinuate or dentate, with petiole shorter than the blade; upper leaves narrower, sessile. Flowers on capillary pedicels much elongating in fruit; petals pink, 4 – 6 mm long. Silicles glabrous, finely veined, elliptical-oblong to suborbicular, strongly compressed, 8 – 14 mm long, 5 – 8 mm wide, with a 2 mm-long truncate apical beak, on spreading capillary pedicels 1.5 – 2 times as long as the fruit. Seeds flat, orbicular, about 3 mm in diameter including the surrounding circular, translucent wing. *Plate 79.*

Habitat: Usually in shallow sand on somewhat rocky ground. Common.

Northern Plains: 1687(BM), al-Hamatiyat, 28 – 50N, 47 – 30E; Jabal al-'Amudah (Dickson 1955); 1171, Jabal al-Ba'al; 1178(BM), Jabal an-Nu'ayriyah; 1263(BM), 15 km ENE Qaryat al-'Ulya; 542, 15 km SE Qaryat al-'Ulya.

Northern Summan: 1402(BM), 42 km W Qaryat al-'Ulya; 2818, 12 km ESE ash-Shumlul; 3232, 74 km S Qaryat al-'Ulya; 3337, 12 km WSW an-Nazim.

Southern Summan: 2102, 49 km S Harad; 2035, 44 km E Khurays; 2061, Jaww ad-Dukhan, 24 – 48N, 49 – 05E; 2118, 3 km SW al-Jawamir (Yabrin area); 2151, 23 – 15N, 48 – 33E; 2157, 23 – 35N, 48 – 39E; 2217, 26 km SSE al-Hunayy; 2251, 26 km ENE al-Hunayy.

Central Coastal Lowlands: 31, 35, 89, 104, Dhahran.

South Coastal Lowlands: 1451(BM), 2 km E Jabal Ghuraymil.

Rub' al-Khali: 8029, al-Mulayhah al-Gharbiyah, 22–53N, 49–28E.

Vernacular Names: QULAYQILĀN (Al Murrah, gen.), QUNQULĀN (Al Murrah variant), QULQULĀN (Qahtan, Bani Hajir). The names are all noun forms derived from the Arabic root verb *qalqala*, 'to shake, move'; they almost certainly refer to the very distinctive quaking motion of this plant's capillary-pedicelled silicles in the slightest wind.

● 13. Lepidium L.

Annual or perennial herbs, glabrous or with simple hairs, with leaves entire or pinnately lobed, often rosulate. Flowers ebracteate, white, rarely pink or yellow, with sepals equal at base and petals sometimes reduced or absent. Stamens 6, 4, or 2. Silicle laterally compressed, 2-locular and dehiscing by 2 valves, sometimes winged or retuse at apex.

Key to the Species of Lepidium

1. Silicles 4.5–6 mm long .... 2. *L. sativum*
1. Silicles 2–4 mm long
    2. Silicles appressed to stem, on flattened pedicels; low desert
      herb .... 1. *L. aucheri*
    2. Silicles spreading on terete pedicels; weed of cultivated
      areas .... 3. *L. virginicum*

Another species, *L. ruderale* L., has been recorded from Kuwait and Iraq and possibly may be found in the Eastern Province. Of the three species keyed it most closely resembles *L. virginicum*, from which it differs in having apetalous flowers and more divided leaves.

1. Lepidium aucheri Boiss.

Annual, branching from base, with stems usually prostrate, 5–30 cm long; sometimes flowering in dwarf condition on drying ground. Basal leaves often rosetted, oblanceolate to spathulate in outline, pinnatifid with rather distant oval to oblong, obtusish dentate lobes; upper leaves nearly entire or serrate. Flowers minute, white. Silicles 2–3.5 mm long, 1–2 mm wide, compressed, with 2 horn-like wings at the emarginate apex, appressed, on flattened pedicels 1–2 mm long.

Habitat: Silt-floored basins. Locally common, particularly where rain pools have lain.

Northern Plains: 584(BM), 40 km SW Hafar al-Batin; 680, Khabari Wadha; 1590(BM), al-Batin/al-'Awja junction; 7558, al-Faw al-Janubi.

Northern Summan: 755(BM), 41 km SW al-Qar'ah wells; 2792 (infested with *Cuscuta*), 26 km SE ash-Shumlul; 3894, 14 km ESE Umm 'Ushar.

Central Coastal Lowlands: 8120, 15 km SSE al-Wannan.

2. Lepidium sativum L.

Erect annual, 20–60 cm high. Leaves pinnatisect with oblong to linear segments; those above entire. Flowers with white petals 2–3 mm long. Silicles broadly oblong to ovate, 5–6 mm long, 4–5 mm wide, compressed, slightly winged-emarginate at apex with style hardly exceeding the notch, on pedicels 2–3 mm long.

Habitat: So far recorded only as a weed of farms and gardens, but might be found on old desert camp sites. Rare.

Central Coastal Lowlands: 3950(BM), Dhahran stables farm.
South Coastal Lowlands: 1147(BM), Abqaiq
Vernacular Names: RASHĀD (gen.).

3. Lepidium virginicum L.

Erect, glabrous to very finely pubescent annual, 10 – 30 cm high. Basal leaves incised to pinnate; upper leaves smaller, lanceolate to linear, entire to incised. Flowers minute, with white petals equalling or somewhat exceeding the sepals. Silicles 2.5 – 4 mm long, somewhat longer than broad, slightly notched at apex.

Habitat: An American weed, recorded only once, at a farm where it was probably introduced with imported grains. Rare.

Central Coastal Lowlands: 3789(BM), Dhahran stables farm.

● 14. Coronopus Zinn

Annual or biennial herbs with leaves more or less pinnately lobed or pinnatisect. Flowers white, in axillary or terminal racemes, with inner sepals non-saccate and petals minute, sometimes absent. Stamens 2 – 6. Silicle somewhat laterally compressed, ovoid to orbicular, entire or emarginate at apex.

Key to the Species of Coronopus

1. Silicle 1.5 – 2.5 mm broad, retuse at apex with style shorter than
   notch .... 1. *C. didymus*
1. Silicle 2 – 4 mm broad, entire at apex with style extended .... 2. *C. squamatus*

1. Coronopus didymus (L.) Sm.

Delicate prostrate to decumbent annual herb with pubescent stems. Leaves 1 – 4 cm long, or those at base somewhat longer, with flattened rhachis and petiole, bipinnatisect with ultimate lobes ovate to linear-oblanceolate, acutish, 0.8 – 1.5 mm wide. Flowers minute, 1 – 1.5 mm broad, when young grouped in dense, corymbose clusters, later in elongating racemes; sepals ovate, somewhat hooding, obtuse; petals usually absent; stamens in 2 groups of 3, 1 group on each side of the ovary, only the central one of each group with an anther, the filament swollen at base. Fruits somewhat compressed, bilobed with emarginate apex, 1 – 1.5 mm long, 1.5 – 2.5 mm broad, finely reticulate pitted, on spreading pedicels 2 – 3 mm long.

Habitat: Known only as a winter-spring lawn weed. Infrequent to rare.

Central Coastal Lowlands: 7054, 8414, Dhahran.

2. Coronopus squamatus (Forssk.) Aschers.

Annual, prostrate to ascending, blue-green herb, 5 – 20 cm high. Leaves pinnatifid, the lower with obovate to oblanceolate segments, the upper with narrower lobes. Flowers in axillary, little-elongated inflorescences; petals white, 1 – 1.5 mm long. Fruits broadly cordate, reticulate and rugose with tubercle-like growths at margins, 2 – 3 mm long, 2 – 4 mm broad, with extending style.

Habitat: A lawn weed. Rare.

Central Coastal Lowlands: 7415(BM), Dhahran.

140

● 15. Cardaria DC.

Perennial herbs with leaves entire, petiolate below, clasping on the upper stem. Flowers small, white, with inner sepals non-saccate. Silicle broadly ovate or cordate, indehiscent, laterally compressed, with each of the 2 locules 1-seeded.

Cardaria draba (L.) Desv.

Erect herb, 15 – 50 cm tall, closely covered with short white curved hairs. Basal leaves obovate-spathulate, dentate; stem leaves oblanceolate-elliptical, dentate, tapering below to a petiole or subsessile, more or less clasping. Inflorescence terminal, corymbose, dense and several-branched. Flowers white. Fruit cordate-reniform, with two rounded lobes, weakly compressed, 2.5 – 4 mm long, 2.5 – 4 mm wide not including the style, with a persistent style half as long, to as long, as the fruit. Fruiting pedicels spreading, about 3 times as long as the fruit.

Habitat: A weed of cultivated areas. Rare to occasional.

Central Coastal Lowlands: 1065(BM), Dhahran Stables farm.

● 16. Horwoodia Turrill

Glabrescent or pubescent annual herb with pinnately lobed leaves. Flowers purplish-pink to violet, with the inner sepals saccate at base. Silicle nearly orbicular, compressed, indehiscent, strongly keeled longitudinally with veins radiating to the winglike margins.

Horwoodia dicksoniae Turrill

Malcolmia musilii Vel. pro parte

Decumbent to ascending annual, 8 – 40 cm high, with older parts glabrescent and young parts densely hirsute with fine spreading white hairs. Leaves ovate to oblong, coarsely serrate or pinnately lobed, on petioles c. ⅓ to ½ as long as the blade. Flowers in terminal racemes on pedicels c. 2 mm long, strongly fragrant, with petals mauve-purple or violet, c. 15 mm long. Silicles orbicular, 12 – 20 mm across, glabrous and glossy when mature, strongly compressed, multiveined from the medial keel toward the winglike margins, with persistent style c. 1 mm long at apex, emarginate at base, on a pedicel 2 – 4 mm long. *Plates 81, 82, Map 8.8.*

The flowers of this plant are highly fragrant, with a sweet, musky scent pleasant to most people but by a few considered disagreeably strong. They give a sweet odor to the milk of camels that feed on it heavily in spring.

Habitat: Stable, well-drained sandy to silty inland soils, often with *Rhanterium*. Common.

Northern Plains: 1173(BM), Jabal al-Ba'al; 1192, Jabal an-Nu'ayriyah; 1301(BM), 15 km ENE Qaryat al-'Ulya; 1719(BM), 28 km E Abraq al-Kabrit; Dickson 155, 496, Abraq al-Khaliqah (Burtt and Lewis 1949, 291).

Northern Dahna: 2320, 22 km ENE Rumah; Dickson 402 (Burtt and Lewis 1949, 291); 3250, S edge of 'Irq Jaham.

Southern Dahna: 2145, E edge ad-Dahna, 23 – 10N; 2036, 9 km E Khurays.

Northern Summan: 3336, 12 km WSW an-Nazim; 7473, Mishash Ibn Jum'ah; 699, 33 km SW al-Qar'ah wells; 772, 69 km SW al-Qar'ah wells; 1395(BM), 42 km W Qaryat al-'Ulya; 2752, 25 – 43N, 48 – 07E; 2847, 26 – 18N, 47 – 56E; 2861, 26 – 00N, 48 – 29E; 2876, 20 km W 'Uray'irah; 3223, 74 km S Qaryat al-'Ulya; 3159(BM), 9 km SSW Hanidh; 7934, 25 – 31N, 48 – 42E.

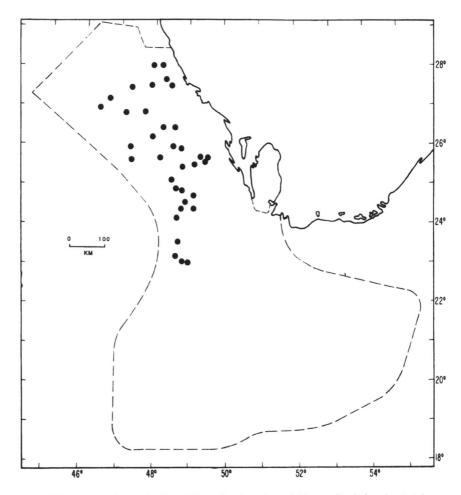

Map 8.8. Distribution as indicated by collection sites of *Horwoodia dicksoniae* (●) in the coverage area of this flora (dashed line), eastern Saudi Arabia. Each symbol represents one collection record.

Southern Summan: 2225, 15 km WSW al-Hunayy; 2221, 5 km S al-Hunayy; 2200, 24–39N, 48–51E; 2033, 44 km E Khurays; 2056, Jaww ad-Dukhan, 24–48N, 49–05E; 2088, 27 km N Harad Station; 2132, 6 km WNW Jabal Dab' (Yabrin area); 2135, 20 km WNW Jabal Dab' (Yabrin area); 2147, 23–15N, 48–33E; 2158, 23–35N, 48–39E; 2171, Railroad KP–349; 2174, 24–25N, 48–41E; 2208, 26 km SSE al-Hunayy.

Central Coastal Lowlands: 193, 423, Na'lat Shadqam; 78(BM), 'Ayn Dar GOSP 3; 7922, Batn al-Faruq, 25–34N, 49–05E.

South Coastal Lowlands: 382, al-Hasa, 1 km N cement plant.

Vernacular Names: KHUZĀMĀ (gen.).

● This striking Arabian endemic was first fully described in 1939 from specimens collected in Kuwait by Mrs Violet (now Dame Violet) Dickson, the specific name

142

acknowledging her very active pioneering work in collecting the northeast Arabian flora. The generic name honors A. R. Horwood (1879–1937), botanist at Kew.

Alois Musil was apparently the first to collect this species, although it is difficult to see how it could have been overlooked by earlier travelers. His specimens were studied and named by Velenovský in Prague. As explained by Rechinger (1962), Velenovský (1912) described the new species *Malcolmia musilii* from Musil's specimens of *Horwoodia* flowers which were, however, mixed with fruits of *Diplotaxis acris*. It was thus improperly placed in *Malcolmia* and invalidly named.

The full range of the species has apparently not yet been mapped, but it extends from the Eastern Province to the northwest across much of northern (and probably central) Saudi Arabia, southern Iraq and, reportedly, parts of Jordan. It extends south of the 23rd Parallel in the southern Summan but does not, apparently, reach Qatar in the east.

● 17. Capsella Medik.

Annual or perennial herbs with leaves entire, dentate or pinnatifid, rosulate below, clasping on the stem. Flowers white or pink with inner sepals non-saccate at base. Silicle obcordate to obtriangular, dehiscent by 2 valves.

Capsella bursa-pastoris (L.) Medik.

Erect annual to about 30 cm high, pubescent below. Leaves mostly in a basal rosette, lyrate, pinnately lobed to parted; upper leaves smaller, clasping. Flowers white, minute. Silicles 4–8 mm long, obcordate to triangular, compressed, with a prominent longitudinal suture on face, minutely reticulate-nerved, spreading; pedicels 5–20 mm long, elongating with age. Seeds numerous. *Plate 83.*

Habitat: A cosmopolitan weed sometimes seen in desert around former Bedouin camp sites, where it is probably introduced with sacked grain. Sometimes locally common.

Northern Plains: 614(BM), al-Batin, 30 km NE Umm 'Ushar.

Northern Summan: 8434, edge of 'Irq al-Jathum, 25–54N, 48–36E.

● 18. Anastatica L.

Small, stellate-pubescent annual with simple leaves. Flowers small, white, nearly sessile, with inner sepals non-saccate at base. Fruit an indurating, small thick silicle with 2 rounded, earlike appendages at apex. Seeds 4.

Anastatica hierochuntica L.

Stellate-pubescent dwarf annual, branched radially from base, prostrate or decumbent and often c. 15 cm across, the branches rolling inward after maturity to form a tight woody ball 4–10 cm in diameter *(Plate 85)*. Leaves oblanceolate to obovate, entire or obsoletely dentate above, to c. 3 cm long, 2 cm wide, tapering at base to a petiole often about ⅔ as long as the blade. Petals white, ovate-oblong, obtuse, clawed, c. 3 mm long. Fruit 4–6 mm long, with a prominent earlike appendage on each side and persistent style about half as long or as long as the fruit body. *Plates 84, 85.*

Habitat: Silty basins or wadi beds. Frequent.

Northern Plains: 1230, Jabal an-Nu'ayriyah; 1494(BM), 30 km ESE al-Qaysumah; Dickson 499, 543, Abraq al-Khaliqah (Burtt and Lewis 1949, 283).

Southern Summan: 2238, 15 km WSW al-Hunayy; 2202, 24 – 39N, 48 – 51E; 2043, 25 – 00N, 48 – 30E; 2068, Jaww ad-Dukhan, 24 – 48N, 49 – 05E; 8342, al-Ghawar, 69 km N Harad.

South Coastal Lowlands: 351, 9 km N al-Mutayrifi, al-Hasa; 2106, 49 km S Harad.

Vernacular Names and Uses: KAFTAH (gen.), KAFF MARYAM ('Mary's hand', gen.), BIRKĀN, BARUKĀN (Shammar), JUMAY· FAṬIMAH ('Fatimah's fist', Rawalah), KAFF AL-'ADHRA' ('virgin's hand', Dickson in Burtt and Lewis 1949), KAFN, QUFAY'AH ('shriveled one', Qahtan), QUNAYFIDHAH (northern, Musil 1926, 357), the last meaning 'little hedgehog' and probably referring to the appearance of the dried plant.

● This plant, inconspicuous and short-lived when green and flowering, is more commonly noticed in the dry season after it has taken its characteristic woody, globose form. Its resemblance to a clutched hand has led to its being likened in Arabian folklore to the hand of the Virgin Mary at childbed. It is thus associated with childbirth and is still used as an herbal remedy popularly believed to ease childbirth if consumed as a tea or used as a charm. It is occasionally seen dried in jars for sale at herbalists' shops. The dried plant's clenched branches expand and straighten when soaked in water, perhaps as a fruit-releasing mechanism, and this action — like that of an opening and closing fist — reinforces the likeness to the human hand.

● 19. **Schimpera** Hochst. et Steud.

Annual herb with dentate to pinnatifid basal leaves; stem leaves nearly entire. Flowers yellow with the sepals equal at base and the petals small. Silicle ovoid, indehiscent, indurating, with a compressed beak diverging from the fruit body.

Schimpera arabica Hochst. et Steud.

Erect annual, often with a branching main stem from a rosette of leaves, 3 – 30 cm high, densely hirsute in young parts, sparsely pubescent below, sometimes viscid near base with adherent sand. Basal leaves lanceolate-spathulate, runcinate-dentate or pinnatifid, sometimes nearly entire in dwarf specimens; stem leaves oblong-linear, nearly entire. Flowers in dense terminal racemes elongating greatly in fruit; petals yellow, c. 2 mm long. Fruit ovoid, indehiscent, 1-seeded, appressed to stem, hardening with age, 2 – 5 mm long, with a 3 – 8 mm-long laterally compressed beak diverging at an angle from the fruit body. *Plates 86, 87.*

Habitat: Most abundant on the northern plains, usually on silty-sand. Common.

Northern Plains: 539, 9 km SE Qaryat al-'Ulya; 671, Khabari Wadha; 692, 20 km WSW Khubayra; 1186, Jabal an-Nu'ayriyah; 1302(BM), 15 km ENE Qaryat al-'Ulya; 1486(BM), 50 km WNW an-Nu'ayriyah.

Northern Summan: 2839, 32 km ESE ash-Shumlul; 714, 33 km SW al-Qar'ah wells; 828, 13 km W Qaryat al-'Ulya; 1331(BM), 42 km W Qaryat al-'Ulya; 3910, 27 – 41N, 46 – 02E.

Northern Dahna: 2325, 22 km ENE Rumah.

Central Coastal Lowlands: 3294, 2 km W Sabkhat Umm al-Jimal.

Southern Summan: 2169, Railroad KP-349; 2180, 24 – 34N, 48 – 45E; 2243, 26 km ENE al-Hunayy.

Schimpera sometimes appears, apparently in cycles of several years, as a widespread vernal dominant on the northern plains. A year in the 1960s when the Dibdibah glowed yellow with a near-continuous cover of the plant is still known among the Bedouins as *'sanat aṣ-ṣuffār'* ('year of the *ṣuffār'*).

Vernacular Names: ṢUFFĀR (gen.), ṢIFĀR (Shammar). The name, 'yellow-weed', refers to the strong yellow color of the flowers. Young plants are sometimes nibbled raw by Bedouin herdsmen; they have a mustardy cabbage-like flavor.

● 20. Neslia Desv.

Pubescent annual herbs with simple leaves. Flowers yellow, with sepals equal at base. Fruit an obovoid or subglobose, somewhat compressed, indehiscent silicle with a single seed.

Neslia apiculata Fisch., Mey. et Avé-Lall.

Erect annual, simple or branched, with branched or stellate pubescence. Leaves lanceolate, acute, 2 – 8 cm long, 0.5 – 1.5 cm wide, entire or obscurely toothed, tapering to an amplexicaul base; upper leaves smaller, almost linear, sagittate with acute auricles at base extending past stem. Inflorescence paniculate, with multiple racemes branching from near top of stem. Flowers shortly pedicellate at top of raceme but pedicels much growing below; petals yellow, c. 2.5 mm long. Fruits slightly compressed, nearly spherical, 1-seeded, reticulate-nerved at sides, apiculate at apex, 1.5 – 2 mm long, on nearly straight, spreading capillary pedicels 5 – 8 mm long.

Habitat: A weed rarely seen in our area. The author's single collection was from an abandoned Bedouin camp where Mediterranean weeds were growing, most probably associated with sacked grain that had been stored there.

Central Coastal Lowlands: 7945, N Batn al-Faruq, 25 – 43N, 48 – 53E.

● 21. Farsetia Turra

Herbs or shrublets, mostly perennials, canescent with appressed branched hairs, and with simple, entire, narrow leaves. Flowers racemed, with sepals equal at base, the petals narrowly oblong, varying in color from whitish or lead-grey to yellowish and brownish pink or purple. Fruit a strongly compressed silicle or silique, oblong to linear, dehiscing by 2 valves. Seeds flat, each surrounded by a scarious wing, 1 – 2-rowed.

Key to the Species of Farsetia
1. Fruit 6 – 12 mm wide, about twice as long as broad . . . . 1. *F. aegyptia*
1. Fruit 1.5 – 4 mm wide, usually more than twice as long as broad
  2. Fruit more than 25 mm long
    3. Fruit 1.7 – 2 mm wide, tapering gradually to an acute
      apex . . . . 3. *F. heliophila*
    3. Fruit 2 – 4 mm wide, constricting more or less abruptly to a
      distinct persistent mucronate style 2 – 3 mm long . . . . . 4. *F. longisiliqua*
  2. Fruit less than 25 mm long
    4. Seeds surrounded by a scarious wing ½ as wide to as wide as
      seed body; diffuse plant with distant leaves . . . . 5. *R. stylosa*
    4. Seeds surrounded by an obscure wing less than half as broad as
      the seed body; low dense plant with approximating leaves 2. *F. burtonae*

1. Farsetia aegyptia Turra
Ascending, many-branched shrublet, 15 – 50 cm high, greyish with indumentum of appressed hairs. Leaves linear, 10 – 40 mm long, 1 – 2 mm wide. Petals 10 – 15 mm

long, lead-grey or whitish to pink, yellowish or purplish. Silicle strongly compressed, oblong with obscure medial-lateral nerve, 12 – 24 mm long, 6 – 12 mm wide, on 3 – 8 mm-long pedicels. Seeds with a broad membranous circular wing. *Plate 88.*

Habitat: Shallow sand, mostly on rocky ground. Common.

Northern Summan: 816, near Wabrah; 2868, 26 – 01N, 48 – 29E.

Southern Summan: 1141(BM), W edge of al-Faruq, Riyadh road.

Central Coastal Lowlands: 410, 431, Dhahran; Dickson 465 (Burtt and Lewis 1949, 290), Dhahran.

Vernacular Names: HALTA ('scratch-bush', Al Murrah), ḤAMĀH ('hot-weed', Qahtan, ad-Dawasir), JURAYBA' ('mangey-bush', Rawalah), ẓABYAH (Bani Khalid). Several of these names refer to the irritating effect obtained by rubbing parts of this very pubescent plant hard against the skin. Bedouin children thus use it in pranks as a kind of 'itching powder'.

2. Farsetia burtonae Oliv.

Rather dense ascending low perennial, often woody at base, densely appressed-pubescent, 3 – 25 cm high. Leaves linear-elliptical to linear-oblanceolate, acute, 10 – 40(50) mm long, 2 – 5 mm wide, tapering to base. Flowers white or pink-purplish, with a sweet, fruity fragrance. Silicles narrowly oblong, compressed, 10 – 18 mm long, 2 – 4 mm wide, on pedicels 2 – 3 mm long, terminating in a mucronate style 2 – 3 mm long. *Plates 89, 90.*

The plant has been collected flowering as hardly more than a seedling, less than 3 cm high. It has also been seen flowering in mid-November, before the first rains of the season.

Habitat: Silty inland soils, often in rocky terrain. Occasional.

Northern Plains: 572(K), 105 km E Hafar al-Batin; 569(BM), 20 km N Khabari Wadha; 1271(BM), 15 km ENE Qaryat al-'Ulya; 1597(BM), 3 km ESE al-Batin/al-'Awja junction; 8748, ravine to al-Batin, 28 – 50N, 46 – 24E.

Northern Summan: 7475, Batn Sumlul, 26 – 26N, 48 – 15E; 505(K), 5 km S Rawdat Ma'qala; 780, 69 km SW al-Qar'ah wells; 2757, 25 – 43N, 48 – 07E; 2843, 26 – 21N, 47 – 55E; 3902, 27 – 41N, 45 – 35E; 7935, 25 – 39N, 48 – 48E; 8608, 26 – 35N, 47 – 32E.

Southern Summan: 8693, N slope of Wadi as-Sahba, 24 – 12N, 48 – 34E.

Rub' al-Khali: 7500(BM), 2 km NW al-Hawk, 22 – 07N, 48 – 36E. This and the preceding records indicate that this plant's distribution is continuous through both the northern and southern Summan regions. The southernmost specimen had both leaves and fruits shorter than northern examples, the fruits approaching those of *F. aegyptia* in shape although smaller.

3. Farsetia heliophila Bge.

   F. hamiltonii auct. non Royle
   F. linearis auct. non Decne.
   F. arabica Boulos

Ascending, many-branched, greyish appressed-pubescent shrublet, 10 – 30 cm high. Leaves in specimens from our area 2 – 10 mm long, 1 – 1.5 mm wide (but may reach greater dimensions under favorable conditions), linear to narrowly spathulate. Petals pinkish, c. 7 – 10 mm long. Siliques linear, 28 – 40(50) mm long, only c. 2 mm wide, compressed, tapering to an acute apex, on pedicels 1 – 3(4) mm long. Seeds 10 – 30(40) in a single row, compressed, surrounded by a scarious wing less than half as wide as the seed body (description based partly on Boulos 1978).

Habitat: Rocky (limestone) terrain. Rare, as far as known, except very locally.

South Coastal Lowlands: 7455(BM), al-'Udayd, 24–28N, 51–20E.

This species has also been recorded from rocky areas on Bahrain. It appears to be a Sudanian associate barely entering our Province in the extreme southeast. Boulos (1978) described specimens from Qatar as a new species, *F. arabica*, treated here as a synonym.

4. Farsetia longisiliqua Decne.
Shrubby perennial, grey with appressed hairs. Leaves narrowly linear. Petals whitish to pale orange, 10–12 mm long. Siliques 25–35 mm long, 2–4 mm wide, linear, slightly curved outward, with persistent style 2–3 mm long, on appressed or ascending 1.5–5 mm-long pedicels.

Habitat: Sandy or somewhat silty terrain. Rare.

Northern Dahna: Vesey-Fitzgerald 13468, 'Dahna sand belt' (Zohary 1957, 637; precise location not given).

Rub' al-Khali: edges of Umm as-Samim, approx. 21–30N, 56–00E (Thesiger 1950, 166, n. 37).

● The author has not collected this species, and its status in the Eastern Province is uncertain. Thesiger's specimen from the far northeastern Rub' al-Khali is on the fringe of our area in Sudanian floristic territory. Zohary's listing from the northern Dahna (which could be either inside our area or outside it to the northwest) is more questionable and could be the result of confusion with *F. stylosa*. *F. stylosa* in our area has fruits that sometimes approach 25 mm long. *F. longisiliqua* is usually described as having flowers twice as large as *F. stylosa* and siliques with linear, rather than undulate, margins.

5. Farsetia stylosa R. Br.
F. ramosissima Fourn.

Ascending, grey-green, rather diffuse plant, often perennial in our area with a somewhat woody base, canescent with very closely appressed fine straight hairs. Leaves linear, sessile, 10–35 mm long, 1–3.5 mm wide. Petals 6–8 mm long, white or sometimes tinged orange to yellow. Siliques compressed, 12–25 mm long, 2.5–4 mm wide with more or less undulate margin, tending to be appressed to stem on pedicels 1–3 mm long, and terminating in a subfiliform style c. 2 mm long. Seeds surrounded by a scarious wing more than half as wide as the seed body.

Habitat: Silt-sand soils, often on somewhat rocky ground. Occasional.

Central Coastal Lowlands: 67(BM), Thaj.

South Coastal Lowlands: 1128(BM), al-'Udayliyah; 386, 1 km N cement plant, al-Hasa; 7601, Jaww ash-Shanayin, 25–46N, 49–22E; 8042, Wadi as-Sahba, 23–59N, 49–10E.

Vernacular Names: ḤISHAM (Al Rashid), ḤAMAH (Qahtan).

● 22. Alyssum L.

Annual or perennial, stellate-pubescent herbs with simple, mostly entire leaves. Flowers yellow or whitish, with inner sepals not saccate at base and petals entire or bilobed. Silicle orbicular to elliptical, compressed or inflated, 2-locular with 1–8 seeds in each.

Key to the Species of Alyssum

1. Silicle orbicular, as broad as long; ovary 1 – 2-ovuled in each
   cell .... 1. *A. homalocarpum*
1. Silicle broadly elliptical, longer than broad; ovary 3 – 6-ovuled in each
   cell .... 2. *A. linifolium*

1. Alyssum homalocarpum (Fisch. et C. A. Meyer) Boiss.

Erect, stellate-pubescent annual, several-branched from near the base, 7 – 15 cm high. Leaves mostly on lower branches, oblanceolate, sessile, long-tapering to base, 1 – 3 cm long, 0.3 – 0.6 cm wide. Flowers racemed with minute, yellowish petals equalling or somewhat shorter than the sepals. Fruiting racemes elongated, rather dense; silicles strongly compressed, nearly orbicular, 4 – 6 mm in diameter, on erect pedicels 1 – 2 mm long. *Plate 91.*

Habitat: Rocky knolls in the north. Occasional.

Northern Plains: 1596(BM), 3 km ESE al-Batin/al-'Awja junction; 8747, ravine to al-Batin, 28 – 50N, 46 – 24E; 8716, 20 km SW ar-Ruq'i wells, 28 – 51N, 46 – 25E.

Northern Summan: 4043, tributary to al-Batin, 27 – 56N, 45 – 23E; 7546, 70 km ESE by E of Umm al-Jamajim.

2. Alyssum linifolium Steph. ex Willd.

Erect or ascending stellate-pubescent annual, 5 – 20 cm high, several-branched from near base. Leaves narrowly-oblanceolate or linear, 8 – 15 mm long, 1 – 2.5 mm wide. Flowers minute, white to cream-colored. Fruiting racemes elongated; silicles elliptical on pedicels 2 – 5 mm long.

Habitat: Rocky knolls in the Summan. Rare.

Northern Summan: 1353(BM), ash-Shayyit al-'Atshan, 42 km W Qaryat al-'Ulya.

● 23. Notoceras R. Br.

Annual herb, canescent with medifixed hairs and with entire leaves. Flowers minute, white or yellow, with sepals equal at base, in racemes elongating in fruit. Silique rigid, linear-oblong, somewhat tetragonous, with the valves keeled and each with a short horn near the apex.

Notoceras bicorne (Ait.) Amo

Prostrate or decumbent annual, branching at base with stems to 20 cm long, covered with very closely appressed straight white-translucent hairs. Leaves oblanceolate, 15 – 30 mm long, 3 – 5 mm wide, tapering to base. Flowers in terminal racemes, with white to yellowish petals less than 2 mm long. Fruits more or less appressed to stem, 6 – 7 mm long, c. 2 mm wide, somewhat compressed, constricted between the seeds, with distinct lateral nerve and 2 diverging horns at apex, on stout pedicels c. 1.5 mm long.

Habitat: Silty desert basins. Frequent. *Plate 80.*

Northern Plains: 589, al-Batin, 16 km ENE Qulban Ibn Busayyis; 1495(BM), 30 km ESE al-Qaysumah; 1543(BM), 36 km NE Hafar al-Batin; 1564(BM), al-Batin/al-'Awja junction; 7569, al-Faw al-Janubi, 2 km N highway.

Northern Summan: 3234, Dahl al-Furayy; 729, 751(BM), 33 km SW al-Qar'ah wells; 7555, 65 km ESE Umm 'Ushar; 2777, 25 – 50N, 48 – 00E; 8623, Dahl Abu Harmalah, 26 – 29N, 47 – 36E.

Southern Summan: 2187, 24–34N, 48–45E; 2052, Jaww ad-Dukhan, 24–48N, 49–05E.
Vernacular Names: 'USHBAT UMM SĀLIM (north; Musil 1927, 598), meaning 'Umm Salim's herb', Umm Salim being the Bedouin name for the hoopoe lark, one of the most common resident birds of the desert.

● 24. Matthiola R. Br.

Annual or perennial herbs, often canescent with stellate or branched hairs, with leaves entire to pinnatifid. Flowers purplish to pink, yellow or white, with inner sepals strongly saccate at base and the petals clawed, sometimes undulate. Siliques narrowly linear, terete or somewhat compressed, with or without 2 apical horns developing from the stigma.

A distinctive genus represented in eastern Arabia by two native species readily recognized by their strongly undulate petals and smooth, terete narrow siliques. The common garden stock, *Matthiola incana* (L.) R. Br., is sometimes cultivated in local gardens.

Key to the Species of Matthiola

1. Lower leaves pinnately lobed; mature fruits with 2 spreading horns at
   apex .... 2. *M. longipetala*
1. Leaves entire; fruit without horns .... 1. *M. arabica*

1. Matthiola arabica Boiss.

M. arabica Vel.

Erect annual, to 50 cm or more high, more or less densely grey-tomentose with stellate hairs. Leaves elliptic-oblanceolate, entire, petiolate or the upper ones subsessile. Flowers sessile or nearly so, with petals 1.5 – 2 cm long, reddish with undulate-wrinkled margins. Siliques terete, to 6 cm long, c. 1 mm wide, on pedicels c. 1 mm long, with somewhat enlarged stigma at apex but without spreading horns. *Map 8.9.*

● Burtt and Lewis (1949) state that Velenovský (1912) applied the same name to 'a different plant'. Velenovský's description, however, fits *M. arabica* Boiss. very well. Rechinger examined Musil's collections and reported (1962) that the specimen from the Nafud described by Velenovský was indeed Boissier's species and that Velenovský's name was thus a homonym and a synonym.

Habitat: Stable sand, primarily in the coastal plain. Occasional.
Northern Summan: between Kuwait and Riyadh (Pelly 1866).
Central Coastal Lowlands: 1807(BM), 5 km NE ar-Khursaniyah; 4136, BM 426, S Abu Hadriyah; 7465, 7 km SW Qannur on Darb al-Kunhuri.
South Coastal Lowlands: Dickson 529(K), SW Abqaiq on al-Hufuf road (Burtt and Lewis 1949, 295); 1448(BM), 2 km E Jabal Ghuraymil.

This species has been described as having a Saharo-Arabian distribution. It appears to be restricted, however, to southern parts of that region or to parts not greatly distant from Sudanian territory. It is thus found in Sinai and parts of Jordan, but not, apparently, in Iraq. In eastern Arabia it is rare or absent north of the 28th Parallel; to the south it is often found in looser sand habitats, while *M. longipetala* (below) prefers silty, gritty soils.

Vernacular Names: SHUQARĀ (Al Murrah, gen.), a name possibly derived from the verb *shaqira* meaning 'to be red, blond' and referring to the reddish flowers.

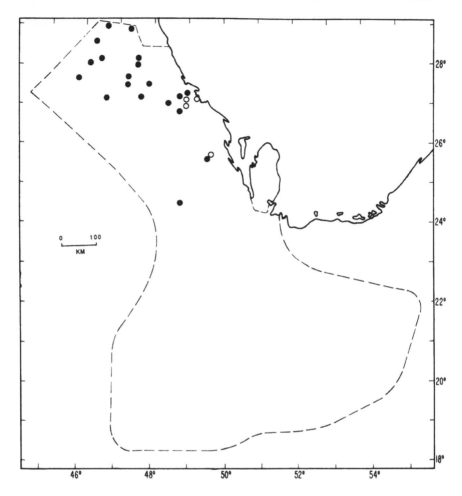

Map 8.9. Distribution as indicated by collection sites of *Matthiola longipetala* (●) and *Matthiola arabica* (○) in the coverage area of this flora (dashed line), eastern Saudi Arabia. Each symbol represents one collection record.

2. Matthiola longipetala (Vent.) DC.

M. oxyceras DC.

Erect or ascending annual, grey-tomentose with stellate pubescence, 8 – 40 cm high. Lower leaves pinnately lobed with obtusish, oblong-triangular segments; upper leaves narrower, sometimes entire. Flowers subsessile or on 1 – 2 mm-long pedicels; petals 15 – 20 mm long, undulate, purple but often tinged yellowish or greenish. Fruits terete, 4 – 6 cm long, subsessile, when mature with 2 prominent, curved, usually 2 – 5 mm-long horns at apex. *Plates 92, 93, Map 8.9.*

Habitat: Silty plains or stable shallow sands. Frequent.

Northern Plains: 556(BM), 655, Khabari Wadha; 1498(BM), 30 km ESE al-Qaysumah; 1553(BM), 36 km NE Hafar al-Batin; 991, 12 km N Jabal al-'Amudah; 1588(BM), al-Batin/al-'Awja junction; 1329(BM), 8 km W Jabal Dab'; 1303, 15 km ENE Qaryat

150

al-'Ulya; 1668(BM), 2 km S Kuwait border at 47 – 14E; 3261, 3285, 8 km W Qatif-Qaisumah Pipeline KP-144; 4047, ad-Dibdibah, 28 – 04N, 46 – 05E.

Northern Summan: 219(BM), 25 km S Qaryat al-'Ulya; 1381, 42 km W Qaryat al-'Ulya; 3906, 27 – 41N, 45 – 48E; 698, 33 km SW al-Qar'ah wells.

Southern Summan: 2182, 24 – 34N, 48 – 45E.

Central Coastal Lowlands: 1757(BM), Abu Hadriyah; 3722, 8 km N Abu Hadriyah; 3322(BM), 12 km S Nita'.

South Coastal Lowlands: 350, 1 km N cement plant, al-Hasa.

Vernacular Names: SHUQARA (Al Murrah, gen.).

● 25. Leptaleum DC.

Small annual herbs with leaves entire or pinnately divided. Flowers bracteate, white or pink, with the inner sepals non-saccate and the petals linear. The longer stamens are connate by the filaments in pairs and often reduced to 2. Siliques dorsally compressed, linear, opening at the apex; seeds in 2 rows.

Leptaleum filifolium (Willd.) DC.

Compact dwarf annual, branched from base, 2 – 15 cm high, shortly pubescent on branches, pedicels, and fruit margins. Leaves 2 – 5 cm long, extremely narrow-linear to filiform, simple or pinnate with filiform segments. Flowers white to pink. Siliques linear, more or less compressed, 15 – 25 mm long, 2 – 3 mm wide, often somewhat deflexed at base, nearly sessile or on pedicels 1 – 2 mm long. Immediately recognizable by its extremely fine, nearly hairlike leaves.

Habitat: Silty wadis and basins. Occasional to frequent.

Northern Plains: 1554(BM), 36 km NE Hafar al-Batin; 1578(BM), al-Batin/al-'Awja junction; 1637(BM), 2 km S Kuwait border at 47 – 14E; 7572, 22 km SW Hafar al-Batin.

Northern Summan: 1364(BM), ash-Shayyit al-'Atshan (42 km W Qaryat al-'Ulya); 3229, 74 km S Qaryat al-'Ulya; 3246, 5 km N Umm al-Hawshat; 8614, near Dahl Umm Hujul, 26 – 35N, 47 – 32E.

Vernacular Names: ḤUWAYWIRAH (Dickson), meaning 'little hotweed'.

● 26. Malcolmia R. Br.

Annual herbs with simple or branched hairs and entire or dentate simple leaves. Flowers purple, pink, yellow or white; sepals equal or the inner pair broader and saccate at base. Stamens connate in pairs by the filaments, or free. Silique linear, terete.

Malcolmia grandiflora (Bge.) O. Kuntze

Strigosella grandiflora (Bge.) Botsch.

Erect annual, single-stemmed or branched from base, 10 – 40 cm high, pubescent in young parts, otherwise glabrescent. Leaves mostly in a basal rosette, oblong to oblanceolate, sinuate to dentate, tapering below to a petiole, the older ones sparsely ciliate at margins, the younger pubescent on faces. Petals 8 – 15 mm long, more than twice as long as the sepals, rather showy pink to purple with darker veins. Siliques 20 – 60 mm long, c. 1 mm wide, spreading-ascending, straight or semi-coiled near the apex, on pedicels 1 – 2 mm long and about as thick as the fruit. *Plate 94, Map 8.10.*

Our plants belong to var. *glabrescens* (Boiss.) Burtt et Lewis.

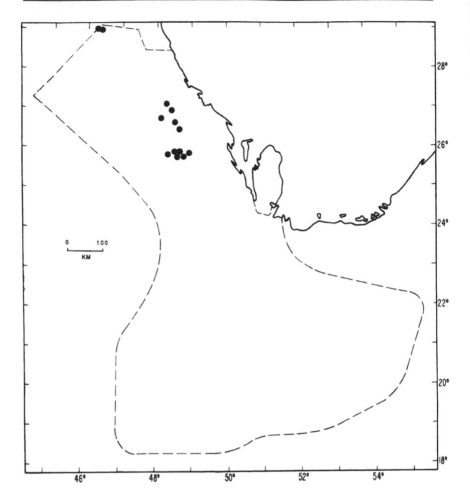

Map 8.10. Distribution as indicated by collection sites of *Malcolmia grandiflora* (●) in the coverage area of this flora (dashed line), eastern Saudi Arabia. Each symbol represents one collection record.

Habitat: Shallow silty sand on limestone. Appears to be somewhat tolerant of saline soils as found once on *sabkhah* edges. Infrequent except locally.

Northern Plains: 8708, 8735, ravine to al-Batin, 29–01N, 46–32E.

Northern Summan: 2860(BM), 26–00N, 48–29E; 3334(BM), 12 km WSW an-Nazim; 3353(BM), 10 km SSW as-Sarrar; 4137, 26–58N, 48–18E; 6829, 1 km S Hamra Judah; 8431, edge of 'Irq al-Jathum, 25–54N, 48–36E; 8436, 25–56N, 48–32E; 8447, 8 km W Nita'; 8557, 13 km W al-'Uyaynah; 8431, edge 'Irq al-Jathum, 25–54N, 48–36E; 8436, 3 km NW Jabal Muqaynimah, 25–56N, 48–32E; 8447, 8 km W Nita'.

Central Coastal Lowlands: 3323(BM), 12 km S Nita'; 8101, 'Uray'irah; 8119, 15 km SSE al-Wannan; 8111, 2 km S Hanidh.

• 27. Eremobium Boiss.

Annual herb, thinly to densely tomentose with stellate hairs; leaves narrow and entire. Flowers white, yellowish, or pink-purple with inner sepals saccate at base. Siliques linear, pubescent, weakly torulose, with the seeds 1-rowed in each locule.

Eremobium aegyptiacum (Spreng.) Boiss.
E. lineare (Del.) Boiss.
E. nefudicum (Vel.) Burtt et Rech.f.
Decumbent to ascending branched, stellate-pubescent annual, 5 – 30 cm high, often viscidulous below with adherent sand. Leaves linear, 5 – 45 mm long, 1 – 3 mm wide, obtuse, sessile. Petals 4 – 10 mm long, varying in color from white to pink or mauve, sometimes tinged with yellow. Siliques linear, weakly compressed, 10 – 35 mm long, 1.5 – 2 mm wide, spreading or ascending, on pedicels 2 – 4 mm long. *Plate 95.*

• Seedlings and young plants have a characteristic habit: one central erect-ascending stem, with 2 lateral decumbent branches from the base. Townsend (in Townsend and Guest 1980, 1038-39) refers to varieties *aegyptiacum* and *lineare* (Del.) Hochr. The characters given, however, do not appear to correlate either in individual plants or geographically in eastern Arabia. This species is notable as one of the few annuals with a distribution extending into at least the northern parts of the Rub' al-Khali. Its flowers there tend to be nearly all white.

Habitat: Deep sands. Common.

Northern Dahna: 7537, 8 km NE Umm al-Jamajim; 3247, S edge of 'Irq Jaham; 2319, 22 km ENE Rumah; 2037, 9 km E Khurays; 792(LE), Hawmat an-Niqyan; 800, 15 km SE Hawmat an-Niqyan.

Northern Summan: 220(BM), 10 km SSW al-'Uwaynah; 7472, Mishash Ibn Jum'ah; 3153(BM), 9 km SSW Hanidh.

Central Coastal Lowlands: 908, Mulayjah; 92, Dhahran; Dickson 471A, 523, Dhahran (Burtt and Lewis 1949, 287); 1812(BM), 49 km SE Abu Hadriyah; 3184(BM), 9 km N Shadqam junction; 3310, 5 km S Nita'.

South Central Lowlands: 548(LE), 5 km E Abu Shidad; 79(BM), 'Ayn Dar GOSP 3; 1050(LE), 24 km N al-'Udayliyah; 1431(BM), 2 km E Jabal Ghuraymil; 2119, 3 km SW al-Jawamir (Yabrin); 2128, Barqa as-Samur (Yabrin).

Southern Summan: 2156, Summan Yabrin, 23 – 27N, 48 – 34E; 2133, 20 km WNW Jabal Dab' (Yabrin); 2089, 27 km N Harad; 2209, 26 km SSE al-Hunayy; 8022, 23 – 11N, 48 – 53E.

Rub' al-Khali: 7671, al-Bahath, 22 – 58N, 50 – 01E; 7673, al-Khuwayran, 22 – 49N, 50 – 11E; 7684, 7 km SW Shalfa, 21 – 51N, 49 – 39E; 7690, Bani Tukhman, 22 – 38N, 49 – 34E; 7694, 8027, al-Mulayhah al-Gharbiyah, 22 – 53N, 49 – 29E.

Vernacular Names: GHURAYRA' (Al Murrah, Qahtan), QURḤAN (Al Murrah). Both names are derived from classical roots meaning 'to have a blaze on the forehead' (horse).

• 28. Maresia Pomel

Dwarf annuals, usually pubescent or tomentose with stellate or branched hairs; leaves entire, dentate, or pinnatifid. Flowers pinkish to white with inner sepals saccate at base. Siliques linear, terete, and somewhat torulose with seeds 1-rowed in each cell.

Maresia pygmaea (DC.) O. E. Schulz

Delicate stellate-pubescent dwarf herb, 3 – 10 cm high, with very fine branches ascending from the base. Leaves mostly in a basal rosette, less than 10 mm long, oblanceolate-spathulate, entire or pinnately lobed, obtuse, long-tapering to a petiole. Flowers equalling or exceeding their pedicels; petals pink, 5 – 8 mm long. Fruits linear-terete, 10 – 20 mm long, 1 – 1.5 mm wide, on subcapillary pedicels 8 – 20 mm long. *Plate 96.*

Habitat: Well drained sands. Occasional.

Northern Plains: 1218(BM), Jabal an-Nu'ayriyah.

Northern Summan: 1361(BM), 42 km W Qaryat al-'Ulya.

Northern Dahna: 4037, ad-Dahna, 27 – 35N, 44 – 51E.

Central Coastal Lowlands: 1411(BM), 7 km NW Abu Hadriyah.

● 29. Sisymbrium L.

Annual, biennial, or perennial herbs, glabrous or pubescent, with leaves lyrate to pinnatifid, rarely entire. Flowers bracteate or ebracteae, yellow, or rarely white or pink, with the inner sepals hardly saccate at base. Siliques narrowly linear, teretish, not beaked; valves 1 – 3-veined. Seeds usually in 1 row.

Key to the Species of Sisymbrium

1. Fruit pedicels 5 – 12 mm long .... 2. *S. irio*
1. Fruit pedicels 1.5 – 4 mm long
   2. Petals less than 3 mm long; siliques tapering toward apex . 1. *S. erysimoides*
   2. Petals 7 – 14 mm long; siliques linear .... 3. *S. septulatum*

1. Sisymbrium erysimoides Desf.

Erect annual, simple or branched, 15 – 60 cm high, sparsely pubescent with fine, white, erect hairs. Basal leaves lyrate-pinnatifid or dentate, with hastate terminal lobe; stem leaves similar, narrower. Flowers minute, with petals c. 2 mm long, yellow, hardly exceeding the calyx. Siliques 20 – 45 mm long, 1 mm wide, erect-spreading, on 1.5 – 4 mm-long pedicels; pedicels pubescent on the upper side.

Habitat: Silty soil in large wadis or Summan basins; very often associated with, and growing beneath, *Ziziphus nummularia*. Locally common.

Northern Plains: 580(BM), 15 km SW Hafar al-Batin; 1592(BM), al-Batin/al-'Awja junction; 8765, ravine to al-Batin, 28 – 49N, 46 – 23E; 8736, ravine near ar-Ruq'i wells, 29 – 01N, 46 – 32E.

Northern Summan: 2780(BM), 25 – 50N, 48 – 00E; 3239, Dahl al-Furayy; 3901, 27 – 41N, 45 – 31E.

● This species appears to occur most frequently along the fringes of Sudanian territory in the western Summan, extending north along al-Batin with its Sudanian-derived associate, *Ziziphus nummularia*.

2. Sisymbrium irio L.

Erect, branched annual, 15 – 50 cm high, pubescent or glabrescent. Lower leaves pinnately divided with hastate terminal lobe. Petals yellow, 2.5 – 4 mm long. Siliques 30 – 45 mm long, 1 mm wide, spreading-ascending, on 5 – 12 mm-long pedicels which are distinctly narrower than the fruit.

Habitat: A weed of gardens or waste ground.

Central Coastal Lowlands: 8097, N Batn al-Faruq, 25 – 43N, 48 – 53E, Bedouin camp ground.

Northern Summan: 8583, Dahl al-Furayy.

Vernacular Names: ḤARRAH (north; Musil 1928a, 697), meaning 'hotweed'.

3. Sisymbrium septulatum DC.

S. bilobum (C. Koch) Grossh.

Erect annual, 20 – 40 cm high, glabrous or sparsely pilose. Basal leaves rosetted, deeply lyrate-pinnatisect with somewhat retrorsely angled, acutish lobes. Upper stem leaves with narrowly linear segments. Petals yellow, 7 – 14 mm long, much exceeding the sepals. Siliques spreading-ascending, linear, 30 – 60 mm long, 1 – 1.5 mm wide, with bilobed stigma, on 2 – 4 mm-long pedicels about as thick as the fruit.

Habitat: Silty soils in the northern parts of the Province. Occasional.

Northern Summan: 621(K), al-Lihabah; 756(BM), 41 km SW al-Qar'ah wells; 3893(BM), 14 km ESE Umm 'Ushar; 3900(BM), 27 – 41N, 45 – 31E; 4040(BM), al-Batin, 27 – 52N, 45 – 15E; 2794(BM), 26 km SE ash-Shumlul; 2829(BM), 28 km N ash-Shumlul; 4018(BM), 27 – 45N, 45 – 32E; 8435, 25 – 56N, 48 – 32E.

● 30. **Neotorularia** Hedge et J. Léonard
Torularia O. E. Schulz

Annual or perennial herbs, usually pubescent, with leaves entire, dentate, or pinnatifid. Flowers pink-violet or white; sepals equal or nearly equal at base. Siliques linear, terete, straight or curved-contorted, dehiscent. Seeds 1-rowed. Léonard (1986b) discusses the establishment of this new generic name, required to replace the long used *Torularia* because of the latter's prior application to an alga.

Neotorularia torulosa (Desf.) Hedge et J. Léonard

Torularia torulosa (Desf.) O. E. Schulz

Ascending, rather rigid annual, branched from the base, 5 – 25 cm high, finely hirsute with erect white hairs. Basal leaves rosulate in young plants, oblong, sinuate to dentate or pinnately lobed, often absent in larger plants; cauline leaves linear, sinuate-dentate or nearly entire, tapering to base. Petals 2 – 3 mm long, white. Siliques in elongated rigid racemes, spreading to somewhat ascending, terete, 12 – 25 mm long, less than 1 mm wide, coiling or (less frequently) quite straight, on thick pedicels c. 1 mm long. Usually easily recognized by its dense racemes of often spirally coiled fruits. *Plate 97.*

Habitat: Silty basins of the northern plains and the Summan. Occasional to frequent.

Northern Plains: 675, Khabari Wadha; 1711(BM), ash-Shaqq, 28 – 30N, 47 – 42E; 1516(BM), 30 km ESE al-Qaysumah; 1544(BM), 36 km NE Hafar al-Batin; 1562(BM), al-Batin/al-'Awja junction; 1639(BM), 1666(BM), 2 km S Kuwait border at 47 – 14E; 4051, ad-Dibdibah, 28 – 04N, 46 – 05E; 7567, al-Faw al-Janubi, 2 km N highway; 8785, 30 km SW ar-Ruq'i Post, 28 – 54N, 46 – 27E.

Northern Summan: 3231, 74 km S Qaryat al-'Ulya; 3245, 5 km N Umm al-Hawshat; 3347, 5 km WSW an-Nazim; 7548, 7552(BM), 65 km ESE by E of Umm 'Ushar; 723, 752, 33 km SW al-Qar'ah wells; 1345(BM), 1369(BM), 42 km W Qaryat al-'Ulya; 2788, 41 km SE ash-Shumlul; 2816, 12 km ESE ash-Shumlul; 8581, Dahl al-Furayy; 8611, near Dahl Umm Hujul, 26 – 35N, 47 – 32E.

Southern Summan: 2072, Jaww ad-Dukhan, 24 – 48N, 49 – 05E.

Vernacular Names: KHUSHSHAYN ('rough-weed', Mutayr).

26. RESEDACEAE

Annual or perennial herbs, or shrubs, with leaves alternate, entire or lobed to divided. Flowers bisexual, or rarely unisexual, zygomorphic, bracteate in narrowly racemose inflorescences. Sepals 4 – 8; petals 4 – 8, or sometimes absent, often laciniate and unequal. Stamens 3 to numerous, inserted on a disc. Ovary superior, on a gynophore or sessile, 3 – 4-carpellate and usually 1-locular. Fruit a 1-celled capsule, 3 – 4-toothed at apex and often gaping, or a berry.

Key to the Genera of Resedaceae

1. Shrub with apetalous flowers; fruit a berry .... 1. *Ochradenus*
1. Herbs with 2 – 8-petalous flowers; fruit a capsule, or of 6 nearly free carpels
   2. Fruits sessile; petals 2; stamens 3 .... 2. *Oligomeris*
   2. Fruits pedicellate; petals 4 – 8; stamens 10 – 20
      3. Fruits stipitate with gynophore as long as or longer than the persistent calyx; carpels nearly free, 1-seeded .... 4. *Caylusea*
      3. Fruit non-stipitate or with gynophore shorter than the calyx; carpels coalescent in a multi-seeded capsule .... 3. *Reseda*

Two basic vernacular names appear to be used indiscriminately for all the Resedaceous plants of eastern Arabia except *Ochradenus*, which is distinctly frutescent. Both are derived from Arabic roots associated with the concept of 'tail' (of an animal) and obviously refer to the spiciform recemose inflorescence characteristic of the family. *Dhanabān* is an adjectival form of the word *dhanab*, 'tail', and the form *dhanabnāb* probably represents a dialectal inversion of the suffix, although it could conceivably be a relic of an early Semitic element related to the root of *nabt* or *nabat*, 'plant'. *Shawlah* is a substantive derived from a root meaning 'to raise the tail', as a scorpion or other creature.

● 1. Ochradenus Del.

Glabrous shrub with simple leaves. Flowers bisexual, or unisexual and the plant polygamous or dioecious. Sepals 5 – 8; petals absent or reduced to rudiments. Stamens 10 to numerous. Ovary 3-carpellate. Fruit open or closed, dry or fleshy and berrylike.

Ochradenus baccatus Del.

Glabrous, erect, branched shrub, nearly always dioecious, 0.5 – 1.5 m high. Leaves single or fascicled, narrowly linear, mostly 2 – 4 cm long, 1 – 1.5 mm wide, deciduous. Flowers in spiciform terminal racemes; sepals 4 – 6; petals 0; stamens 10 – 18. Fruit an ovoid to globose glabrous berry, 4 – 8 mm in diameter, greenish or pinkish when young, ripening to waxy white, on 1 – 2 mm-long pedicels. Seeds subreniform, c. 1.8 mm long, dark brown to nearly black, very finely but distinctly tuberculate. *Plates 98, 99.*

The great majority of our plants are dioecious, but two specimens with flowers apparently functionally bisexual (Nos. 7805, 8419) were found in the al-Hasa area. Shrubs of opposite sex occasionally grow very close together and intertwined, giving the appearance of a single shrub with branches of different sex.

Habitat: Usually on rocky terrain with shallow soil; occasionally on silty basins. Occasional to frequent.

Central Coastal Lowlands: 7751 (staminate), 7752 (pistillate), Dhahran.

South Coastal Lowlands: 7805, 5 km N al-Mutayrifi, al-Hasa; 8416, 8417, 12 km N al-Mutayrifi; 8418, 8419, 7 km N al-Mutayrifi.

Southern Summan: 2154, Summan Yabrin, 23 – 27N, 48 – 34E; 8698 (staminate), 8699 (pistillate), 23 – 53N, 48 – 50E.

Vernacular Names: QIRḌĪ, QURḌĪ (gen.), ʿALQÁ (Bani Hajir).

Another species, *O. arabicus* Chaudhary, Hillcoat et Miller, was described in 1984 from specimens collected in central Saudi Arabia, South Yemen, and Oman (Miller 1984). Its distribution appears so far to be limited to Sudanian floristic territory (*sensu* Mandaville, 1984). This might in future be found to occur in the southwestern or southeastern parts of the Eastern Province. It differs from *O. baccatus* by its rather thick twigs tapering to their apices and becoming spinescent, and by its larger fruits ripening green to yellow and inflated, rather than white and berrylike. The seeds are highly diagnostic, being completely smooth and glossy rather than fine-tuberculate as in *O. baccatus*.

● 2. **Oligomeris** Cambess.

Annual or perennial herbs or subshrubs with entire leaves. Flowers bisexual, or rarely unisexual and the plant polygamous, sessile. Sepals 2 – 6; petals 2, entire or lobed. Disc absent; stamens 3 – 12. Ovary sessile. Capsule 1-celled, gaping at apex; seeds smooth.

Oligomeris linifolia (Vahl) Macbride
O. subulata Webb
O. subulata (Del.) Boiss.

Ascending to erect annual, or sometimes perennial, glabrous herb, 5 – 30 cm high, branched from the base. Leaves sessile, solitary or fascicled, linear, tapering to an acute apex, 1 – 4 cm long, 1 – 3 mm wide. Flowers white, in spiciform terminal racemes; petals 2; stamens 3. Fruit a depressed-globose, sessile capsule, 4-toothed at mouth, c. 3 mm long, 4 mm broad, with numerous seeds.

Habitat: Shallow sand or silty sand, on flatlands or rocky terrain; rarely a weed in cultivation. Frequent.

Northern Plains: 1665(BM), 2 km S Kuwait border in 47 – 14E; 1157, Jabal al-Ba'al; 1207(BM), Jabal an-Nu'ayriyah; 1288(BM), 15 km ENE Qaryat al-'Ulya.

Central Coastal Lowlands: 188(BM), Dhahran; 398, al-Midra ash-Shimali; 1464(BM), 2 km E al-Ajam; 3736, Ras az-Zawr; 3259, 8 km W Qatif-Qaisumah Pipeline KP-144.

South Coastal Lowlands: 354(BM), near al-Fudul, al-Hasa; 2732, ar-Ruqayyiqah, al-Hasa.

Southern Summan: 2152, Summan Yabrin, 23 – 15N, 48 – 33E; 2074, Jaww ad-Dukhan, 24 – 48N, 49 – 05E.

Vernacular Names: DHANABÁN (Al Murrah).

157

● 3. **Reseda** L.

Annual or perennial herbs, sometimes suffrutescent, with leaves usually divided or dissected. Flowers usually bisexual, in spicate racemes. Sepals 4 – 8; petals 4 – 8, white, yellow, or greenish, unequal, with some of them laciniate. Stamens 10 to numerous, inserted asymmetrically on the disc. Ovary often short-stipitate, 3 – 4-carpellate. Capsule 1-celled, gaping at apex.

Key to the Species of Reseda

1. Ovary 4-toothed; leaves pinnately divided with more than 3 pairs of
    lobes .... 2. *R. decursiva*
1. Ovary 3-toothed; leaves ternately cleft-divided, or entire, or if appearing
    pinnately divided then with 3 or less pairs of lobes
    2. Plant densely glandular-papillose-muricate .... 4. *R. muricata*
    2. Plant not papillose-muricate
        3. Some leaves (usually at base) entire; capsules pendulous; petals
            white .... 1. *R. arabica*
        3. All leaves ternate or sometimes (near base) appearing pinnately
            divided with 2 – 3 pairs of lobes; capsules usually erect; petals
            cream to yellow .... 3. *R. lutea*

1. **Reseda arabica** Boiss.

Ascending or erect, branched glabrous herb, 10 – 30 cm high. Leaves entire-oblanceolate or distally ternate, sometimes wavy-margined, up to c. 6 mm wide but in some plants narrowly oblanceolate to linear, in very young plants sometimes obovate, to 12 mm wide. Flowers in elongated terminal racemes, pedicellate with whitish petals and c. 20 stamens. Capsules pendulous, globose-ellipsoid, more or less gaping, 5 – 10 mm long, 3-toothed. *Plate 100.*

Habitat: Sandy or silty soils. Frequent.

Northern Plains: 1245, Jabal an-Nu'ayriyah; 1289, 15 km ENE Qaryat al-'Ulya; 1475(BM), 50 km WNW an-Nu'ayriyah; 1610(BM), 3 km ESE al-Batin/al-'Awja junction.

Northern Summan: 2747, 25 – 28N, 48 – 10E; 1374(BM), 42 km W Qaryat al-'Ulya; 8593, 38 km NE Umm al-Hawshat, 27 – 20N, 47 – 26E.

Central Coastal Lowlands: 191(BM), Dhahran Airport; 63(BM), 68(K), Thaj; 1769(BM), Abu Hadriyah; 3281, 8 km W Qatif-Qaisumah Pipeline KP – 144; 3358, Jabal al-Ahass; 7724, 7734, 2 km S Ras Tanaqib.

South Coastal Lowlands: 387(K), 1 km N al-Hasa cement plant.

Southern Summan: 2054, Jaww ad-Dukhan, 24 – 48N, 49 – 05E.

2. **Reseda decursiva** Forssk.

  R. alba L. subsp. *decursiva* (Forssk.) Maire
  R. propinqua R. Br.

Decumbent to erect glabrous annual, 6 – 30 cm high; stem simple or branched from near the base. Leaves rosetted at base, pinnately divided with narrowly oblong to linear, rather distant, lobes. Flowers subsessile in dense spikelike terminal racemes, sometimes with smaller racemes branching laterally from base of the terminal one; petals white, 3 – 7 mm long, slightly exceeding the sepals; stamens 10 – 13. Capsules 4 – 6 mm long, ovoid, subsessile, 4-toothed at the gaping mouth.

● The majority of authors is followed here in maintaining the species distinct from *R. alba*. Its status is questionable, however, and Abdallah and De Wit in their recent treatment in *Flora of Iraq* (vol. 4) reduce it to the subspecies. *R. alba* (or subsp. *alba*) is separated by its capsules nearly twice as large as those of *R. decursiva* and constricted at base of the mouth, and by having petals considerably longer than the sepals.

Habitat: Silty sands of the northern plains. Occasional.

Northern Plains: 590, al-Batin, 16 km ENE Qulban Ibn Busayyis; 1277(BM), 5 km ENE Qaryat al-'Ulya; 1706(BM), ash-Shaqq, 28 – 30N, 47 – 42E; 2710, 14 km ENE Qaryat al-'Ulya; 1691(BM), 28 – 41N, 47 – 35E; 7563, al-Faw al-Janubi, 2 km N highway.

Northern Summan: 612(BM), 18 km NE Umm 'Ushar.

3. **Reseda lutea** L.

Erect annual or perennial, glabrous, smooth or sometimes moderately papillose, 15 – 70 cm high. Leaves ternately or biternately divided, sometimes near base appearing pinnately divided with 2 – 3 pairs of lobes. Flowers in dense terminal, spiciform racemes on pedicels exceeding the calyx. Petals greenish cream to yellow, 3-partite with lobes entire, the central one much smaller. Stamens 15 – 20. Capsules 5 – 15 mm long, cylindrical, usually erect. Any doubts in separating specimens of *R. lutea* from *R. muricata* or *R. arabica* can be resolved by examining the petals; those of the last two species are multi-partite with more than 3 lobes.

Habitat: Unlikely to be found outside cultivated areas or disturbed sites where grain has been stored or spilled. Rare.

Central Coastal Lowlands: 8085, N Faruq, 25 – 43N, 48 – 53E, Bedouin camp ground.

4. **Reseda muricata** Presl

Ascending to erect perennial herb, 10 – 70 cm high, usually branching from base, densely muricate-papillose, with several erect stems leafy to bases of the racemes. Leaves linear, mostly distally ternate, with linear lobes sometimes wavy at margins. Flowers pedicellate, ascending-spreading, c. 3 mm long, in terminal racemes; petals white; stamens 14 – 17. Capsules erect, obovoid or subglobose, 5 – 8 mm long, 3-toothed at apex, on pedicels slightly or much shorter than the capsules. *Plate 101.*

Habitat: Usually on shallow soil on somewhat rocky terrain. Tolerant of disturbed ground and sometimes seen as a ruderal. Occasional.

Northern Plains: 8781, 18 km WSW ar-Ruq'i Post, 28 – 59N, 46 – 31E.

Northern Summan: 814, 61 km NE Jarrarah; 2869, 1 km N Jabal Mutayrihah, 25 – 59N, 48 – 32E.

Central Coastal Lowlands: Dickson 470(K), Dhahran (Burtt and Lewis 1949, 304).

Southern Summan: 2222, 5 km S al-Hunayy.

Vernacular Names: DHANABAN (gen., Rawalah), DHANABNAB (Shammar), SHAWLAH (Al Murrah).

● 4. **Caylusea** St. Hil.

Annual or perennial herbs with entire leaves. Flowers bisexual, with 5 unequal sepals and 5 unequal and laciniate petals. Stamens 10 – 15. Ovary on a gynophore, with carpels 5 – 6, nearly free. Carpels 1-seeded in fruit, gaping and ciliate.

Caylusea hexagyna (Forssk.) M. L. Green
C. canescens St. Hil.
Annual or perennial ascending herb, usually several-stemmed from the base, 20 – 50 cm high, papillose to pubescent with erect hairs. Leaves entire, narrowly oblong to lanceolate, sessile, 1 – 5 cm long, 0.3 – 1 cm wide, wavy and ciliate at margins. Flowers in dense terminal racemes elongating in fruit, on short pedicels c. 1 mm long; petals 5, laciniate; stamens 10 – 15. Ovary stipitate, with gynophore equalling or exceeding the sepals, 6-toothed, consisting of 6 nearly free, 1-seeded carpels gaping when ripe with woolly mouths. *Plate 102.*

Habitat: Silty soils in beds of large wadis or Summan basins.

Northern Plains: 587(BM), al-Batin, 48 km SW Hafar al-Batin; 3099(BM), al-Batin, 28 – 00N, 45 – 28E.

Northern Summan: 8626, Dahl Abu Harmalah, 26 – 29N, 47 – 36E.

Southern Summan: 2236, 15 km WSW al-Hunayy; 8056, 24 – 06N, 48 – 48E.

Vernacular Names: SHAWLAH (Qahtan) DHANABNÁB (Shammar).

27. PRIMULACEAE

Annual or perennial herbs, rarely shrubs, with alternate or opposite leaves. Flowers bisexual, generally actinomorphic, solitary or in racemose, umbellate, or paniculate inflorescences. Sepals usually 5, connate at least at base; petals usually 5, more or less connate, with corolla rotate, campanulate, or funnel-shaped. Stamens 5 – 9. Ovary superior or rarely half-inferior. Fruit a capsule, circumscissile or opening by teeth or valves.

Key to the Genera of Primulaceae
1. Flowers white, racemed .... 1. *Samolus*
1. Flowers red or blue, solitary in the axils .... 2. *Anagallis*

● 1. Samolus L.

Perennial herbs with alternate, entire leaves. Flowers in terminal racemes, or corymbose, 5-merous. Corolla more or less campanulate, with scalelike staminodes alternating with the stamens in the throat. Ovary half-inferior. Capsule ovoid or globose, opening by 5 valves.

Samolus valerandi L.
Erect, glabrous herb, perennial (or sometimes apparently annual in our area), 10 – 40 cm high and often branching above. Leaves obovate to elliptical-spathulate, obtusish, the lower ones rosetted, tapering to a petiole, those above nearly sessile. Flowers corymbose above, then in rather loose, elongating terminal racemes, on pedicels c. 3 times as long as the flower, with a narrow bract 1.5 – 2.5 mm long inserted at the pedicel knee. Calyx c. 2 mm long, divided c. half-way to base with acute triangular lobes. Corolla white, c. 4 mm long, divided c. half-way to base with 5 obtuse lobes and 5 minute scales inserted on the throat between the lobes. Capsule ovoid-globose, 2 – 3 mm in diameter, dehiscing by 5 valves, enclosed by the calyx except at apex. Seeds 0.3 – 0.4 mm long, angular.

Habitat: Wet shaded ground in the oases or in lawns. Frequent.

Central Coastal Lowlands: 526, Tarut Island; 7412(BM), Dhahran. Often seen in the oases of al-Qatif and al-Hasa.

● 2. Anagallis L.

Glabrous herbs with leaves opposite or alternate. Flowers 5-merous, usually pedicellate, solitary, axillary. Calyx divided nearly to base; corolla lobed, rotate or campanulate. Filaments often bearded. Ovary superior; capsule globose.

Anagallis arvensis L.

Glabrous, ascending annual, 5 – 20 cm high, with quadrangular stems. Leaves opposite or whorled, sessile, ovate to triangular-ovate, 15 – 25 mm long, often dotted dorsally with dark glands. Flowers red or blue, c. 10 mm in diameter, on slender pedicels exceeding the subtending leaf. Capsule globose, circumscissile, 4 – 6 mm in diameter. *Plate 103.*

Two varieties occur in our area:

var. *arvensis* .... corolla scarlet
var. *caerulea* (L.) Gouan .... corolla blue

Further study is required to determine whether the blue-flowered subspecies *foemina* (Mill.) Schinz et Thell. (sometimes treated as the species *A. foemina* Mill.) occurs in our area. No. 8195 (below) had leaves 2 cm long, 1.5 cm wide, on the borderline of another variety, *latifolia*. Both the red and blue-flowered varieties may occur together abundantly in the same field.

Habitat: Usually as a weed in farms or gardens. Dickson (1955) reports it also occurs in the Kuwait area as a desert annual (listed as *A. femina*). No. 7950 (below) was collected in remote desert but on an obvious former Bedouin camp site with a number of Mediterranean weeds.

Central Coastal Lowlands: 437 (var. *arvensis*), Dhahran; 7950 (var. *caerulea*), Batn al-Faruq, 25 – 43N, 48 – 53E.

South Coastal Lowlands: 2729(BM), (var. *caerulea*), 2730 (var. *arvensis*), ar-Ruqayyiqah, al-Hasa; C. Parker 12, 13(BM), farm near al-Hufuf; 8194 (var. *arvensis*), 8195 (var. *caerulea*), near al-Qurayn, al-Hasa Oasis.

Vernacular Names: 'UWAYNAH (gen., 'little eye').

28. CRASSULACEAE

Herbs or shrubs with opposite or alternate, usually succulent leaves. Flowers bisexual, or rarely unisexual and the plant dioecious, actinomorphic, in cymose inflorescences. Sepals 3 – 30, usually free; petals 3 – 30, free or united. Stamens 3 to numerous. Carpels 3 to numerous. Fruit a follicle dehiscing ventrally.

● Crassula L.

Herbs or shrubs with opposite, usually entire leaves. Flowers with (3 –)5(– 9) nearly free sepals and (3 –)5(– 9) white, yellowish or pinkish petals. Stamens equal to or twice the number of sepals. Ovary usually of free carpels. Fruit a follicle with seeds 2 to numerous.

161

Crassula alata Berger
Tillaea alata Viv.
Ascending or erect, glabrous, often reddish dwarf annual, 1.5 – 5 cm high; stems 4-angled with minute longitudinal wings. Leaves opposite, lanceolate, connate at base, 1 – 3 mm long. Flowers 1 – 3 mm long, short-pedicelled in the axils; petals 3 – 4, lanceolate, acuminate; stamens 3 – 4. Carpels 3 – 4, each 2-seeded.

Habitat: Silty soils in basins of the north. Apparently rare, but perhaps often overlooked because of its minute size.

Northern Plains: 1481(BM), 50 km WNW an-Nu'ayriyah; 4012(BM), ad-Dibdibah, 27 – 56N, 45 – 35E.

29. ROSACEAE

Herbs, shrubs or trees with usually alternate, simple or compound leaves. Flowers mostly bisexual, actinomorphic, solitary or in various inflorescences. Sepals (4)5; petals 5(9), rarely absent. Stamens usually twice the sepals in number, or numerous. Ovary superior or inferior, with carpels 1 to numerous, free or united. Fruit of follicles or achenes, or a pome or drupe.

● Neurada L.

Prostrate, woolly annual herbs with lobed leaves. Flowers solitary or paired, with a 5-lobed, cupuliform, persistent, expanding calyx. Epicalyx of bracteoles present. Petals 5, hardly exceeding the calyx. Stamens 10. Ovary inferior, with 5 – 10 carpels connate at base and adnate to the hypanthium. Fruit flat-orbicular, woody, spinescent above. A distinctive genus often placed in a segregate family, the Neuradaceae.

Neurada procumbens L.
Prostrate, grey-green, densely tomentose annual, usually with several branching stems radiating from the base. Leaves ovate-lanceolate in outline, obtuse, unequally pinnately and bluntly lobed, with raised nerves on lower face, on petioles ⅓ to ⅔ the length of the blade. Flowers inconspicuous, short-pedicelled in the upper axils; sepals 5; petals 5, cream, greenish, or pinkish; stamens 10. Fruit flat, discoid, 1.2 – 1.8 cm in diameter, hardening and woody in maturity, smooth and flat below, convex and furnished with prickles 1 – 5(10) mm long above, on a peduncle c. 10 mm long. Plants collected in the northwestern Rub' al-Khali, at al-Mulayhah al-Gharbiyah, were notable for the unusually long prickles — to fully 1 cm long — on their fruits. *Plate 104.*

● The plant may be recognized immediately by its habit and distinctive discoid fruits, which are deciduous and become scattered about the desert after maturity in late spring. They are effective dispersal units, the dorsal prickles clinging equally well to the feet of animals, the shoes of man, and the tires of automobiles.

The seeds of *Neurada* germinate while still contained in the fruit disc, the woody remnants of which are usually persistent at the base of seedlings and often mature plants. Vesey-Fitzgerald (1957, 782) describes how individual *Neurada* seeds germinate only one at a time in the fruit, after different rain showers, a habit he interprets as an insurance mechanism against premature germination. The dry taproots of perished seedlings, he thus suggests, indicate the number of rain showers that fell before soil moisture reached levels sufficient for survival of the first plant. Some form of germination inhibitor may well exist in this species, but seedlings may often in fact be seen to consist of multiple, simultaneous germinations from the same fruit. The number of germinations may perhaps be correlated with soil moisture level.

Habitat: Widespread on both shallow and deep sands; common in the Rhanterietum and one of the few annuals to be found in the Rub' al-Khali, where it is present in at least the northern parts. Common.

Northern Plains: 1262(BM), 15 km ENE Qaryat al-'Ulya; 672, Khabari Wadha; 859, 9 km S 'Ayn al-'Abd; 1008, 12 km N Jabal al-'Amudah; 1244, Jabal an-Nu'ayriyah.

Northern Summan: 3129(BM), 6 km SE ad-Dabtiyah; 3154(BM), 9 km SSW Hanidh.

Central Coastal Lowlands: 156, 183, 414, Dhahran; 1414(BM), 7 km NW Abu Hadriyah; 1465(BM), 2 km E al-Ajam; 7731, 2 km S Ras Tanaqib.

South Coastal Lowlands: 384, 1 km N al-Hasa cement plant; 1433(BM), 2 km E Jabal Ghuraymil; 2104, 49 km S Harad; 2126, 4 km SW al-Jawamir, Yabrin area; 8019, Yabrin area, 23–11N, 48–53E.

Rub' al-Khali: 7666, 23 km W Nadqan; 7696, 8031, al-Mulayhah al-Gharbiyah, 22–53N, 49–29E.

Vernacular Names and Uses: SA'DĀN (gen.; apparently derived from the Arabic root verb *sa'ada*, 'to be auspicious, of good omen'; also cf. *sa'dānah* of the classical dictionaries, meaning 'knot', 'camel's callosity', or 'areola of the nipple', all resembling the peculiar fruits of this plant although the direction of simile is uncertain), NAQQI' (Ahl Wahibah, southeastern Rub' al-Khali; a name more generally applied to *Blepharis*). Dickson (1955, 67) notes that the young fruits of *Neurada* are sometimes eaten by Bedouin children. They are indeed tender and not unpleasant in taste, although somewhat mucilaginous, if plucked well before they begin to harden.

30. LEGUMINOSAE (Fabaceae)

Herbs, shrubs or trees with leaves usually alternate and compound. Flowers mostly bisexual, zygomorphic or actinomorphic, in racemes, panicles or heads. Sepals usually 5, free or connate; petals 5, free or variously united. Stamens often 10, sometimes fewer, or usually numerous in the Mimosoideae, free or connate in a tube. Ovary superior, usually of a single carpel developing into a pod (legume) with the seeds alternately inserted on the opposing margins.

The characteristic flower structure of the majority of our legume species, which belong to the subfamily Papilionoideae, is papilionaceous. This is a specialized irregular flower form in which the upper petal is usually large and erect, forming a 'standard'; the two lateral ones forming 'wings'; and the two lower ones usually connate along the lower margin, forming a 'keel'. The calyx is equally or unequally toothed, and the stamens are usually 10, with the filaments all connate in a tube and 'monadelphous', or with 7–9 connate and 1–3 free and thus 'diadelphous'.

An important family in Arabia, providing dominant perennial genera in Sudanian floristic territory but in the Eastern Province of Saudi Arabia represented mainly in the annual flora. The family is followed here in the wide sense, treating as subfamilies the groups Mimosoideae, Caesalpinoideae, and Papilionoideae. These are often given separate family rank as the Mimosaceae, Caesalpiniaceae, and Papilionaceae.

Key to the Genera of Leguminosae

1. Flowers regular (Mimosoideae)

  2. Flowers in globular heads; stamens numerous; trees with stout stipular
    spines .... 3. *Acacia*

  2. Flowers in cylindrical spikes; stamens 10; prickly or unarmed shrubs
    or trees .... 2. *Prosopis*

1. Flowers irregular

  3. Corolla not papilionaceous; stamens free (Caesalpinoideae) .... 1. *Cassia*

  3. Corolla papilionaceous; at least 7 stamens coalescent in a tube
   (Papilionoideae)

    4. Leaves apparently simple or absent (a few trifoliolate leaves sometimes
     present in *Taverniera*)

     5. Shrubs

      6. Branches spiny; pod curved-linear, subterete .... 19. *Alhagi*

      6. Branches not armed; pod compressed .... 18. *Taverniera*

     5. Herbs

      7. Petiole terminating in paired tendrils above a single pair of
       apparent leaflets; pod strongly compressed .... 21. *Lathyrus*

      7. Petiole without tendrils; leaves simple; pod coiled or
       contorted .... 15. *Scorpiurus*

    4. Leaves compound

     8. Leaves 3-foliolate (or, in *Lotus*, appearing so because the
      lower leaflet pair is on the stem, resembling stipules)

      9. Leaflets dentate or serrate

       10. Flowers pinkish to purple

        11. Flowers in dense heads on peduncles at least twice as
         long as the subtending leaf .... 9. *Trifolium*

        11. Flowers solitary, short-pedicelled or in axillary racemes

         12. Perennial; racemes c. 3 times the length of the
          subtending leaf; leaves distant with leaflets plicate
          on veins .... 12. *Psoralea*

         12. Annual; flowers among the leaves; plant densely
          leafy; leaflets not plicate .... 5. *Ononis*

       10. Flowers yellow or white

        13. Pod spirally twisted into tight, prickly or smooth
         coils .... 7. *Medicago*

        13. Pod not coiled

         14. Pod ovoid, 2–4 mm long .... 8. *Melilotus*

         14. Pod linear and straight, curved, or plicate .. 6. *Trigonella*

      9. Leaflets entire

       15. Calyx teeth equal or subequal; leaves sessile, 5-foliolate but
        often appearing 3-foliolate because lower pair is reduced,
        stipulelike .... 10. *Lotus*

       15. Calyx teeth unequal, the 2 bifid lateral ones much larger
        than the single one below; leaves petioled, clearly
        3-foliolate .... 4. *Lotononis*

     8. Leaves pinnate

      16. Leaves terminating in fine tendrils

       17. Leaves with only a single apparent leaflet pair .. 21. *Lathyrus*

SUBFAMILY CAESALPINOIDEAE

● 1. Cassia L.

Herbs, shrubs or trees with paripinnate leaves and leaflets in 1 to several pairs. Flowers racemed, with a calyx of 5 free, imbricate sepals and a non-papilionaceous corolla of 5 nearly equal petals. Stamens usually 10, free, often unequal. Pod terete or compressed, dehiscent or indehiscent.

Cassia italica (Mill.) F. W. Andr.

C. obovata Collad.

Erect to ascending-spreading branched shrub with blue-green foliage, to c. 1 m high, finely glandular-pubescent particularly on younger parts. Leaves paripinnate with 3 – 6 pairs of oblong to obovate, obtuse, mucronate leaflets which are 1.5 – 3 cm long, 1 – 2 cm wide, sometimes unequal at base, on petiolules c. 1 mm long. Flowers racemed, on pedicels 3 – 5 mm long; petals yellow with darker veins, 1 – 1.7 cm long. Pod flat, curved-oblong, 3 – 5 cm long, 1 – 1.8 cm wide, obtuse, minutely pubescent to glabrescent, with a transverse series of creases and a longitudinal series of short crests on face. *Plates 105, 106.*

Habitat: Bottoms of larger wadis or silty flats; usually in or near Sudanian floristic territory.

South Coastal Lowlands: 2101, 21 km SSE Harad; 1029, 28 km NE Harad; 2947(BM), Railroad KP – 323; 1049, old al-Hasa airport area.

Vernacular Names and Uses: ʿISHRIQ (gen.), SHAJARAT AD-DABB ('snake bush', Al Murrah). Trease (1961, 412) says that the leaves of this species have been imported to Europe from Egypt as the drug 'dog senna'. There is apparently no record of such medicinal use in eastern Arabia, where the leaves, according to some Bedouin reports, are considered toxic to livestock.

SUBFAMILY MIMOSOIDEAE

● 2. Prosopis L.

Shrubs or trees, usually with spines or prickles, with bipinnate leaves. Flowers regular, in spikelike or headlike axillary racemes, yellow or greenish-yellow. Calyx cupuliform or campanulate, 5-toothed. Petals 5, free or connate at base. Stamens 10, free, usually exserted. Pod linear to ovoid-irregular, straight or curved, terete to compressed, indehiscent, hardened and often spongy inside.

Key to the Species of Prosopis

1. Pods ripening straw-yellow, mostly linear; leaflets in 15 – 23
   pairs .... 3. *P. juliflora*
1. Pods ripening brown or reddish brown, torulose or irregularly swollen;
   leaflets in 5 – 14 pairs
   2. Pods 0.4 – 0.8 cm wide, 8 – 25 cm long; brownish .... 1. *P. cineraria*
   2. Pods 0.8 – 3 cm wide, 2 – 10 cm long, reddish brown
      3. Pod ovoid-oblong or irregularly swollen, usually over 1.5 cm broad;
         pinnae in (3)4 – 6 pairs .... 2. *P. farcta*
      3. Pod terete-elongated with 1 or several constrictions, less than 1.5 cm
         broad, pinnae in 1 – 3 pairs .... 4. *P. koelziana*

1. Prosopis cineraria (L.) Druce

P. spicigera L.

Tree or large shrub with scattered prickles. Leaves 2-pinnate, usually with 2 pinnae pairs; leaflets in 7 – 12(14) pairs, oblong, 10 – 18 mm long, 3 – 5 mm wide. Flowers 4 – 5 mm long, in solitary or panicled spikes 7 – 11 cm long. Calyx c. 1.5 mm long, somewhat toothed; corolla yellow, c. 3 mm long. Pods 10 – 20(25) cm long, 0.4 – 0.8 cm wide, somewhat compressed to subterete, torulose, glabrous, ripening brownish, with 10 – 15 longitudinally disposed seeds. (Description after Bhandari 1978 and Léonard 1981 – 1989, fasc. 6).

Habitat: So far known only from shallow sands or steppes on the eastern and southern margins of the Rub' al-Khali. Infrequent to rare.

Rub' al-Khali: Southern and eastern borderlands of our area (Thesiger 1946, 1950).

Vernacular Names: GHAF (gen. in Oman and the southeastern Gulf states), SHIBHĀN (Al Murrah). According to a Rashidi tribesman of the southern Rub' al-Khali, young GHAF trees, 'those with many prickles,' are called HAḌĪB.

● Not yet collected by the author in our area although a probable example was seen from an aircraft in the southern Rub' al-Khali in the vicinity of Ghanim (approx. 20°00'N, 54° 20'E). It is common in the steppes and wadis of neighboring Oman as well as in parts of the United Arab Emirates. Vesey-Fitzgerald (1957b) reported his northwestern-most experience of this species at Ba'ja (in the present United Arab Emirates, 24° 06'N, 51° 06'E), within some 15 km of the Saudi Arabian border. There must now be some doubt, however, as to whether the trees at this site are this species or *P. koelziana*.

2. **Prosopis farcta** (Banks et Sol.) Macbride

P. stephaniana (Willd.) Spreng.

Lagonychium stephanianum (Willd.) M. Bieb.

L. farctum (Banks et Sol.) Bobr.

Straggling, many-branched shrub, 0.4 – 2 m high, densely fine-pubescent on leaves and young stems, with greyish to white woody older branches and scattered prickles. Leaves 2-pinnate with 3 – 6 pinnae pairs; leaflets oblong-elliptical, acutish, 2 – 3(4) mm long, 0.5 – 1 mm wide, in 8 – 14 pairs, approximate and often overlapping. Flowers cream, 3 – 4 mm long, in spikes 4 – 10 cm long. Pod fat, ovoid or irregularly swollen, to c. 5 cm long, dark purplish brown when ripe.

Habitat: A weed shrub of waste land and disturbed ground around agricultural areas.

Northern Summan: 507, Rawdat Ma'qala — an unusual desert record of this species but in a location somewhat disturbed and probably a site of seasonal silt-basin cultivation.

Central Coastal Lowlands: 1098(BM), al-Qatif.

South Coastal Lowlands: Seen in the al-Hasa oasis area.

Vernacular Names: YANBŪT (Al-Hasa farmers), 'AWSAJ (sic, Qatifi gardeners).

3. **Prosopis juliflora** (Sw.) DC.

P. chilensis Stuntz

Straggly shrub to large, wide-crowned tree, more or less prickly, with glabrous foliage. Leaves 2-pinnate with 1 – 3 pinnae pairs; leaflets oblong, 3 – 10(12) mm long, in 15 – 23 pairs. Flowers greenish-yellow, c. 4 – 5 mm long, in dense cylindrical pendulous spikes mostly 6 – 9 cm long, 1 cm broad. Stamens well exserted; ovary and ventral surface of the petals densely pubescent. Pod compressed, straight or somewhat curved, 8 – 22 cm long, c. 10 mm wide, with margins linear or obscurely constricted between the seeds, a slight beak at apex, green, ripening to straw-yellow.

All plants seen in our area belong to var. *juliflora*, with closely-spaced oblong leaflets and short fruit beak.

Habitat: Cultivated, but frequently escaped around urban and village areas; always on somewhat disturbed ground. Locally frequent.

Central Coastal Lowlands: 1082(BM), Dhahran.

Vernacular Names and Uses: No local name has been noted for this introduced American native, the mesquite. It is widely planted in the oil company community at Dhahran, where with frequent irrigation it reaches great stature, 12 – 15 m or more, with a thick trunk often branching about 2 – 3 m above ground level. It has also been planted along city streets and roadsides in al-Khubar, ad-Dammam, and other towns. It has the virtues of being extremely tough and hardy, quick in growth, and requiring virtually no care. Its vices are the production of messy litter from fallen leaves, an apparent inhibiting effect on the growth of lawns and other plants beneath it, and a reputation for causing hay fever by its abundant wind-borne pollen.

The tree has blooming periods in both spring and autumn and tends to drop its leaves in late winter, bringing out new foliage in the spring before blooming. Some trees at Dhahran, however, may be seen blooming and fruiting at nearly any time of year. It sometimes escapes from cultivation, apparently by two means: the spread of pods by flooding after rainstorms, and by being carried to dump areas after land clearing operations. It can self-propagate on disturbed ground but is never found in truly natural desert habitats.

4. **Prosopis koelziana** Burkart

Shrub to large tree, 2 – 12 m high, scrubby and straggling or erect with well-defined trunk and somewhat pendulous branches, with sharp prickles or unarmed, finely and shortly pubescent at least on young parts. Larger trees have trunks with grey bark, somewhat fissured below and smoothish above. Leaves 2-pinnate, with 1 – 3 pinnae pairs; leaflets oblong, 3 – 8 mm long, 1.5 – 2.5 mm wide, mostly in c. 12 pairs. Flowers yellow or greenish-yellow, c. 4 – 5 mm long, in dense cylindrical spikes 3.5 – 6(9) cm long, 8 – 11 mm in diameter. Calyx c. 1.5 mm long, faintly toothed, concolorous with the corolla; petals glabrous, lanceolate, acute, 3 – 4 mm long, recurved at apex; stamens slightly exserted, shorter than the exserted style. Pods 3 – 9(11) cm long, 0.8 – 1.2(1.5) cm wide, nearly terete, constricted between the seeds, apiculate at apex, yellowish or pinkish when immature, ripening a strong reddish brown, often with only 1 – 3, obliquely disposed, seeds. Seeds more or less oval in outline, somewhat compressed, 4 – 6 mm long, smooth reddish brown, dullish to weakly shining, sometimes concave on each face; each face with a peculiar horseshoe-shaped fissile line. (Description based on our al-Hasa area specimens). *Plates 107, 108.*

The author is indebted to Prof. J. Léonard, of the Belgian National Botanical Garden and the University of Brussels, for identifying our specimens and establishing these first records of the species from Saudi Arabia. The regional distribution of *P. koelziana* is very imperfectly known, probably due in part to confusion with *P. cineraria*. First described from southern Iran, it was hitherto known in the Arabian Peninsula only in the far south, in Hadhramaut (Léonard 1985). Léonard (1986b) describes a new variety, *puberula*, of this species from our east Arabian specimens. It differs from the typical variety in having the rhachises and lower faces of the leaflets puberulous.

Habitat: Usually found near ruin sites or villages, on disturbed or undisturbed sandy ground.

Central Coastal Lowlands: 3949(BM), Thaj; 8638, 1 km S Thaj.

South Coastal Lowlands: 1048(BM), 2896(BM), 4085, 4086, near al-Mutayrifi, al-Hasa; 8141, 8420, 8644, 8645, 1 km N al-'Uyun, al-Hasa; 8761, 8 km S al-'Uyun.

Vernacular Names: GHAF (villager of al-'Uyun, al-Hasa; the name generally used in Oman for *P. cineriaria*), SHIBHĀN (Al Murrah, cf. classical SHABAHĀN), ṬIRF (al-'Awazim at Thaj).

● In our area this species is found as a well-developed tree only around the edges of the northern al-Hasa Oasis, where good examples may be seen in numerous small groves around the northwestern side of the town of al-'Uyun (al-Muhtaraqah), on both sides of the al-'Uqayr road. It has been observed there flowering profusely in the first week of April. Otherwise, in more remote locations, it hardly develops beyond a straggly shrub or stunted tree 2 – 4 m high. The northernmost record is at the Hellenistic-period ruins at Thaj, where it conceivably could have persisted since an introduction through early caravan trade from southeastern Arabia. Another stand is found west of the *sabkhah* about 5 km south of Thaj. In the south, the author has noted small stands of what appear to be severely cut examples of this species near the base of Qarn Abu Wail south of Salwah, and in the northeastern part of the Yabrin depression. Cheesman, describing his 1924 visit to Yabrin, published a photograph taken there (1926, Plate 46) which appears to include some rather well-developed specimens of a *Prosopis* that may be this species.

• 3. Acacia Mill.

Trees or shrubs, usually spinescent, with pinnately compound leaves or (but not in Middle East natives) leaves modified to phyllodes. Flowers small, usually bisexual, actinomorphic, grouped in dense spicate racemes or globular heads. Calyx campanulate with 4 – 5 lobes or teeth. Petals 4 – 5, free or united. Stamens numerous, exserted. Pod linear to ovate, often compressed but sometimes terete, straight, curved or contorted, with margins parallel or constricted between the seeds.

Trees or large shrubs, those in our area with straight white spines often alternating with shorter curved spines. In the Eastern Province *Acacia* is restricted with some exceptions to the borders of Sudanian floristic territory, of which the genus is an important indicator.

Key to the Species of Acacia

1. Pinnae not exceeding 1 – 2 pairs .... 1. *A. ehrenbergiana*
1. Pinnae exceeding 2 pairs
    2. Pod falcate, not contorted, linear; leaflets in (8)10 – 18(23)
       pairs .... 2. *A. gerrardii*
    2. Pod contorted, constricted between the seeds; leaflets in (5)6 – 9(12)
       pairs
        3. Pods glabrous, with distinct multiple nerves on the valves; young
           branches glabrous .... 3. *A. raddiana*
        3. Pods finely pubescent with obscure nerves; young branches
           pubescent .... 4. *A. tortilis*

1. Acacia ehrenbergiana Hayne

A. flava (Forssk.) Schweinf.

Large shrub or small tree, glabrescent, usually with multiple ascending branches from the base, 2 – 4 m high. Spines mostly straight, whitish, 2 – 5 cm long. Leaves with only 1 – 2 pairs of pinnae; leaflets (5)6 – 9(10) pairs, 2 – 3 mm long. Flowers yellow, in globular heads on peduncles 1 – 2 cm long, solitary or clustered in the axils. Pods up to c. 10 cm long, falcate, constricted between the seeds. *Plates 109, 110, Map 8.12.*

Habitat: Silty basins in the larger wadis or the Summan. Often in groups. Occasional to locally frequent in Sudanian territory. The most common *Acacia* of the Eastern Province.

South Coastal Lowlands: 1026, near Wadi as-Sahba, 24 – 01N, 49 – 09E; 7906, edge of Na'lat Shadqam, 25 – 43N, 49 – 29E; frequent along the bed of Wadi as-Sahba, in some basins of southern Ghawar, extending north into the al-Hasa Oasis region, where it is frequent on silty soil south of the oasis complex; rarer individual trees may be found as far northeast as the northeastern edge of Na'lat Shadqam.

Northern Summan: 2748(BM), 25 – 31N, 48 – 09E.

Southern Summan: 2113(BM), 4 km N Jabal Madba'ah, near Yabrin; 2153(BM), Summan Yabrin, 23 – 15N, 48 – 33E.

Vernacular Names: SALAM (gen.), ḤARDHÁ, pl. ḤARADHĪ (tribes of the S Rub' al-Khali, Al Rashid).

2. **Acacia gerrardii** Benth. subsp. **negevensis** Zoh.

  A. iraqensis Rech. f.

Tree, usually with distinct trunk, 3 – 10 m high. Spines straight, whitish, 2 – 5 cm long, sometimes reduced to paired hornlike spinelets 3 – 5 mm long. Leaves usually finely puberulent-tomentose, rarely glabrescent, with 3 – 9 pairs of pinnae; leaflets in (8)10 – 18(23) pairs, 3 – 4 mm long. Flowers pale yellow to white, in solitary or clustered globular heads, with peduncles c. 15 mm long. Pods falcate, not constricted, compressed, linear or slightly sinuate margined, 6 – 12 cm long, 0.6 – 1.2 cm wide, finely pubescent-tomentose, sometimes becoming glabrescent with age *Map 8.11.*

● It proved difficult to assign some of our specimens clearly to subsp. *negevensis* as described by Townsend (in Townsend and Guest 1974). Our duplicates of Nos 586 and 3081 (which may have been taken from the same tree), for example, have rachises up to 35 mm long and petioles over 5 mm. The eccentricity of leaflet attachment is slight.

Dorothy Hillcoat, who has had much experience with specimens of this taxon from Arabia in her work at the British Museum (Natural History), suggests (personal communication) that a good case can be made for giving preference to the name *Acacia pachyceras* Schwartz for our plants. Schwartz's specimens from Yemen appear to be conspecific with our northern material. The decisive question, of course, is whether *A. pachyceras* and our plant are a subspecies of the African *gerrardii* or are distinct from that one at the species level. Ours do have fruits quite close to those of African *gerrardii* specimens seen by the author at BM, yet they differ rather strongly in some characters, such as leaflet dimensions (our Arabian ones being notably smaller). Further study over a broad specimen base is indicated.

This is one of two large-trunked *Acacias*, the other being *A. raddiana*, that are known in the vernacular as *ṭalḥ* and that are important constituents of the woody vegetation along wadis of the Tuwayq uplands in central Arabia. *A. gerrardii* reaches more northern latitudes than other *Acacias* in central and eastern Arabia, and it is most probably the *ṭalḥ* that helps indicate the northern edge of Sudanian territory along the southern and southeastern edges of the Great Nafud. Its extension into Iraq, where it is extremely rare, may have occurred with caravan traffic along Darb Zubaydah, the millenium-old pilgrim track. The author has collected *A. gerrardii* along the Darb at the ruined caravan station of Zubalah; Musil (1928b) reported *ṭalḥ* north of the Birkat al-Haytham cistern in Iraq on the same route. In the Eastern Province it penetrates Saharo-Arabian territory along the al-Batin entrant, and the single occurrence of what is probably this species in northern Kuwait (Dickson 1955, 11) can probably be attributed to spreading along the same wadi route.

The finest known stand of *A. gerrardii* in the Eastern Province was found at an-Na'ayim, where a grove of some 30 well developed trees was situated in a tributary to al-Batin.

Northern Plains: 586(BM), al-Batin, 65.km SW Hafar al-Batin (possibly the same individual tree as No. 3081, following); 3081, al-Batin, 28 – 01N, 45 – 29E; 3189, an-Na'ayim, 27 – 57N, 45 – 28E.

Northern Summan: 3833, ash-Shayyitat, 27 – 29N, 47 – 22E (probable, requires confirmation); 2751, Rawdat az-Zu'ayyini, 25 – 38N, 48 – 04E; 2786, 25 – 51N, 47 – 59E; 2327, al-Farshah, 21 km ESE by E of Rumah.

Vernacular Names and Uses: ṬALḤ (Mutayr, gen.). It might be found to be a rather copious producer of gum Arabic (*ṣamgh*), although no local exploitation of this product is known to the author. The writer collected large tears of fine, clear, light amber, tasteless gum from what appeared to be this species in Wadi Hanifah, Najd.

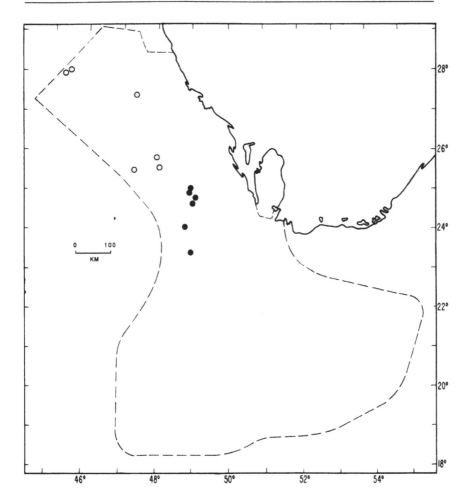

Map 8.11. Distribution as indicated by collection sites of *Acacia gerrardii* (○) and *Acacia raddiana* (●) in the coverage area of this flora (dashed line), eastern Saudi Arabia. Each symbol represents one collection record.

3. **Acacia raddiana** Savi

A. tortilis (Forssk.) Hayne subsp. *raddiana* (Savi) Brenan

Tree, usually with distinct trunk and rounded irregular crown, seldom exceeding 4 m in the Eastern Province but reaching 7 m or more in favorable central Arabian wadi situations. Young branches and leaflets glabrous; rachis sometimes sparsely pubescent. Spines 2 – 5 cm long, white. Leaves with pinnae in 2 – 6 pairs; leaflets in (5)6 – 10(12) pairs, 2 – 3 mm long. Flowers pale yellow to white. Pods usually strongly contorted and sometimes spirally twisted, glabrous, light reddish to yellowish brown, 5 – 12 cm long (when straightened), 6 – 9 mm wide, somewhat compressed, slightly constricted between the seeds, with distinct multiple veination on the valve faces. Seeds oblong, c. 6 mm long, 5 mm wide. *Plate 111, Map 8.11.*

171

Habitat: In our area usually seen as isolated individuals on silty depressions of the Summan. Occasional to rare.

Southern Summan: 1878(BM), Mishash Jaww Dukhan, 24 – 42N, 49 – 04E; 8384, Darb Mazalij, 24 km ENE al-Hunayy; 8386, Darb Mazalij, 25 – 02N, 49 – 00E.

South Coastal Lowlands: 2112(BM), 4 km N Jabal Madba'ah, Yabrin area; 2162(BM), 24 – 08N, 48 – 43E.

Vernacular Names: ṬALḤ (gen.).

4. Acacia tortilis (Forssk.) Hayne

A. spirocarpa Hochst. ex A. Rich.

Large shrub or small tree, usually flat-topped with several main branches ascending from base. Young branches and leaves, and often spines, fine pubescent-tomentose. Spines often in alternating pairs of longer (1 – 3 cm) straight ones and shorter curved ones. Leaves with pinnae mostly in 4 – 6 pairs; leaflets in 6 – 10 pairs. Flowers in pale yellow globular heads. Pods more or less spirally coiled and contorted, weakly compressed, torulose, constricted between the seeds, 3 – 9(10) cm long (when straightened), c. 5 mm broad, densely fine pubescent-tomentose with obscure parallel veins on valve faces. *Plates 112, 113, Map 8.12.*

Habitat: Usually on elevated, well-drained ground. Occasional in Sudanian territory of the south.

Southern Summan: al-Manakhir, 24 – 12N, 48 – 38E (pods collected s.n., 16 September 1977); sight records in vicinity of Bir Harad, 24 – 20N, 49 – 10E; 8381, in Summan 9 km ESE Qirdi GOSP.

South Coastal Lowlands: 2131, Barqa as-Samur, Yabrin area.

Vernacular Names: SAMUR (gen.).

SUBFAMILY PAPILIONOIDEAE

● 4. Lotononis (DC.) Eckl. et Zeyh.

Herbs or shrubs with trifoliolate leaves. Flowers papilionaceous, racemed or rarely solitary. Calyx teeth very unequal, with the lateral pair at each side largely connate, enlarged, much larger than the lower one. Stamens monadelphous. Pod compressed or cylindrical, dehiscent.

Lotononis platycarpa (Viv.) Pichi-Serm.

L. dichotoma (Del.) Boiss.

Prostrate to decumbent, appressed-pubescent, greyish-green annual with stems up to 15 cm long. Leaves 3-foliolate on petioles longer than the blade; leaflets entire, obovate to elliptical, 4 – 7 mm long. Flowers solitary or clustered, 6 – 7 mm long, subsessile in the axils; corolla yellow with cream keel. Pods oblong, 5 – 8 mm long, appressed pubescent, slightly exceeding the calyx. Seeds nearly spherical, c. 1 mm in diameter, smooth greenish-brown. *Plate 114.*

Various authors describe the corolla of this plant as cream-colored. Our specimens have the conspicuous standard clearly yellow; only the less prominent keel is paler.

Habitat: Sand or silt in depressions or large wadis of the southeast. Locally frequent.

Central Coastal Lowlands: 406(US), al-Midra ash-Shimali, near Dhahran.

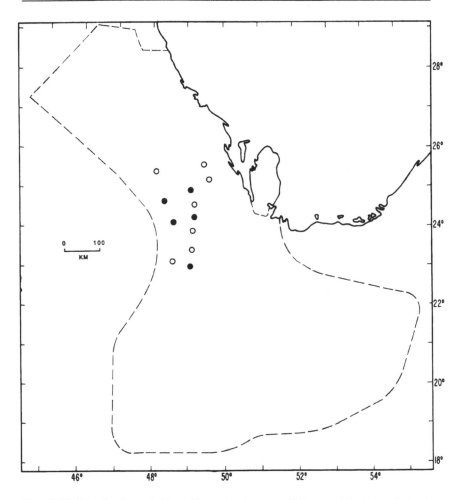

Map 8.12. Distribution as indicated by collection sites of *Acacia tortilis* (●) and *Acacia ehrenbergiana* (○) in the coverage area of this flora (dashed line), eastern Saudi Arabia. Each symbol represents one collection record.

South Coastal Lowlands: 7704, 23 – 52N, 49 – 11E; 8044, Wadi as-Sahba, 23 – 59N, 49 – 10E.
Southern Summan: 8007, 24 – 00N, 49 – 00E; 8062, 24 – 06N, 48 – 48E; 8072, Wadi as-Sahba, 24 – 12N, 48 – 41E.
● *Lotononis* is clearly associated with Sudanian floristic territory, the northernmost record being the single plant from the Dhahran area listed above. It is fairly common in Qatar and the Gulf states to the east. In our area it is frequent, as far as known, only around Wadi as-Sahba and the immediately adjacent country in the south.

● **5. Ononis L.**

Annual or perennial herbs or shrubs with leaves usually trifoliolate, the leaflets dentate

173

or serrate. Flowers papilionaceous, axillary, solitary or in racemose inflorescences. Calyx campanulate or tubular with teeth usually equal. Stamens monadelphous. Pod linear or oblong, terete, dehiscent by 2 valves.

Ononis serrata Forssk.

Ascending annual, often with erect central stem and decumbent lateral stems from the base, 5 – 20 cm high, glandular-hairy and viscid, often with adherent sand. Leaves 3-foliolate; leaflets oblong, serrate, 6 – 10 mm long, 2 – 4 mm wide. Flowers 5 – 6 mm long, solitary in the upper axils, short-pedicelled; corolla pink with white keel, about equalling the calyx. Pod oblong, glabrous, 4 – 5 mm long, not or hardly exceeding the calyx. Seeds subglobose or somewhat flattened, fine-tuberculate or nearly smooth, yellow to brown, c. 1.3 mm in diameter.

Habitat: Sands. Common, particularly in coastal districts.

Northern Plains: 998, 12 km N Jabal al-'Amudah; 1722(BM), 40 km E Abraq al-Kabrit; 1747(BM), Jabal al-'Amudah.

Northern Dahna: 3881, 25 km W Umm 'Ushar.

Central Coastal Lowlands: Dhahran; 1412(BM), 7 km NW Abu Hadriyah; 1763(BM), Abu Hadriyah; 1797(BM), 5 km NE al-Khursaniyah; 1980, ad-Dawsariyah; 3298, 2 km W Sabkhat Umm al-Jimal; 7732, 2 km S Ras Tanaqib, 27 – 49N, 48 – 53E.

Vernacular Names: UMM UDHN ('earweed', N, Musil 1927, 628). Usually referred to by most east Arabian tribesmen by the generic term for annual legumes, ḤURBUTH.

● 6. Trigonella L.

Annual or perennial herbs, often with fragrant foliage; leaves trifoliolate and leaflets usually dentate. Flowers papilionaceous, yellow, rarely white or bluish, in axillary or sometimes headlike racemes. Calyx campanulate or tubular, the 5 teeth usually unequal. Stamens diadelphous. Pod linear to ovate, terete or compressed, usually dehiscent.

Key to the Species of Trigonella

1. Flowers in dense headlike racemes, on peduncles subequalling or exceeding the subtending leaf
    2. Flowering pedicels equalling or shorter than the bracts; pedicels of lower flowers and of fruits strongly deflexed .... 4. *T. hamosa*
    2. Flowering pedicels clearly longer than the minute bracts, not deflexed .... 2. *T. aurantiaca*
1. Flowers sessile or subsessile, numerously clustered or 1 – 3 in the axils
    3. Flowers 13 – 18 mm long; cream-colored .... 3. *T. foenum-graecum*
    3. Flowers less than 8 mm long, yellow
        4. Pod 25 – 60 mm long, 1 – 3 in the axils .... 5. *T. monantha*
        4. Pod 3 – 8 mm long, numerously clustered
            5. Pods wavy, bent or folded sharply from side to side ... 1. *T. anguina*
            5. Pods curved simply .... 6. *T. stellata*

1. Trigonella anguina Del.

Prostrate annual, glabrous or very slightly pubescent on vegetative parts. Petioles about 5 – 10 mm long; leaflets obovate or rounded-cuneate, obtuse, serrate, mucronate, 5 – 9 mm long, 3 – 7 mm wide. Flowers sessile, clustered in the axils, 3 – 4 mm long, with yellow corolla slightly exceeding the calyx. Pods sessile, clustered, slightly compressed, nerved, finely pubescent, to c. 8 mm long, wavy, strongly bent or folded from side to side, resembling a miniature wriggling worm or snake.

Habitat: Silt-floored desert basins. Occasional. Also may be found rarely as a weed on cultivated ground.

Northern Plains: 1584(BM), al-Batin/al-'Awja junction.

Northern Summan: 2880, Rawdat Musay'id, 26 – 02N, 48 – 43E; 3884, al-Batin, 5 km SW Umm 'Ushar.

Central Coastal Lowlands: 7417(BM), Dhahran, weed in new cultivation.

Southern Summan: 8080, al-Ghawar, 24 – 33N, 49 – 06E.

Vernacular Names: NAFAL (gen.), SHAMAṬIRĪ (Qahtan).

2. Trigonella aurantiaca Boiss.

Prostrate to ascending pubescent annual with stems to c. 20 cm long. Petioles 2 – 6 mm long; leaflets obovate to cuneate, sharply dentate, 3 – 10 mm long, 2 – 7 mm wide, appressed-pubescent beneath. Flowers yellow, 4 – 9 mm long, subsessile in dense 3 – 10-flowered headlike racemes, the inflorescence mostly equalling or exceeding the subtending leaf, on a peduncle 6 – 15(20) mm long. Pods linear, somewhat compressed, 15 – 33 mm long, 1 – 1.5 mm wide, straight or gently curved, thinly pubescent and reticulately nerved, on straight pedicels c. 1 mm long.

Habitat: Not a native plant but may occur rarely on disturbed sandy ground. The only record is from an abandoned Bedouin camp site where numerous Mediterranean weeds were growing, probably seeded from stored grain.

Central Coastal Lowlands: 7953, 8086(K), N Batn al-Faruq, 25 – 43N, 48 – 53E.

3. Trigonella foenum-graecum L.

Erect herb, to c. 30 cm high, glabrous or pubescent. Leaflets obovate to oblong, 10 – 30 mm long, 5 – 15 mm wide, dentate above. Flowers sessile or subsessile, solitary or paired in the axils, 13 – 18 mm long with white to cream corolla. Pod erect, 6 – 10(15) cm long, 3 – 5 mm wide, somewhat compressed, straight or somewhat curved, glabrous or pubescent, tapering to a straight beak 2 – 3 cm long.

Habitat: Cultivated; rarely found as an escape.

Central Coastal Lowlands: 378, Ras Tanura, as a garden weed.

Vernacular Names and Uses: ḤULBAH (gen.). Sometimes cultivated on Eastern Province farms as a kitchen herb or as a medicinal. Dickson (1955, 93) notes that it was cut young in Kuwait and used in meat and fish stews.

4. Trigonella hamosa L.

Ascending or decumbent annual, glabrescent or sparingly pubescent, 10 – 30 cm high. Petioles 4 – 15 mm long; leaflets obovate to oblong-obovate, obtuse, somewhat truncate and mucronate at apex, dentate, 5 – 15 mm long, 3 – 8 mm wide. Flowers 3 – 5 mm long with yellow corolla, in dense headlike racemes, the inflorescence borne on a peduncle 15 – 30 mm long, equalling or exceeding the subtending leaf. Pod linear, subterete, fine aristate at apex, 8 – 10(12) mm long, 1 – 2 mm wide, curved, transversely and reticulately veined, pubescent when young, at length glabrescent. *Plate 115.*

Habitat: Sands and silts. Found in both natural desert situations and on disturbed or cultivated ground as a weed. Frequent.

Northern Plains: 1558(BM), al-Batin/al-'Awja junction.

Northern Summan: 3344, 5 km WSW an-Nazim; 4017, 27 – 45N, 45 – 32E; 1340(BM), 42 km W Qaryat al-'Ulya; 7481, Batn Sumlul, 26 – 26N, 48 – 15E; 8597, Bahrat al-Kilab, 26 – 33N, 47 – 53E; 8604, Wadi Mibhil, 26 – 40N, 47 – 34E.

Southern Summan: 2179, 24 – 25N, 48 – 41E; 2235, 15 km WSW al-Hunayy.

Central Coastal Lowlands: 116, 1826(BM), 7886, Dhahran; 7410(BM), Dhahran, hotel garden weed; 1062, Dhahran stables farm.

South Coastal Lowlands: 376, Bani Ma'n, al-Hasa; 8307, al-Hufuf.

Vernacular Names and Uses: DARJAL (Hasawi gardener), QURRAYS (N, Musil 1927, 609; 1928a, 700; 1928b, 350). Musil (1928a, 95) reported that Bedouins eat it raw as a salad herb.

5. Trigonella monantha C. A. Meyer

Prostrate to ascending annual, branched from the base, sparsely pubescent, 5 – 20 cm high. Leaflets obovate to cuneate, dentate, 4 – 12 mm long, 2 – 8 mm wide. Flowers sessile, 1 – 2 in the axils or less frequently 3 together, 4 – 7 mm long; corolla yellow. Pods linear, somewhat compressed, straight or curved or sometimes hooked near apex, 30 – 60 mm long, 1 – 2 mm wide, sparsely appressed-pubescent, reticulately nerved, ripening to brownish purple.

Habitat: A weed to be expected only around cultivation or desert campsites where weeds are introduced with sacked grain. Rare.

Central Coastal Lowlands: 8087(K), N Batn al-Faruq, 25 – 43N, 48 – 53E, Bedouin campsite.

6. Trigonella stellata Forssk.

Glabrescent, prostrate annual, many-branched from the base with stems up to c. 35 cm long. Petioles 10 – 50 mm long; leaflets obovate to obcordate, dentate above, entire below, obtuse, 4 – 12 mm long, 3 – 10 mm wide. Flowers appearing clustered and sessile, 3 – 4 mm long, in very short subsessile axillary racemes; corolla yellow. Pods stellate-spreading, terete, curved, 4 – 8 mm long, 1 – 2 mm wide, thinly pubescent or glabrous. *Plate 116.*

Habitat: Widespread and probably the most common of the desert *Trigonellas*, especially in silty soils of the northern plains.

Northern Plains: 565(BM), 663, Khabari Wadha; 1496(BM), 1527(BM), 30 km ESE al-Qaysumah; 693, 20 km WSW Khubayra; 824, 13 km W Qaryat al-'Ulya; 1539(BM), 36 km NE Hafar al-Batin; 1661(BM), 2 km S Kuwait border in 47 – 14E; 1211, Jabal an-Nu'ayriyah; 1256(BM), 15 km ENE Qaryat al-'Ulya; 1310(BM), 8 km W Jabal Dab'; 1698(BM), ash-Shaqq, 28 – 30N, 47 – 42E.

Northern Summan: 1371(BM), 42 km W Qaryat al-'Ulya; 709, 33 km SW al-Qar'ah wells; 781, 69 km SW al-Qar'ah wells.

Central Coastal Lowlands: 1774(BM), Abu Hadriyah; 1824(BM), Dhahran; 3271, 8 km W Qatif-Qaisumah Pipeline KP – 144; 3774, Ras az-Zawr; 7747, 5 km NW Ras Tanaqib, 27 – 53N, 48 – 49E.

Southern Summan: 2081, al-Ghawar, 24 – 27N, 49 – 05E.

Vernacular Names and Uses: NAFAL (gen.). Dickson (1955, 93) says that Shammar Bedouin women use it to prepare a sweet-scented hairdressing. Excessive grazing on this plant (and perhaps other species of *Trigonella*) is reported by Bedouins to cause bloat in livestock.

● 7. Medicago L.

Annual or perennial herbs, rarely shrubs, with alternate trifoliolate leaves and leaflets usually dentate. Stipules often dentate or laciniate. Flowers papilionaceous, racemed or in heads, yellow, or rarely violet. Calyx with 5 subequal teeth. Stamens diadelphous. Pod more or less tightly spiral-coiled, cylindrical, ovoid, or orbicular in outline, with margins often prickly or tuberculate.

A genus readily recognizable by its characteristic spirally coiled pod, *Medicago* includes the important cultivated plant alfalfa or lucerne, *Medicago sativa* L., as well as a widespread and abundant desert annual, *M. laciniata*.

Key to the Species of Medicago

1. Flowers violet to purple .... 3. *M. sativa*
1. Flowers yellow
   2. Pod spheroid or ovoid with spines strongly interlacing; desert
     annual .... 1. *M. laciniata*
   2. Pod flat-discoid with marginal prickles not or hardly interlacing;
     weed .... 2. *M. polymorpha*

A fourth species, *M. orbicularis* (L.) Bartal., has been collected as a weed in central Arabia (Chaudhary and Zawawi 1983) and may be found to occur around cultivation in the Eastern Province. It has yellow flowers and a distinctive flat, smoothish, orbicular pod, without prickles, 10 – 18 mm in diameter.

1. Medicago laciniata (L.) Mill.

M. aschersoniana Urb.

Prostrate to decumbent annual, branched from the base with stems 5 – 30 cm long, sparingly pubescent. Leaves 3-foliolate; leaflets cuneate to obovate, truncate or emarginate-apiculate at apex, variably dentate to laciniate, 3 – 10 mm long, 2.5 mm wide. Flowers c. 4 mm long, 1 – 3 on peduncles shorter or longer than the subtending leaf; corolla yellow. Pod spheroid or ovoid, 3 – 6 mm in diameter, densely furnished with interlacing prickles from the margins of the coils. *Plate 117*.

● This taxon has long been described by various authors in two species: *M. laciniata* (L.) Mill., with stipules laciniate, leaflets deeply serrate-laciniate, and peduncles much longer than the subtending petioles; and *M. aschersoniana* Urb., with stipules deeply dentate, leaflets serrate and peduncles equalling or shorter than the petiole. Townsend (in Townsend and Guest 1974) treats these, respectively, as var. *laciniata* and var. *brachyacantha* Boiss. Most of our specimens resemble the latter more closely, but many appear to be intermediate and it does not seem profitable now to attempt a varietal segregation. There appears to be little doubt about their being conspecific.

Habitat: Sand or silts. Very common and widespread.

Northern Plains: 1512(BM), 30 km ESE al-Qaysumah; 870, 9 km S 'Ayn al-'Abd; 997, 12 km N Jabal al-'Amudah; 1577(BM), al-Batin/al-'Awja junction; 1745(BM), 1746(BM), Jabal al-'Amudah; 1174(BM), Jabal al-Ba'al; 1195(BM), 1219(BM), Jabal an-Nu'ayriyah; 1258(BM), 15 km ENE Qaryat al-'Ulya; 1682(BM), 2 km S Kuwait border in 47 – 14E; 3063(BM), ash-Shaqq, 28 – 25N, 47 – 47E.

Northern Summan: 1356(BM), 42 km W Qaryat al-'Ulya; 713, 33 km SW al-Qar'ah wells; 2754, 25–43N, 48–07E; 2834, 26–21N, 47–25E; 3177(BM), 13 km SSW Hanidh.

Southern Summan: 1816(BM), 15 km WSW al-Hunayy; 1828(BM), 26 km ENE al-Hunayy; 2063, Jaww ad-Dukhan, 24–48N, 49–05E; 2172, Railroad KP–349; 2192, 24–34N, 48–45E.

Central Coastal Lowlands: 907, Mulayjah; 1036, 7407(BM), Dhahran; 3265, 8 km W Qatif-Qaisumah Pipeline KP–144; 3296, 2 km W Sabkhat Umm al-Jimal.

Vernacular Names: NAFAL (Al Murrah, Qahtan), ḤASAK (Rawalah, Qahtan), ḤUSAYKAH (Al Murrah).

2. Medicago polymorpha L.

Glabrescent decumbent annual, branched from the base with stems up to 50 cm long. Leaves 3-foliolate; leaflets obcordate-cuneate, denticulate (sometimes rather obscurely so), 5–20 mm long. Flowers 3–4 mm long, mostly grouped 2–6 on peduncles subequalling the leaf; corolla yellow. Pod flat-discoid, 4–8 mm in diameter, reticulately veined on the faces and bearing prickles around the margin in a single plane; prickles often hooked distally.

Habitat: A weed of garden, farm. Infrequent.

Central Coastal Lowlands: 7409(BM), Dhahran (in newly established lawn); 8088, N Batn al-Faruq, 25–43N, 48–53E (Bedouin camp site).

3. Medicago sativa L.

Erect or ascending branched annual or perennial herb, 20–90 cm high, sparingly pubescent on leaves and stems. Leaves 3-foliolate; leaflets broadly oblanceolate to oblong-linear, denticulate around the apex, obtuse or acute, apiculate, 10–35 mm long, 4–14 mm wide; petiole 5–20 mm long; petiolule of terminal leaflet 2–10 mm long. Flowers in dense spiciform racemes up to 60 mm long, 30 mm wide, in the upper axils on peduncles 20–40 mm long and greatly exceeding the subtending leaves. Flowers lilac, violet to purple (in the specimen below a rich dark purple), c. 12 mm long. Pod compressed, 4–8 mm in diameter, usually tightly coiled, auger-shaped, without prickles or spines, appressed pubescent and finely reticulate-veined on the faces.

This is alfalfa or lucerne, the cultivated forage plant.

Habitat: Rare at any distance from cultivated plots, but spring roadside accidentals may be found after good rains.

Southern Summan: 8077, S al-Ghawar, 24–33N, 49–06E (roadside).

Vernacular Names and Uses: QATT (gen.; often colloquially pronounced ʻJITTʼ). A very important fodder crop in most parts of Arabia and often sold bundled in local markets.

● 8. Melilotus Mill.

Annual herbs, sometimes perennating, with trifoliolate leaves and leaflets usually dentate. Flowers papilionaceous, racemed, yellow or white. Calyx with 5 nearly equal teeth. Stamens diadelphous or monadelphous. Pod ovoid or globose, smooth and often veined, 1–4-seeded.

56 *Calligonum comosum* x 1/25

57 *Calligonum comosum* (fruits, red form)
x 1 1/4

58 *Calligonum comosum* (fruits, yellow form)
x 1/3

59 *Calligonum crinitum*
x 1 3/5

60 *Calligonum tetrapterum*
x 5/8

63 *Corchorus depressus* x 1 3/5

61 *Limonium axillare* x 1/2

64 *Althaea ludwigii* x 1

62 *Limonium thouinii* x 1

66 *Frankenia pulverulenta*
x 3/4

65 *Tamarix arabica* x 3/8

68 *Citrullus colocynthis* x 1/4

67 *Citrullus colocynthis* x 3/4

69 *Capparis spinosa* x 1 2/5

70 *Capparis spinosa* (opened fruit) x 4/5

71 *Cleome glaucescens* x 1/4 **72** *Diplotaxis acris* x 1/5 **74** *Diplotaxis harra* x 1/5

73 *Diplotaxis acris* x 1 1/3

75 *Cakile arabica* x 2/5

76 *Zilla spinosa* x 2/5

78 *Carrichtera annua* x 1 1/4

77 *Carrichtera annua* x 2/5

80 *Notoceras bicorne* x 1/2

79 *Savignya parviflora* x 1/2

81 *Horwoodia dicksoniae*
x 1/5

82 *Horwoodia dicksoniae*
x 4/5

83 *Capsella bursa-pastoris*
x 2/3

4 *Anastatica hierochuntica* x 2/5

85 *Anastatica hierochuntica* (dried plant) x 1/2

87 *Schimpera arabica*   x 1 1/3

86 *Schimpera arabica*   x 1/5

88 *Farsetia aegyptia*   x 2/3

90 *Farsetia burtonae* (flowering
in its first season)   x 1

91 *Alyssum homalocarpum*
x 1

89 *Farsetia burtonae*   x 3/5

93 *Matthiola longipetala*
x 1/4

92 *Matthiola longipetala*   x 2

94 *Malcolmia grandiflora*   x 1

95 *Eremobium aegyptiacum*
x 3

97 *Neotorularia torulosa*
x 1/4

96 *Maresia pygmaea* x 1

99 *Ochradenus baccatus*
(pistillate plant) x 1/5

98 *Ochradenus baccatus* (staminate plant) x 1/4

100 *Reseda arabica* x 1/3

101 *Reseda muricata* x 1/3

102 *Caylusea hexagyna* x 1/3

103 *Anagallis arvensis* var. *caerulea* x 2 1/2

104 *Neurada procumbens* x 1/3

106 *Cassia italica* (pods) x 1/2

105 *Cassia italica* x 1

107 *Prosopis koelziana*
(6 m high)

08 *Prosopis koelziana* x 1/3

10 *Acacia ehrenbergiana* x 2

109 *Acacia ehrenbergiana*
x 1/2

111 *Acacia raddiana* (6.3 m high)

112 *Acacia tortilis* (3.5 m high)

113 *Acacia tortilis* x 1

115 *Trigonella hamosa* x 1/2

114 *Lotononis platycarpa* x 1

116 *Trigonella stellata* x 1/3

Key to the Species of Melilotus
1. Flowers white, 3.5 – 4 mm long .... 1. *M. alba*
1. Flowers yellow, 2 – 3 mm long .... 2. *M. indica*

1. **Melilotus alba** Medik.

Ascending to erect, branched glabrescent annual, sometimes biennial. Leaves 3-foliolate; petioles to c. 20 mm long; leaflets obovate to oblong, serrulate, 10 – 20 mm long, 5 – 10 mm wide. Flowers 3.5 – 5 mm long, white, in dense racemes to 30 mm or more long and c. 8 mm broad, on peduncles 15 – 30 mm long. Pods ovoid to subglobular, 3 – 5 mm long, reticulately nerved, maturing brown.

Habitat: A weed of lawns, gardens, sometimes growing with the much more common *M. indica*. Infrequent.

Central Coastal Lowlands: 7413(BM), Dhahran; 7888, al-Khubar.

2. **Melilotus indica** (L.) All.

Erect, or rarely decumbent, glabrous annual, 10 – 40 cm high, sometimes to 80 cm in favorable situations. Leaves 3-foliolate, on petioles to 40 mm long; leaflets oblong-elliptic, obtuse, denticulate, 8 – 22 mm long, 3 – 12 mm wide. Flowers 2 – 3 mm long, yellow, in dense racemes up to 4.5 cm long, to 5 – 6 mm wide in fruit, usually longer than the subtending leaf. Pods ovoid, reticulate-nerved, c. 3 mm long, somewhat pendulous, on deflexed pedicels c. 1 mm long. Foliage sweet-scented of coumarin. *Plate 118.*

Habitat: A weed of lawns, gardens, farms, where it is common. Sometimes also seen in the desert around Bedouin camp sites and other disturbed grounds.

Northern Plains: 1686(BM), al-Hamatiyat, 28 – 50N, 47 – 30E.

Central Coastal Lowlands: 394, Dhahran.

South Coastal Lowlands: 2736, ar-Ruqayyiqah, al-Hasa; 7792, adh-Dhulayqiyah, al-Hasa.

● 9. Trifolium L.

Annual or perennial herbs with leaves palmately 3-foliolate, or sometimes 5 – 7-foliolate. Flowers papilionaceous, racemed or capitate, rarely solitary, of various colors. Calyx with 5 equal or unequal teeth, sometimes inflated in fruit. Stamens diadelphous. Pod ovoid or oblong, often enclosed in the persistent calyx, indehiscent and mostly 1 – 4-seeded.

Trifolium resupinatum L.

Prostrate or ascending branched glabrescent annual, to c. 30 cm high. Leaves 3-foliolate, subsessile above and with petioles up to 30 mm long below; leaflets obovate-oblanceolate, finely and very sharply serrate, mucronate, 5 – 10 mm long. Flowers 4 – 6 mm long, sessile in dense, hemispherical, peduncled heads, the peduncle clearly longer than the subtending leaf, up to 25 mm long; calyx growing and inflating in fruit, becoming membranous and pubescent, with reticulate nervation, and furnished above with two awn-like teeth or setae 1 – 2 mm long. Corolla pink. Pod ovoid, enclosed in the calyx.

Habitat: A weed of gardens, farms. Rare.

Central Coastal Lowlands: 1063(BM), Dhahran stables farm; 7392, al-Jubayl, lawn weed.

● 10. Lotus L.

Annual or perennial herbs or shrublets with leaves usually 5-foliolate, the lower pair of leaflets stipulelike on the stem. Flowers papilionaceous, in umbels or heads, rarely solitary, yellow or pink, red to purple, rarely white. Calyx 5-dentate with the teeth equal or unequal, sometimes bilabiate. Stamens diadelphous. Pod linear to oblong, straight or curved, teretish, rarely compressed.

Key to the Species of Lotus

1. Shrublet with minute leaflets 1 – 2 mm long; flowers white to
    pink .... 2. *L. garcinii*
1. Herbs with leaflets mostly 5 mm long or longer; flowers yellow
    2. Flowers on peduncles more than 20 mm long; corolla 9 – 15 mm
        long; weed of lawns, gardens .... 1. *L. corniculatus*
    2. Peduncles 10 – 15 mm long; corolla 4 – 7 mm long; desert
        annual .... 3. *L. halophilus*

1. Lotus corniculatus L.

Prostrate to ascending, glabrescent or pubescent annual, 5 – 30(80) cm high. Leaves 5-foliolate, the lower pair of leaflets stipulelike on the stem. Leaflets of variable shape and pubescence, generally obovate to lanceolate, obtuse or acute, 5 – 20 mm long, 3 – 10 mm wide. Flowers strong yellow, sometimes red-flushed, umbelled mostly 3 – 5 on axillary peduncles 30 mm or more long. Pod linear, straight, 12 – 30 mm long, 2 – 3 mm wide.
Habitat: A weed of lawns or gardens. Infrequent.
Central Coastal Lowlands: 1822(BM), Dhahran.

2. Lotus garcinii DC.

Erect grey-canescent shrublet with flexible branches, to c. 75 cm high. Leaves sessile, 3 – 4 foliolate, the leaflets minute, only 1 – 2 mm long. Flowers solitary in the axils, 4 – 7 mm long, subsessile or on pedicels c. 1 mm long; corolla hardly exceeding the equally toothed calyx, white to pink with fine darker red veins. Pod oblong, subcylindrical, reddish, 5 – 6 mm long, 2.5 – 3 mm wide, faintly constricted between the seeds, mostly 3-seeded. Seeds spherical, 1.3 mm in diameter, smooth, brown with black flecks.
Habitat: Sands along the coast, usually on somewhat saline ground around *sabkhah* edges. Occasional.
Central Coastal Lowlands: 914, Sayhat; 638(BM), 5 km N al-Khubar; 1468(BM), 2 km E al-Ajam; 10 km N al-Jubayl (Dickson 1955, 61).
South Coastal Lowlands: 7457(BM), al-'Udayd.

3. Lotus halophilus Boiss. et Sprun.
L. pusillus Viv.
L. villosus Forssk.

Prostrate or decumbent appressed-pubescent annual, branched from the base with stems up to c. 25 cm long. Leaves sessile, 5-foliolate, the lower pair of leaflets stipulelike; leaflets obovate to elliptical, 3 – 9 mm long, 2 – 4 mm wide. Flowers yellow, 5 – 6 mm long, solitary or paired on peduncles to 16 mm long. Pod glabrous, linear-cylindrical, slightly curved, 15 – 25 mm long, 1.2 – 1.6 mm wide, ripening a shining dark brown.
Plate 119.

Habitat: Shallow or deep sands. Very common; one of the most abundant spring annuals of sandy habitats.

Northern Plains: 846, 9 km S 'Ayn al-'Abd; 1007, 12 km N Jabal al-'Amudah; 1220, Jabal an-Nu'ayriyah; 1261(BM), 15 km ENE Qaryat al-'Ulya; 1621(BM), 3 km ESE al-Batin/al-'Awja junction.

Northern Summan: 710, 33 km SW al-Qar'ah wells; 1367(BM), 42 km W Qaryat al-'Ulya; 2832, 26–21N, 47–25E.

Northern Dahna: 2323, 22 km ENE Rumah.

Central Coastal Lowlands: 83, Dhahran; 368, Umm ar-Rus, Dhahran; 1409(BM), 7 km NW Abu Hadriyah; 1767(BM), Abu Hadriyah; 1790(BM), 5 km NE al-Khursaniyah; 1985, ad-Dawsariyah; 3295(BM), 2 km W Sabkhat Umm al-Jimal; 4059, az-Zulayfayn, 26–51N, 49–51E; 7723, 2 km S Ras Tanaqib, 27–49N, 48–53E.

South Coastal Lowlands: 1429(BM), 2 km E Jabal Ghuraymil.

Vernacular Names: ḤURBUTH (Al Murrah).

● 11. Indigofera L.

Herbs or shrubs, often silvery appressed-pubescent, with leaves imparipinnate, digitate, or rarely simple. Flowers papilionaceous, racemed, with the calyx campanulate or tubular, 5-toothed. Corolla pink to red. Stamens diadelphous. Pod linear to oblong, rarely globose, straight or curved, teretish or compressed.

A Sudanian genus well represented in southern and far-eastern Arabia but with only one species reaching the extreme southeastern edge of our area.

Indigofera intricata Boiss.

Much branched, intricate, densely appressed silvery-canescent shrublet, 10–30 cm high. Leaves subsessile, imparipinnate, c. 5 mm long, with 1–3 pairs of leaflets. Leaflets 1.5–2.5 mm long, obovate. Flowers 4–5 mm long; calyx with linear-acute lobes; corolla red. Pods subsessile, cylindrical, mucronate, pubescent, 3–4 seeded, 7–12 mm long.

Habitat: The single collection of this plant was from shallow sand on gypsum. Rare, except perhaps locally in the extreme southeast.

South Coastal Lowlands: 7066, coastal road east of Qatar, 24–10N, 51–29E. Also collected by the author in southern Qatar.

● 12. Psoralea L.

Perennial herbs or shrubs with leaves digitate, of 3 or more leaflets. Flowers papilionaceous, in various inflorescence types, rarely solitary, blue or violet to pink and white. Calyx campanulate with 5 teeth equal, or the upper teeth shorter and connate. Stamens diadelphous. Pod more or less ovoid, indehiscent, 1-seeded, sometimes beaked.

Psoralea plicata Del.

Ascending grey-green, appressed-pubescent shrublet with scattered white to yellowish glands; stems finely striate, to c. 0.5 m high. Leaves 3-foliolate, subsessile or on 1–2 mm-long petioles; leaflets narrowly oblong, tapering toward the base, plicate on the nerves and repand-undulate margined, the terminal larger, 4–10 mm long. Flowers 3–5 mm long, subsessile in open, spicate racemes 2–5 cm long, the tapering axis hardening and becoming weakly spinescent after fall of the fruits; corolla hardly exceeding the calyx, with white standard and violet-tinged wings and keel; calyx growing in fruit. Pod ovoid, enclosed in the calyx, brown, appressed-pubescent, 3.5 – 4 mm long. **Habitat:** Silt basin floors. Rare.

Southern Dahna: 477, 7013(BM), Rawdat at-Tawdihiyah. This only known record site is at the western edge of the Dahna and the western limit of our area. Batanouny (1981) reports the plant from Qatar, and populations are to be expected along the connecting entrant of Wadi as-Sahba.

Vernacular Names: SHAJARAT AN-NA'AM ('ostrich bush', Qahtan).

● 13. Sesbania Adans.

Herbs, shrubs or trees with paripinnate leaves and numerous leaflets. Flowers papilionaceous, in axillary racemes, yellow to blue or white. Calyx 5-lobed or virtually entire. Stamens diadelphous. Pod linear, rarely oblong, somewhat compressed and torulose, dehiscent.

Sesbania sesban (L.) Merrill
S. aegyptiaca Pers.

Erect shrub or slender small tree, generally 3 – 5 m high, pubescent in very young parts, at length glabrous. Leaves paripinnate, 6 – 8 cm long, on 5 – 7 mm-long petioles; leaflets in (8)10 – 13(15) pairs, oblong with parallel margins, subtruncate at apex, 3 – 4.5 mm wide. Flowers 15 – 18 mm long, racemed; corolla yellow, flecked with purple. Pods subterete, very long and narrow, up to 25 cm long, 2 – 3.5 mm wide, straight or gently curved, pendulous.

Habitat: Shaded roadsides in oases or along edges of date palm plots. Frequent in oases. Probably originally an escape from cultivation but now self-propagating in the larger oases.

Central Coastal Lowlands: 344(K), al-Qatif Oasis, roadside.

Vernacular Names: NAWWAMAH ('sleeper tree', farmers in al-Qatif; possibly a reference to its drooping branches and pods).

● 14. Astragalus L.

Annual or perennial herbs or shrubs, sometimes spinescent, with leaves paripinnate or imparipinnate. Flowers in racemes or heads, or solitary. Calyx with 5 equal or unequal teeth, sometimes inflating. Corolla papilionaceous; stamens diadelphous. Ovary sessile or stipitate. Pod variable, linear to globose, compressed or more or less terete, usually 2-valved.

Astragalus, the largest known genus of flowering plants, is represented in eastern Saudi Arabia by ten known species, the majority of which are small, low-growing, annuals. The number of species found declines steadily from north to south with greater distance from the Middle Eastern diversity center of the genus in Iraq and Iran, and with increasing aridity.

Key to the Species of Astragalus
1. Spiny perennials
   2. Fruiting calyx greatly inflated, 10 mm or more in diameter, enclosing the pod; leaflets in 4 – 6 pairs .... 9. *A. spinosus*

2. Fruiting calyx not inflated; pod long-exserted; leaflets in 10 – 25
   pairs .... 8. *A. sieberi*
1. Unarmed annuals or perennials
   3. Pod enclosed in the inflated calyx; leaflets 10 – 25 mm
      wide .... 6. *A. kahiricus*
   3. Pod exserted, exposed; leaflets less than 10 mm wide
      4. Mature pods glabrous
         5. Leaflets in 1 – 4 pairs; pod purple-spotted, smooth ... 1. *A. annularis*
         5. Leaflets in 5 or more pairs; pod not purple-spotted but with
            raised reticulate nerves .... 3. *A. corrugatus*
      4. Pods pubescent
         6. Pods 6 – 9 mm broad; perennial somewhat woody at
            base .... 2. *A. bombycinus*
         6. Pods 2 – 4 mm broad; annuals
            7. Leaflets retuse; pods not dilated at base, in short racemes
               8. Leaflets mostly in 1 – 3(5) distant pairs; teeth of flowering
                  calyx shorter than the tube .... 5. *A. hauarensis*
               8. Leaflets mostly in 6 – 7 approximate pairs; teeth of flowering
                  calyx equal to or exceeding the tube .... 4. *A. eremophilus*
            7. Leaflets entire at apex, sub-acute; pods dilated at base, in
               capitate clusters of 2 – 6
               9. Pods 3 – 7 mm long .... 10. *A. tribuloides*
               9. Pods 12 – 20 mm long .... 7. *A. schimperi*

Another species, *A. hamosus* L., has been reported from Kuwait by Dickson (1955) and from Qatar by Batanouny (1981). It has not apparently been recorded to date from eastern Saudi Arabia but would emerge close to *A. schimperi* in the key above. It can be recognized by the presence of asymmetrically medifixed hairs on the pod and leaflet undersides and by its longer pods (20 – 40 mm), strongly curved and in peduncled headlike racemes.

Few of the annual species of *Astragalus* have specific vernacular names. They are collectively known as QAF'A', the diminutive QUFAY'A', or (mainly among southern tribes) ḤURBUTH. These same names are sometimes applied to small annuals of other leguminous genera, such as *Lotus* and *Hippocrepis*.

1. Astragalus annularis Forssk.

Prostrate, decumbent or ascending annual, grey-green pubescent, branched from the base with stems 2 – 15 cm or more long. Leaves imparipinnate with 1 – 3 leaflet pairs; leaflets obovate-elliptical with rounded obtuse apices, mostly 5 – 10 mm long. Flowers strong pink to red-purple, in racemes subequalling or exceeding the subtending leaf; peduncle prolonged beyond point of fruit insertion as an aristate point. Pods often solitary, linear, curved, tapering at apex, strongly compressed dorso-ventrally and nearly flat both above and below, pubescent when young but glabrous or nearly so when mature, irregularly spotted and streaked with red-purple, 2 – 5 cm long, c. 3 mm wide. *Plate 120.*

Habitat: Sands or silty sand on many terrain types. Common.

Northern Plains: 1694(BM), 28 – 41N, 47 – 35E; 867, 9 km S 'Ayn al-'Abd; 889, Jabal al-'Amudah; 934, 10 km N Khafji-Saffaniyah junction; 999, 12 km N Jabal al-'Amudah;

1284, 15 km ENE Qaryat al-'Ulya; 1487(BM), 50 km WNW an-Nu'ayriyah; 1653(BM), 2 km S Kuwait border at 47–14E; 1709(BM), ash-Shaqq, 28–30N, 47–42E; 1189(BM), Jabal an-Nu'ayriyah.

Central Coastal Lowlands: 39, 58, 81, 182A, 1836(BM), 7884, Dhahran; 186, N Dhahran Airport; 909, Mulayjah; 1413(BM), 7 km NW Abu Hadriyah; 1768, Abu Hadriyah; 1799(BM), 5 km NE al-Khursaniyah; 3264, 8 km W Qatif-Qaisumah Pipeline KP–144; 3315, 5 km S Nita'; 3357, Jabal al-Ahass; 4064, az-Zulayfayn, 26–51N, 49–51E; 8109, 9 km SSE Hanidh.

Vernacular Names: ABŪ KHAWĀTĪM ('father of signet rings', Al Murrah), ḤURBŪTH (Al Murrah, Qahtan; a generic name), QAF'A' (Bani Hajir; generic). Dickson (1955) lists the name AṢĀBI' AL-'ARŪS, 'bride's fingers'.

2. Astragalus bombycinus Boiss.

Prostrate perennial or, reportedly, sometimes annual, branching from base with densely pubescent stems to 30 cm or more long. Leaves imparipinnate with 5–8 pairs of leaflets; leaflets obovate to cuneate, obtuse and often retuse, 2–8 mm long, densely white pubescent below, glabrescent above, on petiolules 0.5–0.7 mm long. Flowers white to pinkish, in racemes usually longer than the subtending leaf. Pods subsessile, solitary or in racemes of 2–4, somewhat inflated, tapering to a beaklike apex, densely pubescent with fine white, spreading hairs, 15–25 mm long, 5–9 mm wide, sometimes obscurely reticulate. *Plate 121.*

Habitat: Silty basins and plains of the north. Occasional to rare.

Northern Plains: 559(BM), Khabari Wadha; 854, 9 km S 'Ayn al-'Abd; 1518(BM), 30 km ESE al-Qaysumah.

Northern Summan: 8629, Dahl Abu Harmalah, 26–29N, 47–36E; 8631, 2 km NE Dahl Abu Harmalah, 26–30N, 47–37E.

3. Astragalus corrugatus Bertol.

A. tenuirugis Boiss.

Procumbent to ascending annual, glabrous or very sparsely pubescent, branched from the base with stems up to about 25 cm long. Leaves imparipinnate with 5–9 pairs of leaflets; leaflets oblong, truncate-retuse at apex, 4–10 mm long, 2–3 mm wide. Flowers pink or pink-violet to cream, or blue-violet with white, in racemes subequalling the subtending leaf. Pods subsessile, curved linear, 20–35 mm long, 2–3 mm wide, glabrous and distinctly reticulate-rugose. *Plate 122.*

Habitat: Silty basins and plains, sometimes on somewhat saline soils. Occasional.

Northern Plains: 591(K), al-Batin, 16 km ENE Qulban Ibn Busayyis; 1557(BM), al-Batin/al-'Awja junction; 1649(BM), 2 km S Kuwait border in 47–14E; 4016, ad-Dibdibah, 27–56N, 45–35E; 7570, 22 km SW Hafar al-Batin; 7983, 24 km W al-Khafji; 8714, 20 km SW ar-Ruq'i wells; 8731, ar-Ruq' wells.

Northern Summan: 737, 33 km SW al-Qar'ah wells; 1362(BM), 1389(BM), 42 km W Qaryat al-'Ulya; 8589, 2.5 km NE Umm al-Hawshat, 27–03N, 47–17E.

Southern Summan: 2060, Jaww ad-Dukhan, 24–48N, 49–05E; 2097, 8 km N Harad Station.

Central Coastal Lowlands: 8106, 9 km SSE Hanidh.

4. Astragalus eremophilus Boiss.

Prostrate annual, branched from the base with stems up to c. 20 cm long, pubescent with both appressed and spreading hairs. Leaves 20 – 50 mm long, imparipinnate with (5)6 – 7 leaflet pairs; leaflets obovate to oblong, mostly obtuse and minutely retuse, 3 – 8 mm long, 2 – 6 mm wide, pubescent at least beneath. Flowers 1 – 3 in axillary racemes, somewhat inconspicuous, c. 5 mm long with cream to pinkish corolla. Pods linear, nearly terete, with spreading or appressed hairs or both, mostly curved nearly or fully to a half-circle, 20 – 30 mm long, 2 – 3 mm broad. *Plate 123.*

Habitat: Silty or somewhat sandy ground in Summan basins or wadi bottoms of the south. Locally frequent. Apparently a Sudanian associate and not to date recorded north of Wadi as-Sahba.

South Coastal Lowlands: 8041, Wadi as-Sahba, 23 – 56N, 49 – 15E.

Southern Summan: 8003(K), 24 – 00N, 49 – 00E; 8055, 24 – 06N, 48 – 48E.

5. Astragalus hauarensis Boiss.

A. gyzensis Del.

Procumbent to decumbent annual, branched from the base with stems 5 – 20 cm long, more or less pubescent. Leaves imparipinnate with 1 – 4, sometimes 5, leaflet pairs; leaflets oblong-obovate, obtuse and retuse, 4 – 15 mm long, 3 – 9 mm wide, appressed pubescent on both surfaces. Flowers pink or pink-violet, sometimes tinged with yellow, in racemes shorter than the subtending leaf. Pods subsessile, linear and strongly curved, pubescent, 20 – 30 mm long, c. 3 mm wide. *Plate 124.*

Our specimens show a considerable degree of variation in density of pubescence; two from the northern plains (No. 7991) are quite glabrous and attributable to var. *glaber* as described by Townsend (in Townsend and Guest 1974). Specimens from a population at Dhahran nearly all had leaflets in 4 or 5 pairs. Otherwise they are generally in 1 – 3 pairs, as is usual in the species.

Habitat: Sand or silt in a wide variety of terrain types. Perhaps the most common *Astragalus* of sandy habitats.

Northern Plains: 1696(BM), 28 – 41N, 47 – 35E; 1004, 999A, 12 km N Jabal al-'Amudah; 1260(BM), 15 km ENE Qaryat al-'Ulya; 1317, 8 km W Jabal Dab'; 1615(BM), 3 km ESE al-Batin/al-'Awja junction; 1652(BM), 2 km S Kuwait border in 47 – 14E; 1182(BM), Jabal an-Nu'ayriyah; 1705(BM), ash-Shaqq, 28 – 30N, 47 – 42E; 1723(BM), 40 km E Abraq al-Kabrit; 7984, 24 km W al-Khafji; 7989, 7991, 28 – 12N, 48 – 07E.

Northern Summan: 2809, 12 km ESE ash-Shumlul; 7529, 64 km E Umm 'Ushar.

Northern Dahna: 8563, E Dahna, 25 – 25N, 47 – 37E.

Central Coastal Lowlands: 182, 411, 1838(BM), 7756, 7885, 7887, Dhahran; 7957, Batn al-Faruq, 25 – 43N, 48 – 52E; 910, Mulayjah; 3297(BM), 3304(BM), 2 km W Sabkhat Umm al-Jimal.

Southern Summan: 2168, Railroad KP – 349; 2212, 26 km SSE al-Hunayy.

6. Astragalus kahiricus DC.

Procumbent or decumbent perennial, branched from the base with pubescent stems 15 – 50 cm long. Leaves 10 – 25 cm long, pinnate with 5 – 9 leaflet pairs; leaflets orbicular with apiculate apex, 10 – 25 mm in diameter, glabrous above but woolly-tomentose below. Flowers yellow, c. 25 mm long, subsessile in racemes about half the length of the subtending leaf; calyx white-lanate, inflated in fruit and enclosing the pod, 10 – 15 mm wide. *Plates 125, 126.*

Habitat: Silty soils of the northern plains and Summan. Infrequent except locally, as in the vicinity of Qaryat al-'Ulya.

Northern Plains: ar-Ruq'i (Dickson 1955, 21).

Northern Summan: 537(BM), 15 km NW Qaryat al-'Ulya.

Vernacular Names: UDHUN AL ḤIMĀR, 'donkey's ears' (Musil 1926, 356; 1928b, 348; 1927, 606).

7. Astragalus schimperi Boiss.

Prostrate to decumbent pubescent annual, branched at the base with stems 5 – 30 cm long. Leaves imparipinnate with leaflets in (4)5 – 8 pairs; leaflets elliptical, obtuse or acute, 5 – 10 mm long, 2 – 3 mm wide, appressed pubescent. Flowers often with pink standard and whitish wings and keel, in headlike, peduncled or subsessile inflorescences. Pods often in radiate clusters, linear, subterete, sessile, moderately curved or sometimes straight, bigibbous at base, pubescent with appressed or spreading hairs or both, 15 – 20 mm long, 3 – 4 mm wide. *Plate 127.*

No attempt has been made to segregate our specimens into var. *schimperi*, with distinct peduncles, and var. *subsessilis* Eig, with sessile or short-peduncled inflorescences. Both, and intermediate, forms are found. Forms with subsessile inflorescences may resemble large-fruited specimens of *A. tribuloides* rather closely. Separation by dimensions of the pod appears to be generally reliable.

Habitat: Sands and silty terrain. Frequent.

Northern Plains: 1608, 1622(BM), 3 km ESE al-Batin/al-'Awja junction; 1324, 8 km W Jabal Dab'; 1265(BM), 15 km ENE Qaryat al-'Ulya; 1485(BM), 50 km WNW Qaryat al-'Ulya; 1656(BM), 2 km S Kuwait border in 47 – 14E; 1695(BM), 28 – 41N, 47 – 35E; 1710, ash-Shaqq, 28 – 30N, 47 – 42E; 1005(BM), 12 km N Jabal al-'Amudah; 3065(BM), ash-Shaqq, 28 – 25N, 47 – 47E; 7968, Jabal al-'Amudah; 7987, 28 – 12N, 48 – 07E; 8730, ar-Ruq'i wells.

Northern Summan: 618(BM), 20 km N al-Lihabah; 8552, 10 km W Jabal Dawmat al-'Awdah, 27 – 11N, 48 – 03E.

Northern Dahna: 8567, E Dahna, 25 – 25N, 47 – 37E.

Central Coastal Lowlands: 1837(BM), 7755, Dhahran; 1750(BM), Abu Hadriyah; 3263(BM), 8 km W Qatif-Qaisumah Pipeline KP – 144; 3305(BM), 2 km W Sabkhat Umm al-Jimal.

Southern Summan: 8050, 24 – 06N, 48 – 48E.

8. Astragalus sieberi DC.

A. zubairensis Eig

Cushion-like, antrorsely pubescent perennial, 10 – 20 cm high, with lower leaves becoming spinescent with falling of the leaflets. Leaves linear, 8 – 11 cm long, with (10)15 – 22(26) pairs of ovate to suborbicular leaflets 3 – 6 mm long. Flowers bright sulfur yellow, about 20 cm long, short pedicelled. Pod 20 – 40 mm long, 4 – 7 mm wide, oblong-lanceolate, finely appressed pubescent, tapering at apex to a rigid sharp beak. *Plates 128, 129.*

Habitat: Silty basins and bottoms of larger wadis. Occasional.

Northern Plains: 681, Khabari Wadha; 3097(BM), al-Batin, 28 – 00N, 45 – 28E;

2973(BM), 28 – 49N, 46 – 58E; 3850, 55 km ESE al-Qaysumah; 8705, 62 km ENE al-Qaysumah.

Northern Summan: 347, 16 km N Harad; 2047(BM), Jaww ad-Dukhan, 24 – 48N, 49 – 05E; 2239(BM), 15 km WSW al-Hunayy; 7511, Khashm az-Zaynah; 7706, 23 – 52N, 49 – 11E.

South Coastal Lowlands: 349(K), 385, 1 km N al-Hasa cement plant.

Vernacular Names: MISHṬ ADH-DHĪB, 'wolf's comb' (Qahtan).

9. Astragalus spinosus (Forssk.) Muschl.

Ascending branched, extremely spiny shrub, 20 – 70 cm high, appressed pubescent on young parts. Apical leaves 3 – 5 cm long with 4 – 5 pairs of oblong-elliptic, mucronate leaflets 4 – 7 mm long, 2 – 3 mm wide, the rachises soon dropping the leaflets and becoming spines; leaves on older wood below much smaller, with approximate leaflets and not or hardly spinescent. Flowers usually solitary in the axils, about 20 mm long on 5 – 9 mm-long pedicels; corolla whitish tinged with pink, the conspicuous part of the flower being the inflated calyx which is 15 – 18 mm long, 10 – 13 mm wide, pinkish, becoming cream-white, with 30 – 40 obscure longitudinal nerves. Pod enclosed in the calyx, appressed pubescent, oblong, c. 8 mm long, 3 mm wide, with a beak c. 2 mm long. *Plate 130.*

Habitat: Rocky or silty ground, often in overgrazed areas. Occasional to locally frequent.

Northern Summan: 753, 33 km SW al-Qar'ah wells.

Southern Summan: 2203, 26 km SSW al-Hunayy.

South Coastal Lowlands: 381, 5 km S al-'Uyun, al-Hasa.

Vernacular Names and Uses: KIDAD (gen.; derived from the Arabic root verb *kadda*, one meaning of which is 'to comb' and perhaps referring to the spines), SHAWṬ (Harb), KUDAYYIDĀN (Qahtan). The extremely dense armament of this plant makes it a strong range increaser, and its presence often indicates overgrazing conditions. In times of extreme forage shortage, Bedouins reportedly burn off the spines and thus make it more palatable to livestock.

10. Astragalus tribuloides Del.

Prostrate to slightly ascending appressed pubescent annual, 3 – 10 cm high. Leaves 1 – 4 cm long, imparipinnate, with 5 – 8(10) pairs of acutish, elliptical leaflets 2 – 7 mm long, 1 – 2.5 mm wide. Flowers c. 5 mm long, with white to pink corolla, in sessile clusters of 2 – 6. Pods in sessile stellate clusters of (2)3 – 5(6), 2.5 – 6 mm long, 1.5 – 2.5 mm wide, oblong, obtusish, dilated at base, with appressed hairs or both spreading and appressed hairs. After poor rains, particularly on silt basins in the Summan, this plant may take extreme dwarf forms with leaflets hardly greater than 1 mm long.

Our specimens all appear to be attributable to var. *minutus* (Boiss.) Boiss., as described by Townsend (in Townsend and Guest 1974).

Habitat: Silty basins of the northern plains or the northern or southern Summan. Frequent.

Northern Plains: 600(K), al-Batin, 16 km ENE Qulban Ibn Busayyis; 879, 9 km S 'Ayn al-'Abd; 1525(BM), 30 km ESE al-Qaysumah; 1536(BM), 36 km NE Hafar al-Batin; 1575(BM), al-Batin/al-'Awja junction; 1630(BM), 2 km S Kuwait border in 47 – 06E; 1655(BM), 2 km S Kuwait border in 47 – 14E; 1283(BM), 15 km ENE Qaryat al-'Ulya; 7985, 24 km W al-Khafji.

Northern Summan: 617(BM), 20 km N al-Lihabah; 724, 750, 33 km SW al-Qar'ah wells; 2820, 3 km ESE ash-Shumlul; 8574, 12 km WNW Ma'qala.

Northern Dahna: 2322, 22 km ENE Rumah.

Southern Summan: 2167, Railroad KP – 349; 2234, 15 km WSW al-Hunayy; 8078, al-Ghawar, 24 – 33N, 49 – 06E.

Central Coastal Lowlands: 1771(BM), Abu Hadriyah.

Vernacular Names: ḤURBUTH (Al Murrah, Qahtan, Bani Hajir), QAFʿAʾ (az-Zafir).

● 15. Scorpiurus L.

Annual herbs with simple, entire leaves. Flowers papilionaceous, yellow to reddish, in racemes or umbellate inflorescences. Calyx cleft with 5 teeth, the upper 2 connate well above the base. Stamens diadelphous. Pod terete, indehiscent, more or less coiled, strongly constricted into several prickly-ribbed segments.

Scorpiurus muricatus L.

Ascending, several-branched annual, 8 – 25 cm high, somewhat pubescent particularly on younger parts. Leaves simple, entire, oblanceolate, long-tapering below to the stem, acute or rarely subacute, 3 – 12 cm long, to c. 2.3 cm wide. Flowers c. 8 mm long, on peduncles mostly exceeding the subtending leaf; corolla yellow. Pods usually coiled and strongly contorted, consisting of arched segments, each with a series of prickly, often blackish, raised longitudinal lines at sides and back.

Habitat: Strictly a weed. Rare.

Central Coastal Lowlands: 7416(BM), Dhahran, in a new hotel lawn.

● 16. Hippocrepis L.

Annual or perennial herbs or shrubs with imparipinnate leaves. Flowers papilionaceous, umbelled or axillary and subsessile, yellow or whitish. Calyx 5-toothed, the upper 2 teeth more or less connate. Stamens diadelphous. Pod usually strongly compressed, linear in outline, straight or curved, divided into horseshoe-shaped joints by a number of deep, rounded sinuses in one margin.

Key to the Species of Hippocrepis
1. Flowers and pods sessile in the axils . . . . 3. *H. unisiliquosa*
1. Flowers and pods on peduncles 0.5 cm long or longer
    2. Pods spiraled and contorted; margins of horseshoe-shaped sinuses
       extended into linear projecting horns . . . . 1. *H. bicontorta*
    2. Pods straight or slightly curved, not spiraled; margin ends of sinuses
       rounded, flush with margin of the pod . . . . 2. *H. constricta*

1. Hippocrepis bicontorta Lois.

Decumbent annual, 5 – 20 cm high, glabrescent or sparsely appressed-pubescent. Leaves imparipinnate with 3 – 5 pairs of leaflets; leaflets 8 – 15 mm long, 1 – 3 mm wide, linear or (more rarely) oblong, emarginate. Flowers umbelled, (2)3 – 4(5) together, on elongating peduncles sometimes exceeding the leaves; corolla yellow, c. 6 mm long, twice as long as the calyx. Pod contorted and coiled with horseshoe-shaped sinuses bordered by horn-like processes. *Plate 131, Map 8.13.*

Habitat: Shallow or deep sand. One of the most common annuals of the Central Coastal Lowlands.

Northern Plains: 869, 9 km S 'Ayn al-'Abd; 887, Jabal al-'Amudah; 1010, 12 km N Jabal al-'Amudah; 1194(BM), Jabal an-Nu'ayriyah; 1264, 15 km ENE Qaryat al-'Ulya; 8794, 30 km SW ar-Ruq'i Post, 28 – 54N, 46 – 27E.

Northern Summan: 1355(BM), 1375(BM), 42 km W Qaryat al-'Ulya; 7471, Mishash Ibn Jum'ah; 7482, Batn Sumlul, 26 – 26N, 48 – 15E.

Northern Dahna: 4032, 27 – 35N, 44 – 51E.

Central Coastal Lowlands: 32, 40, 88, 1056, 1839(BM), Dhahran; 1419(BM), 7 km NW Abu Hadriyah; 2721, 23 km N Jabal Fazran; 3180(BM), 9 km N Shadqam junction; 3292, 2 km W Sabkhat Umm al-Jimal.

South Coastal Lowlands: 1437(BM), 2 km E Jabal Ghuraymil.

Southern Summan: 2262, 29 km ENE al-Hunayy.

Vernacular Names: ḤURBUTH (Qahtan, Al Murrah), KHURAYMA' ('pierced-weed', N, Musil 1927, 605), QARNAH ('horn-wort', N, Musil 1927, 630; 1928b, 368), UMM AL-QURAYN (N, Dickson 1955, 50; Dickson explains this name, 'horn-weed', as likening the fruit to the Bedouin hair braids known as 'horns', but it perhaps as likely refers to the horn-like projections on the pod).

2. Hippocrepis constricta Kunze

Prostrate, sparsely pubescent to glabrescent annual, branched from base with stems up to c. 30 cm long. Leaves imparipinnate with 2 – 3 leaflet pairs; leaflets obovate to oblong, 3 – 8(15) mm long, 2 – 4(6) mm wide, with obtuse, entire or emarginate apices. Flowers 1 – 4 in peduncled capitate clusters; peduncles 5 – 10 mm long; corolla c. 5 mm long, cream to pale yellow with reddish veins. Fruiting peduncles growing to 15 – 20 mm; pods linear in outline, strongly compressed, straight or gently curved, shortly pubescent or glabrescent, 15 – 30 mm long, 2 – 4 mm wide with fine curved mucro at apex and c. 5 – 9 deep horseshoe-shaped sinuses opening to one edge, the pod disarticulating transversely when ripe into 5 – 9 segments. *Map 8.13.*

Habitat: Silty depressions in the Summan or wadi bottoms. Locally frequent.

South Coastal Lowlands: 8039, Wadi as-Sahba, 23 – 56N, 49 – 15E.

Southern Summan: 8009, 24 – 00N, 49 – 00E; 8057, 24 – 06N, 48 – 48E; 8073, Wadi as-Sahba, 24 – 12N, 48 – 38E.

● This species is distributed as a Sudanian vicariant of *H. bicontorta*, replacing it in areas south of southern Ghawar and particularly around Wadi as-Sahba, where it is usually associated with two other subtropical legumes, *Astragalus eremophilus* and *Lotononis platycarpa*.

3. Hippocrepis unisiliquosa L.

Glabrescent annual, 5 – 20 cm high. Leaves imparipinnate with 3 – 7 pairs of leaflets; leaflets cuneate to linear-oblong, 5 – 12 mm long, 1.5 – 4 mm wide, emarginate. Flowers sessile or (rarely) near-sessile in the axils, 1 – 2 together. Corolla yellow, c. 6 mm long, twice as long as the calyx. Pod straight or slightly curved, linear in outline, 20 – 40 mm long, 3.5 – 5 mm wide, with numerous horseshoe-shaped sinuses bordered by blunt processes not projecting outward. *Map 8.13.*

Habitat: So far known only from shallow sand in a rocky ravine leading to al-Batin. Rare.

Northern Plains: 3082(BM), al-Batin, 28 – 01N, 45 – 29E.

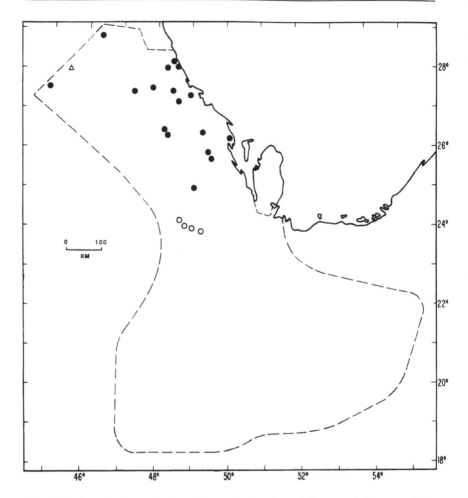

Map 8.13. Distribution as indicated by collection sites of *Hippocrepis bicontorta* (●), *Hippocrepis constricta* (○) and *Hippocrepis unisiliquosa* (△) in the coverage area of this flora (dashed line), eastern Saudi Arabia. Each symbol represents one collection record.

● This species, collected only once, appears to be very rare in our area. Dickson (1955) reported it from Kuwait as 'only found on Faylaka Island'. Good (1955) reported it from Bahrain, where it was less frequent than *H. bicontorta*. Batanouny (1981) says it is 'fairly common' in Qatar; it appears therefore to be more frequent in Sudanian territory, if not simply associated with disturbed ground.

● 17. Onobrychis Mill.

Annual or perennial herbs or shrubs with leaves imparipinnate and leaflets numerous. Flowers papilionaceous, in axillary racemes. Calyx campanulate with 5 subequal or unequal teeth. Stamens diadelphous. Pod flattened, more or less orbicular, indehiscent, prickly at crests and margins, 1–2(3)-seeded.

Onobrychis ptolemaica (Del.) DC.

Ascending perennial, branched from the base with stems 15–30 cm long, densely covered with fine white erect hairs; stems slightly striate-angled. Leaves imparipinnate, 10–15 cm long including the petiole, with 4–6 rather distant pairs of lanceolate to elliptical acute leaflets, 12–20 mm long, 4–8 mm wide, appressed-pubescent. Inflorescence a raceme 8–10 cm long including the peduncle; flowers short-pedicellate, 10–15 mm long; corolla cream with reddish veins. Fruiting racemes elongating to 25 cm; pods flat-orbicular, 8–13 mm in diameter, with short prickles at margins and faces, white silky-pubescent with dense, erect radiating hairs at margins. *Plate 132.*

Our specimens are assignable to subsp. *ptolemaica*, with wings of the corolla very short and with lamina nearly suppressed (Townsend in Townsend and Guest 1974).
Habitat: Silty soils of the northern plains and the Summan. Infrequent.
Northern Plains: 1623(BM), 3 km ESE al-Batin/al-'Awja junction; 2963(BM), 100 km ESE al-Qaysumah; 3100(BM), al-Batin, 28–00N, 45–28E; 7524, 20 km WNW al-Wari'ah; 7994, 50 km ESE as-Su'ayyirah; 8750, al-Batin, 28–50N, 46–24E.
Northern Summan: 708(BM), 33 km SW al-Qar'ah wells; 7549, 65 km ESE Umm 'Ushar; 8620, Rawdat Ma'qala, 26–26N, 47–27E.
Vernacular Names: TUMMAYR (Shammar).

● 18. Taverniera DC.

Shrubby perennials with leaves trifoliolate or unifoliolate. Flowers papilionaceous, pink or white, in axillary racemes. Calyx campanulate with teeth subequal or unequal. Stamens diadelphous. Pod compressed, indehiscent, constricted into 1 to several suborbicular chained joints, sometimes prickly, breaking into segments.

Taverniera spartea DC.

Erect, branched, silvery-grey canescent shrub, 0.5–1.5 m high. Leaves 1-foliolate or 3-foliolate, subsessile or on 2 mm-long petioles; leaflets obovate, obtuse, entire or faintly retuse at apex, 5–8 mm long, 3–5 mm wide, closely appressed-pubescent. Flowers solitary or paired in the axils, on peduncles c. 3 mm long; calyx 3–4 mm long, cleft c. half-way into narrowly triangular acute lobes; corolla c. 10 mm long, pink with red-mauve veins. Pod 5–17 mm long, 3–4 mm wide, strongly compressed, constricted laterally into (1)2–3(4) rounded joints, rounded-obtuse at apex or acute or with minute beak, faintly nerved transversely, densely covered with short, fine white hairs and longer, pinkish-yellow bristles. Seedlings of this plant have been observed to have both 1-foliolate and 3-foliolate leaves, with petioles equalling or exceeding the blade, the terminal leaflet to 15 × 20 mm. *Plate 133.*
Habitat: Coastal sites, both on low-lying saline soils and on higher stony ground. Occasional to locally common.
Central Coastal Lowlands: 125, Safwa; 162, 7088, al-Ajam; 915, Sayhat-Ras Tanura road junction; 1467(BM), 2 km E al-Ajam; 4074, 3 km W Shi'ab, 26–48N, 49–55E; al-'Aziziyah (Hill and Dickson, Dickson 1955, 89); 8445, near al-'Aba, 26–44N, 49–43E.
South Coastal Lowlands: 7459, al-'Udayd, 24–37N, 51–26E.
Vernacular Names: ALAL (al-Ajam villagers), 'ALQA (sic, Bani Hajir), ẒABYAH (Bani Khalid), NAZZA' (Al Wahibah).

● 19. **Alhagi** Adans.

Shrubby perennials with spinescent branches and simple entire leaves. Flowers axillary, solitary or racemed. Calyx campanulate, 5-toothed. Corolla papilionaceous, red to purple. Stamens diadelphous. Pod linear, terete, often constricted between the seeds.

Alhagi maurorum Medik.
A. graecorum Boiss.
A. mannifera Desv.
Erect to ascending many-branched shrublet, 0.3 – 1 m high, glabrous to short-pubescent, with lateral twigs becoming spines up to 5 cm long. Leaves simple, obovate, obtuse, shortly petiolate, up to 20 mm long. Flowers pink or reddish to purple, c. 1 – 1.2 cm long, in axillary racemes of about 3. Ovary pubescent, maturing to a glabrescent or shortly pubescent linear-cylindrical pod 10 – 30 mm long, 2 – 3 mm broad; pod usually somewhat curved and more or less constricted between the seeds.
Habitat: Known mainly as a weed or a plant of disturbed waste ground. Occasional.
Central Coastal Lowlands: Dhahran.
South Coastal Lowlands: 1908(BM), al-Mutayrifi; 2906(BM), al-'Arfaj farm, al-Hasa.
Vernacular Names and Uses: ʿAQŪL (gen.). The plant is recognized by some to have uses in traditional medicine, the roots being collected to make an infusion used for 'liver or kidney ailments'.

● 20. **Vicia** L.

Annual or perennial herbs with paripinnate leaves ending in simple or branched tendrils. Flowers papilionaceous, usually blue to purple, or yellowish, solitary or in axillary racemes or clusters. Calyx teeth equal or the upper pair shorter. Stamens diadelphous. Pod compressed, coriaceous, dehiscent; seeds mostly globose.

Key to the Species of Vicia
1. Pods on peduncles 0.8 – 4 cm long .... 1. *V. monantha*
1. Pods sessile or subsessile .... 2. *V. sativa*

1. Vicia monantha Retz.
Ascending or decumbent annual, 5 – 50 cm high, thinly pubescent at least in young parts. Leaves ending in fine tendrils, paripinnate with 4 – 8 leaflet pairs; leaflets linear-oblong, mostly obtuse, 8 – 20 mm long, 2 – 5 mm wide. Flowers 8 – 15 mm long, solitary or grouped 2 – 3, racemed on peduncles 1 – 3 cm long and shorter than the subtending leaf; corolla violet to blue. Pod strongly compressed, oblong, glabrous, light brown, finely veined, 20 – 30 mm long, 5 – 9 mm wide, tapering abruptly and obliquely at apex to a fine, acuminate beak.
Habitat: A weed of farms, gardens. Rare.
Central Coastal Lowlands: 7405(BM), Dhahran.

2. Vicia sativa L.
Sparsely pubescent decumbent or scrambling annual, 5 – 50 cm high, branching from near the base. Leaves paripinnate, ending in fine tendrils; leaflets in 3 – 5(8) pairs, oblong-linear, usually obtuse, mucronate, of much variable size, 3 – 25 mm long,

1 – 10 mm wide. Flowers solitary or paired, sessile or subsessile in the axils, 7 – 20 mm long; corolla blue-violet. Pods linear, compressed, glabrous or very finely pubescent, 20 – 35 mm long, 4 – 8 mm wide, acuminate at apex, ripening to a very dark shining brown. Seeds nearly spherical, 3 – 5 mm in diameter, smooth, grey-brown, mottled darker.

Habitat: A weed of farmlands. May occur rarely in desert at campsites where grain is spilled. Rare.

Central Coastal Lowlands: 8083(K), N Batn al-Faruq, 25 – 43N, 48 – 53E, Bedouin camp site.

● 21. Lathyrus L.

Annual or perennial herbs with paripinnate leaves usually ending in a tendril; leaflets few, or replaced by a pair of leaflike stipules. Flowers papilionaceous, solitary or in axillary racemes. Calyx teeth equal or unequal. Stamens diadelphous. Pod compressed, linear or oblong, dehiscing by 2 valves, with seeds usually numerous and globose.

Lathyrus aphaca L.

Glabrous herb with striate, angled stems. Leaves reduced to a naked tendril, twining at apex, but subtended at base by paired, leaflike, acute hastate stipules, 10 – 30 mm long, 6 – 20 mm wide. Flowers pale yellow, mostly solitary, on peduncles equalling or exceeding the stipules. Pod glabrous, linear-oblong, compressed, abruptly narrowed at apex to an acute beak.

Habitat: A weed of farm, garden. Rare.

Central Coastal Lowlands: 7411(BM), Dhahran.

● 22. Sophora L.

Trees, shrubs or perennial herbs with imparipinnate leaves. Flowers papilionaceous, panicled or racemed, white, yellow or blue. Calyx campanulate, with short subequal teeth, often gibbous above at the base. Filaments all free, or the upper 2 – 3 free and the others connate for up to one half their length. Pod more or less constricted between the seeds, indehiscent or late-dehiscing.

Sophora gibbosa (DC.) Yakovl.

Ammothamnus gibbosus (DC.) Boiss.

Erect shrublet, strongly woody at base, 30 – 70(90) cm high and silvery-grey canescent with dense, silky appressed hairs. Leaves imparipinnate, to c. 15 cm long with (4)6 – 9(10) pairs of leaflets, the rachises sometimes persistent after fall of the leaflets, indurating and pungent; stipules linear-subulate, 3 – 4 mm long. Leaflets obovate to cuneate or suborbicular, 5 – 15 mm long, 4 – 12 mm wide, rounded-entire or sometimes emarginate at apex. Flowers in 10 – 25 cm-long, c. 2.5 cm-wide terminal racemes; bracts linear-subulate, reaching about middle of the calyx; calyx c. 7 mm long, apressed-pubescent, gibbous above at base, with acute teeth ¼ to ⅓ as long as the tube, asymmetrically inserted on a pedicel 2 – 3 mm long; corolla about twice as long as the calyx, pale yellow, the standard often darkening with age from the apex; all petals appressed-pubescent at back. Upper 2 – 3 filaments free, the others connate for ⅓ to ½ their length; style with a tuft of hairs at the stigma; ovary densely appressed-pubescent. Pod strongly compressed, densely appressed-silky-pubescent, linear and 2 – 3 mm wide except where swollen around the seed, to c. 60 mm long (measured straightened), contorted or weakly coiled with age; only 1 or 2 seeds developing.

Habitat: Low red sand dunes or shallow red sand over limestone. Overall very rare but common and dominant at the only known record site.

This species, collected by the author in 1964 in the Sakaka basin, northwestern Arabia, is so far known in our eastern territory only at one locality on the western edge of the Northern Dahna. It grows there, spaced on the order of 5 – 12 m, in an almost pure stand of several square kilometers' area northwest of Umm al-Jamajim village. There is some indication that it may have taken dominance here from overgrazed *Rhanterium epapposum*. Around the fringes of this stand it is associated with *Rhanterium epapposum*, *Artemisia monosperma* and *Convolvulus oxyphyllus*.

Northern Dahna: 8824, 2 km NNW Umm al-Jamajim village, 26 – 53N, 45 – 19E; 8830, 3 km WNW Umm al-Jamajim village.

Vernacular Names and Uses: UMM QUṬAYNAH, 'cotton bush' (Mutayr); SHUHAYBĀ', 'grey-bush' (Mutayr). Herdsmen near our record locality say that the shrub is avoided by livestock. This is confirmed by our observation that it is invariably untouched even where palatable plants such as *Rhanterium* and *Panicum* at the same site are heavily cut back by grazing.

31. HALORAGACEAE

Herbs, generally aquatic, with opposite or whorled leaves. Flowers usually unisexual, actinomorphic, minute, with 2 – 4 sepals, 0 – 4 petals, 2 – 8 stamens, and ovary inferior, of 4 carpels. Fruit nutlike, a drupe, or a schizocarp.

● Myriophyllum L.

Monoecious perennial aquatic herbs with whorled, pinnate, filiform leaves; flowers whorled in the axils, the staminate above with 4-merous calyx and corolla and 8 stamens, the pistillate below with minute perianth and 4-celled ovary.

Myriophyllum spicatum L.

Glabrous aquatic perennial with creeping rhizome rooted on the bottom and ascending submerged shoots to 1 m or more long. Leaves whorled, 1 – 2 cm long, pinnatisect with filiform segments. Flowers whorled in interrupted bracteate spikes emergent above the water surface, the staminate ones above, with 8 stamens. Fruit ovoid, about 2 mm long, of 4 nutlets.

Habitat: Freshwater pools or canals.

Central or South Coastal Lowlands: Not seen by the author but reported as a weed, presumably in irrigation canals of al-Hasa or Qatif Oases, by Chaudhary, Parker and Kasasian (1981).

32. LYTHRACEAE

Herbs or shrubs with leaves usually opposite. Flowers bisexual, mostly actinomorphic, in cymes or panicles, sometimes solitary or clustered. Sepals usually 4 – 6(8), sometimes alternating with 4 – 6 lobes of an epicalyx. Petals usually 4 – 6(8), sometimes absent. Stamens 4 – 12, or numerous. Fruit a capsule.

Key to the Genera of Lythraceae
1. Herb with alternate leaves .... 1. *Lythrum*
1. Shrub or small tree with opposite leaves .... 2. *Lawsonia*

● 1. **Lythrum** L.

Herbs or small shrubs with leaves opposite, alternate, or both. Flowers purple or pink with a tubular calyx or hypanthium, the calyx segments alternating with epicalyx lobes. Petals 4 – 6; stamens 2 – 12. Fruit a 2-celled capsule with numerous seeds.

Lythrum hyssopifolia L.
Erect or ascending glabrous annual or perennial, to c. 50 cm high. Leaves alternate, or the lower sometimes opposite, 1 – 2.5 cm long, 0.2 – 0.8 cm wide, obtuse and oblong to linear-oblong below, linear-lanceolate and generally acute above. Flowers 4 – 8 mm long, subsessile and solitary in the axils. Calyx (with the epicalyx) 10 – 12-toothed; petals 6, pink-lilac; stamens 4 – 6, included. Capsule enclosed in the calyx, cylindrical.
Habitat: Wet or swampy ground in the oases. Infrequent.
South Coastal Lowlands: near Umm Sab' spring, al-Hasa (Dickson 1955, 65).

● 2. **Lawsonia** L.

Shrub or small tree with opposite leaves and flowers in paniculate cymes, or corymbose. Calyx campanulate, 4-lobed. Petals 4. Stamens 8(9), inserted in pairs. Ovary globose, 4-locular, with style persistent. Capsule globose, 2 – 4-celled; seeds angular.

Lawsonia inermis L.
L. alba Lam.
Shrub or small tree, to about 7 m high with branches sometimes somewhat spinescent. Leaves opposite, elliptical-oblanceolate, mostly acute, 15 – 35 mm long, 5 – 13 mm wide, sessile or subsessile. Flowers c. 10 mm across, in dense paniculate corymbs at ends of the branches; sepals 4, acute; petals 4, greenish yellow, wrinkled and reflexed, c. 4 – 5 mm long; stamens 8 (rarely 9), in 4 pairs. Fruit a globose, many-seeded capsule, 5 – 7 mm in diameter, ripening brown.
Habitat: Not infrequently cultivated as an ornamental and sometimes found apparently spontaneous on waste ground or along roadsides in oasis areas. Its self-propagating ability in our area, however, is questionable.
South Coastal Lowlands: 3943, 2.5 km ENE al-Hufuf.
Vernacular Names and Uses: ḤINNA' (gen.). The powdered leaves of this, the henna plant, have been used widely as a cosmetic dye by women, particularly of the village populations but also among the Bedouins. It is applied to the hands and feet as well as other parts of the body in decorative patterns, giving a characteristic brownish-orange color. Elderly men sometimes use it to dye their beards, and the villagers of al-Qatif and al-Hasa apply it decoratively to their white donkeys. Much of this dyestuff is probably imported rather than derived from local plants. Indian material is commonly used in the Eastern Province. Henna from Medina is also sold in local markets and is specially esteemed. Henna, as well as having these cosmetic uses, is believed to toughen the skin of the hands and feet and thus help protect against chafing and abrasions. The dye is used by mixing a paste of the ground leaves with plain water or with an

aqueous infusion of crushed dried limes; the latter is said to strengthen the coloring action. The paste is applied in patterns and left to 'set' for several hours or even overnight, until the paste dries. Recent practice is to add a small amount of motor gasoline to the paste, which reportedly intensifies the color.

33. SANTALACEAE

Partially parasitic herbs, shrubs, or trees with chlorophyll. Leaves generally alternate, entire. Flowers bisexual or unisexual with a uniseriate perianth usually with 4–5 lobes; stamens mostly 4–5; ovary inferior. Fruit indehiscent, 1-seeded.

● Thesium L.

Annual or perennial root parasites with alternate, narrowly linear leaves. Inflorescence a raceme or panicle of few-flowered cymes. Flowers bisexual, bracteate, with 4–5-merous perianth and 4–5 stamens. Fruit a 1-seeded nutlet with the perianth persistent above.

Thesium humile Vahl

Erect or ascending glabrous, pale green or yellow-green annual herb, 5–35 cm high, with numerous striate-angled stems. Leaves numerous, narrowly linear or filiform, 1.5–5 cm long, c. 1 mm wide. Flowers 1.5–2 mm long, rotate, c. 2.5 mm in diameter, subsessile in the upper axils and exceeded by subtending, leaflike bracts; perianth lobes whitish to greenish. Nutlet ovoid-ellipsoid, 2–3 mm long, with the persistent, inconspicuous perianth at apex. (Description largely following Townsend in Townsend and Guest 1980).

Habitat: A root parasite on cultivated plants, particularly cereals.

● Not seen or collected by the author but reported by Chaudhary and Zawawi (1983) as 'a cool-season weed causing heavy infestations in nurseries and vineyards around Al-Kharj', near the western edge of our territory. Chaudhary, Parker and Kasasian (1981) list it as a weed of the Eastern Province.

34. CYNOMORIACEAE

Parasitic, leafless, polygamodioecious herbs without chlorophyll. Flowers in a dense, clublike terminal spadix with perianth lobes 3–6 or 0; stamens 1. Fruit nutlike.

● Cynomorium L.

A monotypic genus with characters as for the family.

Cynomorium coccineum L.

Fleshy, reddish, club-shaped leafless perennial herb, to c. 30 cm high, parasitic on the roots of desert shrubs and visible above ground only during its spring flowering period. Stem simple, erect, cylindrical, succulent, c. 2 cm in diameter, with scales imbricated below, distant and deciduous above. Flowers c. 5 mm long, densely packed over the surface of the obtuse, very dark red terminal spadix, which is 10–20 cm long, 3–6 cm wide, with flat-topped peltate bracts scattered over the surface; perianth segments 1–5, with a single anther in the staminate and bisexual flowers. *Plate 134.*

● A highly specialized plant readily recognized by its unusual form and habit as a parasite on the roots of various shrub species. The subterranean connection with roots of the host may often be demonstrated by careful excavation. The flowering stems may emerge from the ground singly but more often are grouped several together. Rings or circular clumps of up to 40 flowering heads are sometimes seen, in which cases the spadices are generally smaller. The stems may often be found to be connected and branching below ground at a depth of c. 30 cm.

The spadix is mildly fetid and appears to attract flies, which may act as agents of pollination. The author, however, has never observed (or recognized) growth from seed. The flowering stems wither and decay by late spring, after which the plant aestivates as subterranean 'tubers', which have been observed in August at a depth of 20–30 cm on the roots of heavily parasitized *Zygophyllum qatarense* near Dhahran. These were subspheroid to ovoid or oblong bodies, 0.7–2.5 cm long, with the host root attached at one end. The opposite end closely resembles the rounded, club-like basal end of the spring season flowering stems. An iodine test of the sliced, white, potato-like interior of one of these tubers proved the presence of starch granules. Stored sugars were indicated by a sweetish taste, and a considerable amount of stored water was also present. With respect to life-form, it thus appears that *Cynomorium* should be classed as a geophyte, avoiding desert environmental stress by remaining below ground except for a relatively brief flowering period after winter rains. Propagation by seed would be thought to have adaptive value as a dispersal mechanism, and further study of this plant's life history would be of interest. It might be expected that seed germination would lead to points of host root infestation rather than the development of a plant at ground surface.

Habitat: Sandy ground, parasitic on the roots of saltbushes including *Haloxylon salicornicum* and *Haloxylon persicum;* also frequently seen on *Zygophyllum*. Locally frequent.

Central Coastal Lowlands: 8652, S of Dhahran, on *Zygophyllum qatarense* (tubers in liquid); numerous sight records on *Haloxylon* and *Zygophyllum*.

Rub' al-Khali: 7685, 7 km SW Shalfa, 21–51N, 49–39E, on *Haloxylon persicum*.

Vernacular Names and Uses: ṬURṬHŪTH (gen.). After good rains the young fleshy stems of this plant, particularly the below-ground parts, are sometimes sweet tasting and edible raw, with a pleasant crisp, succulent texture. The tubers are prepared simply by washing and peeling off the outer skin to expose the white fleshy interior. The sweet plants, however, are usually found growing, apparently randomly, among others which are bitter and astringent; finding edible individuals is thus largely a matter of chance. Bedouin elders of the Dhahran area say that Qatif oasis villagers in pre-oil days used to go out in the neighboring desert in spring and bring back donkey loads of *ṭurṭhūth* to sell in local markets as a seasonal delicacy or tonic. The crimson pigment of the plant's epidermis and flowering parts stains fast and was used in earlier times by Bedouin women of the Manasir tribe as a dyestuff. It was employed to dye clothing the dark red color known as *damī*, or 'blood-colored'. The phalloid form of this plant has led to a rich Bedouin repertoire of associated ribald names and folklore.

35. SALVADORACEAE

Shrubs or small trees with simple opposite leaves and unisexual or bisexual flowers. Flowers mostly with 4-merous perianth and 4 stamens; ovary superior. Fruit a 1-seeded berry or drupe.

● Salvadora L.

Glabrous shrubs or trees with leathery leaves and bisexual or unisexual flowers. Calyx 4-lobed; petals and stamens 4. Fruit a berry-like drupe.

Salvadora persica L.

Large shrub with opposite branches, 1–3.5 m high, sometimes growing as dense thickets on sand hummocks. Leaves elliptical to broadly lanceolate, entire, acute or obtuse, 2–5 cm long, 1–2 cm wide, on petioles c. 5 mm long. Flowers c. 3 mm long, in paniculate racemes, on 1–2 mm-long pedicels, with coffee-like odor. Fruits globular, fleshy, reddish, 3–6 mm in diameter.

Habitat: Coastal or inland sands. Rare.

Central Coastal Lowlands: 2959(BM), ar-Rakah, 26–49N, 48–41E.

South Coastal Lowlands: al-'Uqayr, Abu Zahmul.

This plant, not uncommon in the Sudanian floristic territory of southeastern, southern and western Arabia, is known from only three small stands in our area. Each of these is near an archaeological site, and they may well have been introduced by man in earlier times, accidentally, or deliberately as a source of *masāwīk* (see below). There is a well-established grove at al-'Uqayr, on a sand ridge near a ruined well overlooking the harbor and on the old trail inland to al-Hasa. A second stand, known in living memory but destroyed by mid-20th Century, was at the coastal settlement still known as ar-Rakah (from the name of the shrub), between al-Khubar and Dammam. The third is inland to the north, southwest of Thaj.

Vernacular Names and Uses: RĀK, ARĀK (gen.). This shrub is well known in all parts of Arabia as the source of the twigs used as *masāwīk* (sing. *miswāk*), the fibrous toothbrushes widely used by both townsmen and Bedouins. These often may be seen for sale in markets or by streetcorner vendors, in unfinished form as washed and dried straight lengths of branches or roots about 15–20 cm long and 1–1.5 cm in diameter. The user prepares these by cutting back the epidermis and cortex from the end for about 2 cm. The fibrous stele thus exposed, after soaking and preparatory rubbing, becomes the 'brush' at the end of the stick 'handle'. Natural compounds in the wood, which have a mildly astringent medicinal flavor, act as a dentrifice and mouth cleanser. This traditional use has been the subject of recent study by dental specialists; by 1988 at least one commercial toothpaste containing *Salvadora* was being marketed in Saudi Arabia under the tradename 'Fluoroswak'.

36. EUPHORBIACEAE

Herbs, shrubs or trees with alternate or opposite leaves. Flowers generally unisexual, the plant monoecious or dioecious. Sepals 5 or more; petals 5 or 0; stamens several or reduced to 1. Fruit generally a dehiscent trilocular capsule.

Key to the Genera of Euphorbiaceae

1. Flowers in cyathia, with a common involucre containing several
   1-stamened male flowers around a single female flower consisting of
   a stalked ovary; plants with milky juice (latex) .... 4. *Euphorbia*
1. Flowers not in cyathia and without common involucre; plants non-
   lactiferous

2. Prostrate herb; leaves 10 mm or less long .... 1. *Andrachne*
2. Erect herbs or shrubs; leaves much longer than 10 mm
   3. Coarse herbs or shrubs 20 – c.100 cm high with dense stellate
      indumentum; capsules scurfy .... 2. *Chrozophora*
   3. Glabrous coarse treelike herb, commonly over 2 m high, with palmately
      lobed leaves and prickly capsules .... 3. *Ricinus*

● 1. Andrachne L.

Monoecious, commonly prostrate, perennials with simple, alternate leaves. Calyx 5-parted; staminate flowers with 5 – 6 stamens; pistillate flowers with 3-celled ovary and 2 ovules per cell.

Andrachne telephioides L.

Glabrous, prostrate perennial, herbaceous but sometimes woody at base, glaucous and often bluish-green, with stems spreading from base. Leaves alternate to subopposite, obovate to suborbicular, mostly obtuse, 2 – 6 mm long, 2 – 5 mm wide, on petioles 1 – 4 mm long. Flowers 2 – 3 mm in diameter, 1 – 3 in the axils on pedicels 1 – 3 mm long; petals shorter than the sepals; pistillate flowers growing in fruit. Capsules depressed-globose, obscurely 3-lobed, 1 – 3 mm in diameter, with 6 triquetrous seeds.

Habitat: Silty basins. Locally frequent.

Northern Summan: 215, 25 km S Qaryat al-'Ulya; 809, 37 km NE Jarrarah; 2879, Rawdat Musay'id, 26 – 02N, 48 – 43E.

● 2. Chrozophora A. Juss.

Monoecious herbs or shrubs, often with dense stellate indumentum; leaves alternate. Staminate flowers racemose above the pistillate flowers, with 5 petals and 3 – 15 stamens; pistillate flowers with petals 5 or 0, sometimes with 5 staminodes. Capsule 3-seeded, often ovoid and scurfy.

Key to the Species of Chrozophora

1. Herb to c. 35 cm high; leaves grey-green, obtuse, ovate to rhombic,
   undulate-margined but not serrate .... 2. *C. tinctoria*
1. Suffrutescent plant, 40 – 100 cm high; leaves grey-white, mostly acute,
   lanceolate to triangular, the younger repand-serrate .... 1. *C. oblongifolia*

1. Chrozophora oblongifolia (Del.) A. Juss. ex Spreng.

Shrubby, many-branched, ascending grey to grey-white, stellate-canescent perennial, somewhat woody below, 40 – 100 cm high. Leaves generally acute, elliptical-lanceolate and tapering at base with repand-serrate margins, to broadly ovate-triangular with truncate base, 2 – 5(6) cm long, 1 – 3(5) cm wide, on petioles shorter than, equalling, or exceeding the blade. Flowers racemose, 3 – 5 mm long, the pistillate growing in fruit, with narrowly linear, stellate-ciliate calyx lobes and inconspicuous petals. Capsules depressed-globular with 3 rounded lobes, silvery-scurfy tubercled, 5 – 7 mm in diameter, on pedicels growing to as long as 20 mm. Seeds ovoid, brown, 3 – 4 mm long, with faintly tubercled or pitted surface.

Habitat: Silty inland soils, sometimes on disturbed ground. Infrequent.

Northern Plains: 3204, al-Batin, 35 km SW Hafar al-Batin.
Northern Summan: 509(BM), Wadi an-Najabiyah.
Southern Summan: 1869, 8 km S Jaww al-Hawiyah.
South Coastal Lowlands: 7806, 5 km N al-Mutayrifi, al-Hasa.
Vernacular Names and Uses: TANNŪM (gen., 'Ujman, al-Hasa villagers). *Chrozophora* was formerly used in dyeing and as a medicinal in Europe (Radcliffe-Smith in Townsend and Guest 1980, 322). Dickson (1955, 32) notes that an indigo blue juice comes from the root and that oil is obtained from the fruit. In the author's experience, the plant is traditionally associated with ink manufacture in earlier times in Arabia, although the method of preparation is not now generally known. Juice from the capsules, apparently with oxidation, frequently leaves blue-black or reddish stains on herbarium dryers.

2. Chrozophora tinctoria (L.) Raf.

Erect or ascending, green or grey-green annual with stellate indumentum, to c. 35 cm high. Leaves ovate or deltoid to rhomboid, mostly obtuse, with entire or gently undulate margins, 3 – 8 cm long, 2 – 8 cm wide, often nearly as broad as long, on petioles 1 – 7 cm long. Inflorescence racemose in the upper axils; flowers 2 – 3 mm long, subsessile or (particularly the pistillate) on elongating pedicels; sepals narrowly linear; petals minute. Capsule depressed-globose with 3 rounded lobes, silvery-scurfy or scaly, 6 – 8 mm in diameter. *Plate 135.*
Habitat: Sandy wadi bottoms or banks. Rather rare except locally.
Northern Summan: 3174(BM), 13 km SSW Hanidh.
Southern Summan: 8349, Wadi as-Sahba, 10 km W Harad.
Central Coastal Lowlands: 8126, 10 km SSE al-Wannan.
Vernacular Names and Uses: This is the tournsole of southern Europe and the Mediterranean, long used as a source of red and blue dyes. There do not appear to be any records of such use in eastern Arabia.

● 3. Ricinus L.

Large shrublike or treelike monoecious herb with alternate, palmately lobed leaves. Flowers apetalous, in paniculate racemes, the staminate with numerous connate and branched stamens, the pistillate with a 3-celled ovary developing into a 3-lobed capsule covered with thick prickles. A monotypic genus.

Ricinus communis L.

Erect, glabrous, shrublike herb up to 5 m tall, with stems branched above. Leaves alternate, peltate, 10 – 50 cm across, palmately 5 – 11-lobed, the lobes acute, ovate-lanceolate and dentate; petioles about equalling the blade. Flowers c. 2 cm wide, greenish-yellow, in axillary and terminal racemes; the pistillate producing ovoid capsules 1 – 3 cm long and covered with rather thick, at first soft, prickles or spines. Seeds highly variable in size and markings, oval, somewhat compressed, 0.5 – 2 cm long, smooth, greyish and variously dark-mottled.
Habitat: Apparently sometimes cultivated as an ornamental or a windbreak but often spontaneous around gardens or farms. Locally frequent.
Central Coastal Lowlands: sight records at Dhahran and in al-Qatif Oasis areas.
South Coastal Lowlands: 3922, near 'Ayn al-Khadud, al-Hasa.

Vernacular Names and Uses; KHAṢAB (al-Hasa farmer), KHIRWAʿ (Bani Hajir), a name also noted in Iraq (Radcliffe-Smith in Townsend and Guest 1980). This is the well-known castor oil plant of commerce, producing the oil used as a laxative in medicine and in industry as a machine lubricant. The oil is free of toxins, but all other parts are poisonous to man and animals to some degree. Cases of human poisoning are apparently rare in eastern Arabia, but because of its high toxicity and occurrence near inhabited areas, *Ricinus* must be respected as one of the more hazardous poisonous plants of our region. The seeds are particularly dangerous, containing a relatively high concentration of the toxin, ricin.

● 4. Euphorbia L.

Monoecious herbs, shrubs or trees with milky sap and alternate or opposite leaves. Flowers in a specialized inflorescence, a cyathium, consisting of several 1-stamened staminate flowers surrounding a single pistillate flower, all within a common cuplike, gland-bearing involucre. Ovary 3-celled; fruit a 3-seeded capsule; seeds often carunculate. The size, shape and surface characteristics of the seed are useful delimiting characters in this genus.

Key to the Species of Euphorbia

1. Leaves all opposite, stipulate; plants often prostrate
    2. Cyathia in peduncled head-like or umbel-like clusters; some leaves more than 1.5 cm long
        3. Cyathia in dense heads; capsules hairy .... 7. *E. hirta*
        3. Cyathia in umbel-like groups; capsules glabrous .... 8. *E. nutans*
    2. Cyathia solitary although sometimes crowded; all leaves less than 1.5 cm long
        4. Leaves ovate-subcordate; plant glabrous .... 11. *E. serpens*
        4. Leaves oblong or oblong-elliptical; plants pubescent at least in part
            5. Leaves 3–3.5 times as long as broad, each usually with a distinct central, elongate red-purple blotch above .... 12. *E. supina*
            5. Leaves mostly 2–2.5 times as long as broad, without distinct central blotch although sometimes reddish overall ... 4. *E. granulata*
1. Leaves alternate at least on lowest part of stem (although often opposite in upper parts), exstipulate; plants erect
    6. Some leaves dentate or serrate, at least in part
        7. Stems leaves obovate-spathulate; caruncle less than ¼ length of the seed; weed .... 6. *E. helioscopia*
        7. Stem leaves linear; caruncle more than ¼ length of the seed; desert plants
            8. Leaves of the inflorescence ovate and subauriculate at base, acuminate-caudate at apex; seeds with caruncle shorter than the seed or absent .... 10. *E. retusa*
            8. Leaves of the inflorescence linear-lanceolate, not caudate; seeds with caruncle about as long as seed or longer ...... 5. *E. grossheimii*
    6. Leaves all entire
        9. Leaves linear to long-lanceolate .... 2. *E. dracunculoides*

201

9. Leaves ovate
   10. Leaves cuspidate with a sharp apical point; faces of seed with series of
       transverse grooves .... 3. *E. falcata*
   10. Leaves obtuse; seeds longitudinally grooved, or rugulose, or smooth
       11. Lobes of capsule smooth-rounded; seeds irregularly rugulose;
           desert plant .... 1. *E. densa*
       11. Lobes of capsule longitudinally 2-ridged; seeds regularly grooved
           and pitted; weed .... 9. *E. peplus*

1. **Euphorbia densa** Schrenk

Ascending glabrous annual, many-branched, to about 12 cm high, sometimes broader than tall. Main stems with a few alternate leaves near base, but those above nearly all opposite, oblong-ovate or obovate, obtuse, 8 – 15 mm long, sessile above, sometimes short-petiolate below. Cyathia at or near the stem tips; glands oblong, each with 2 distinct horns, or horns 1 or nearly absent. Capsules 2 – 3 mm long, with lobes smooth-rounded at back. Seeds quadrangular, irregularly rugulose on faces, nearly smooth at angles, pale grey to whitish but often darker in channels between the rugulosities; caruncle distinct, rounded-yellowish, or indistinct, or absent entirely.

Habitat: Shallow sand, often over rocky ground. Frequent.

Northern Plains: 7963, Jabal al-'Amudah.

Northern Summan: 1346(BM), ash-Shayyit al-'Atshan, 42 km W Qaryat al-'Ulya.

Central Coastal Lowlands: 1783(BM), Abu Hadriyah; 8107, 9 km SSE Hanidh.

2. **Euphorbia dracunculoides** Lam.

Glabrous ascending to erect annual, sometimes perennating, with several to many stems branching from the base, to 25 cm high. Stem leaves linear or linear-lanceolate, sessile, mostly acute, 1 – 2(4) cm long, 1 – 2 mm wide, tending to be lanceolate above. Capsule c. 3.5 mm long, 3-lobed with lobes smooth-rounded at back or very faintly keeled. Seeds c. 2.2 mm long, cylindrical or ovoid-cylindrical, finely tuberculate-rugose, light grey-brown and often mottled darker. Caruncle apparently present in some specimens, absent in others.

Habitat: Silty basins of the Summan. Rare.

Northern Summan: 2885(BM), Rawdat Musay'id, 26 – 02N, 48 – 43E.

Southern Summan: 8051, 24 – 06N, 48 – 48E.

3. **Euphorbia falcata** L.

Glabrous, often much-branched erect annual up to c. 30 cm high. Lowest stem leaves obovate to oblanceolate, obtuse or acutish, entire, up to 2 cm long; upper leaves ovate or subovate, acute and usually cuspidate with a sharp, narrow point at apex 1 – 3 mm long, entire or weakly undulate, sometimes faintly denticulate under magnification. Capsules subovoid to oblong, glabrous, c. 2 – 2.3 mm long, with obtusely angled lobes. Seeds oblong, somewhat compressed and only weakly faceted, 1.5 – 1.8 mm long, reddish brown maturing to a pale grey, with 2 series of transverse grooves on each of the front and back sides.

Aspect of a small-leaved *E. peplus* but recognizable by the cuspidate leaves.

Habitat: A weed of farms and gardens.

● Not seen by the author in the Eastern Province but collected by S. A. Chaudhary near the western limits of our area at al-Kharj (RIY 8069, seen by the author) and likely to be found in the east as a weed.

4. Euphorbia granulata Forssk.

Prostrate, sparsely to densely pubescent annual or perennial herb, usually greyish-green with yellowish stems, sometimes greenish or reddish throughout, spreading from the base with many-branched stems to 20 cm or more long. Leaves opposite, mostly unequally oblong and obtuse, strongly asymmetrical at base, sometimes acutish, 3 – 6(8) mm long, 1.5 – 3 mm wide, entire, subsessile or shortly petiolate. Cyathia axillary; glands with minute whitish petaloid appendages; capsule 1.5 – 2 mm long, distinctly or weakly keeled at backs of lobes, of variable pubescence with appressed straight to spreading and interlaced hairs, or rarely glabrescent. Seeds 1 – 1.3 mm long, angled, grey to brown, ecarunculate, usually more or less rugulose or transversely wrinkled.

This species is quite variable in form, density of indumentum and overall coloration. Nearly all of our specimens exhibit pubescence in at least some parts; both the normal green and the red anthocyanin-pigmented forms have been collected. As a weed it tends to occupy habitats drier than those of *E. serpens*.

Habitat: Occurs both in undisturbed desert, usually on silty basins or shallow sands, and as a ruderal or weed. A red-pigmented form with appressed-hairy capsules is common in Dhahran in the edges of lawns and walks. Frequent.

Northern Plains: 673, Khabari Wadha.

Northern Summan: 3130(BM), 6 km SE ad-Dabtiyah; 3150(BM), 9 km SSW Hanidh; 3175(BM), 13 km SSW Hanidh.

Central Coastal Lowlands: 1749(BM), Abu Hadriyah; 3214, an-Nu'ayriyah-as-Saffaniyah junction; 3788(BM), 7762, Dhahran.

South Coastal Lowlands: 7842, 5 km SE al-Hufuf, 25 – 20N, 49 – 37E; Dhahran to al-Hufuf road (Dickson 1955, 40); 7808, adh-Dhulayqiyah, al-Hasa.

Southern Summan: 2247(BM), 26 km ENE al-Hunayy; 1130(BM), al-'Udayliyah; 1872, 30 km S Jaww al-Hawiyah; 8004, 24 – 00N, 49 – 00E; 8046, Wadi as-Sahba, 24 – 00N, 49 – 08E.

Vernacular Names: LABNAH (Al Murrah), ḤILLAB ('Ujman). Both names may be translated approximately as 'milkweed' and refer to the characteristic milky sap of the plant.

5. Euphorbia grossheimii Prokh.

E. isthmia V. Täckh.

Glabrous annual or perennial, branching from the base, 8 – 30 cm high. Cauline leaves linear to long-lanceolate, acute, entire or nearly so, sessile, 2 – 3(5) cm long, 3 – 6 mm wide; upper leaves linear to oblong, acute or obtuse, denticulate particularly near apex, 1 – 3 cm long. Cyathia in the upper axils; capsules c. 6 – 7 mm long with rounded, sometimes faintly keeled lobes. Seeds ovoid to spherical, 2.5 – 3 mm long (not including caruncle), smooth grey mottled darker, with a yellowish conical caruncle 3 – 4 mm long and longitudinally hollowed at back.

Habitat: Silty basins or plains. Rare.

Northern Plains: 1555(BM), al-Batin/al-'Awja junction.

6. Euphorbia helioscopia L.

Ascending to erect, glabrous or sparsely pubescent annual, simple or branched, 10 – 40 cm high. Cauline leaves obovate-spathulate, tapering to base, obtuse and serrulate at apex, 2 – 4 cm long; floral leaves ovate, obtuse to acute. Capsules 3 – 3.5 mm long, smooth with rounded lobes. Seeds c. 2 mm long, ovoid, dark brown, reticulate-pitted, with small flattish caruncle.

Habitat: A weed of gardens, farms. Infrequent.

South Coastal Lowlands: Chaudhary/Aramco H – 26, al-Hasa Oasis.

7. Euphorbia hirta L.

Erect or decumbent annual herb, 10 – 30 cm high; stems with spreading yellowish hairs. Leaves opposite with pairs rather distant below, ovate-lanceolate, acute, with oblique base, serrate at least on upper margins, mostly 1 – 3.5 cm long, 0.5 – 2 cm wide, very short-petiolate, finely appressed-pubescent at least on underside. Cyathia in dense, headlike, subglobose axillary clusters to c. 12 mm in diameter, with peduncles 5 – 10(15) mm long. Capsules c. 1.5 mm long, 1.5 mm broad, with somewhat angular lobes, appressed-pubescent. Seeds c. 0.8 mm long, quadrangular with facets faintly wrinkled transversely, pale reddish brown, ecarunculate.

Habitat: A weed of farms and gardens. Infrequent.

Central Coastal Lowlands: 8802, nursery 3 km E Dhahran. Also collected by S. A. Chaudhary, whose specimen from al-Qatif (RIY 8070, seen by this writer), provided the first record of the species in our territory.

8. Euphorbia nutans Lag.

Ascending, branching annual, glabrescent (as in ours) or sparingly pubescent, with stems up to 40(55) cm long. Leaves elliptic to lanceolate-linear, 1 – 3(3.5) cm long, 0.4 – 1 cm wide, obtuse to acutish, finely serrate at least on upper margins, more or less oblique at base, on petioles 1 – 2 mm long. Inflorescence branching, umbelliform, on common peduncles 3 – 12 mm long. Glands white to purplish. Capsule glabrous, c. 1.5 mm long and broad. Seed c. 1 mm long, brown to blackish, transversely wrinkled, ecarunculate.

Habitat: So far found, and only to be expected, as a weed of nurseries or other cultivated ground. Infrequent.

Central Coastal Lowlands: 8803, nursery 3 km E Dhahran.

9. Euphorbia peplus L.

Erect glabrous annual, simple or several-branched from the base, 10 – 40 cm high. Lower stem leaves obovate, entire, tapering at base, petiolate, 1 – 2 cm long; upper leaves ovate to rhomboid or triangular, entire, obtuse or acutish, sessile. Capsule c. 2 mm long with 2 longitudinal ridges at back of each lobe. Seeds 1 – 1.5 mm long, 6-sided, pale grey, with a longitudinal groove on each of the 2 inner facets and a longitudinal row of 3 – 4 pits on the 4 outer faces.

Habitat: A weed of farms and gardens. Frequent.

Central Coastal Lowlands: 3840, Sayhat.

South Coastal Lowlands: 'Ayn Najm, al-Hasa (Dickson 1955, 43); 7797, 5 km SE al-Hufuf, al-Hasa; 8197, near al-Qurayn, al-Hasa Oasis.

● The few records of *E. peplis* L. in our area are probably name confusions and attributable to this species.

10. **Eurphorbia retusa** Forssk.

E. cornuta Pers.

Erect glabrous, often reddish perennial herb, many stemmed from the base, often growing as dense rounded shrublets 10 – 50(60) cm high. Cauline leaves linear, sessile, entire or (especially near the apex) denticulate, acute or truncate-denticulate at apex, 1 – 3(5) cm long, 2 – 3 mm wide; floral leaves broadened and rounded at base, ovate with acuminate-caudate apex, denticulate. Glands with 2 entire or weakly lobed horns. Capsules 5 – 6 mm long with lobes smooth-rounded. Seeds ovoid, smooth, pale grey, c. 3 mm long (excluding caruncle), with a fluted caruncle 1 – 2 mm long or with caruncle sometimes absent.

Habitat: Silty basins. Infrequent.

Northern Summan: 789(BM), 142 km SW al-Qar'ah wells; 2750, Rawdat az-Zu'ayyini, 25 – 38N, 48 – 04E.

Southern Summan: 2067, Jaww ad-Dukhan, 24 – 48N, 49 – 05E; 472, 12 km NE Harad; 8350, 10 km W Harad.

Vernacular Names: 'IḌAT AL-ḤA'ISH (Qahtan), GHAZALAH (Dickson 1955, 40).

11. **Euphorbia serpens** Kunth

Prostrate, slender, glabrous annual; stems spreading, often with 2 – 4 minute, fleshy, conical protuberances on the underside of each node. Leaves opposite, ovate-subcordate, laterally entire, sometimes weakly retuse, 3 – 7 mm long, 2 – 6 mm wide, on petioles less than 1.5 mm long. Cyathia axillary; glands reddish with narrow white margins. Capsules c. 1.5 mm long and wide, glabrous. Seeds ecarunculate, c. 1 mm long, brownish, with smooth faces.

Like the following species this is an American weed, and was apparently introduced since about 1960 – 70. Like *E. granulata* (above), it is often seen around lawn and garden edges, but tends to occupy somewhat wetter sites. It may be recognized by its leaf form and wholly glabrous state.

Habitat: A town or farm weed of damp places, lawn or garden edges. Locally frequent.

Central Coastal Lowlands: 7762, 8392, 8833, Dhahran; 8399, Jubayl Industrial City, residential area.

12. **Euphorbia supina** Raf.

Prostrate, more or less villous annual, possibly sometimes perennating, with stems spreading up to c. 40 cm long. Leaves opposite, short-petiolate, oblong or elliptical-oblong, 4 – 10(15) mm long, 1.5 – 4 mm wide, sometimes acutish or weakly mucronate, entire or serrulate above, the rather narrow base nearly symmetrical or somewhat oblique, the blade usually with an elongate, central, red-purple blotch on the upper surface. Cyathia axillary, mostly crowded on lateral branchlets; glands with narrow white appendages. Capsules c. 1.5 mm long, 1.5 mm broad, angled at backs of the 3 lobes, usually with appressed whitish hairs. Seeds ecarunculate, 0.7 – 1 mm long, pale greyish-brown, quadrangular with more or less distinct transverse ridges on the facets.

A common weed of the eastern United States accidentally introduced at Jubayl around 1984, apparently with American landscape plants. It has so far been collected only as a walk-side weed in the Jubayl Industrial City residential area and in 1984 was apparently unknown elsewhere in Arabia. Its range may be increasing.

Habitat: A weed of disturbed ground around inhabited areas. Rare except locally.
Central Coastal Lowlands: 8398(K), Jubayl Industrial City, residential area.

37. RHAMNACEAE

Shrubs, trees or climbers, often spinescent. Leaves simple, 3 – 5-nerved. Flowers actinomorphic, bisexual or unisexual; sepals 4 – 5; petals 4 – 5, or 0; stamens 4 – 5. Ovary 2 – 4-locular, free or immersed in the disc; fruit a drupe or a capsule.

● Ziziphus Mill.

Shrubs or trees with both straight and hooked stipular spines usually present. Leaves alternate. Flowers bisexual with 5 sepals, 5 petals, and 5 stamens. Fruit a globular or ovoid drupe.

Key to the Species of Ziziphus
1. Tree with leaves mostly 3 – 6 cm long . . . . 2. *Z. spina-christi*
1. Shrub with leaves 0.8 – 2 cm long . . . . 1. *Z. nummularia*

1. Ziziphus nummularia (Burm. f.) Wight et Arn.
Ascending, many-branched spiny shrub, more or less rounded in outline and 1 – 3 m high, tomentose on younger parts, with somewhat zig-zag ultimate branchlets; stipular spines dimorphic, with 1 straight, to c. 1 cm long, the other hooked. Leaves ovate to ovate-orbicular, often mucronate, with 3 main nerves from the base, fine-tomentose especially below, 0.8 – 2 cm long, 0.5 – 1.8 cm wide, on petioles 2 – 4 mm long. Flowers axillary, greenish-yellow, 3 – 4 mm long, with a 10-lobed disc. Drupe globose, reddish, 7 – 8 mm in diameter, pedicelled. *Plate 136.*

Habitat: Silty basins of the Summan. Locally frequent. One of the characteristic species of the Summan, where it may form relatively dense stands in basins, often sheltering an undergrowth of delicate annuals.

Northern Summan: 504, Rawdat Ma'qala; 2746, 5 km SSW BM – 38, 25 – 20N, 48 – 14E.

Southern Summan: 8383, Darb Mazalij, 16 km WSW al-Hunayy.

Vernacular Names and Uses: SIDR (gen.). This shrub, although one of the few woody plants occurring in its rocky Summan habitat, is seldom used for firewood. The wood is tough, elastic and fibrous, and thus difficult to break. Dickson (1951, 652) says that chiefs of the Mutayr tribe used to guard *sidr* against firewood gathering and that the wood was used to make the hook-ended camel stick known as the *mish 'āb*. According to Musil (1928a, 416), tribes occupying the northern ranges of this shrub have a tradition that *sidr* thickets are haunted and that spirits (*jinn*) 'have their gardens' where the *sidr* grows. The shrubs were therefore not used as fuel.

2. Ziziphus spina-christi (L.) Willd.
Tree up to c. 12 m high, spiny or unarmed, glabrous or somewhat pubescent in young parts, with pale grey branchlets. Leaves ovate to oblong, obtuse or sometimes weakly retuse, entire or with crenulate margins, with 3 main nerves from the base, the blade 2.5 – 6 cm long, 1.5 – 4 cm wide, on petioles 0.5 – 1.7 cm long. Flowers c. 3 – 8 together in axillary cymes, yellowish-green, 4 – 6 mm in diameter, with calyx lobes acute; petals

obtuse, shorter than the sepals, inserted between the calyx lobes beneath the stamens. Drupe short-pedicelled, ovoid or globular, 0.8–1.5 cm in diameter, yellowish.

The majority of trees in northeastern Arabia have spines absent, or weak and deciduous, and may thus be attributed to var. *inermis* Boiss.

Habitat: Occurs wild in some parts of Arabia within Sudanian territory but in the Eastern Province known as a common cultivated tree with only one uncertain record of wild occurrence. Frequently planted around towns and villages.

Central Coastal Lowlands: 7004(BM), 6 km NW Ghunan. This is the only possible record known to the author of a wild occurrence of this species in our area. It was a small tree in open sandy desert distant from present habitation; it could have grown from a dropped seed or as a relict of a sand-drifted abandoned settlement.

Vernacular Names and Uses: SIDR (the tree, gen.), NABAQ, NABIQ (the fruit, gen.). An important cultivated tree and one of the few truly Arabian native tree species still grown in towns and villages along with the many exotics now introduced. The fruits are edible, and local gatherings are sold in village fruit markets. The *sidr* tree is mentioned several times in the *Qur'an*: in a description of paradise where is found 'the *sidr* without thorns' (LVI, 27), in a description of an impoverished garden (XXXIV, 15), and as a landmark in an earthly or heavenly landscape associated with a vision of the Prophet (LIII, 14, 16). J. M. Rodwell, in a footnote to his translation of the *Qur'an*, describes without source a legend of a *sidr* tree that has leaves inscribed with the names of all living persons, the fallen leaves of which foretell the deaths in the coming year of the souls represented by each. The story appears to have no basis in orthodox Islamic traditions and is not generally current in Arabia today.

● The dried, powdered leaves of *sidr* have long been used as a hairwash in eastern Arabia, and it is still sold for such use in the older markets. Mixed with water, the powder forms a paste that lathers and is reputed to both soften the hair and strengthen its roots. This application is noted also in Kuwait (Dickson, 1955). Carter (1917, 195) and Dickson (1955) report its use also in washing the dead. The author has experimented with this product as a shampoo and has found it very effective, even in rather hard water. The following notes are based on personal experience: Finely powdered leaves should be chosen in the market; flat leaf bits are more difficult to rinse out of the hair and provide less lather. A half cup of leaf powder well covered with warm water will begin to produce thin suds within a few minutes with little stirring. This mixture has a vegetable odor, but none is left in the hair. Two liberal applications, well rubbed in and rinsed between, leaves the hair soft and lustrous. It is superior to chemical shampoos in leaving the hair straight and more manageable.

38. RUTACEAE

Herbs, shrubs or trees with alternate or opposite, gland-dotted and often strong-odored leaves. Flowers generally bisexual, actinomorphic, with 4–5 sepals and 4–5 petals; stamens equal to or twice the petals in number. Ovary superior. Fruit fleshy, capsular, or a samara.

Key to the Genera of Rutaceae
1. Petals fimbriate; leaves pinnatipartite; weed .... 1. *Ruta*
1. Petals entire; leaves entire; desert plant .... 2. *Haplophyllum*

● 1. Ruta L.

Perennial, strong-odored herbs with alternate divided leaves. Flowers yellow with 4 sepals and 4 petals except in terminal flowers, which have 5 each. Stamens 8(10). Ovary and capsule 4–5-lobed.

Ruta chalepensis L.

Glabrous ascending perennial herb, sometimes somewhat woody below, 20–50 cm high. Leaves petiolate, or sessile above, 2–10 cm long, oblong in outline, once or more pinnately divided into unequal entire or somewhat dentate lanceolate segments up to 2 cm long. Inflorescence corymbose with cordate to ovate sessile bracts. Flowers c. 12 mm in diameter with yellow petals fringed above at margins and narrowing below to a claw. Capsule c. 7–8 mm long, divided about half-way to base into 4(5) acute lobes.

Habitat: not a native species and recorded by the author only once, as an escape or weed. Rare.

Central Coastal Lowlands: 3842, Sayhat, weed at a plant nursery.

Vernacular Names and Uses: This rue is commonly known in Arab countries as SADHAB, and that name was recorded by the author in the 'Asir highlands of southwestern Saudi Arabia, where the plant occurs as a ruderal.

● 2. Haplophyllum A. Juss.

Perennial, strong-odored herbs with alternate entire or divided leaves. Sepals and petals 5; stamens 10. Fruit a 5-celled dry capsule.

Haplophyllum tuberculatum (Forssk.) A. Juss.

H. arabicum Boiss.
H. longifolium Boiss.
H. obovatum (Hochst. ex Boiss.) Hand.-Mazz.

Erect or ascending pubescent perennial, usually several-branched from a somewhat woody base, 15–50 cm, or rarely approaching 100 cm high. Leaves sessile to short-petiolate, highly variable in shape and size, from suborbicular and a few mm across to obovate or narrowly linear, up to c. 3(5) cm long, dotted with glands which may be punctate, or craterlike with raised circular margin, often somewhat crisped-undulate or revolute at margins. Flowers c. 8 mm broad, bright sulfur yellow, in a terminal corymbose, flat-topped inflorescence. Petals 3–5 mm long, exceeding the sepals; stamens dilated and pubescent below. Capsule c. 4 mm long, glabrous or pubescent, with glands. *Plate 137.*

This plant's identity is confirmed immediately by the powerful skunklike odor emitted by a bit of crushed foliage.

Habitat: Occurs in a variety of shallow sand or silty habitats. Frequent to common.

Northern Plains: 855, 'Ayn al-'Abd; 902, Jabal al-'Amudah; 1692(BM), 28–41N, 47–35E; 831, 31 km E Qaryat as-Sufla.

Northern Summan: 3138(BM), 3139(BM), 6 km SE ad-Dabtiyah; 3145(BM), 9 km SSW Hanidh.

Central Coastal Lowlands: 131, 1033, Dhahran; Dickson 522(K), Dhahran (Burtt and Lewis 1954, 409); 1776(BM), Abu Hadriyah; Dickson 580(K), 607(K), W Abu Hadriyah (Burtt and Lewis 1954, 409).

Southern Summan: 1871, 25 km S Jaww al-Hawiyah; 2224, 5 km S al-Hunayy.

Vernacular Names and Uses: Some of the various vernacular names for *Haplophyllum* are as strong as its odor, the following having been collected by the author, loosely translated as indicated: MUSAYKAH ('muskweed', gen., the most commonly used name in eastern Arabia), ZUQAYQAH ('turdlet weed', Al Murrah), FURAYTHAH ('gutweed', Musil 1927, 599), ZIFRAH ('filthweed', Shammar). As noted by Dickson (1955, 48 – 49), the dried powdered leaves have been used by Bedouins in a poultice to treat scorpion stings.

39. ZYGOPHYLLACEAE

Shrubs or herbs with opposite or alternate stipulate leaves. Flowers generally bisexual and pentamerous, actinomorphic. Sepals and petals 4 – 5; stamens (4)8 – 10(15). Ovary 4 – 5-celled, often with angular or lobed carpels. Fruit a capsule or a schizocarp.

A family including important, drought-adapted components of the east Arabian desert vegetation, exhibiting a high degree of variability with a development of endemic species or forms in two genera, *Tribulus* and *Zygophyllum*.

Key to the Genera of Zygophyllaceae
1. Stipules spinescent, at least at their apices .... 2. *Fagonia*
1. Stipules unarmed
   2. Leaves pinnately compound with 4 or more leaflet pairs; hairy
     plants .... 5. *Tribulus*
   2. Leaves simple, entire or dissected, or compound with 1 – 3 leaflets;
     plants glabrous or fine-tomentose
      3. Leaves 3-foliolate; petals 0 .... 3. *Seetzenia*
      3. Leaves simple or 1 – 2 foliolate; petals present
        4. Leaves mostly opposite, modified to highly succulent terete,
          club-like or thickened structures .... 4. *Zygophyllum*
        4. Leaves alternate or clustered, flat, entire or much dissected
          5. Leaves entire; shrub, often with spinescent twigs ..... 6. *Nitraria*
          5. Leaves irregularly dissected with linear lobes; coarse, non-
            spinescent herb .... 1. *Peganum*

● 1. Peganum L.

Perennial herbs or shrubs with alternate, irregularly dissected leaves. Flowers bisexual with 5 sepals and 5 petals; stamens 12 – 15. Ovary with 3(4) carpels. Fruit a dehiscent or indehiscent capsule with numerous angular seeds.

Peganum harmala L.
Glabrous perennial herb to c. 50 cm high, somewhat woody at base. Leaves sessile, to 6(10) cm long, irregularly pinnatifid into linear-lanceolate, acute, entire lobes. Flowers in a terminal cymose or corymbose inflorescence, pedicellate; sepals narrowly linear, 10 – 20 mm long; petals about equalling the sepals, elliptical-lanceolate, white, sometimes streaked yellowish or green; stamens 12 – 15. Fruit depressed-globose, somewhat 3-lobed, glabrous, 6 – 10 mm in diameter.

Habitat: A weed or ruderal, nearly always on disturbed ground. Rare.

Central or South Coastal Lowlands: Not seen by the author in our area but recorded by Chaudhary, Parker and Kasasian (1981) as a weed of the Eastern Province. It is well known in central and northern Arabia and Iraq but rare in the east; Dickson (1955, 72) notes that it occurs in Kuwait but is rare there.

Vernacular Names: KHIYYAYS ('stinkweed', as-Sulubah), SHAJARAT AL-KHUNAYZIR ('piglet bush', north, Musil 1927, 621), ḤARMAL (origin of Linnaeus's specific epithet, northern Arab countries). The plant has a long history of uses in traditional medicine in some Arab lands. It is known to contain toxic alkaloids but is not, apparently, a serious problem to livestock.

● 2. Fagonia L.

Spiny perennial herbs or subshrubs with opposite, unifoliolate or trifoliolate leaves and 4 spiny stipules at each node. Flowers solitary, axillary, with 5 sepals, 5 pink to purple petals, and 10 stamens. Fruit a pyramidal capsule with the carpels forming 5 prominently angled lobes, often on a deflexed pedicel.

Key to the Species of Fagonia

1. Leaves all 3-foliolate (except sometimes at the distal nodes); central leaflet obovate . . . . 2. *F. glutinosa*
1. Leaves 1-foliolate at least in upper parts of the stem; leaflets all elliptical-lanceolate, or subovate
    2. Leaves 3-foliolate below, 1-foliolate in upper stems . . . . 1. *F. bruguieri*
    2. Leaves all 1-foliolate
        3. Middle and upper stems covered with minute glands, finely striate but terete
            4. Leaflets narrowly elliptical to lanceolate-linear, 2 – 3.5 mm wide; sepals persistent on capsules . . . . 3. *F. indica*
            4. Leaflets elliptical to subovate, 4 – 8(13) mm wide; sepals often deciduous in fruit . . . . 4. *F. ovalifolia*
        3. Stems glabrous or with only sparse, scattered glands, the middle and upper ones sulcate and hence quadrangular . . . . 5. *F. olivieri*

1. Fagonia bruguieri DC.

Procumbent, minutely glandular subshrub; stems sulcate, angled, many-branched, spreading horizontally from a woody base. Leaves 3-foliolate in lower stems, the lowest on petioles up to c. 5 mm long but becoming subsessile above; central leaflet larger, 4 – 10(12) mm long, linear-lanceolate to oblong-ovate, acute; lateral leaflets absent in upper parts of stem. Stipular spines 3 – 12 mm long. Flowers solitary, very fragrant, c. 10 – 15 mm across, on pedicels 1 – 4 mm long; petals pale pink to purple, or rarely white, twice as long as the sepals. Capsule pyramidal, 5-angled, 3 – 5 mm long, with a persistent style; fruiting pedicels growing, somewhat deflexed. *Plate 138, Map 8.14.*

One population in the Northern Summan (No. 8598, below) had flowers all white; elsewhere there is some evidence that the corollas become paler with age.

Habitat: Shallow sand or silt, often over a rocky substrate. Frequent.

Northern Plains: 1689(BM), 28 – 41N, 47 – 35E; 573, 105 km E Hafar al-Batin; 1149, Jabal al-Ba'al; 3113(BM), al-Batin, 28 – 00N, 45 – 28E; 3849, Qatif-Qaisumah Pipeline KP – 380; 4077, Jabal an-Nu'ayriyah.

Northern Summan: 7483, Batn Sumlul, 26 – 26N, 48 – 15E; 3915, 27 – 41N, 46 – 24E; 2852, 26 – 18N, 47 – 56E; 2758, BM – 30, 25 – 49N, 48 – 01E; 763, 69 km SW al-Qar'ah; 1377(BM), 42 km W Qaryat al-'Ulya; 8598, 17 km ESE Jabal Burmah, 26 – 41N, 47 – 45E; 8613, near Dahl Umm Hujul, 26 – 35N, 47 – 32E.

Central Coastal Lowlands: 555, Rish Ibn Hijlan; Dickson 555(K), Abu Hadriyah (Burtt and Lewis 1954, 394); 8639, 4 km SW Thaj.

South Coastal Lowlands: 551, 8 km N al-Mutayrifi, al-Hasa; 7909, Na'lat Shadqam, 25 – 43N, 49 – 29E; 8323, 18 km NW al-Hufuf.

Vernacular Names: JANBAH (Rawalah, gen. N, perhaps from the sense of the Arabic root meaning 'to avoid, shun' and referring to the very spiny nature of the plant), ḤULAYWAH ('sweetweed', referring to the sweet fragrance of the flowers, Bani Hajir), ḤULĀWÁ (a variant of the preceding name, Bani Hajir), DARMA' (Al Rashid), DURAYMA' (diminutive form of the preceding, Al Murrah, Bani Hajir), 'ALQÁ (Bani Hajir).

2. Fagonia glutinosa Del.

Prostrate perennial, glandular-viscid and often with adherent sand, often reddish-tinged, with stems spreading up to c. 20 cm from the base. Leaves on petioles 1 – 4 mm long, virtually all trifoliolate, only those at extreme stem apices sometimes with lateral leaflets absent; the central leaflet larger, obovate, rarely elliptical-oblanceolate, c. 4 – 10 mm long, 3 – 8 mm wide, with a sharp white mucro. Stipular spines shorter than or about equal to the petioles, greenish and somewhat soft except at the sharp white apex. Flowers solitary, 3 – 5 mm long, short-pedicellate, with mauve petals. Capsule 3 – 5 mm long, about as broad as long, glandular-viscid and often with larger, non-viscid whitish hairs; style persistent.

Some specimens from the southern Summan and adjoining southern regions have been seen to have unusually narrow leaflets, approaching those of *F. bruguieri.*

Habitat: Silts or shallow sands. Frequent.

Northern Plains: 3121(BM), tributary to al-Batin, 28 – 01N, 45 – 33E; 679, Khabari Wadha; 842, 9 km S 'Ayn al-'Abd; 1170(BM), Jabal al-Ba'al; 1688(BM), 28 – 41N, 47 – 35E.

Northern Summan: 620, 20 km N al-Lihabah; 1358(BM), 42 km W Qaryat al-'Ulya; 7544, 65 km ESE Umm 'Ushar; 3916, 27 – 41N, 46 – 24E.

Central Coastal Lowlands: 66, Thaj; 1780(BM), Dickson 554(K), Abu Hadriyah (Burtt and Lewis 1954, 395).

South Coastal Lowlands: 2100, 21 km SSE Harad; 2105, 49 km S Harad; 7701, 11 km S Wadi as-Sahba, 23 – 52N, 49 – 11E.

Southern Summan: 2223, 5 km S al-Hunayy; 2032, 44 km E Khurays; 2160, 23 – 35N, 48 – 39E.

Vernacular Names: UMM AT-TURAB (N, Musil 1927, 628, meaning 'earth-weed' and probably referring to the granules of soil almost always adherent on this viscid plant). Philby's record (1933) of *birkān* as a name for this plant is probably based on confusion with *Limeum arabicum,* which resembles this *Fagonia* in some respects and which is generally known by that name.

3. Fagonia indica Burm. f.

F. parviflora Boiss.

Minutely and densely glandular perennial, with several to many stems ascending from a more or less woody base. Stems finely striate but terete; only the highest internodes sulcate. Leaves all unifoliolate, the leaflets elliptical, or sometimes linear-lanceolate, mostly acute, mucronate, often 8 – 15 mm long; petioles variable, mostly 1 – 3 mm long. Stipular spines 5 – 20 mm long, shorter than or exceeding the leaves. Flowers solitary in the axils, fragrant, to c. 12 mm in diameter with pink to purple petals, on pedicels 3 – 8 mm long, the pedicels growing and strongly deflexed in fruit. Capsule 3 – 5 mm long, 2 – 5 mm wide, short-pubescent, with persistent style. *Map 8.14.*

Habitat: Shallow sands, often over limestone or gravels, both in desert and on disturbed ground along roadbanks. Basically a Sudanian species and most frequently seen in the southern half of our area. Locally frequent to common.

Central Coastal Lowlands: 518, 4081, 4083, Dhahran; 3791(BM), 5 km W Dhahran; 3212(BM), Jalmudah, N al-Jubayl.

South Coastal Lowlands: 1024, 23 – 36N, 49 – 34E.

Southern Dahna: 3821(BM), 24 – 26N, 48 – 05E.

Southern Summan: 7512, Khashm az-Zaynah.

Rub' al-Khali: 8037, al-Jawb, 23 – 07N, 49 – 53E; 7506, 13 km SE Niqa Fardan, 21 – 53N, 48 – 34E; 7682 (seedlings), 7 km SW Shalfa, 21 – 51N, 49 – 39E; 8425, near G – 1 airstrip, 20 – 44N, 55 – 09E.

Vernacular Names: DURAYMA' (Al Murrah), ḤULAYWAH (Bani Hajir, Al Murrah).

4. Fagonia ovalifolia Hadidi

Procumbent annual or perennial with branches spreading from base, densely glandular and sometimes viscid when young. Stipular spines 8 – 15 mm long. Leaves elliptical to subovate, mostly 10 – 20(27) mm long, 4 – 8(13) mm wide, often 3-nerved on lower face, on petioles mostly 2 – 4 mm long. Flowers c. 8 mm in diameter, rose, with sepals often deciduous in fruit. Capsules c. 4 mm long, 5 mm broad, on deflexed pedicels.

A species rather close to *F. indica* (above). Our specimens often, but by no means invariably, have the sepals deciduous in fruit.

Habitat: Sands, often over limestones. Fairly frequent in the southeast.

Central Coastal Lowlands: 1081(BM), 8136, Dhahran.

Rub' al-Khali: 7667, 23 km W Nadqan; 7692, al-Mulayhah al-Gharbiyah.

5. Fagonia olivieri DC.

Perennial, glabrous to very sparsely and minutely glandular, with a main stem ascending from a woody base and then often branching repeatedly in a horizontal plane. Lowest stems terete but the middle and upper internodes distinctly sulcate-quadrangular. Leaves all unifoliolate, the leaflet subsessile, elliptical-lanceolate, acute, mucronate, 5 – 12(15) mm long. Stipular spines 3 – 14 mm long, mostly shorter than the leaves but sometimes exceeding them. Flowers solitary in the axils, fragrant, c. 10 mm in diameter; sepals acute, c. half as long as the pink-mauve petals. Capsules c. 3 – 4 mm long and broad, short-pubescent, with persistent style, on deflexed pedicels c. 3 mm long.

Habitat: Shallow sand, often over limestone. Infrequent except in a few, generally coastal, areas.

Central Coastal Lowlands: 518(K), 1075(BM), 1817(BM), 4082, 8137, Dhahran; 520(K), Jabal al-Barri; 7712, Ras al-Ghar, 27 – 32N, 49 – 12E.

Southern Dahna: 2141, E edge of ad-Dahna, 23 – 07N.

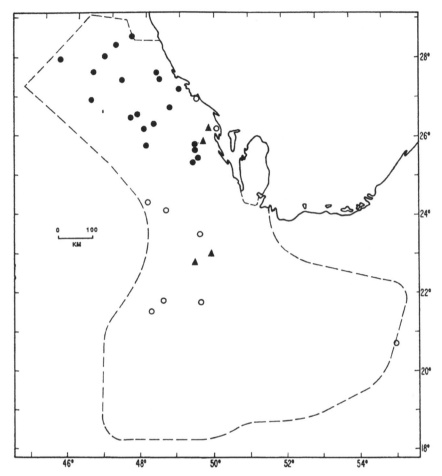

Map 8.14. Distribution as indicated by collection sites of *Fagonia bruguieri* (●), *Fagonia indica* (○), and *Fagonia ovalifolia* (▲) in the coverage area of this flora (dashed lines), eastern Saudi Arabia. Each symbol represents one collection record.

Southern Summan: 2226, 15 km WSW al-Hunayy.
Vernacular Names: DURAYMA' (Al Murrah), ḤULAYWAH (Al Murrah, Bani Hajir).

● 3. Seetzenia R. Br.

Perennial herb with opposite, 3-foliolate leaves. Stipules unarmed. Flowers bisexual, apetalous, with 5 sepals and 5 stamens. Ovary 5-celled. Fruit an ovoid capsule, dehiscing between the 5, 1-seeded carpels.

Seetzenia lanata (Willd.) Bullock
S. orientalis Decne.
Glabrous, light green prostrate perennial with finely glandular-muricate stems branching from base and above to c. 30 cm long. Leaves opposite, 3-foliolate, on petioles c. 2 mm

213

long; central leaflet obovate-deltoid, obtuse, 4 – 8 mm long, often with a fine white mucro; lateral leaflets oblique, subreniform, 3 – 6 mm long, mucronate. Flowers solitary in the axils on pedicels 3 – 5 mm long; sepals oblong, 2 – 3 mm long; petals 0. Capsule ellipsoid-ovoid, 6 – 10 mm long, 4 – 6 mm wide, yellowish with 5 green herbaceous longitudinal stripes, on growing pedicels to c. 7 – 10 mm long, at length semi-reflexed at insertion with the pedicel. Seeds dark brown to black, somewhat compressed, elliptical, 4 – 5 mm long, 2 – 3 mm wide, covered in a glue-like exudate remaining thick-viscous well after drying of the capsule. *Plate 139.*

Habitat: Silty floors of basins or wadi bottoms. Locally frequent.

South Coastal Lowlands: 383, 1 km N cement plant, al-Hasa; 463(BM), al-Mustannah, 60 km SSE al-Hufuf; 8040, Wadi as-Sahba, 23 – 56N, 49 – 15E; 8327, 25 km NW al-Hufuf.

Southern Summan: 2146(BM), E edge of ad-Dahna, 23 – 10N; 474(BM), 12 km NE Harad.

● 4. Zygophyllum L.

Herbs or shrubs with opposite simple or pinnate, often succulent leaves. Flowers bisexual, 5-merous. Stamens 5 – 10, with a scale-like appendage near the base. Fruit a rounded, angled, or winged capsule, fleshy to maturity, dehiscing longitudinally between the carpels when dry.

Fully mature fruits are generally required for species identifications of *Zygophyllum*, and this stage is generally marked in our plants by a color change in the living capsules from green to yellowish. Leaf characteristics vary according to age and growth conditions. Seedlings and young plants of normally unifoliolate species are often predominantly bifoliolate. The same phenomenon may be seen sometimes in lush growths of normally unifoliolate plants under good soil moisture conditions. Ismail (1983) described seasonal changes in the proportion of bifoliolate leaves in *Z. qatarense* and measured associated changes in transpiration rates and thus water conservation efficiency.

Key to the Species of Zygophyllum
1. Leaves sessile, simple; low herb .... 1. *Z. simplex*
1. Leaves with succulent petiole, 1- or 2-foliolate; shrubs
    2. Leaves all or predominantly 1-foliolate, glabrous or glabrescent when mature
        3. Capsules cylindrical with truncate apex, 5 – 10 mm long . 2. *Z. qatarense*
        3. Capsules clavate to ellipsoid, rounded at apex, (10)12 – 25 mm long .... 3. *Z. mandavillei*
    2. Leaves all or mostly 2-foliolate, more or less finely grey-tomentose .... 4. *Z. migahidii*

1. Zygophyllum simplex L.

Dense, many-branched, succulent procumbent summer annual or perennial herb, spreading to c. 30 cm across. Leaves simple, sessile, succulent, ovoid to cylindrical, obtuse, 3 – 15(20) mm long, 2 – 3 mm wide. Flowers mostly solitary in the axils, short-pedicelled, c. 4 – 5 mm in diameter, with yellow petals and 10 stamens. Capsules obovoid to subglobose, becoming somewhat 5-angled or keeled, 2 – 3 mm long, short-pedicelled. *Map 8.15.*

Habitat: Silty soil or shallow sand over silt. Infrequent except locally. A Sudanian species found mainly in the southern parts of our area.

Central Coastal Lowlands: 8100, 'Uray'irah; sight records at Dhahran, where it appeared for the first time in 1982, possibly associated with imports of clay soil from the al-Hasa area.

South Coastal Lowlands: 372(K), Jabal Sha'ban, al-Hasa; 511, al-'Uqayr; 7912, Na'lat Shadqam, 25–43N, 49–29E; 3179(BM), 9 km N Shadqam-'Ayn Dar junction; 4134, Ras al-Qurayyah; 1873, 24 km N Harad; 2121, 3 km SW al-Jawamir, Yabrin; 3813, 7 km SE al-Hufuf; 3837, 3838, 3839, Wadi as-Sahba at Harad Agricultural Project.

Vernacular Names: QARMAL (gen.).

2. Zygophyllum qatarense Hadidi
Z. coccineum auct. non L.

Ascending, many-branched, highly succulent shrublet to c. 75 cm high, with glabrous to sparsely tomentose foliage sometimes turning yellow or reddish. Leaves 1-foliolate, or sometimes 2-foliolate in seedlings or young plants, or sometimes in older shrubs during the rainy season; leaflets succulent, terete, cylindrical to ovoid or even globose, sometimes somewhat flattened, 3–8 mm long, on succulent petioles about equalling or exceeding the leaflet. Flowers solitary on 2–3 mm-long pedicels; petals whitish, c. 5 mm long; stamens 10. Capsules cylindrical or obscurely obconical, glabrous, weakly 5-angled, truncate at apex, 4–8(10) mm long, 3–5(6) mm wide, on pedicels 2–3 mm long. *Plate 142, Map 8.15.*

A rather variable species that was often attributed incorrectly to *Z. coccineum* L. in earlier literature on our area. It is the characteristic *Zygophyllum* of Gulf coastal districts and tends to develop smaller, nearly spherical leaflets in shoreside habitats. The foliage sometimes turns red to dark purple. Ismail (1983) provides useful data on leaf morphology and transpiration rates.

Habitat: Sands, generally shallow and overlying saline ground. Common.

Central Coastal Lowlands: 233, 7759, Dhahran; 7496(BM), 7497(BM), 8 km S Dhahran; 7492(BM), Safwa; 7049, Abu 'Ali Island.

South Coastal Lowlands: 1944, 28 km S al-'Uqayr; 7460(BM), al-'Udayd; 8289, 6 km N al-Mutayrifi; 8319, 5 km SE al-Hufuf; 8329, 25 km NW al-Hufuf.

Vernacular Names and Uses: HARM (gen.). A useful summer grazing plant for camels, and those animals accustomed to its salinity and purgative effects are called *hawārim* (sing. *hārimah*), a term derived from the plant name. Excessive grazing on it is considered unhealthy. The author has experimentally produced useful amounts of potable water by distillation of the crushed succulent foliage of this plant in a simple solar ground still (Mandaville 1972).

3. Zygophyllum mandavillei Hadidi

Ascending, many-branched, highly succulent shrub, glabrous except sometimes on youngest growth where fine-tomentose, to c. 80 cm high. Younger branches smooth, glossy, light yellow, swollen at the nodes. Leaves predominantly 1-foliolate but sometimes 2-foliolate in seedlings or fresh growth under good moisture conditions; blades subcylindrical, highly succulent, obtuse, 4–15(18) mm long, 2–3 mm wide, on a succulent cylindrical petiole of about equal or sometimes greater length. Flowers c. 4.5 mm long, solitary, on pedicels c. 3 mm long; sepals obtuse, hooded, slightly shorter than the white, spathulate, 4 – 5 mm-long petals; stamens 10(11), c. 3 mm long, with winged appendage on the filament for c. half its length. Capsules clavate, circular in cross section, rounded at apex, or rarely subtruncate, (10)12 – 20(22) mm long, 3 – 7 mm wide, on pedicels 3 – 5 mm long. *Plate 140, Map 8.15.*

● The specimen figured with El Hadidi's (1977) description of this plant is somewhat atypical in showing more 2-foliolate than 1-foliolate leaves. The proportion of bifoliolate leaves probably changes seasonally and with varying moisture conditions, as in *Z. qatarense* (Ismail 1983); much of the time, however, the species is predominantly or entirely unifoliolate.

The fruit form in typical specimens of this species is certainly distinct from that of *Z. qatarense*. In the author's experience however, intermediates may be found, and the apparent presence of these along with typical plants suggests that hybridization could be occurring. Further investigations along points of contact of the two species would be of interest. El Hadidi considers both species to be distinct from Schweinfurth's *Z. hamiense.*

Specimens of what appears to be *Z. mandavillei* collected by the author outside the western edge of the Rub' al-Khali at al-Aflaj are notable in having unusually thick fruits and distinctly greyish tomentose leaves. No. 7759 is an anomalous plant from Dhahran with very narrow leaves and fruits; the capsules are linear, somewhat angular in section, 12 – 16 mm long and only 1.5 mm wide with a more or less truncate apex and persistent filiform style. In dimensions it would belong to this species although the truncate fruit apex suggests a form of *Z. qatarense.*

Habitat: Sands; often in the northern and eastern Rub' al-Khali concentrated on the lower edges of large dunes overlying gravels or saline flats. Locally common or dominant.

Central Coastal Lowlands: 7447(BM), 5 km W Dhahran; 7759, Dhahran.

South Coastal Lowlands: 8415(BM), 32 km NW Nadqan.

Rub' al-Khali: 2892(holotype BM, CAI), Camp Shaybah 9, 22 – 32N, 54 – 03E; 2891(BM), 7859, Camp Ramlah 1, 22 – 10N, 54 – 21E; 3999(BM), Shaybah Camp; 4006(BM), Camp S – 3, 21 – 38N, 53 – 21E; 7388, Camp Tumaysha 1, 21 – 09N, 54 – 52E; 7029, 21 – 29N, 55 – 24E; 7051, Rawakib 1 Camp; 7055, 23 – 07N, 50 – 56E; 7085, 22 – 15N, 54 – 16E; 7607, 5 km W 'Amad 2 Camp; 7609, 7611, al-Jawb, 23 – 10N, 49 – 52E; 7668, 22 km N Nadqan, 23 – 08N, 49 – 53E; 7683, al-Mulayhah al-Gharbiyah, 22 – 53N, 49 – 29E; 7882, 21 – 36N, 54 – 46E; 8034, al-Jawb, 23 – 07N, 49 – 53E; 8423, ES-562 airstrip, 21 – 45N, 55 – 31E.

Vernacular Names and Uses: HARM (Al Murrah, Al Rashid, gen.). A useful grazing plant for camels accustomed to feeding on it, but generally considered unhealthy for prolonged grazing unless other forage plants are available to temper its salinity and purgative effects.

4. Zygophyllum migahidii Hadidi

Ascending, much-branched succulent shrublet to c. 75 cm high, with greyish-tomentose foliage, sometimes becoming glabrescent with age. Leaves all or predominantly 2-foliolate; leaflets succulent, cylindrical or obscurely compressed, to ovoid, 2 – 15 mm long, on a succulent petiole about equal to or distinctly longer than the leaflets. Flowers all solitary, or rarely 2 at some nodes, 4 – 5 mm long, on 2 – 12(15) mm-long pedicels; petals white or yellowish. Capsule cylindrical to somewhat obconical, 5-angled, truncate or sometimes somewhat retuse and weakly lobed at apex, 8 – 13 mm long. *Plate 141, Map 8.15.*

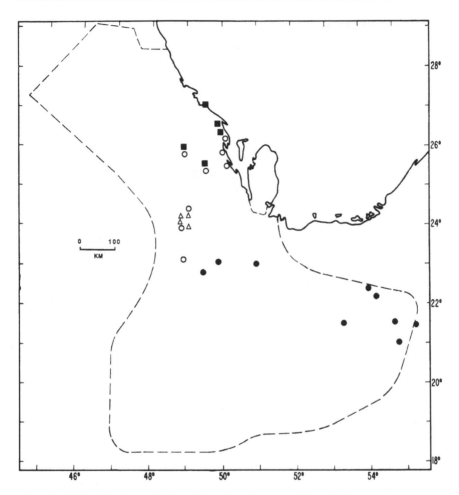

Map 8.15. Distribution as indicated by collection sites of *Zygophyllum qatarense* (■), *Zygophyllum simplex* (○), *Zygophyllum migahidii* (△), and *Zygophyllum mandavillei* (●) in the coverage area of this flora (dashed line), eastern Saudi Arabia. Each symbol represents one collection record.

● This species is closely related to *Z. propinquum* Decne., differing from it in having solitary rather than clustered flowers and fruits. Our specimens from around 'Uray'irah and Hanidh have smaller leaflets and a rather high proportion of unifoliolate leaves.
Habitat: Silts or sandy silts of wadi bottoms or alluvial plains; may be found on disturbed ground along roadsides. Infrequent except locally.
Central Coastal Lowlands: 7440(BM), 5 km SE 'Uray'irah; 7441(BM), 7442(BM), 'Uray'irah.
Northern Summan: 7445(BM), Hanidh.
Southern Summan: 7015, Wadi as-Sahba, 24–11N, 48–49E; 7490(BM), Wadi as-Sahba at Harad Agricultural Project; 8077, Mishash al-'Ashawi, 24–17N, 48–53E; 8369, al-Ghawar, 28 km N Harad; 8367, N bank Wadi as-Sahba in 48–50E; 8696, 23–53N, 48–50E.

217

● 5. Tribulus L.

Perennial or annual herbs, sometimes frutescent, with opposite, pinnate leaves and one leaf of each pair often reduced in size. Flowers bisexual, solitary in the axil of the reduced member of the leaf pair. Sepals and petals 5; stamens (5 – 8)10. Ovary 5-loculed. Fruit dry, woody, with carpels often winged or spinescent, dehiscing between the 4 – 5 carpels.

The delimitation and nomenclature of Arabian species in this genus have long been complicated by the high degree of variability in the taxa concerned. These troubles have been compounded by confusion in the many names that have been given to multiple forms that may be conspecific. Our plants with winged carpels (Section *Alata* Hadidi) fall into two main groups: one with small, mostly pentandrous flowers, the other with flowers much larger and decandrous. These two forms are clearly distinct, but difficulties arise in attempts to subdivide them. Some of these problems are indicated in notes with our species or species groups below. Better taxonomic solutions may require multi-disciplinary, including cytogenetic, techniques.

Key to the Species of Tribulus
1. Fruit carpels with terete, tapering spines .... 1. *T. terrestris*
1. Fruit carpels spineless, but furnished with dentate or subentire wings, or rarely unappendaged.
    2. Flowers 6 – 12 mm in diameter; stamens 5(6) .... 2. *T. pentandrus* agg.
    2. Flowers 15 – 40 mm in diameter; stamens (8)10 .... 3. *T. arabicus* s.l.

1. Tribulus terrestris L.
Prostrate annual with stems spreading from the base up to c. 100 cm long; stems sparsely tomentose with scattered longer white, erect hairs. Leaves (in ours) mostly 20 – 40 mm long, with one of each pair reduced; leaflets in 4 – 7(8) pairs, oblong, appressed pubescent below, nearly glabrous above except along the midrib, 4 – 8(12) mm long, 2 – 4 mm wide. Flowers solitary; petals yellow, 4 – 6 mm long, exceeding the sepals. Fruiting pedicels 4 – 10 mm long. Fruits globose, c. 5 – 7 mm in diameter excluding the spines, furnished with a pair of diverging terete, 3 – 6 mm-long spines on each carpel. **Habitat:** Usually a weed or ruderal but occasionally found in desert around Bedouin camping grounds. Infrequent.
Northern Plains: 599(BM), al-Batin, 16 km ENE Qulban Ibn Busayyis; 3198, al-Batin, 5 km NE Dhabhah; 7450, Qaisumah Pump Station.
Northern Summan: 3151(BM), 9 km SSW Hanidh; 3161(BM), 11 km SSW Hanidh.
South Coastal Lowlands: 1862(BM), Abqaiq; 7795, adh-Dhulayqiyah, al-Hasa.
Vernacular Names: SHIRSHIR ('cutweed', gen.), SHARSHĪR (Rawalah), ḌURAYSAH ('toothweed'), QAṬB (Rawalah), BAQL (Qahtan).

2. Tribulus pentandrus Forssk. agg.
Prostrate perennial, with stems up to c. 50 cm long spreading from the base; stems more or less tomentose and often with some erect pubescence. Leaves appressed-pubescent, (10)15 – 30(40) mm long, with one of each pair smaller; leaflets in (4)5 – 7(8) pairs, oblong, mostly acute, 5 – 9(11) mm long, 2 – 4(5) mm broad. Flowers solitary, 6 – 10(12) mm in diameter; petals pale yellow, obovate and sometimes indistinctly toothed at apex, c. 4 mm long, equalling or somewhat exceeding the lanceolate sepals;

stamens 5(6); ovary densely hirsute. Flowering and fruiting pedicels 5 – 10(12) mm long. Fruits globular, more or less hirsute between the wings, 8 – 11(13) mm in diameter, the margins of the carpels furnished with 2 – 4 mm-broad, dentate, finely tomentose or glabrescent longitudinal wings, or wings sometimes nearly absent or reduced to a few teeth. *Plate 145.*

● It would appear unwise, pending a badly needed full regional revision of *Tribulus*, to attempt assigning a specific name to this taxon. Dr H. Hosni at the University of Cairo, one of the most recent researchers to work on the genus, calls our Dhahran specimens *T. megistopteris* Kralik and our inland plants *T. macropterus* Boiss. (personal communication). The first differ from the second in being overall less pubescent and in having carpel backs shorter and smaller overall in relation to the wings; the wings thus tend to surround the carpel entirely rather than just border it laterally. Also, the fruit as a whole tends to be evenly spheroid rather than oblately so as in the inland plants. These differences, however, seem to the present writer to be ones of degree within a single polymorphic species. The informal group name adopted here is based on Forsskal's early epithet, *sensu* Hadidi (1978), who has examined the type and reports it to have winged carpels.

Habitat: Found both on disturbed ground and as a desert native on shallow sand over limestone or silts. At Dhahran, as a ruderal it is a habitat vicariant of *T. terrestris*, occupying roadsides and waste ground where the latter would be expected elsewhere. Locally frequent, as at Dhahran, but overall rather restricted in distribution.

Central Coastal Lowlands: 137, 337, 409, 517(K), 653(BM), 923(BM), 7753, 8134, 8335, Dhahran.

Southern Summan: 7493, Wadi as-Sahba near Khashm az-Zaynah.

Southern Dahna: 8361, 8363, 8 km ENE Rawdat at-Tawdihiyah. Both of these plants have the wings of the fruits reduced, tending to become a few acute, triangular teeth or sometimes only one, retrorse tooth near the base of the carpel.

Vernacular Names: ZAHR (Al Murrah).

3. Tribulus arabicus Hosni s.l.

Ascending to decumbent greyish-green perennial herb, densely pubescent with both appressed and erect white hairs, 20 – 70(100) cm high. Leaves 10 – 40 mm long, the reduced member of each pair usually less than half this long; leaflets in (4)5 – 8(10) pairs, oblong-elliptical, acutish, 4 – 8(9) mm long, 2 – 4 mm wide. Flowering and fruiting pedicels 10 – 18(22) mm long, usually exceeding the (reduced) subtending leaf. Flowers solitary, (12)15 – 30(40) mm in diameter; petals obovate, bright yellow, about twice as long as the narrowly lanceolate sepals; stamens (8)10. Ovary densely hirsute. Fruit globose-ovoid, 9 – 12 mm long, hairy between the wings, the carpels with subentire to dentate, tomentose to glabrescent, 1.5 – 2.5 mm-broad wings. *Plates 143, 144.*

● This apparent Arabian endemic is highly variable in such features as flower size and carpel wing margins; as described above it includes a series of forms widely distributed in the southern and central sands of our area. Very young plants have decidedly smaller flowers, but the author has collected mature plants growing within a few meters of each other with flowers greatly differing in size. Features of the carpel wings are also variable; some on mature specimens have nearly entire or sinuate margins while others bear distinct acute teeth. The wings may be nearly glabrous or densely tomentose. There also appears to be much variation in vegetative features, with overall pubescence ranging from moderate to dense. Leaflets may be approximate and even imbricate, or more widely spaced. Several of our specimens (2889, 7831, 7495) have been ascribed to *T. omanensis*, described by Hosni (in El Hadidi 1978), on the basis of their glabrescent, entire-margined carpel wings and deciduous calyces. The many intermediate forms seen in the field, however, leave the distinction of that taxon in doubt; it would seem preferable to treat the complex provisionally as a single polymorphic species.

Habitat: Usually in deep, even semi-mobile, sand. Very common in the eastern and southeastern Rub' al-Khali, where it attains maximum stature and subdominant status. Its distribution extends through the Jafurah sands to our northernmost record at al-Juhaymi.

Central Coastal Lowlands: 7495, 1.5 km S al-Juhaymi, 26−44N, 49−20E.

South Coastal Lowlands: 194(BM), 550, 7446(BM), 5 km E Jabal Abu Shidad; 425, 7395, al-Hasa/'Ayn Dar junction; 7915, 3 km S Jabal Ghuraymil, 25−46N, 49−32E; 7833, 8138, Burq Aba ad-Dalasis, 25−43N, 49−31E; 8312, 3 km S adh-Dhulayqiyah.

Rub' al-Khali: 1014, 23−03N, 49−55E; 1955, 1 km S Jirwan; 2889, 7831, Camp Ramlah 1, 22−10N, 54−21E; 4003, 21−38N, 53−21E; 7017, ST−38, 22−17N, 53−22E; 7028, Camp S−3, 21−28N, 55−23E; 7670, al-Jawb, 23−06N, 49−58E; 7686, Shalfa, 21−52N, 49−43E; 7687, 7 km W Bir Fadil, 22−06N, 49−42E; 7881, Camp S−3, 21−36N, 54−46E; 8025, 8030, al-Mulayhah al-Gharbiyah, 22−53N, 49−28E; 8032, BM T−42, 23−07N, 49−56E; 8033, 2.5 km N al-Mulayhah al-Gharbiyah; 8036, al-Jawb, 23−07N, 49−53E; 7857, Camp S−3, 22−11N, 54−55E.

Vernacular Names and Uses: ZAHR (Al Murrah, Al Rashid, al-Manahil). This name, widely known and used quite specifically among the southern tribes of our area, means simply 'flower' in standard Arabic. This unusual use of an anatomical term as a specific plant name is perhaps understandable in view of the fact that *Tribulus* is virtually the only plant over thousands of square kilometers in the Rub' al-Khali that has conspicuous flowers.

● This plant is one of the main camel grazing species for tribes in the eastern and southeastern Rub' al-Khali, and its importance is reflected in the specialized Bedouin terminology applied to its different growth stages. Among Al Rashid tribesmen seedlings of *zahr* are called ZURAYQA', while plants in the second and third years of growth are known as 'UTHWAH. The plant is valued for grazing, but tribal lore ascribes two deleterious effects to excessive feeding on it by camels. These would appear to be physiological disorders associated with mineral or trace element deficiencies or excesses. Grazing on *Tribulus* along with *Cornulaca arabica* without other plants for long periods leads to a condition called *qiswār*. The animals lose appetite, weaken, cease drinking, and may die. Long pasturage on *Tribulus* alone reportedly may lead to an ailment called *zahr ḥumrah*. Grazing on *Tribulus* along with *Cyperus* and *Zygophyllum* is considered healthy and safe.

● 6. Nitraria L.

Shrubs with alternate or clustered simple leaves. Flowers bisexual, in cymes, with 5-merous calyx and 5 hooded petals; stamens 15. Ovary trilocular. Fruit a conical, 1-seeded fleshy drupe with wrinkled stony endocarp.

Nitraria retusa (Forssk.) Aschers.

Ascending, stiff-branched shrub, appressed pubescent-tomentose particularly in young parts, 1 – 2 m high, often broader than high in hummocks, with grey woody twigs at length becoming spinescent. Leaves alternate or fascicled, obovate-deltoid, obtuse, truncate or faintly retuse at apex, tapering at base, subsessile or shortly petiolate, 8 – 15(20) mm long, 4 – 10(15) mm wide. Flowers 5 – 6 mm long, in cymes on terminal and lateral shoots from woody twigs; petals hooded, hispidulous at backs, about 2 – 3 times as long as the appressed-hispid sepals, greenish white to yellowish; stamens (12) – 15. Fruit an ovoid, somewhat trigonous drupe 5 – 8(10) mm long, red, with a 3-angled, wrinkled, stony endocarp. *Plate 146.*

Various descriptions of this plant in the literature indicate that there is considerable geographical variation in leaf shape, leaf apices, and petiole length. Shrubs from the single known station for *Nitraria* in our area have nearly sessile leaves with entire to faintly retuse apices.

Habitat: Shallow sand hummocks on *sabkhah*; to be expected only on saline ground near the coast.

Northern Plains: 7977, Khawr al-Khafji. So far recorded only from this one station in the extreme northeast of our area. This is apparently the southernmost record of it on the Gulf coast of Arabia.

Vernacular Names: GHARDAQ (Kuwait, Dickson 1955). This is obviously a form derived by metathesis from the classical name GHARQAD. The plant description under this name by Abu Hanifah ad-Dinawari and other early Islamic writers matches this species very closely.

40. GERANIACEAE

Annual or perennial herbs with alternate or opposite stipulate leaves. Flowers bisexual with sepals 5, petals 5, and stamens 5 – 15 with filaments more or less connate at base, sometimes not all bearing anthers. Ovary superior, 5-carpellate, developing into a 5-lobed schizocarp often with elongated beak.

Key to the Genera of Geraniaceae
1. Stamens with anthers 5; leaves not plicate, with pinnate nerves ... 1. *Erodium*
1. Stamens with anthers 15; leaves subplicate with subpalmate
   nerves .... 2. *Monsonia*

● 1. Erodium L'Hér.

Annual or perennial herbs, sometimes slightly woody at base, with alternate or opposite pinnately lobed, or pinnately compound, rarely entire, leaves. Flowers usually umbelled, bracteate, with reflexed pedicels, 5-merous with 5 anther-bearing stamens and 5 staminodes. Fruit a 5-carpellate schizocarp with elongated beak, splitting longitudinally into 5, 1-seeded, awned, achenes, the awns often twisting hygroscopically after dehiscence.

Key to the Species of Erodium
1. Achene without pits at apex; awns plumose
   2. Outer sepals 3-nerved; petals c. 6 mm long; filaments ciliate
      below .... 1. *E. bryoniifolium*

2. Outer sepals 5-nerved; petals 10 – 12 mm long; filaments
   glabrous .... 4. *E. glaucophyllum*
1. Achene with 2 pits at apex near base of the awn; awns appressed pilose
   3. Leaves all pinnatisect or bipinnatisect
      4. Beaks 2.5 – 3.5 cm long, slender; fruiting sepals 5 – 7(8) mm
         long .... 3. *E. deserti*
      4. Beaks 5 – 10 cm long, stout; fruiting sepals 11 – 15 mm
         long .... 2. *E. ciconium*
   3. Leaves shallowly to deeply lobed, sometimes divided nearly to base
      5. Concentric furrow present immediately below the pits at apex of
         achene; beak 2 – 2.5(3) cm long .... 6. *E. malacoides*
      5. Furrow absent below the pits; beaks 3 – 4(5) cm long .. 5. *E. laciniatum*

1. Erodium bryoniifolium Boiss.

Prostrate to decumbent, rarely ascending annual or rarely perennating, grey-green, sometimes pinkish, with leaves mostly in a basal rosette, some opposite, on stems spreading to c. 15 cm long. Leaves ovate-cordate, 10 – 20(25) mm long, 8 – 15(20) mm wide, with nerves impressed above, obscurely 3 – 5 lobed, the lobes crenulate; basal leaves on petioles exceeding the blade, the upper with petioles shorter than the blade to subsessile. Flowers umbelled 3 – 4(5) in the upper axils, on peduncles exceeding the (reduced) subtending leaf; pedicels shorter than or sometimes equalling the calyx. Sepals c. 5 mm long, mucronate, tomentose, yellowish-margined with 3 green nerves; petals c. 6 mm long, pink. Fruit beak 5 – 8 cm long, pinkish-grey canescent; achenes narrowly obconical, c. 4 mm long, light brown, antrorsely hairy; pits absent but a very faint concentric groove present around apex at insertion of the awn; awn plumose with fine white hairs.

Habitat: Shallow silty soil in basins associated with stony terrain. Infrequent.

Northern Plains: 1594(BM), 3 km ESE confluence of al-Batin with Wadi al-'Awja; 1690(BM), 28 – 41N, 47 – 35E; 8738, ravine to al-Batin, 28 – 50N, 46 – 24E; 8771, 49 km NE Hafar al-Batin.

Northern Summan: 7956, Batn al-Faruq, 25 – 43N, 48 – 52E.

Southern Summan: 7920, Batn al-Faruq, 25 – 34N, 49 – 05E; 8070, Wadi as-Sahba, 24 – 12N, 48 – 41E.

Vernacular Names: TUMMAYR (N Arabia, Musil 1927, 628; 1928a, 710).

2. Erodium ciconium (L.) L'Hér.

Procumbent to ascending glandular-pubescent annual to c. 30 cm high, usually lower. Leaves ovate-oblong in outline, the blade 4 – 10 cm long, 1 – 4 cm wide, 1 – 2-pinnatisect into crenate to incised lobes, the rachis winged-dentate between the primary segments; basal leaves on 2 – 4 cm-long petioles, those above shorter. Umbels 2 – 4(6)-flowered, on peduncles 4 – 12 cm long, the pedicels exceeding the calyx. Sepals 5-nerved, growing markedly in fruit to 12 – 15 mm long; petals obovate, 9 – 20 mm long, pink to violet. Fruiting beak 5 – 11 cm long, very stout; achenes c. 10 mm long, obconical with erect hairs, with 2 pits near apex below insertion of the appressed-hairy awn but without concentric furrow.

Easily recognized by the large flowers and very large fruits, much stouter and longer than in our other *Erodiums* with dissected leaves.

Habitat: Shallow sand on limestone. Rather rare.

Northern Plains: 8762, ravine to al-Batin, 28 – 49N, 46 – 23E.

Northern Summan: 1344(BM), ash-Shayyit al-'Atshan, 42 km W Qaryat al-'Ulya.

Northern Dahna: 4039, edge of the Dahna, 27 – 37N, 44 – 53E.

Vernacular Names: RAQAM (Kuwait, Dickson 1955), TUMMAYR (N Arabia, Musil 1927, 628; 1928a, 710).

3. Erodium deserti (Eig) Eig

Procumbent or decumbent-ascending, pubescent annual, with stems spreading from base to c. 30 cm long. Leaves oblong in outline, 2 – 8 cm long, 1- or 2-pinnatisect with oblong-elliptical, incised segments, the rachis pubescent, not winged between the segments, the lower segments of larger leaves distant, alternate. Umbels 3 – 8(10)-flowered, on long peduncles exceeding the subtending leaf and up to 10(20) cm long; pedicels c. equalling or somewhat exceeding the sepals. Sepals c. 4 – 5 mm long in fruit, white-pubescent at base, less so above, obscurely subreticulate-nerved, with a purplish mucro; petals oblong, pink-mauve, c. 1.5 – 2 times as long as the sepals. Fruit beaks 25 – 35(40) mm long; achenes 4 – 5 mm long, obconical, brown, antrorsely pubescent, with 2 pits at apex, each with 1 distinct concentric furrow below; awn appressed-pilose. *Plates 147, 148.*

This species is rather close to *E. cicutarium* (L.) L'Hér., from which it differs in having the filaments laterally toothed at base, sepals without the white bristle terminating the mucro, and pits of the achene glandular.

Habitat: Silty soils of desert basins, wadi bottoms, sometimes on disturbed ground. Frequent.

Northern Plains: 1559(BM), confluence of al-Batin with Wadi al-'Awja; 588(BM), 48 km SW Hafar al-Batin; 4009, ad-Dibdibah, 28 – 10N, 45 – 55E; 7526, 27 km WSW al-Qaysumah; 7560, al-Faw al-Janubi; Dickson s.n., Jabal al-'Amudah (Burtt and Lewis 1954, 401); 8744, ravine to al-Batin, 28 – 50N, 46 – 24E.

Northern Summan: 702, 746, 33 km SW al-Qar'ah wells; 2793(BM), 26 km SE ash-Shumlul; 2812(BM), 12 km ESE ash-Shumlul; 3224, 74 km S Qaryat al-'Ulya on Riyadh track; 8596, Bahrat al-Kilab, 26 – 33N, 47 – 53E.

Southern Summan: 2045(BM), 25 – 00N, 48 – 30E; 2190(BM), 24 – 34N, 48 – 45E.

Vernacular Names: RAQAM (N).

4. Erodium glaucophyllum (L.) Ait.

Stout, ascending, branched perennial herb to c. 75 cm high, glaucous, glabrescent, with thin short pubescence on young sepals and bracts. Leaves ovate or subcordate to oblong, opposite at least above, obscurely lobed, dentate, somewhat coriaceous, to c. 4 cm long, 3 cm wide, on petioles longer than or equalling the lower blades, shorter above. Umbels 2 – 4-flowered on peduncles exceeding the reduced subtending leaf; pedicels about equalling the calyx, reflexed in fruit. Sepals in fruit c. 6 mm long including the 1 mm-long mucro, puberulent, obscurely nerved; petals 10 – 12 mm long, rather showy, bright purple, fugaceous. Fruit beaks 6 – 7 cm long; achenes c. 5 mm long, light brown-grey, pubescent, with 2 – 3 concentric grooves around apex at insertion of the awn; defined achene pits absent although faint wrinkling sometimes present; awns yellowish-plumose in upper two-thirds. *Plate 149.*

Habitat: Often in rocky terrain on shallow sand or silts. Locally frequent.

Northern Plains: 1595(BM), 3 km SE al-Batin/al-'Awja confluence.

Northern Summan: 2849, 26–18N, 47–56E; 8619, Rawdat Ma'qala, 26–26N, 47–20E.

Central Coastal Lowlands: 390, 1047, 8135, Dhahran; 1782(BM), Abu Hadriyah; Dickson 553, as-Sudah, 140 miles S Kuwait (Burtt and Lewis 1954, 403).

Vernacular Names: KIRSH ('paunch', Bani Hajir), DABGHAH ('tanweed', Rawalah). Some Bedouins report that overgrazing on this plant can lead to bloat in camels and other ruminants. The name *kirsh*, above, refers to this.

5. Erodium laciniatum (Cav.) Willd.

E. pulverulentum (Cav.) Willd.

Procumbent to ascending annual, greyish-green and short-pubescent to canescent, with stems to c. 30 cm long. Leaves ovate in outline, 2–4 cm long, 1–3 cm wide, varying greatly in lobation from subentire-dentate to pinnately divided nearly to base, often with 3–5 main lobes; petioles on lower leaves exceeding the blades, those above shorter. Umbels 2–8-flowered, on peduncles exceeding the subtending leaf, to c. 6 cm long; pedicels 1–3 times the length of the sepals. Fruiting sepals 4–5 mm long, mucronate, pubescent, 5-nerved; petals 5–6 mm long, pink, rarely white. Fruit beaks 30–40(45) mm long; achenes c. 4 mm long, brown, short-pubescent, with 2 pits at apex at insertion of the awn, grooves or furrows absent; awn sparsely appressed-pubescent.

The majority of our specimens would appear to be assignable to var. *pulverulentum* (Cav.) Boiss (= *E. pulverulentum* (Cav.) Willd.). As noted by Rechinger (1964), the characters distinguishing *E. pulverulentum* as a species appear to lack correlation in the eastern part of its range, and it would seem preferable not to attempt a segregation on the basis of available material.

Some plants in the southern Summan around Wadi as-Sahba have very shallowly lobed, dentate, leaves and sparse pubescence; one individual in that area was seen to have white rather than pink flowers.

Habitat: Sands and silty soils in a wide variety of terrain types. The most widespread and common *Erodium* in our area.

Northern Plains: 659, Khabari Wadha; 895, Jabal al-'Amudah; 864, 9 km S 'Ayn al-'Abd; 536(BM), 25 km W Qaryat al-'Ulya; 683, 20 km WSW Khubayra; 1154(BM), 1164(BM), Jabal al-Ba'al; 1250(BM), Jabal an-Nu'ayriyah; 1275, 15 km ENE Qaryat al-'Ulya; 1599(BM), 3 km ESE al-Batin/al-'Awja confluence; 1681(BM), 2 km S Kuwait border in 47–14E; 2717, 3 km NW Jabal Dab'; 7986, 28–12N, 48–07E; 8739, ravine to al-Batin, 28–50N, 46–24E.

Northern Dahna: 27–35N, 44–51E.

Northern Summan: 7485, Batn Sumlul, 26–26N, 48–15E; 701, 33 km SW al-Qar'ah wells; 1391(BM), 42 km W Qaryat al-'Ulya; 2778, 25–50N, 48–00E; 2797, 15 km ESE ash-Shumlul; 2842, 32 km ESE ash-Shumlul; 3908, 27–41N, 46–02E; 3220, 3227, 74 km S Qaryat al-'Ulya on Riyadh track; 8439, edge of 'Irq al-Jathum, 25–54N, 48–36E.

Central Coastal Lowlands: 38(BM), 1814(BM), 1830(BM), Dhahran; 931, 10 km S Khafji-Saffaniyah junction; 3729, Ras az-Zawr; 1466(BM), 2 km E al-Ajam; 1801(BM), 1806(BM), 5 km NE al-Khursaniyah; 3306, 2 km W Sabkhat Umm al-Jimal; 7041, Karan Island.

South Coastal Lowlands: 1436(BM), 2 km E Jabal Ghuraymil; Dickson 559, Abqaiq-al-Hasa road (Burtt and Lewis 1954, 406).

Southern Summan: 2259, 29 km ENE al-Hunayy; 2183, 24–34N, 48–45E; 2075, Jaww ad-Dukhan, 24–48N, 49–05E; 8054, 24–06N, 48–48E.

Vernacular Names: QARNUWAH ('hornweed', prob. derived from *qarnā'*, 'horned', and referring to the beaked fruits, Bani Hajir), KIRSH ('paunch, rumen', Al Murrah), SAMNAH ('fatweed', Bani Hajir), RAQAM (N Arabia, Musil 1927, 596), DAHMĀ' ('greyling', Musil loc. cit.).

6. Erodium malacoides (L.) L'Hér.

Stout ascending, pubescent annual, 10–40 cm high. Leaves ovate to subcordate, shallowly lobed, crenate-dentate, 2–6(10) cm long, 1–4(6) cm wide; petioles up to 6 cm long, shorter above. Umbels 2–8-flowered, on peduncles mostly exceeding the subtending leaves; pedicels 1–3 times the length of the sepals. Fruiting sepals 5–6 mm long, 5-nerved, mucronate; petals oblong, somewhat exceeding the sepals, lilac. Fruit beaks often only 2 cm long, rarely longer; achenes c. 5 mm long, pubescent, with 2 pits at apex at insertion of the awn, each pit with a narrow concentric groove below; awns sparsely pubescent to glabrescent.

Habitat: A weed, and to be expected only in farms or gardens. Rare.

Not seen by the author but reported as a weed of the Eastern Province by Chaudhary, Parker and Kasasian (1981).

● 2. Monsonia L.

Ascending annuals or perennials with alternate or opposite petiolate leaves. Aspect of *Erodium*, but differing in having 15 functional stamens grouped in 3's opposite the petals. Fruit a beaked schizocarp like that of *Erodium*.

Monsonia nivea (Decne.) Decne. ex Webb

Ascending to erect silver-grey canescent perennial, 8–30 cm high. Leaves ovate to oblong-elliptical, 15–30 mm long, 7–15 mm wide, crenulate, subplicate on the impressed, subpalmate nerves; petioles of lower leaves exceeding the blade, those above shorter. Umbels 2–6-flowered, on peduncles to c. 15 cm long; fruiting pedicels mostly 15–20 mm long. Sepals c. 5 mm long, obtuse, mucronate, tapering to base, densely covered with grey, mostly appressed, pubescence; petals pink, slightly exceeding the sepals, fugaceous. Fruit beaks 35–40 mm long; achenes 4–5 mm long, brown, pubescent, with 2 faint concentric grooves at apex near insertion of the awn; awn plumose with fine white hairs.

Batanouny (1981, 115) reports *Monsonia heliotropoides* (Cav.) Boiss. from southern and western Qatar. No plants seen by us in eastern Saudi Arabia, however, can be attributed to that species, which appears to be restricted to the Peninsula's Sudanian floristic territory.

Habitat: Deep or shallow sands, particularly in the south. Frequent.

Northern Dahna: 794, 15 km SE Hawmat an-Niqyan; the Dahna (Pelly 1866, 81).

Northern Summan: 2848, 26–18N, 47–56E.

Central Coastal Lowlands: 102, 130, 1042, 8338, Dhahran; Dickson 572, W al-Jubayl (Burtt and Lewis 1954, 407); 7947, 7958, al-Faruq, 25–43N, 48–53E.

Southern Summan: 7919, Jabal ar-Ruhayyah, 25–34N, 49–05E; 8012, 24–00N, 49–00E; 8075, Wadi as-Sahba, 24–12N, 48–50E; 2163, 24–08N, 48–43E.

South Coastal Lowlands: 8038, Wadi as-Sahba, 23–52N, 49–22E.

Rub' al-Khali: 7689, Bani Tukhman, 22 – 38N, 49 – 34E; 7697, al-Mulayhah al-Gharbiyah, 22 – 53N, 49 – 29E.

Vernacular Names: QARNUWAH (Al Murrah).

41. OXALIDACEAE

Predominantly herbs, with alternate, usually 3-foliolate leaves. Flowers bisexual, 5-merous and often heterostylous, with 10 stamens connate at base. Ovary superior, (3)5-carpelled. Fruit a loculicidal capsule.

● Oxalis L.

Annual or perennial caulescent or acaulescent herbs with sour juice and leaves mostly 3-foliolate. Sepals 5; petals 5, in ours yellow. Stamens 10, the 5 outer with shorter filaments. Fruit a 5-celled capsule.

Key to the Species of Oxalis

1. Corolla 5 – 10 mm long, c. twice as long as the calyx; plant with creeping stem .... 1. *O. corniculata*
1. Corolla 15 – 30 mm long, c. three times as long as the calyx; stemless plant .... 2. *O. pes-caprae*

1. Oxalis corniculata L.

Annual or perennial pubescent herb with procumbent or ascending creeping stems rooting at the nodes, 5 – 30 cm high. Leaves 3-foliolate, digitate, 1 – 4(6) cm long, on petioles exceeding the blade, with obcordate leaflets 5 – 10 mm long, usually broader than long. Peduncles equalling or exceeding the leaves, 1 – 4-flowered. Flowers 5 – 10 mm long with yellow petals c. twice as long as the calyx. Capsule 8 – 25 mm long, c. 3 mm broad, puberulent, prismatic-cylindrical, acute, on a reflexed pedicel shorter than or equalling the fruit.

Habitat: An introduced weed of greenhouses, nurseries, and some shaded habitats around cultivated grounds. Rare except perhaps very locally.

Central or South Coastal Lowlands: Not observed by the author but reported by Chaudhary and Zawawi (1983) as a weed, at least in central Saudi Arabia and likely to be found occasionally in the east.

● A specimen with small, hardly reflexed fruits and obsolete stipules, collected by the author among potted plants in Dhahran (No. 8124), may be attributable to *O. europaea* Jord.

2. Oxalis pes-caprae L.

O. cernua Thunb.

Ascending stemless perennial, glabrescent or with sparse scattered hairs, 10 – 40 cm high with bulbiferous rhizome. Leaves digitate, 3-foliolate, 2 – 5 cm long, broader than long, on 10 – 20 cm-long petioles arising directly from the base; leaflets obcordate or 2-lobed about half-way to the base, 1 – 2 cm long, 1.5 – 3 cm wide, sometimes reddish-spotted. Flowers c. 8 – 20 together, umbelled on rather thick, scapelike peduncles exceeding the leaves; pedicels c. 2 – 3 cm long; corolla yellow, funnel-shaped, 15 – 20(30) mm long, 3 – 4 times as long as the calyx. Fruits (in our specimens) not known.

Our specimens appear to be the short-styled, sterile form, also prevalent in Egypt and Palestine, of this heterostylous species.

Habitat: A garden weed of sheltered, partially shaded spots, probably introduced with nursery stock from abroad. Rare.

Northern Plains: 7556, Qaisumah Pump Station. Weed.

42. POLYGALACEAE

Herbs or shrubs with usually alternate, simple, entire leaves. Flowers bisexual, zygomorphic. Sepals imbricate, the 2 inner ones frequently larger, petaloid; petals 3 – 5, unequal; stamens 8. Ovary superior, usually 2-celled; fruit a capsule or drupe.

● Polygala L.

Herbs or small shrubs with leaves usually alternate. Flowers racemed, compressed, with sepals 5, the two inner ones much enlarged and forming 2 distinctive petaloid wings enclosing the other flower parts. Petals 3; stamens 8. Fruit a compound 2-celled capsule; seeds caruncled.

Polygala erioptera DC.

Erect herb or straggling shrublet, finely short-pubescent, usually woody at base, to c. 60 cm high. Leaves of variable shape and size, in ours linear to narrowly lanceolate, usually obtuse at apex, 10 – 25 mm long, 0.5 – 3 mm wide, subsessile or on petioles 1 – 2 mm long. Flowers compressed, 3 – 5 mm long, racemed, with reflexed pedicels c. 2 mm long; 2 inner sepals much enlarged, growing to form 2 oblong wings c. 5 mm long, 3 mm broad, finely nerved, fine-pubescent, white to cream and sometimes green- or reddish-veined or marked, enclosing the other flower parts. Petals fugacious, rose toward the apices, the large central one with white-fringed appendage at back. Capsule compressed, oblong, retuse, ciliate, 2-seeded, 4 – 5 mm long, enclosed in the wings. Seeds oblong, teretish, 3 – 3.5 mm long including the distinct caruncle, somewhat compressed, with black body densely covered in pale, appressed, ascending hairs.

This species is often described as an annual, but our plants appear to be perennial. It is a Sudanian species so far not collected north of Dhahran, where it is seen on limestone.

Habitat: Shallow sand or silt, often over limestone. Infrequent to rare.

Central Coastal Lowlands: 339, 1037, 8273, Dickson 467 (Burtt and Lewis 1949, 307), Dhahran.

Southern Summan: 8011, 24 – 00N, 49 – 00E.

43. UMBELLIFERAE (Apiaceae)

Annual or perennial, often aromatic herbs, sometimes shrubby, with alternate, usually divided, leaves. Inflorescence usually a compound umbel, the primary (umbel) rays subtended or not by an involucre of bracts, the secondary umbel (umbellule or umbellet) with raylets or pedicels sometimes subtended by an involucel of bracteoles. Flowers bisexual, or unisexual and then usually monoecious, 5-merous with sepals often inconspicuous; petals sometimes very unequal. Ovary inferior, nearly always 2-locular, with 2 styles mounted on a broadened or conical base (the stylopodium). Fruit a schizocarp separating into 2, 1-seeded, more or less ribbed mericarps.

Key to the Genera of Umbelliferae

1. Leaves entire .... 3. *Bupleurum*
1. Leaves divided or absent
  2. Shrub, leafless or with a few filiform leaf rudiments .... 7. *Deverra*
  2. Herbs, with well-developed leaves
    3. Involucel present
      4. Outer flowers of umbellules sterile, when mature on indurated pedicels standing parallel to each other and to their supporting primary ray .... 1. *Anisosciadium*
      4. Outer flowers fertile; pedicels spreading, non-parallel
        5. Fruits smooth, glabrous; outer flowers with radiate petals .... 2. *Coriandrum*
        5. Fruits tuberculate or pubescent; flowers non-radiate
          6. Fruits velvety-hirtulous; rays 3 – 7 cm long ...... 10. *Ducrosia*
          6. Fruits finely glandular-tuberculate; rays 0.5 – 2.5 cm long .... 5. *Trachyspermum*
    3. Involucel absent
      7. Flowers white
        8. Calyx and fruit densely hirsute .... 8. *Pimpinella*
        8. Calyx and fruit glabrous .... 4. *Apium*
      7. Flowers yellow or yellow-green
        9. Fruit 4 – 6 mm long, distinctly compressed and somewhat winged laterally, on more or less equal pedicels ...... 9. *Anethum*
        9. Fruit 1.5 – 2 mm long, weakly compressed and unwinged, on unequal pedicels .... 6. *Ridolfia*

● 1. Anisosciadium DC.

Desert annuals with pinnatisect leaves. Umbels with persistent bracts and bracteoles, the central fertile and bisexual flowers of each umbellule sessile or subsessile, surrounded by sterile pedicellate flowers. Indurated umbellules detaching as a unit at maturity.

Anisosciadium lanatum Boiss.

Ascending or decumbent pubescent annual, to c. 40 cm high. Leaves ovate-oblong in outline, 2 – 3-pinnatisect into linear lobes; petioles below equalling or exceeding the blades, those above shorter. Umbels axillary, on elongated peduncles, (5)8 – 18-rayed, the rays somewhat flattened and glandular hispid, subtended by coriaceous, ciliate, sharp, tapering bracts to c. 8 mm long. Pedicels c. 10 – 18, at first spreading, later in maturity becoming straight-standing hardened columns nearly parallel to each other and to their supporting primary ray, surrounding a few fertile flowers, the umbellule forming a peculiar indurated cylindrical dispersal unit c. 6 – 10 mm long, 6 mm wide; calyx teeth of the outer, sterile flowers persistent, indurated, acute. Petals white, the outer ones sometimes enlarged-radiate; styles in fertile flowers much elongated, yellow, indurated, diverging above. *Plate 150.*

This plant is by far the most common of our desert umbellifers and is easily recognized by its dome-shaped, dense umbels of white flowers and the peculiar structure of its umbellules.

Habitat: Usually shallow sand over rocky ground or silty substrates. Common.

Northern Plains: 1490(BM), 50 km WNW an-Nu'ayriyah; 872, 9 km S 'Ayn al-'Abd; 884, 1735(BM), Jabal al-'Amudah; 1001, 12 km N Jabal al-'Amudah; 1225, 1238(BM), Jabal an-Nu'ayriyah; 1281(BM), 15 km ENE Qaryat al-'Ulya; 1702(BM), ash-Shaqq, 28–30N, 47–42E; 3116(BM), al-Batin, 28–00N, 45–28E; 3753, 10 km W as-Saffaniyah; 8769, 49 km NE Hafar al-Batin; 8793, 30 km SW ar-Ruq'i Post, 28–54N, 46–27E.

Northern Summan: 218, 25 km S Qaryat al-'Ulya; 2878, 20 km W 'Uray'irah; 697, 33 km SW al-Qar'ah wells; 760, 69 km SW al-Qar'ah wells; 2756, 25–43N, 48–07E.

Central Coastal Lowlands: 422, 7905, Na'lat Shadqam; 7748, 5 km NW Ras Tanaqib, 27–53N, 48–49E.

Southern Summan: 2220, 26 km SSE al-Hunayy; 2266, 29 km ENE al-Hunayy.

Vernacular Names: BASBĀS (Al Murrah, gen.).

Another species, *A. isosciadium* Bornm., is known from the Syrian Desert and adjacent regions in the north. It differs in having ovate to orbicular bracteoles and calyx lobes.

● 2. Coriandrum L.

Annual or biennial glabrous, aromatic herbs with pinnatisect leaves and white to pink flowers with some petals radiate. Bracts absent or 1; involucel present. Fruit ovoid to globular, with wavy ribs alternating with straight ones; mericarps not separating.

Coriandrum sativum L.

Erect, glabrous, rather fetid annual, 50–90 cm high in cultivation, often shorter as an escape. Leaves of several kinds may be present: the lowest long-petioled and ternate, lower stem leaves pinnatisect with 2–4 pairs of incised segments, and the higher leaves finely pinnatisect with linear-subfiliform lobes. Umbels peduncled, mostly 4–8-rayed, with a single inconspicuous bract or involucre entirely absent; umbellules with involucel, 5–11-flowered. Flowers with persistent calyx lobes; outer petals radiate, 3–4 mm long, pale pink or violet, or whitish. Fruit 3–5 mm long, smooth, subglobose, with persistent calyx lobes at apex and with weak, straight to wavy ribs when dry.

Habitat: A rare escape along roadsides or on other disturbed ground.

Central Coastal Lowlands: 8000, al-Khubar, roadside; 8160, near al-Jishsh, al-Qatif Oasis (specimen from cultivation); 8647, 8648, Ras Tanura.

Vernacular Names and Uses: KUZBARAH (gen. name in many Arab countries). This is the coriander used as a condiment and spice; it is sometimes cultivated in oasis gardens, as in al-Qatif, and appeared to be becoming more common in the early 1980s because favored by Egyptian farm laborers.

● 3. Bupleurum L.

Annual herbs (ours) with entire sessile leaves. Umbels compact, subtended by leafy bracts or bracteoles. Flowers (in ours) inconspicuous, yellowish-green.

Bupleurum semicompositum L.

Glabrous, ascending dwarf annual herb with fine, almost capillary, many-branched stems, to c. 20 cm high. Leaves sessile, narrowly linear-lanceolate, acute, 0.5–3(5) cm long, 1–2 mm wide. Umbels axillary or terminal, 3–5-rayed; bracteoles linear-lanceolate, acute, 1.5–2 times the flowering umbellules. Umbellules 3–8-flowered;

fruits ovate to short-oblong, somewhat compressed, c. 1 mm long and broad, the ribs not visible, whitish granular-tuberculate.

Habitat: Northern coastal sands. Apparently rather rare, but an inconspicuous plant that may be more widespread than generally appreciated.

Central Coastal Lowlands: 7743, 2 km N Umm Judhay', 27 – 53N, 48 – 49E.

● 4. Apium L.

Annual or perennial glabrous herbs with ternate or pinnately divided leaves. Petals (in ours) white; stylopodium depressed. Fruit somewhat laterally compressed, broadly ovate to subglobular, distinctly filiform-ribbed.

Apium graveolens L.

Annual or biennial, erect to ascending glabrous, aromatic herb, 15 – 50(80) cm high. Leaves 1.5 – 6 cm long, often broader than long, mostly ternate with rhomboid to cuneate, dentate to laciniate, sometimes parted, lobes; petioles exceeding the blades below, shorter above. Umbels axillary, 5 – 9-rayed, mostly peduncled; umbellules with 8 – 15 minute, white-petaled flowers c. 0.5 mm long. Fruits c. 1 mm long, somewhat angled on the ribs.

Habitat: A weed of wet shaded oasis fields, gardens and lawns. Locally frequent.

Central Coastal Lowlands: 918, al-Qatif; 3918, 7758, Dhahran.

South Coastal Lowlands: 8193, near al-Qurayn, al-Hasa Oasis.

Vernacular Names: KARFAS (Qatifi farmers).

● 5. Trachyspermum Link

Annual herbs with leaves multi-pinnatisect into linear lobes. Involucre and involucel usually present. Flowers bisexual, with calyx teeth obsolete and petals white, 2-lobed, medially inflexed. Stylopodium capitate-conical. Fruit laterally compressed, papillose to short pubescent; mericarps with 5 prominent ribs.

Trachyspermum ammi (L.) Sprague

T. copticum (L.) Link

Annual, many-branched, ascending or erect glabrescent herb, 10 – 40(60) cm high. Lower leaves petiolate, 2 – 3-pinnatisect with filiform or linear lobes; upper leaves ternate-pinnatisect, 2.5 – 9 cm long, with linear lobes 0.5 – 1 mm broad. Umbels terminal and axillary, with (6)8 – 12(20) rays 7 – 15(30) mm long, subtended by linear-oblong bracts; umbellules subtended by linear-oblong bracteoles. Flowers 12 – 16, on pedicels (in our specimen) c. 3 mm long. Petals white, with medial portion strongly inflexed and puberulent without. Fruits c. 2 mm long, ovate, brown, somewhat laterally compressed, with conspicuous paler ribs, distinctly granular-papillose.

Habitat: A weed around cultivated areas. Rare.

South Coastal Lowlands: 3921(E), 'Ayn al-Khadud, al-Hasa; 8225, near al-Qurayn, al-Hasa Oasis.

● 6. Ridolfia Moris

Annual glabrous herbs with filiform leaf lobes. Involucre and involucel absent. Flowers bisexual, minute, with yellowish petals. Fruit oblong-prismatic.

Ridolfia segetum (Guss.) Moris

Glabrous erect annual herb with anise-like odor, up to 80(120) cm high. Lower leaves long-petioled, to c. 30 cm long, ovate in outline with sheathing petioles, multi-pinnatisect into filiform lobes; upper leaves shorter-petiolate, filiform-lobed. Umbels on peduncles c. 3 – 10 cm long, 10 – 15-rayed, the rays to c. 5 cm long. Flowers mostly 10 – 15 in each umbellule, on unequal pedicels; petals yellow, minute; fruits 1 – 2 mm long, oblong-cylindrical, prismatic or weakly compressed, glabrous, surmounted by a disciform depressed stylopodium.

Habitat: A Mediterranean weed of grain fields, so far recorded in our territory only as a roadside city weed. Rare.

Central Coastal Lowlands: 8001, al-Khubar.

● 7. **Deverra** DC.
Pituranthos Viv.

Shrubby, nearly leafless perennials with small terminal or axillary umbels and deciduous bracts and bracteoles. Flowers bisexual. Fruit ovoid to globular, weakly compressed laterally, usually densely pubescent.

Deverra triradiata Hochst. ex Boiss. subsp. **musilii** (Chrtek, Osbornova et Sourkova) Pfisterer et Podlech

Pituranthos triradiatus (Hochst. ex Boiss.) Aschers. et Schweinf.

D. musilii Chrtek, Osbornova et Sourkova

Aromatic shrub with ascending, glabrous, virtually leafless wandlike stems up to c. 1.7 m high. Leaves absent or only sheath bases remaining on older branches, but rudimentary, simple or divided, filiform sessile lobes sometimes present on young growth. Umbels at or near the branch tips, mostly 3 – 4 rayed, the rays rigid, 0.7 – 2.5(3) cm long; bracts and bracteoles caducous. Umbellules mostly 5 – 9-flowered. Fruits c. 2.5 mm long, ovoid, densely whitish-hirsute without visible ribs, with glabrous yellow stylopodium and ascending diverging styles at apex, on unequal pedicels exceeding or shorter than the fruit, some subsessile. *Plate 151.*

Our plant, formerly accepted by some as a segregate species, *D. musilii*, is now generally treated as a subspecies of *triradiata* (I. Hedge, personal communication 1989).

Habitat: Sandy bottoms or banks of rocky wadis and ravines, often standing within other shrubs. Infrequent.

Northern Summan: 704, 33 km SW al-Qar'ah wells; 768, 69 km SW al-Qar'ah wells; 817, near Wabrah; 2955(BM), 6 km SE ad-Dabtiyah; 7399, Batn Sabsab.

Southern Summan: 2269, 30 km ENE al-Hunayy; 1868, Jaww al-Hawiyah, 24 – 42N, 49 – 08E.

Central Coastal Lowlands: 7904, Na'lat Shadqam, 25 – 43N, 49 – 29E.

Vernacular Names and Uses: ḤAZZA' (Shammar), sūs ('liquorice', 'Ujman). Many Bedouin herdsmen note the camel's particular fondness for this aromatic shrub.

● 8. **Pimpinella** L.

Annual or perennial herbs with divided leaves, or rarely the lower leaves undivided, and white flowers. Involucre and involucel (in ours) absent. Fruit weakly compressed laterally, ovoid to subglobular, with ribs of the mericarps distinct or obsolete.

Pimpinella puberula (DC.) Boiss.

Erect puberulent annual herb, 15 – 40 cm high. Leaves variable, 1.5 – 4 cm long, the lowest often trisect with rounded laciniate segments, those above becoming increasingly divided, 2-trisect or pinnatisect with narrower lobes, the highest sometimes with ultimate lobes filiform. Umbels on peduncles 1.5 – 2.5 cm long, (8)10 – 20-rayed, the rays 10 – 15 mm long. Flowers with densely hirsute ovary and white petals puberulent at back, somewhat radiating. Fruits about 1 mm long, ovoid, with ribs obsolete, densely hirsute, with depressed stylopodium and persistent styles longer than the fruit body.

Habitat: Sandy or silty ground. Infrequent except rather locally.

Northern Plains: 7576, Jabal Dab'.

Northern Summan: 3891, 14 km ESE Umm 'Ushar; 2783(BM), 25 – 50N, 48 – 00E.

Central Coastal Lowlands: 3718, 8 km N Abu Hadriyah; 3765, 13 km N Abu Hadriyah.

Southern Summan: 2085, al-Ghawar, 24 – 27N, 49 – 05E (a sterile specimen probably attributable to this species, and apparently marking the known southern limit of its range).

Vernacular Names: KUSAYBIRAH (N, Musil 1927, 609; 1928a, 700). This name is certainly a variant on the diminutive form of the name *kuzbarah* (coriander).

● *Pimpinella eriocarpa* Banks et Sol. has been recorded in southern Iraq, outside our area to the northwest and might be found to occur within our boundaries. All specimens collected to date, however, would appear to be referable to *P. puberula*. *P. eriocarpa* has narrower elliptical fruits with elongate conical stylopodium, shorter styles, and usually only 5 – 7 rays in the umbel.

● 9. Anethum L.

Tall aromatic herbs with leaves highly dissected into linear or filiform lobes. Flowers with yellow, notched, suborbicular petals; stylopodium depressed-conical. Fruit strongly dorsally compressed, the mericarps with dilated, winglike lateral margins.

Anethum graveolens L.

Erect, branching, glabrous, strongly aromatic annual with striate stems, 15 – 60 cm high. Leaves 10 – 30 cm long, multi-pinnatisect with filiform lobes, petiolate at least below, the petioles somewhat sheathing at base. Umbels 10 – 30-rayed; rays 2 – 5(7) cm long, mostly equal; involucre and involucel absent. Fruit elliptical, 3 – 6 mm long, compressed, the marginal ribs winged.

Our specimens have rather poorly developed marginal wings on the fruits.

Habitat: A weed or escape from cultivation in the oases. Infrequent.

Central Coastal Lowlands: 1027, Abu Ma'n; 8262, 5 km S al-Qatif town.

South Coastal Lowlands: al-Hasa Oasis, reported by Cheesman (1926, 417), who collected a cultivated plant flowering in November in or near al-Hufuf.

Vernacular Names and Uses: ḤULWAH (Qatifi farmers). Leaves and seeds used for their aniselike flavoring. Cheesman (1926, 417) says that the seeds were sprinkled as a spice on packed dried dates in al-Hasa.

● 10. Ducrosia Boiss.

Glaucous perennial herbs with much-divided leaves and white to yellowish flowers. Petals entire, inflexed. Fruits compressed with thickened margin; mericarps 5-lined or ribbed.

Ducrosia anethifolia (DC.) Boiss.

Perennial herb of strong, unpleasant scent, branching mostly from the base, with glabrous stems and leaves, 15 – 30 cm high. Leaves ovate-oblong to rhomboid in outline, 2 – 6 cm long, sometimes broader than long, trisect, with divided lobes, the ultimate segments linear-lanceolate, acute; petioles 5 – 18 cm long. Umbels long-peduncled, 10 – 15-rayed, the rays up to 7 cm long, subtended by an involucre of small bracts c. 2 mm long. Umbellules with involucel of acute bracteoles. Flowers c. 12 – 18 per umbellule, yellowish, with pubescent ovary and petals puberulent at their backs. Fruit 7 – 8 mm long, ovate to elliptical, hirtulous, with smooth dilated margin. *Plate 152.*

Habitat: Silty soils of the northern plains and the northern Summan. Infrequent and usually found as single individuals.

Northern Plains: 570(K), 105 km E Hafar al-Batin; 3084(BM), al-Batin, 28 – 01N, 45 – 29E.

Northern Summan: 790, 142 km SW al-Qar'ah wells; 805, 37 km NE Jarrarah; 2770, 25 – 49N, 48 – 01E; 2870, 25 – 59N, 48 – 32E; 7400, Batn Sabsab; 8630, 2 km NE Dahl Abu Harmalah, 26 – 30N, 47 – 37E.

Vernacular Names: ḤAZZAZ (Mutayr).

44. GENTIANACEAE

Glabrous herbs with opposite entire leaves. Flowers actinomorphic, mostly bisexual and (4)5 – 8-merous with deeply lobed tubular calyx and sympetalous rotate, campanulate, or funnel-shaped corolla. Stamens 5 – 8, inserted on the corolla tube. Ovary superior, 1-celled; fruit a septicidal capsule with numerous seeds.

● Centaurium Hill

Glabrous annuals with tetragonal stems and opposite, sessile leaves. Flowers 5-merous with calyx deeply 5-cleft and corolla with cylindrical tube. Stamens 5, inserted in upper part of corolla tube, exserted. Capsule 2-valved.

Key to the Species of Centaurium

1. Flowers in forked cymes, not appressed to stem; calyx lobes shorter than the corolla tube .... 1. *C. pulchellum*
1. Flowers in elongate spicate racemes, appressed to the stem; calyx lobes equalling or exceeding the corolla tube .... 2. *C. spicatum*

1. Centaurium pulchellum (Sw.) Druce

Erythraea pulchella (Sw.) Fries

Erect glabrous annual to c. 50 cm high. Leaves ovate-elliptical to lanceolate, entire, 8 – 25 mm long, 3 – 8 mm wide. Flowers both pedicellate and subsessile in dichotomously branching, open, terminal cymes; calyx divided nearly to base, with linear-lanceolate lobes reaching about the middle of the corolla tube; corolla 10 – 16 mm long, pink. Capsule cylindrical, 8 – 10 mm long, c. 1.5 mm wide, with 2 mucronate valves dehiscing longitudinally from the apex. Seeds numerous, minute (c. 0.2 mm long), brown, finely reticulate. *Plate 153.*

Habitat: A weed of wet, often shaded ground in the oases. Occasional.

Central Coastal Lowlands: 1087(BM), Abu Ma'n, al-Qatif Oasis area.

South Coastal Lowlands: 3920, 'Ayn al-Khadud, al-Hasa Oasis; 8226, near al-Qurayn, al-Hasa Oasis.

2. Centaurium spicatum (L.) Fritsch

Erythraea spicata (L.) Pers.

Erect glabrous herb, simple or branched from near base and at base of inflorescence above, 20 – 50 cm high. Cauline leaves lanceolate, sessile, entire, acute, 15 – 30 mm long, 3 – 11 mm wide, becoming smaller above in the inflorescence. Flowers in several simple or branched terminal spicate racemes up to c. 15 cm long, sessile or short-pedicellate, appressed to the stem. Calyx deeply cleft, with unequal linear, acute, keeled lobes equalling or exceeding the corolla tube. Corolla 9 – 12 mm long, pink or white. Capsule cylindrical-ellipsoid, terete, tapering at apex, membranous, shining brown, 6 – 8 mm long, c. 1.5 mm wide, with numerous minutely reticulate seeds c. 0.3 mm long.

Habitat: A weed of wet places in the oases, sometimes found with the preceding species. Occasional.

Central Coastal Lowlands: 1088(BM), Abu Ma'n, al-Qatif Oasis area; 1915(BM), al-Jarudiyah, al-Qatif Oasis; 7771, 4 km S al-Qatif town, flowers white.

Vernacular Names: KHIFJI ('tremble-weed', Qahtan).

45. APOCYNACEAE

Shrubs, trees or herbs with white milky sap and opposite or (rarely) alternate leaves. Flowers bisexual, actinomorphic, usually 5-merous. Calyx deeply lobed; corolla gamopetalous; stamens equalling the corolla lobes in number. Ovary usually superior, 2-carpelled. Fruit usually a follicle with seeds often winged or comose.

Key to the Genera of Apocynaceae
1. Leaves alternate; plant of dry desert . . . . 1. *Rhazya*
1. Leaves opposite; plant of saline marshes . . . . 2. *Trachomitum*

● 1. Rhazya Decne.

Shrubs with alternate entire leaves. Flowers 5-merous, with salverform corolla much longer than the small 5-lobed calyx; stamens inserted in upper part of the tube. Ovary superior, 2-carpelled. Fruit a glabrous terete follicle.

Rhazya stricta Decne.

Glabrous shrub with many branches ascending from the base, to c. 80 cm high and often broader than tall. Leaves linear-oblong or elliptical, entire, acute, tapering at base to a short petiole, or subsessile, c. 5 – 10 cm long, 1 – 2 cm wide. Flowers in dense terminal cymes; calyx c. 2 – 3 mm long with acute triangular lobes. Corolla c. 12 – 15 mm long, with brownish-green tube expanded somewhat above the middle and longer than the salverform limb, partly occluded by bristles at the throat; lobes of the limb broadly obovate, obtuse, mucronate, white inside and often bluish on back. Follicles linear, terete, tapering toward apex, often densely grouped, erect-standing, 5 – 10 cm long, c. 0.4 cm wide. *Plate 154.*

Habitat: Silty soil or shallow sand over silts. Locally frequent and sometimes forming pure stands.

South Coastal Lowlands: sight records south of al-Hasa.

Southern Summan: sight records around Harad, Wadi as-Sahba.

Southern Dahna: 2138, eastern Dahna, 23 – 05N.

Vernacular Names: ḤARMAL (gen.). This plant is generally acknowledged to be somewhat toxic, but livestock are said usually to avoid it and it does not have a reputation as a serious threat. It has traditional medicinal uses, and inhalation of the smoke of the dried leaves (using a tobacco pipe) is considered by some elder Bedouins to be an effective treatment for rheumatism.

● 2. **Trachomitum** Woodson

Erect shrubby perennial herbs with rhizomatous roots. Leaves opposite. Calyx 5-partite; corolla campanulate with open throat, the stamens inserted at base of the tube, alternating with scales. Ovary superior, 2-carpellate; fruit a terete follicle; seeds with apical hair tuft.

Trachomitum venetum (L.) Woodson

Apocynum venetum L.

Shrubby perennial herb with ascending-erect stems up to c. 1.5 m tall. Leaves lanceolate-elliptical, acute or obtuse, mucronate, with denticulate cartilaginous margins, 2 – 5(6) cm long, 0.8 – 2 cm wide, on petioles 1 – 4 mm long. Inflorescence bracteate, terminal, branched. Flowers short-pedicelled, fragrant; calyx lobes 1 – 2 mm long, acute, finely pubescent; corolla c. 4 – 6(8) mm long, finely glandular-pubescent, white, with lobes about equalling the tube. Follicles paired on a short deflexed peduncle, terete-linear, tapering gradually to an acute apex, 8 – 17 cm long, 0.2 – 0.4 cm wide. Seeds windborne by an apical tuft of fine white hairs c. 20 mm long.

Habitat: Wet ditches or swampy saline ground with irrigation run-off in and around the oases. Often associated with *Juncus*. Locally frequent.

Central Coastal Lowlands: 459(K), al-Khubar; 7775, Sayhat.

South Coastal Lowlands: 1888(BM), NE al-Mutayrifi, al-Hasa Oasis; 3925, 1 km SW Jabal al-Qarah, al-Hasa Oasis; 8665, 6 km NE al-Hufuf, al-Hasa Oasis.

46. ASCLEPIADACEAE

Herbs or shrubs with opposite leaves and often milky sap. Flowers actinomorphic, 5-merous, gamopetalous, the filaments usually connate and the anthers adnate to the pistil forming a gynostegium. Corona of 5 or 10 segments present at base of the stamens. Fruit a follicle, often in pairs, elongate or inflated. Seeds compressed and often comose.

Key to the Genera of Asclepiadaceae

1. Twining plants with cordate or triangular-sagittate leaves
    2. Follicles lanceolate-ovoid, spiny-tubercled .... 4. *Pergularia*
    2. Follicles narrow-fusiform, not tubercled; green oasis weed .. 5. *Cynanchum*
1. Non-twining plants with leaves not as above, or leaves absent
    3. Leafless shrub with follicles narrowly linear-fusiform .... 3. *Leptadenia*
    3. Leafy shrub or herb with follicles swollen, not fusiform
        4. Large woody shrub with leaves 10 – 25 cm long .... 2. *Calotropis*
        4. Low herb with leaves 1 – 3 cm long .... 1. *Glossonema*

● 1. **Glossonema** Decne.

Perennial herbs with flowers in axillary cymes. Corolla deeply divided. Follicles inflated, ovoid-acuminate.

Glossonema varians (Stocks) J. D. Hooker
G. edule N. E. Br.

Ascending-spreading, branched, rounded perennial lactiferous herb, grey-green and densely pubescent with short straight white hairs, 8 – 20(30) cm high. Leaves opposite or sometimes appearing alternate, ovate to suborbicular or deltoid, obtuse, repand-undulate, rounded or truncate at base, mostly 1 – 2 cm long and broad, on petioles ⅓ – ⅔ as long as the blades. Flowers mostly about 5 together, in axillary umbellate cymes, on pedicels c. 2 mm long. Calyx c. 2 mm long, deeply divided with lanceolate-triangular acute lobes, hairy; corolla c. 4 mm long, divided slightly more than half way to base with yellow-brown, elliptical acute lobes obtusely keeled at back and sometimes sparsely pubescent on the keel; corona 5-lobed, pale, somewhat shorter than the calyx. Follicle inflated, ellipsoid, 3 – 5(6) cm long, 1.5 – 2.5 cm wide, somewhat attenuate at apex, smooth-glaucous, bearing soft conical tubercles c. 6 mm long. Seeds comose. *Plate 155.*

Habitat: Sand in cracks in limestone, often on rocky elevated ground. So far not found north of Dhahran, where it is occasional on limestone. Infrequent.

Central Coastal Lowlands: 29, 142, 1034, 1823(BM), Dhahran.

South Coastal Lowlands: 7841, 8318, 5 km SE al-Hufuf.

Southern Summan: 1877, 14 km N Harad.

Vernacular Names and Uses: The plant: KURRAYSH ('paunch-weed', probably referring to the baglike fruits, Al Murrah), ʿITR (Al Rashid, Bani Hajir), KUBBAYSH (Qahtan), KABŪSH (Yemeni tribesmen), ʿANTAYR (Sulubah, Hutaym). The fruits: ʾITRĪ (Qahtan), KABASH (Al Rashid), JARW ('cubs', Bani Hajir). This plant is edible, and the author has found the raw young fruits quite palatable, with a flavor somewhat like sweet cabbage. With maturity they become practically inedible because of their tougher texture and comose seeds. The very young leaves are also edible.

● 2. **Calotropis** R. Br.

Shrubs with large, broad subsessile fleshy leaves. Flowers in umbellate cymes. Calyx 5-partite; corolla 5-lobed. Fruit a smooth, ovoid, inflated spongy follicle.

Calotropis procera (Ait.) Ait. f.

Ascending to erect glaucous shrub with copious latex, woody below with pale corky bark, coarsely succulent-herbaceous above, 1.5 – 4 m high, sometimes assuming treelike form up to 5 m high. Leaves opposite, oblong or obovate, sessile, obtuse with acute mucro, weakly auriculate at base, 10 – 25 cm long, 8 – 17 cm broad, fine mealy-tomentose when young, becoming glabrous. Inflorescence an umbellike cyme on stout primary peduncles c. 5 cm long from the upper axils. Flowers 1.5 – 2 cm across (3 cm with corolla lobes spread) on pedicels 2 – 2.5 cm long. Calyx c. ¼ to ⅓ as long as the corolla, divided more than half way to base with triangular acute lobes c. 4 mm long; corolla subcampanulate, c. 1.6 cm long, divided about half-way to base with ovate-triangular lobes pale greenish white outside, purple flushed in the upper half within. Stigmatic cap pale, pentagonal, with anthers adnate at the angles. Corona of 5, laterally compressed segments, shiny purple above, radiating from the gynostegium. Fruit a smooth or somewhat wrinkled, inflated, ovoid follicle 8–13 cm long. Seeds comose, wind dispersed. *Plates 156, 157.*

Easily recognizable by its great glaucous leaves, unlike those of any other non-cultivated plant of our area, and by its copious latex bleeding from the slightest wound. **Habitat:** Often on disturbed waste ground; rarer in desert, on silty soil sometimes with thin sand cover. Occasional around Dhahran and other towns on vacant lots and roadsides.

Northern Summan: sight record at Umm 'Ushar, 27–43N, 45–04E.

Central Coastal Lowlands: 1080(BM), Dhahran.

Vernacular Names and Uses: 'USHAR (gen.). This plant is often considered to be poisonous, but available evidence from experiments on small mammals (Verdcourt and Trump 1969) suggest that the reported toxicity of the raw latex may be somewhat exaggerated. There are, however, records of bark extracts being used in some tropical countries in the preparation of arrow poisons. Various alkaloids or other principles have been isolated from the plant, and these in concentrated form are known to be toxic. A powerful cardiac poison, gigantin, has been extracted from a related species, *P. gigantea*, of India (UNESCO 1960). Bedouin tradition in eastern Arabia does hold it to be poisonous, and livestock are said to avoid it. There is evidence for this in the plant's persistence on ruderal sites around villages and towns where household goats and sheep would otherwise be expected to browse it to destruction soon. Small doses of the latex are sometimes used medicinally by Bedouins, although its specific indications are usually rather vague. An elder of the Bani Hajir tribe told the author that a safe but effective dose can be obtained by scooping out the seeds and pulp from a halved ripe follicle and drinking a full measure of sheep, goat or camel milk from the resulting hollow, cuplike, still-green skin. Enough of the active principle is said to be absorbed from the fruit wall to be effective, and not enough of the latex will be consumed to pose any danger. *'Ushar* is also well known to many Bedouins who remember the shrub as the source of wood used in preparation of the best charcoal for the manufacture of black gunpowder.

● 3. Leptadenia R. Br.

Virtually leafless shrubs with flowers in short axillary clusters. Corolla rotate; corona of short lobes. Follicles linear-terete.

Leptadenia pyrotechnica (Forssk.) Decne.

Ascending, dense, many-branched shrub with clear yellowish juice and green, wandlike branches, 1.5–3(5) m high, virtually leafless except occasionally for small, soon-deciduous linear-lanceolate rudiments on young spring growth. Flowers subsessile, clustered in short axillary cymes; calyx fine-tomentose with triangular lobes c. ⅓ the length of the corolla limb; corolla rotate, c. 6 mm in diameter, yellow-green, with elliptical-triangular acute lobes convex above, concave and weakly revolute-margined at back, fine tomentose. Corona of very short lobes alternating with the petals. Follicles terete, linear, striate, tapering toward apex, 9–13 cm long, c. 0.8 cm wide. Seeds comose. **Habitat:** Usually on the sandy flanks of limestone hills. More frequent south of Dhahran but occasionally seen at coastal points to the north.

Central Coastal Lowlands: 235, Dhahran; al-Jubayl (Dickson 1955, 59); Ras Tanaqib (sight record).

South Coastal Lowlands: 1957, 21 km S Jal al-Wutayd; 8 km SW al-'Udayliyah (sight record).

Vernacular Names and Uses: MARKH (gen.). The young flowers and young fruits, MA'ALĪT (Bani Hajir), are considered edible by southern tribesmen. The shrub is also known for having provided in earlier days a fine silky tinder much used for catching the sparks of flint and steel in firemaking. The dried hair tufts from the seeds were employed for this, and this use was presumably the basis for Forsskal's specific epithet.

● 4. Pergularia L.

Twining shrubs with cordate or reniform leaves. Flowers in axillary umbellate cymes, with rotate corolla and double corona, the outer one 10-lobed, the inner of 5 segments. Follicles inflated-lanceolate, tapering to apex.

Pergularia tomentosa L.
Daemia cordata (Forssk.) R. Br. ex Schult.
Greyish-tomentose twining lactiferous shrub. Leaves opposite in somewhat distant pairs, distinctly cordate, acute, apiculate, 1.5 – 3(5) cm long, on petioles 8 – 15 mm long. Flowers in axillary pedunculate umbels, on pedicels longer than the flower; calyx lobes acute, hirsute; corolla whitish, c. 10 mm in diameter, with oblong acute lobes, longitudinally downfolded, ciliate. Follicles lanceolate-ovoid, tapering to apex, spiny-tubercled, fine tomentose, 4 – 5 cm long, 1.5 – 2 cm wide. Seeds ovate, compressed, brown with paler margins, 7 – 10 mm long, furnished at apex with a pappus of fine white silky hairs about 15 mm long.
Habitat: Silty basins. Rare to locally occasional.
South Coastal Lowlands: 352, 1053, 9 km N al-Mutayrifi, al-Hasa.
Southern Summan: 348, 16 km N Harad; 1875, 32 km S Jaww al-Hawiyah; 25 – 00N, 48 – 30E (sight record).
Vernacular Names and Uses: GHALQAH (gen.). Formerly used to remove the hair from hides before tanning.

● 5. Cynanchum L.

Perennial herbs with twining or erect stems. Leaves simple, often cordate or sagittate at base. Inflorescence an umbel-like cyme. Calyx divided with lanceolate lobes. Corolla rotate, deeply 5-partite. Follicles fusiform, usually solitary.

Cynanchum acutum L.
Perennial, widely climbing and scrambling herb, hairy in parts but overall appearing glabrous, with numerous twining stems. Leaves mostly triangular, acute, sagittate at base with the basal lobes rounded, in our specimens 4 – 8(11) cm long, 2 – 6(8) cm wide, on petioles ⅓ to ½ the length of the blade. Flowers in peduncled, umbel-like cymes; sepals 1 – 1.5 mm long, narrowly lanceolate, acute, hirsute; corolla (in ours) white, c. 4 – 5 mm long. Follicle (not yet seen by the author in the few specimens from our area) fusiform, long-tapering at apex, 9 – 15 cm long, 0.4 – 0.8 cm wide.

Habitat: Wet ground in oases, particularly along canal or irrigation ditches. Apparently rare.

South Central Lowlands: 8653 (legit S. Collenette), al-Hasa Oasis; 8672, 1 km N 'Ayn Luwaymi, al-Hasa Oasis.

● This plant appears to have been introduced as an oasis weed as recently as the 1980s, and its distribution is so far restricted. It may be expected to spread, however, and it has the potential to become a troublesome weed of canal banks and perhaps some crops. It has been seen climbing on date palms in al-Hasa Oasis.

47. SOLANACEAE

Herbs or shrubs, sometimes toxic or narcotic, with leaves alternate, fascicled, or rarely the upper ones opposite. Flowers bisexual, usually actinomorphic, 5-merous with 5-lobed calyx and rotate, campanulate or tubular 5-lobed corolla. Ovary superior, 2-carpellate. Fruit a capsule or a berry.

Key to the Genera of Solanaceae
1. Shrub over 1.5 m high with rigid woody twigs becoming
   spinescent . . . . 1. *Lycium*
1. Herbs, rarely shrubby, under 1.5 m, unarmed
   2. Flowers 5 cm or more long; fruit a spiny capsule . . . . 5. *Datura*
   2. Flowers shorter than 3 cm; fruit not spiny
      3. Fruit a berry
         4. Berry enclosed in the inflated calyx . . . . 3. *Withania*
         4. Berry exposed, much longer than the calyx . . . . 4. *Solanum*
      3. Fruit a capsule enclosed in the persistent calyx . . . . 2. *Hyoscyamus*

● 1. Lycium L.

Shrubs or small trees with rigid woody branchlets becoming thorns, and clustered leaves. Flowers solitary or clustered, with cuplike or tubular 5-dentate calyx and tubular or funnel-shaped corolla with spreading limb. Stamens 5, inserted near middle of the corolla tube. Fruit a berry.

Key to the Species of Lycium
1. Corolla funnel-shaped, c. 8 – 10 mm long, with limb ½ to 1 times as
   long as the exposed tube; stamens nearly equal, clearly
   exserted . . . . 1. *L. depressum*
1. Corolla tubular, c. 12 – 15 mm long, with limb ¼ to ⅓ length of the tube.
   Stamens unequal, included or some shortly exserted, others
   included . . . . 2. *L. shawii*

1. **Lycium depressum** Stocks
   L. barbarum sensu Boiss.
Ascending-erect dense glabrous shrub, 1.5 – 3.5 m high, with many rigid branchlets becoming more or less spinescent. Leaves clustered, obovate-oblong to spathulate, obtuse or acute, tapering to a short petiole or subsessile, 1 – 3(4) cm long, 0.3 – 0.8(1) cm wide. Flowers usually in clusters of 3 – 8 on pedicels 3 – 10(15) mm long. Calyx

cuplike or short-tubular, with unequal teeth, 1 – 3 mm long; corolla 8 – 10 mm long, funnel-shaped with lobes of the limb ½ to 1 times as long as the tube, pale violet or whitish with branched purple lines on each lobe. Stamens equal or subequal, exserted from the corolla to a distance about equal to the limb. Berries globose, orange-red, smooth, 4 – 6 mm in diameter. *Plate 158.*

● This species is so far recorded only from a few locations in the Summan but may be more widespread. In addition to the key characters above, the following points will assist in separating it from the more common *L. shawii:* Leaves and pedicels virtually glabrous, or with a few scattered minute glands, while *L. shawii* nearly always is shortly, finely, but definitely tomentose, at least toward the leaf bases. The flowers are clustered 3 – 5(8) together, while *L. shawii* has flowers solitary or, less frequently, 2 together. It also has a somewhat denser habit, with shorter internodes, than does *L. shawii.* The two species have been collected growing together (Nos 3126 above and 3127 below) in 28 – 26N, 48 – 36E. *L. depressum* occurred here as an isola*t*ed stand of about 9 shrubs surrounded for a great distance all around by scattered individuals or groups of *L. shawii.*

Habitat: Silty soil, often over rocky substrate, usually well inland. Occasional.

Northern Summan: 503, Rawdat Ma'qala; 3126(BM), 6 km SE ad-Dabtiyah, 28 – 26N, 48 – 36E; 7528, 64 km E Umm 'Ushar (a questionable record, somewhat intermediate between this and the following species).

Vernacular Names and Uses: ʿAWSAJ (gen.). Other notes for *L. shawii,* below, apply also to this species.

2. **Lycium shawii** Roem. et Schult.

L. persicum Miers
L. arabicum Schweinf. ex Boiss.

Ascending, dense, intricately rigid-branched shrub, 1.5 – 2.5 m high, fine tomentose at least on younger parts, rarely glabrescent. Leaves elliptical-oblanceolate to spathulate, obtuse or acute, tapering at base to a short petiole or subsessile, 1 – 2(3) cm long, 0.3 – 0.8(1) cm wide. Flowers solitary or rarely 2 together, 13 – 16 mm long; calyx tubular, c. 3 – 4 mm long, 5-toothed; corolla c. 15 mm long, narrowly tubular, with the obtuse lobes of the limb ¼ to ⅓ the length of the tube, variable in color from white through pink to purple, sometimes lined. Stamens distinctly unequal, all included or 2 somewhat exserted. Berries globose, red, c. 4 – 5 mm in diameter. *Plate 159.*

Habitat: Stable coastal and inland sands and silts, tolerant of disturbed ground and sometimes seen around abandoned village and ruin sites. Frequent.

Northern Plains: 3193, al-Batin, 28 – 00N, 45 – 27E; 3196, al-Batin, 27 – 56N, 45 – 23E; 1205(BM), Jabal an-Nu'ayriyah.

Northern Summan: 7469, al-Lahy ar-Rayyan, 26 – 37N, 48 – 09E; 3127, 6 km SE ad-Dabtiyah, 28 – 26N, 48 – 36E.

Central Coastal Lowlands: Dhahran; frequent in the coastal regions to the north.

South Coastal Lowlands: 1108(BM), 10 km W Abqaiq.

Southern Summan: 2175, 24 – 25N, 48 – 41E.

Vernacular Names and Uses: ʿAWSAJ (gen.), ʿAWSHAJ, ʿAWSHAZ (N dialectal forms). The berries, DAWM (Bani Hajir), of *Lycium* are edible when ripe and somewhat sweet to the taste. According to an old superstition among some Bedouins ʿawsaj shrubs are an abode of *jinn* (malevolent or mischievous spirits). Musil (1928) reports that for this reason the plants are not cut for fuel by the Rawalah. Dickson (1955) says that Bedouins of the Kuwait area avoid cutting the plant because of the *jinn* association. H. Dickson (1951, 537 – 38) writes also that in the Kuwait area *Lycium* bushes are sometimes seen surrounded by stones at the base because of some Bedouins' habit of appeasing the *jinn* by throwing a stone into the bush while passing and repeating some protective phrase such as *bismallah* ('in the name of God'). Bedouins in our area also say the shrub is not used for fuel; it is not very suitable for this in any case, being rather thorny and difficult to break. An elderly Bani Hajir acquaintance of the author was able to recount a number of campfire anecdotes and folk stories relating to *'awsaj* bushes and *jinn*. *Lycium* is frequently seen around ruin sites or desert burial grounds in the central part of our area, and this may account at least partially for the apparent wealth of superstition associated with it.

● 2. Hyoscyamus L.

Annual or perennial herbs, usually pubescent, with simple alternate leaves. Flowers axillary, usually in leafy, one-sided racemes or spikes. Calyx tubular, 5-dentate, often indurate in fruit; corolla funnel-shaped to campanulate, more or less zygomorphic with 5 unequal lobes. Stamens inserted at base of the corolla. Fruit a circumscissile capsule enclosed in the calyx.

Hyoscyamus pusillus L.

Annual herb, simple or branched from the base with stems erect or ascending, cobwebby-pubescent, 3 – 20(30) cm high. Leaves oblanceolate-elliptical in outline, usually coarsely dentate or pinnately lobed with acute lobes, or subentire, 1 – 6 cm long, 0.5 – 3 cm wide, smaller above. Flowers subsessile in 1-sided, spicate racemes. Calyx small in flower, much growing in fruit, becoming tubular-campanulate with spreading, acute, somewhat prickly teeth, distinctly veined, indurate-coriaceous, c. 1.5 – 2 cm long and 0.8 – 1 cm wide at mouth. Corolla 8 – 15 mm long, distinctly irregular, pale yellow, dark purple in the throat, with 5 lobes towards one side. Capsule enclosed in the calyx and about ½ its length, opening from above by a circumscissile, dome-shaped lid. Seeds c. 1.2 mm long, brown, minutely short-tuberculate. *Plate 160.*

Habitat: Sands or silty sands. Sometimes around desert wells. Rare to occasional.

Northern Plains: 1626(BM), 3 km ESE al-Batin/al-'Awja confluence; 8715, 20 km SW ar-Ruq'i wells, 28 – 51N, 46 – 25E.

Northern Summan: 625(BM), Wabrah; 3166(BM), 11 km SSW Hanidh; 3346, 5 km WSW an-Nazim; 8585, Dahl al-Furayy, 26 – 47N, 47 – 05E.

Central Coastal Lowlands: 76(BM), Thaj; 7468, al-Hinnah.

● 3. Withania Pauquy.

Shrubby plants with entire, alternate or opposite leaves. Calyx campanulate, 5 – 6-toothed, much enlarging in fruit and enclosing the berry. Corolla narrow-campanulate, 3 – 6-lobed with stamens inserted near base. Fruit a berry.

Withania somnifera (L.) Dun.

Ascending to erect shrub, more or less greyish-tomentose, to c. 1.5 m high. Leaves sometimes opposite above, ovate, acute, entire, 3 – 10 cm long, 1 – 7 cm wide, on petioles

c. ⅕ to ¼ as long as the blade. Flowers in often dense axillary clusters, short-pedicelled; calyx cupuliform, tomentose, with linear lobes, growing to c. 18 mm long in fruit; corolla yellowish-green, tomentose, campanulate, exceeding the calyx, with spreading lobes. Fruit a red berry, c. 5 mm in diameter, enclosed in the much enlarged calyx, the latter with connivent linear lobes 2 – 3 mm long at apex. *Plate 161.*

Habitat: A weed of waste ground and cultivated areas. Occasional.

South Coastal Lowlands: 2900(BM), al-'Arfaj farm, al-Hasa; 8664, 2 km E al-Hufuf, al-Hasa Oasis.

Vernacular Names and Uses: ḤAML BALBŪL ('bulbul fruit', Hasawi gardeners). *Withania* is a sedative narcotic, the principal active alkaloid being somniferine (Emboden 1972), and it has a wide variety of medicinal uses in India. It is not, apparently, highly toxic but should be treated with some respect.

● 4. Solanum L.

Herbs (ours), shrubs or trees with simple to pinnate leaves and axillary corymbose or umbellate inflorescence. Flowers nearly always actinomorphic with lobed calyx and 5-lobed rotate or campanulate corolla. Stamens inserted at the corolla base. Fruit a globose or somewhat elongated berry.

Solanum nigrum L.

Ascending or decumbent annual herb, glabrous or pubescent, to c. 50 cm high. Leaves ovate to rhomboid-lanceolate, acute or (less often) obtuse, entire or sinuate-dentate, mostly 2 – 8 cm long, 1 – 5 cm wide, tapering below to a petiole shorter than the blade. Flowers 3 – 6 in peduncled umbellate inflorescences; pedicels c. 3 – 5 mm long, elongating in fruit to c. 10 mm; corolla white, rotate, 2 – 3 times as long as the calyx, c. 7 mm in diameter. Fruit globose, black, or in some varieties red, c. 5 – 8 mm in diameter.

Habitat: A weed of farm and gardens. Occasional.

Northern Plains: 7557, Qaisumah Pump Station.

Central Coastal Lowlands: 7414(BM), Dhahran; 7773, 4 km S al-Qatif town.

South Coastal Lowlands: 8673, 3 km NE al-Hufuf, al-Hasa Oasis.

Vernacular Names and Uses: SHAJARAT AL-BALBŪL ('bulbul bush', referring to the oasis bird's supposed fondness for the berries, Qatifi farmers). This nightshade is sometimes listed as a poisonous plant, but the author has seen farmers in al-Qatif pluck the ripe berries and eat them raw, laughing at the suggestion that they might be dangerous. The plant's content of the toxic alkaloid, solanine, may vary by variety and environment; it would be inadvisable to eat the berries in any quantity.

● 5. Datura L.

Strong-odored narcotic and toxic herbs, rarely shrubs or trees, with simple alternate leaves. Flowers very large, solitary, axillary, with tubular 5-dentate calyx and funnel-shaped, 5 – 10-lobed corolla. Stamens equal, included, inserted near base of the corolla. Fruit a spiny capsule.

Key to the Species of Datura

1. Mature fruit deflexed; corolla 12 – 18 cm long .... 1. *D. fastuosa*
1. Mature fruit erect; corolla 5 – 10 cm long .... 2. *D. stramonium*

1. **Datura fastuosa** L.

D. metel L.

Coarse, glabrous branched annual or perennial, sometimes exceeding 1 m in height. Leaves ovate to sublanceolate, acutish, more or less coarsely toothed, often asymmetrical at base, 10 – 20 cm long, 5 – 15 cm wide, on petioles half or more as long as the blade. Calyx tubular, dentate at apex, 6 – 10 cm long; corolla funnel-shaped, (in ours) white, 12 – 18 cm long, the limb with 5 small, acute tails between the 5 main lobes. Capsule subglobose, deflexed when mature, 2.5 – 4 cm in diameter, tubercled to spiny, usually dehiscing irregularly. Seeds ear-shaped or reniform, compressed, 5 – 6 mm long, smooth-wrinkled, brown.

Habitat: A weed of gardens, waste ground. Rare.

Central Coastal Lowlands: 8426, Dhahran; 8650, Ras Tanura. Seen by the author only as a sporadic in town gardens.

• A poisonous plant with narcotic and intoxicant properties similar to, but apparently less dangerous than, those of *D. stramonium* (q.v. below).

• Another *Datura* with flower dimensions similar to *D. fastuosa* might be encountered as a weed in our area: *D. innoxia* L. This also has deflexed, irregularly dehiscing fruits but may be recognized by its dense, short, greyish pubescence. The flowers are white.

2. **Datura stramonium** L.

Stout annual with ascending to erect branched stems, glabrous to partly pubescent, ranging in height from c. 20 cm to over 1 m. Leaves ovate to lanceolate-oblong, acute, irregularly and coarsely sinuate-dentate, petiolate, 6 – 17 cm long, 5 – 13 cm wide. Flowers erect; calyx tubular, 3.5 – 4.5 cm long, with triangular-acuminate lobes; corolla funnel-shaped, 5 – 10 cm long, white or purple with 5 broad lobes. Capsule erect, ovoid, 3 – 4 cm long, covered with slender spines c. 5 – 7 mm long, short-pedicelled, dehiscing regularly from the apex by 4 valves. Seeds blackish, reniform, with minutely pitted faces, 2.5 – 3.5 mm long.

Habitat: A weed around gardens or farms, or on waste ground. Rare, except perhaps locally.

Central Coastal Lowlands: One sight record of a *Datura* by the author at Dhahran may have been this species. It is included here primarily on the basis of its listing by Chaudhary, Parker and Kasasian (1981) as a weed observed on one or more croplands in eastern Saudi Arabia.

• Jimson weed: a dangerously poisonous plant but seldom ingested by man or livestock because of its strongly unpleasant scent and bitter principles. It contains several toxic alkaloids, including hyoscyamine, hyoscine, and atropine. In small doses it is an intoxicant; larger amounts lead to convulsions, stupor and death. The seeds are more strongly poisonous than other parts, and the alkaloids in all parts remain active after drying (Verdcourt and Trump 1969).

48. CONVOLVULACEAE

Annual or perennial herbs, climbers, or shrubs, sometimes twining, with alternate simple leaves. Flowers bisexual, 5-merous, with sepals free or connate and corolla usually funnel-shaped with entire or lobed limb. Stamens free. Ovary superior, 2-loculed; fruit a capsule or indehiscent.

Key to the Genera of Convolvulaceae
1. Both flowers and leaves under 7 mm long; styles 2; stigmas capitate . 1. *Cressa*
1. Flowers and leaves longer; style 1, bifid with filiform or linear
   stigmas .... 2. *Convolvulus*

● 1. Cressa L.

Perennial herbs or subshrubs with small entire leaves. Sepals free, subequal; corolla campanulate with 5 reflexed lobes. Stamens exserted; ovary 2-loculed with 2 styles, each with a capitate stigma. Capsule dehiscing by 2 or 4 valves, usually 1-seeded.

Cressa cretica L.
Erect or ascending, rarely prostrate perennial, grey-green with both appressed and spreading hairs, 10–30(40) cm high. Leaves ovate to lanceolate, sometimes subcordate, sessile or very short-petioled, acute, 3–6 mm long, 1–4 mm wide, often with scattered crystals of exuded salt. Flowers in short, dense spikelike racemes at branch apices; pedicels c. 0.5–1 mm long. Sepals obovate, c. 3 mm long, acutish; corolla c. 5–6 mm long, divided about half-way to base into oblong to long-triangular acutish, reflexed, white to cream, somewhat membranous lobes, pubescent at tips. Stamens exserted. Styles usually exserted, with capitate stigmas.

Habitat: A weed of saline ground in and around agricultural areas or salt marshes, or around *sabkhahs*. Common.

Central Coastal Lowlands: 23(BM), Dammam; 225, Abu Ma'n; 1099(BM), 7767, al-Qatif; 2958(BM), Bakha.

Northern Summan: 2883, Rawdat Musay'id, 26–02N, 48–43E.

South Coastal Lowlands: 1910, al-Mutayrifi, al-Hasa.

Vernacular Names: SHUWWAYL (Bani Hajir, Bani Khalid, gen.).

● 2. Convolvulus L.

Annual or perennial herbs or shrub, sometimes twining, with simple leaves. Flowers solitary or in cymose inflorescences; sepals free, somewhat unequal; corolla usually funnel-shaped, entire or lobed, often with 5 pubescent longitudinal bands. Stamens included, inserted near base of the corolla. Fruit a 2-locular, 1–4-seeded capsule.

Key to the Species of Convolvulus
1. Shrublet with rigid spinescent branches; flowers sessile,
   axillary .... 6. *C. oxyphyllus*
1. Unarmed shrublets or herbs; inflorescence peduncled
     2. Leaves sagittate, hastate or cordate at base, abruptly petiolate
       3. Leaves shallowly lobed around all of the margin; corolla 10 mm
         or less long .... 5. *C. fatmensis*
       3. Leaves entire except at base; corolla 15 mm or more
         long .... 1. *C. arvensis*
     2. Leaves not sagittate, hastate or cordate at base, sessile or tapering
       gradually to a petiole
       4. Corolla 15–20 mm long .... 3. *C. cephalopodus*
       4. Corolla 8–14 mm long

5. Sepals acuminate, villous, with hairs mostly spreading and longer than
   width of the sepals or nearly so
   6. Corolla 8 – 11 mm long; stem hairs mostly erect, spreading, with
      appressed hairs below them sparse or absent .... 2. *C. cancerianus*
   6. Corolla 12 – 14 mm long; stem with distinctly appressed hairs, these
      overtopped by very sparse to rather abundant spreading hairs . 4. *C. deserti*
5. Sepals oblong-ovate, acute but short-pointed, with hairs mostly appressed and
   shorter than width of the sepal .... 7. *C. pilosellifolius*

Chaudhary (1983) records *C. pentapetaloides* L. from the al-Hasa Oasis. This species
has a yellow and blue corolla with a distinctly lobed limb. It has not been collected
by the author, and there may be some doubt about the identity of the specimen referred
to, which does not appear to have the nearly glabrous, submembranous sepals typical
of *C. pentapetaloides*.

Another species possibly present in the Eastern Province is *C. siculus* L., with small
blue flowers; the author has seen a plant somewhat resembling it growing as a weed
in the al-Hasa Oasis area.

1. **Convolvulus arvensis** L.

Perennial herb, glabrous or sometimes pubescent on young shoots, prostrate or climbing
with twining stems up to 1 m long. Leaves petiolate, sagittate or hastate with 2 small
obtuse lobes at base, triangular to linear-oblong, 1.5 – 4(5) cm long, 0.5 – 1.5(2) cm
wide, becoming smaller above. Peduncles exceeding the leaves, usually 1-flowered,
with pedicels equalling or exceeding the calyx. Sepals oblong to obovate, obtuse,
glabrous, c. 4 mm long; corolla 15 – 20(25) mm long, 20 – 25 mm in diameter, white
to pale pink or sometimes bluish, glabrous, with reddish, glossy, longitudinal stripes
at back. Capsule glabrous, globose, c. 6 mm long.

Habitat: A weed of gardens and farms, where it is a serious and persistent agricultural
pest climbing over and choking other plants. Common in cultivation throughout the
Eastern Province.

Central Coastal Lowlands: 433, Dhahran stables farm.

Vernacular Names: FAḌAKH (Hasawi farmers), FADGHAH (Qatifi farmer), KHAṬMĪ
(Shammar).

2. **Convolvulus cancerianus** Abdallah et Sa'ad

Somewhat shrubby, procumbent to ascending perennial, branching from near the base
with stems to c. 75 cm long. Branches short-villous with spreading white to brownish
hairs, sometimes with sparser somewhat appressed pubescence below them. Leaves
sessile, lanceolate to oblong, rather weakly acute, those on the branches to c. 35(40)
mm long, 8(10) mm wide, becoming gradually smaller above, all with sparse to
moderately dense spreading or weakly appressed hairs. Flowers mostly 2 – 6 together,
crowded at the ends of short-villous peduncles mostly well exceeding the subtending
leaves below but becoming shorter above. Sepals (in ours) mostly 5 – 6 mm long, ovate
to elliptical, acuminate, villous and ciliate with whitish hairs. Corolla white to pale
pink, (in ours) mostly 8 – 11 mm long.

This one of our specimens (No. 8375), appearing marginally distinct from the complex
included below under *C. deserti* s.l., is assigned provisionally to this species on the

basis of its predominantly spreading indumentum, larger leaves, and somewhat smaller flowers. A critical revision of Arabian material will be required to determine, with any certainty, the status of this taxon in our area.

Habitat: Silty soil or thin sand over silty terrain. Apparently rather rare as recognized to date.

Southern Summan: 8375, 38 km S 'Uthmaniyah GOSP 13, along road.

3. Convolvulus cephalopodus Boiss.

    C. sericeus Burm. f. non L.

    C. buschiricus Bornm.

Ascending or decumbent, rarely prostrate shrublet, many-branched from the base with stems more or less white-woolly and villous, to c. 60 cm high. Leaves linear-oblong to lanceolate, appressed-pubescent, the lateral nerves usually distinct and depressed and the margins sometimes somewhat repand-wavy, the lowest long-tapering to a petiole and obtusish, 2 – 6(8) cm long, (0.5)0.7 – 1.2(2) cm wide, becoming shorter, lanceolate-elliptical, sessile and acute above. Flowers clustered several together, but usually opening singly, on peduncles somewhat shorter than or exceeding the subtending leaf. Sepals lanceolate-acuminate, white-villous, 5 – 9 mm long; corolla rather showy, c. 15 – 20 mm long, hardly lobed, pink or sometimes nearly white; ovary white-villous at apex. *Plates 162, 163.*

• As pointed out by Dorothy Hillcoat (personal communication), *C. buschiricus* Bornm. is almost certainly a synonym of the earlier *C. cephalopodus* Boiss. The type of *C. cephalopodus* (duplicate seen by the author at BM) has tightly clustered, small leaves, but this is a variable seasonal or edaphic effect. A specimen at BM from Sharjah (Guichard No. KG/83/Oman) has some leaves like those of this type, and others up to 2.5 cm long quite like those of our usual form.

Certainly the showiest of our *Convolvulus*, presenting an impressive morning display of delicate pink flowers in the sands of the coastal lowlands.

Habitat: Stable sands, perhaps most commonly in coastal districts but also inland on somewhat silty ground. Frequent. Tolerant of disturbed ground and sometimes at roadsides.

Northern Plains: 3123(BM), al-Batin, 30 km SW Hafar al-Batin.

Northern Dahna: 3882, 25 km W Umm 'Ushar.

Northern Summan: 2872, 5 km NE al-Jawwiyah wells; 3167(BM), 11 km SSW Hanidh.

Central Coastal Lowlands: 913(BM), Jabal Haydaruk; 133, 408, 552(K), 1031(BM), 1083(BM), Dhahran; 1785(BM), 2 km SE Abu Hadriyah; 1787(BM), 13 km E Abu Hadriyah; 2887, 25 – 51N, 49 – 08E; 3731, Ras az-Zawr; 8127, al-Wannan; 4057, az-Zulayfayn, 26 – 51N, 49 – 51E.

Southern Summan: 8016, 24 – 00N, 49 – 00E; 8052, 24 – 06N, 48 – 48E; 8069, Wadi as-Sahba, 24 – 12N, 48 – 41E.

Vernacular Names: RUKHAYMĀ (Al Murrah, Al Rashid), RUKHĀMĀ (Qahtan), RUKHKHĀMĀ (Bani Khalid, Shammar, Rawalah).

4. Convolvulus deserti Hochst. et Steud. s.l.

Shrubby, greyish to pale green, procumbent to ascending perennial, of variable size, sometimes rather diffuse and somewhat straggling, with stems spreading from the base, 20 – 100 cm long. Stems with appressed hairs predominating but these sometimes overtopped by rather abundant spreading ones. Leaves linear to oblong or lanceolate, acute or obtuse, mostly sessile, the lowest tapering to a petiole, appressed-pubescent, up to c. 25 mm long, 5 mm wide below, smaller above and often smaller throughout. Flowers 1 – 3(4) together on straight, rather rigid, rarely branching peduncles mostly well exceeding the subtending leaf and up to c. 4 cm long. Sepals elliptical, more or less acuminate, c. 5 – 6 mm long, villous with spreading white or brownish hairs mostly longer than the width of the sepal; corolla white to pale pink, c. 12 – 14 mm long, with villous longitudinal bands outside, drying pinkish-brown.

● The specimens assigned here provisionally to this species form a confusing complex of apparently intergrading forms that requires more detailed study. Sa'ad's (1967) monograph does not seem to be of much help in segregating the plants in our area, and the presence of closely allied species in this group will have to be tested through careful review of a large Arabian specimen base. The leaves in our specimens vary from linear-lanceolate-acute to oblong-obtuse, and the stem pubescence ranges from closely appressed throughout through various proportions of mixed appressed and spreading hairs.

Habitat: Silty ground in desert basins, wadi bottoms; also on disturbed ground. Occasional from Dhahran to the south.

Central Coastal Lowlands: 641(BM), 1030(BM), Dhahran.

South Coastal Lowlands: 7836, 10 km SE al-Hufuf; 7050(BM), 34 km N al-Hufuf; 8288, 6 km N al-Mutayrifi.

Southern Summan: 8008, 24 – 00N, 49 – 00E; 8376, al-Ghawar, 31 km S 'Uthmaniyah GOSP 13.

5. **Convolvulus fatmensis** Kunze

Prostrate annual or perennial herb with trailing stems up to c. 40 cm long, sparingly pubescent at least on young parts. Leaves deltoid-oblong to ovate in outline, sagittate-cordate at base, with all of the margin shallowly lobed or dentate, mostly 1 – 2 cm long, 0.5 – 1 cm wide, on petioles ½ to ¾ as long as the blade. Flowers mostly solitary, axillary, on peduncles somewhat shorter than or equalling the subtending leaf. Outer sepals c. 4 mm long, oblong, with truncate mucronulate apex, scarious-margined. Corolla c. 8 mm long, pale pink. Ovary glabrous.

Habitat: Silt-floored basins, also apparently sometimes as a weed. Rare.

Northern Summan: 2884(BM), Rawdat Musay'id, 26 – 02N, 48 – 43E. Also noted by Chaudhary, Parker and Kasasian (1981) as a weed, probably in the al-Hasa area.

6. **Convolvulus oxyphyllus** Boiss. subsp. **oxycladus** Rech. f.

Ascending, rounded shrublet, 15 – 65 cm high, with rigid woolly-tomentose main branches arising from the base, bearing numerous straight, rigid, lateral branchlets, all at length becoming pungent-spinescent at apices. Leaves elliptical-oblanceolate to linear-spathulate, acute or obtuse, greyish-green and more or less woolly-tomentose, those below larger and tapering to a petiole, up to c. 4 cm long, 0.7 cm wide, those above smaller and sessile, grading above into small elliptical-ovate bracts; lateral nerves of leaves either distinct and somewhat depressed-plicate, or obsolete. Flowers mostly on the lateral branchlets, solitary in the axils, sessile, 8 – 10 mm long. Sepals c. 5 mm long, 2 mm wide, oblong-triangular, appressed-pubescent, somewhat unequal with the outer ones wider, acute or obtusish, greenish toward apex, paler below. Corolla

8 – 10 mm long, not or hardly lobed, white or near-white but often drying somewhat pinkish, with longitudinal pubescent bands. Stamens 5 mm long, inserted near base of the corolla; anthers c. 1.2 mm long. Ovary pubescent. *Plate 164.*

The assignment of our plant to this taxon should be considered provisional pending a detailed review of the group. Specimens have been described by some workers as being close to *C. harmrinensis* Rech. f.

Habitat: Silty inland soils; bottoms of the larger wadis; also sometimes in sands, as in the southern Dahna, but then always where sands are shallow and overlying silts. Rare to occasional overall, but sometimes frequent locally.

Northern Plains: 7850, al-Fulayj al-Janubi, 28 – 22N, 46 – 01E; 3124, 30 km SW Hafar al-Batin; 3208(BM), 28 km SW Hafar al-Batin; 545, 2 km N Qaryat as-Sufla.

Northern Summan: 217(BM), 25 km S Qaryat al-'Ulya.

Northern Dahna: 2328, al-Farshah, 21 km ESE Rumah.

Southern Dahna: 8360, 8 km ENE Rawdat at-Tawdihiyah; 8380, 7 km SE Qirdi GOSP.

Southern Summan: 2271, 16 km E al-'Udayliyah; 8366, S bank Wadi as-Sahba, 32 km W Harad.

Vernacular Names and Uses: 'ADRIS (Qahtan, gen.), 'UDRIS (Al Murrah). According to Dickson (1955, 33), who lists what most probably is this plant from Kuwait, Bedouin children sometimes suck a gum which exudes from its branches.

7. Convolvulus pisellifolius Desr. in Lam.

Prostrate or ascending perennial herb, more or less appressed-pubescent with some spreading hairs, with stems spreading from the base to c. 80 cm long. Leaves pale green, sometimes repand-wavy, the lower ones oblong-lanceolate, up to 7(8) cm long, 1 cm wide, tapering to an obscure petiole, those above considerably smaller, lanceolate, sessile. Flowers mostly 1 – 3 together on peduncles shorter than, or sometimes well exceeding, the subtending leaf. Sepals c. 5 mm long, elliptical-ovate, greenish above, paler below, acute but short-pointed, mostly appressed-hairy with hairs shorter than the width of the sepal, some spreading hairs at margins. Corolla 10 – 13 mm long, pink, or sometimes near-white. Capsule glabrous, ovoid, c. 5 mm long, with 3 – 4 seeds. Seeds dark brown, c. 2 mm long, somewhat trigonous, short-pubescent to glabrous.

Habitat: Silty basins and bottoms of the larger wadis. Also sometimes a weed around cultivated plots. Occasional to locally frequent.

Northern Plains: 3086(BM), al-Batin, 28 – 01N, 45 – 29E; 609(BM), Qulban Ibn Busayyis.

Northern Summan: 506, Rawdat Ma'qala; 7402, Batn Sabsab; 2881, Rawdat Musay'id, 26 – 02N, 48 – 43E; 8553, W edge Humr Mathluth, 26 – 57N, 47 – 53E; 8610, near Dahl Umm Hujul, 26 – 35N, 47 – 32E.

South Coastal Lowlands: 1096(BM), Abqaiq; 2909(BM), al-'Arfaj farm, al-Hasa (a non-typical form with identity somewhat questionable).

49. CUSCUTACEAE

Twining, virtually leafless parasites without chlorophyll. Flowers minute, 4 – 5-merous, with calyx lobes 4 – 5 and 4 – 5-lobed sympetalous corolla. Stamens 4 – 5, inserted between the corolla lobes with petaloid scales below. Ovary 2-loculed; styles 2 or 1. Fruit a capsule.

● Cuscuta L.

Annual threadlike, often yellowish, virtually leafless parasites adhering to their hosts by haustoria. Flowers minute, white or reddish-white, usually in headlike clusters. Sepals 4 or 5, connate at base; corolla 4 – 5-lobed, with 4 or 5 scales below and between the lobes. Stamens 4 – 5. Capsule ovoid to globose with 4 or less seeds.

Key to the Species of Cuscuta

1. Flowers mostly 4-merous .... 2. *C. pedicellata*
1. Flowers all 5-merous
   2. Stigmas capitate or subglobose; flowers short-pedicellate ... 1. *C. campestris*
   2. Stigmas filiform; flowers sessile .... 3. *C. planiflora*

1. **Cuscuta campestris** Yuncker

Twining parasite with stems c. 0.3 – 0.5 mm in diameter when dry, yellowish. Leaves absent; sometimes a few minute bracts present at base of the inflorescence. Flowers 2 – 3 mm long and broad, on pedicels c. 0.5 – 1 times as long as the flower, 5-merous, in rather poorly defined clusters. Calyx ovate-orbicular, obtuse; corolla lobes triangular, somewhat exceeding the calyx. Stamens exserted. Styles 2, persistent in fruit, with capitate-subglobose, often darker, stigmas. Capsule c. 3 mm in diameter, depressed-globose, pale, about half immersed in the perianth. Seeds light brown, c. 1 – 1.5 mm long.

Habitat: A parasitic weed on cultivated plants, found on a variety of hosts (see below with specimen records). Occasional to locally frequent.

Central Coastal Lowlands: 3795, on *Vinca*, Dhahran; 8130, on *Clerodendrum inerme*, Dhahran; 7772, on *Corchoris olitorius*, 4 km S al-Qatif town; 8655, on *Alhagi maurorum*, W edge of al-Qatif Oasis.

South Coastal Lowlands: 1902(BM), on *Ocimum basilicum*, ash-Shuqayq, al-Hasa; 2902(BM), on *Convolvulus arvensis*, al-'Arfaj farm, al-Hasa; 2904(BM), on *Citrus*, al-'Arfaj farm, al-Hasa; 3927, on *Conyza linifolia*, near 'Ayn al-Khadud, al-Hasa; 3940, on *Convolvulus arvensis*, 2 km ENE al-Hufuf.

Vernacular Names: SŪYAH (Qatifi farmer, a name apparently derived from the root verb *sa'a*, 'to be bad, evil'). This weed is a serious agricultural pest of the oases, sometimes forcing farmers to pull out and burn parts of crops in desperate (and often futile) attempts to stamp out the infestation.

2. **Cuscuta pedicellata** Ledeb.

Twining parasite with stems c. 0.3 mm wide when dry, yellow or purplish. Flowers in umbellate groups of 3 – 8, 4-merous throughout, or with a few scattered 5-merous ones. Calyx with ovate-triangular, mostly acute, lobes, slightly shorter than the corolla tube; corolla lobes rather erect, triangular and acute, somewhat acuminate, well exceeding the calyx and about as long as the corolla tube. Stamens exserted. Capsule depressed-globose, c. 2 mm in diameter, enclosed in the perianth to near the top. Seeds light brown, c. 0.75 – 1 mm long.

Habitat: A twining parasite. Rare.

Northern Dahna: 2329(BM), on *Trigonella anguina*, al-Farshah, 21 km ESE of Rumah. Also noted outside our area to the west, on *Fagonia*. Records to date have been deep inland, and the species may be more frequent in central Arabia than in the east.

Vernacular Names: The plant would probably be known under the same assemblage of vernacular names used for *C. planiflora*, below, q.v.

3. Cuscuta planiflora Ten.

Twining filiform leafless parasite growing upon, sometimes virtually covering, other plants with threadlike, usually yellow stems 0.2 – 0.3 mm in diameter when dry. Flowers 5-merous, sessile, c. 2 – 2.5 mm in diameter, in dense, 4 – 10 mm-broad globose clusters. Calyx with oblong, obtusish lobes about as long as the corolla tube; corolla limb with more or less triangular, subacute, white or reddish-white lobes somewhat hooded at apex. Stamens exserted. Stigmas, parts of the anthers, and the stems sometimes purplish. Capsule depressed-globose, about ¾ enclosed in the perianth. Seeds brown, c. 0.5 – 0.7 mm long. *Plates 165, 166*.

Habitat: A twining parasite on a wide variety of wild desert plants. The common *Cuscuta* of the desert and perhaps sometimes to be expected on cultivated plants. Occasional to locally frequent in the spring.

Northern Plains: 1704(BM), on *Astragalus schimperi*, ash-Shaqq, 28 – 30N, 47 – 42E; 4011, on *Helianthemum*, ad-Dibdibah, 27 – 56N, 45 – 35E; 540(BM), on *Horwoodia* seedlings, 9 km SE Qaryat al-'Ulya; 7053, on *Rhanterium*, near an-Naqirah wells.

Northern Summan: 8601, on *Fagonia bruguieri*, 17 km ESE Jabal Burmah, 26 – 41N, 47 – 45E.

Central Coastal Lowlands: 3267, on *Astragalus*, 8 km W Qatif-Qaisumah Pipeline KP – 144; 7038, Karan Island.

Vernacular Names: All variants on the root verb *shabaka*, 'to be intricate, entangled, netlike', referring to the netlike tangled stems enveloping the host: SHUBAYKAH (Al Murrah, Qahtan, al-Hasa farmers), SHABBAKAH, SHUBBĀK (Bani Hajir).

50. BORAGINACEAE

Herbs or shrubs (or rarely trees), usually with rough indumentum and leaves mostly alternate and entire. Flowers in spikelike or racemelike monochasial cymes, regular or sometimes irregular, generally 5-merous. Calyx usually divided to near base; corolla tubular to funnel-shaped, campanulate or rotate, sometimes with scales or hairs in throat. Stamens inserted on the corolla. Pistil 1 with 2 – or 4-lobed superior ovary. Fruit usually of 4 nutlets, or sometimes fewer by abortion, each with 1 seed.

Key to the Genera of Boraginaceae

1. Inflorescence not leafy and without bracts; style terminal; shrubs or shrubby herbs with flowers all white or all yellowish . . . . 1. *Heliotropium*
1. Inflorescence among leaves or bracteate; style arising basally between the carpels
    2. Shrublets or perennial herbs
        3. Corolla with yellowish hairs closing throat; leaves less than 5 mm long and less than 3 mm wide . . . . 4. *Echiochilon*
        3. Corolla without hairs in throat; leaves larger . . . . 2. *Moltkiopsis*
    2. Annual herbs
        4. Flowers yellow . . . . 6. *Arnebia*
        4. Flowers reddish, purple, blue or white

5. Corolla 1.5 cm or more long, strong red and purple, irregular ..... 7. *Echium*
5. Corolla 1 cm or less long, not strong red, regular
   6. Flowers in dense, crowded, 1-sided spicate inflorescence, pale violet,
      drying whitish .... 6. *Arnebia*
   6. Flowers solitary, axillary, blue or white (rarely pink in *Gastrocotyle*)
      7. Leaves wider than 5 mm; fruiting calyx lobes 2–3 mm
         long .... 8. *Gastrocotyle*
      7. Leaves narrowly linear, 1–3 mm wide, fruiting calyx lobes 3–7 mm long
         8. Corolla with internal appendages forming a ringlike structure in
            mouth of tube; nutlets spiny or prickly at outer angles ... 5. *Lappula*
         8. Corolla without appendages in mouth; nutlets evenly tuberculate
            without spiny projections at margins .... 3. *Ogastemma*

● 1. Heliotropium L.

Annual or perennial herbs, or shrublets. Leaves often with nerves depressed above, elevated below. Flowers usually sessile in terminal, ebracteate, leafless, helicoid cymes. Calyx 5-lobed; corolla 5-lobed, sometimes with 5 intermediate teeth. Stamens included. Nutlets 4 or 2.

Key to the Species of Heliotropium

1. Corolla yellow, with lobes abruptly constricting to a narrowly linear
   extension .... 4. *H. digynum*
1. Corolla white, with lobes rounded
   2. Leaves narrowly elliptic-oblanceolate, all acute, less than 1 cm wide
      3. Nutlets 2 (2 connate pairs) .... 2. *H. bacciferum*
      3. Nutlets 4 .... 3. *H. ramosissimum*
   2. Leaves oblong-ovate, over 1.5 cm broad, those above
      obtuse .... 1. *H. lasiocarpum*

1. Heliotropium lasiocarpum Fisch. et C. A. Meyer

Erect, somewhat shrubby, softly pubescent annual to c. 40 cm high. Leaves elliptical-oblong to ovate, obtuse at least above, petiolate, 1.5–5 cm long 1–2 cm wide, soft-pubescent at least on the nerves. Flowers sessile in dense spicate helicoid cymes to c. 10 cm long; calyx c. 2.5 mm long with lanceolate obtusish, hirsute lobes; corolla white, c. 1.5 times as long as the calyx, pubescent outside, with minute lobes. Stigma elongate-conical, sessile or subsessile on the ovary. Nutlets 2–2.5 mm long, ovoid, with muricate glabrescent ventral face, rounded and rather densely velutinous at back.
Habitat: A ruderal in the oases. Rare in desert around Bedouin camp grounds.
Northern Plains: 1586(BM), al-Batin/al-'Awja confluence.
South Coastal Lowlands: 2905(BM), al-'Arfaj farm, al-Hasa.

2. Heliotropium bacciferum Forssk.

H. undulatum Vahl var. *tuberculosum* Boiss.
H. tuberculosum (Boiss.) Boiss.
H. kotschyi Bge.
H. crispum Desf.

Ascending, dark green or greyish green many-branched perennial, hard-herbaceous above, woody at base, 15 – 75(100) cm high, rough with hard appressed hairs and larger, often tubercle-based bristles. Leaves narrowly elliptic, oblanceolate to linear, acute, tapering at base, sessile or indistinctly petiolate, somewhat revolute and repand-undulate or crenulate at margins, 0.3 – 3.5(4) cm long, 0.2 – 0.5(0.8) cm wide, smaller above, often larger on seedlings and young plants after good rains. Flowers in simple or forked terminal helicoid cymes; inflorescence often straightening and elongating in fruit. Calyx usually 2 – 2.5 mm long with lanceolate lobes; corolla white, 3 – 4 mm long, with ovate to oblong, rounded obtuse lobes somewhat eroded at margins. Stigma elongate-conical, on a style shorter than it. Maturing gynoecium more or less globose or depressed-globose, sometimes swollen with suberous callosities, c. 2 – 3 mm long, 2 – 5 mm broad, separating into 2, 2-seeded, pubescent or glabrescent nutlets (each consisting of 2 fused nutlets); commissural face of nutlets sometimes slightly winged at margins. *Plate 167.*

● The corky swellings often present on the nutlets of this plant occur more or less at random, although they seem to be more common on coastal specimens, perhaps on more saline soils. They may be found on some fruits, but not others, of the same plant and in varying degree, either in small patches or virtually covering the surface of both nutlet pairs. This tissue appears pale greenish or whitish when fresh but dries brownish, with a pithy texture.

See the following species for notes on the nomenclature of this plant and its relationship to *H. ramosissimum.*

Habitat: Shallow sands over limestone or silty soil. Common in coastal regions and inland about as far as the al-Hasa Oasis area. Often seen colonizing vacant ground around towns and along roadsides.

Central Coastal Lowlands: 1043, 1054(BM), 8257, Dhahran; 7729, 2 km S Ras Tanaqib, 27 – 49N, 48 – 53E; 1470(BM), 2 km E al-Ajam.

South Coastal Lowlands: 8277, 6 km N al-Mutayrifi; 8328, 8330, 25 km NW al-Hufuf.

Vernacular Names and Uses: RAMRĀM (gen.). The notes on uses of the following species, *H. ramosissimum,* apply also to this plant.

3. Heliotropium ramosissimum (Lehm.) DC.

Ascending, often intricately branched, densely hairy to bristly shrublet, 15 – 50 cm high. Leaves narrowly elliptical or oblanceolate to linear, 0.5 – 3.5 cm long, 0.2 – 0.7 cm wide but often in the smaller range, with crenulate-dentate margins; all or those above sessile, the lower ones sometimes tapering to a petiole up to half as long as the blade. Flowers in simple or forked terminal helicoid cymes; inflorescence often straightening and elongating in fruit. Calyx usually c. 2 mm long with lanceolate lobes; corolla white, 3 – 4 mm long, with ovate to oblong, rounded obtuse lobes; stigma elongate-conical. Maturing gynoecium more or less globose, broad, separating into 4 pubescent nutlets.

● Chaudhary (1985) treats *H. ramosissimum* as a synonym of *H. crispum* Desf. and uses the latter name for this plant. There appears to be some evidence, however, that *H. crispum* is a synonym of *H. bacciferum* (see Léonard 1985, 27; 1987, 95), and that view is followed provisionally here. This species is closely related to *H. bacciferum* and seems to replace it inland west of the coastal plain. Some of our desert specimens collected not far from al-Hasa Oasis appear to show some intermediate characters and may represent hybrid forms along the line of contact of the two species (if they are distinct species). In general *H. bacciferum* is the stouter plant, usually having a rougher indumentum with tubercle-based bristles, while *H. ramosissimum* more often has softer,

252

non-tubercle-based hairs. Under some conditions, however, the latter may be rough-hairy. Dwarfed forms of *H. ramosissimum* with minute leaves and smaller flowers are found on some inter-dune, silty terrain of the northern Rub' al-Khali. Chaudhary (1985) describes forms of *H. ramosissimum* (as *H. crispum*) having corky callosities on the nutlets, but we have not observed these so far in our eastern specimens.

Habitat: Often on soils with a silt component but also on shallow sands. Widespread and fairly common in many terrain types.

Northern Plains: 3110(BM), al-Batin, 28–00N, 45–28E; 1693(BM), 28–41N, 47–35E; 3209, 5 km ESE al-Qaysumah.

Northern Summan: 727, 33 km SW al-Qar'ah wells; 808(BM), 37 km NE Jarrarah; 3134(BM), 6 km SE ad-Dabtiyah; 7474, Batn Sumlul, 26–26N, 48–15E; 8826, as-Summan, 27–45N, 45–40E.

Southern Summan: 2198, 24–39N, 48–51E; 2210, 26 km SSE al-Hunayy; 8368, 8374, al-Ghawar, 25 km N Harad; 8348, Wadi as-Sahba, 10 km W Harad.

South Coastal Lowlands: 8037, Wadi as-Sahba, 23–52N, 49–22E; 7705, 11 km S Wadi as-Sahba, 23–52N, 49–11E; 7663, 23–54N, 49–08E; 2109, 49 km S Harad.

Rub' al-Khali: 7509(BM), 22–07N, 48–36E.

Vernacular Names and Uses: RAMRAM (gen.). This plant has a reputation as a medicinal in eastern Arabia and at least in earlier times was used as a remedy for snake bite. According to folklore the desert monitor lizard *(Varanus)* gains immunity from venomous snakes by rolling in the shrub's branches and eating its leaves (H. Dickson 1951, 467), and human snake bite victims used to be given tea from the leaves while a leaf poultice was applied to the bite. There is, of course, no medical evidence that such treatment has more than possible psychological value, and persons bitten by the small common sand viper *(Cerastes cerastes)* usually recover without treatment in any case. Dickson *(op. cit.,* 160) also reports its use as an infusion or paste to treat mouth sores.

4. Heliotropium digynum (Forssk.) Aschers. ex C. Christ.

H. luteum Poir. in Lam.

Ascending to erect diffusely branched shrublet, rather softly pubescent with stems white at least above, 15–50 cm high. Leaves greyish green, ovate to oblong, obtuse or acute, subsessile above and short-petiolate below, 0.5–1.5 cm long, 0.3–1.2 cm wide, sometimes weakly undulate-margined. Flowers sessile in terminal, spicate, helicoid cymes elongating in fruit to up to 6 cm long. Calyx 2.5–3 mm long, hirsute; corolla c. 5 mm long, yellow, hirsute outside, the lobes narrowing abruptly above to linear-triangular with apices often inflexed. Nutlets 2–3 mm long, pubescent or nearly glabrous. *Plate 168.*

Habitat: Stable sands. Occasional.

Central Coastal Lowlands: 7713, 2 km E Ras al-Ghar, 27–31N, 49–14E; 159, al-Ajam; 906, Mulayjah; 1469(BM), 2 km E al-Ajam.

Northern Summan: 2743, 25–18N, 48–20E.

Southern Dahna: 2140, E edge of the Dahna, 23–05N.

South Coastal Lowlands: 380, 5 km S al-'Uyun, al-Hasa. 1438(BM), 2 km E Jabal Ghuraymil; Abqaiq (Dickson 1955, 49); also sight records by author.

Rub' al-Khali: 7630, 19–31N, 48–13E; 7632, 'Uruq al-Awarik, 19–25N, 47–53E.

Vernacular Names and Uses: KARY (Al Murrah, gen. S). According to Al Murrah Bedouins, milk camels have a great fondness for the plant.

● 2. Moltkiopsis I. M. Johnston

Perennial herb or dwarf shrublet, hirsute-strigose with white stems. Flowers in helicoid cymes with corolla tubular or narrowly funnel-shaped, exceeding the 5-parted calyx, the throat without scales. Stamens unequal. Nutlets 4, or sometimes 1–3, triquetrous-pyramidal.

Moltkiopsis ciliata (Forssk.) I. M. Johnston
Lithospermum callosum Vahl

Erect to ascending dwarf shrublet, 10–30 cm high, both appressed pubescent and strigose with hard, broadly calcareous-based bristles; stems whitish, and often pinkish above in fresh growth. Leaves mostly oblong-lanceolate, sometimes linear or subovate, acute, sessile, ciliate at margins with hard bristles, 10–25 mm long, 2–8 mm wide. Flowers subsessile in helicoid cymes spiralled at apex, straightening below. Calyx c. 5 mm long with lanceolate-linear strigose lobes, deciduous at maturity; corolla tubular-cylindrical, pubescent outside, about twice as long as the calyx and often of different colors in the same inflorescence: blue-violet, red, and white, or bluish limb with red tube; one specimen with all flowers white has been collected. Stamens about equalling the corolla, sometimes slightly exserted. Style often exserted. Nutlets often only 1 or 2, glossy smooth greyish-tan, 2–2.5 mm long, ovoid with a blunt horn at apex, acutely keeled ventrally, rounded at back. *Plate 169.*

● A highly drought-adapted perennial, dying back to a clump of nearly dry stems and leaves in the hot season and putting forth fresh growth in the winter and spring. An important constituent of the *qara'ah* terrain in the northern plains, where larger shrubs are entirely absent; also very frequent in both shallow and deep sands of the coastal lowlands, the Dahna, and the northern and western Rub' al-Khali.

Northern Plains: 832, 30 km SW al-Mish'ab; 894, Jabal al-'Amudah; 993, 12 km N Jabal al-'Amudah.

Northern Dahna: 793, 15 km SE Hawmat an-Niqyan; 2321, 22 km ENE Rumah.

Northern Summan: 2742, 25–18N, 48–20E; 3157(BM), 9 km SSW Hanidh.

Central Coastal Lowlands: 111, 391, 1045, Dhahran; 930, 10 km S al-Khafji/as-Saffaniyah junction; 3302, 2 km W Sabkhat Umm al-Jimal.

South Coastal Lowlands: 7903 (flowers all white), Na'lat Shadqam, 25–43N, 49–29E; 1439(BM), 2 km E Jabal Ghuraymil.

Southern Dahna: 2137, E edge of the Dahna, 23–05N.

Rub' al-Khali: 1020, 22–45N, 50–20E; 7669, al-Jawb, 23–06N, 49–58E; 7683, 7 km SW Shalfa, 21–51N, 49–39E.

Vernacular Names: Generally known as ḤALAM among southern tribes such as Al Murrah, Bani Hajir and Al Rashid. To the north it is called ḤAMAṬ, a name used by Bani Hajir only in referring to it in the dry state less palatable to livestock.

● 3. Ogastemma Brummitt

Dwarf annual with narrow leaves. Flowers solitary, axillary or in loose bracteate spicate inflorescences. Calyx divided nearly to base, somewhat growing in fruit; corolla tubular-campanulate, white with naked throat. Nutlets 4, ovoid, rounded at back, evenly tuberculate-granular. See Brummitt (1982) for the change in name from *Megastoma.*

Ogastemma pusillum Brummitt, Bonnet et Barratte
Megastoma pusillum Coss. et Dur.

Erect-ascending dwarf annual, 3 – 10(15) cm high, usually branched, appressed pubescent and setose with tubercle-based bristles. Leaves linear, 5 – 25 mm long, 1 – 2 mm wide. Flowers solitary, subsessile in axils of bracts or upper leaves; fruiting calyx 3 – 4 mm long with unequal lobes; corolla minute, 2 – 3 mm long with oblong obtuse lobes, white, without scales in throat. Nutlets c. 1.5 mm long, shorter than the fruiting calyx, ovoid, ventrally keeled, rounded at back, evenly granular-tuberculate.

Not to be confused with *Lappula spinocarpos* (*q.v.* below), a plant of similar habit and with which it is sometimes found growing.

Habitat: Shallow sands or silty basins and plains. Occasional.

Northern Plains: 1315(BM), 8 km W Jabal Dab'; 2720, 3 km NW Jabal Dab'; 1662(BM), 2 km S Kuwait border in 47 – 14E; 1257(BM), 15 km ENE Qaryat al-'Ulya; 1483(BM), 50 km WNW an-Nu'ayriyah.

Northern Summan: 2819, 3 km ESE ash-Shumlul; 8575, 14 km S Dahl al-Furayy, 26 – 39N, 47 – 05E.

Central Coastal Lowlands: 135(BM), 189, 1831(BM), Dhahran; 1759(BM), Abu Hadriyah; 3766, 13 km N Abu Hadriyah; 2714, 18 km SW as-Saffaniyah.

● 4. **Echiochilon** Desf.

Mostly small shrublets with firm leaves and flowers solitary or in elongating spicate inflorescences. Calyx divided nearly to base; corolla more or less zygomorphic, the limb spreading with upper lip 2-lobed, the lower 3-lobed. Stamens included, inserted above middle of the corolla tube.

Echiochilon kotschyi (Boiss. et Hohen.) I. M. Johnston
Lithospermum kotschyi Boiss.

Greyish ascending, many-branched, densely appressed-pubescent shrublet, 15 – 40 cm high. Leaves linear-elliptical or linear-oblong, sessile, mostly 3 – 5 mm long, only 1 mm wide, somewhat longitudinally folded, rounded at back and channeled above, acutish, often with a minute mucro at apex, densely covered with hard whitish appressed hairs. Flowers solitary, axillary, subsessile, 5 – 6 mm long; calyx 3 – 4 mm long, divided ¾ to base with oblong to linear-oblong lobes somewhat unequal in width. Corolla slightly exceeding the calyx, with whitish to cream spreading limb, the lobes nearly equal, the throat closed by yellowish hairs. Stamens inserted in upper part of tube; anthers c. 1 mm long. Nutlets minute, rough, dark brown.

Habitat: So far known only from elevated rocky, limestone terrain at Dhahran. Rare to locally occasional.

Central Coastal Lowlands: 114, 343, 519(K), 1040, Dhahran.

● A southern species originally described from southern Iran and the Makran. It is to be expected south of Dhahran toward the southern Gulf coast, and has been recorded in Qatar by Batanouny (1981). This is almost certainly the *'Sericostoma'* collected by Dickson (1955) at Dhahran in 1942 and the *'Sericostoma'* reported by Good (1955) from Bahrain.

● 5. **Lappula** Gilib.

Annual, or rarely perennial, herbs with linear leaves. Flowers small, blue or white,

255

with calyx divided nearly to base and corolla campanulate to funnel-shaped, with 5 scales in the throat. Stamens included. Nutlets 4, triquetrous-pyramidal, with acute, often prickly outer angles.

Lappula spinocarpos (Forssk.) Aschers.

Erect to ascending-decumbent, appressed-pubescent, dwarf annual, 3 – 15 cm high, usually branched from the base. Leaves linear to linear-spathulate, 1 – 3(5) cm long, 1 – 2(3) mm wide, obtusish, tapering toward base. Flowers solitary, or in very loose racemes, axillary, subsessile or short-pedicellate; calyx 4 – 5 mm long with linear, somewhat unequal lobes; corolla 3 – 4 mm long, shorter than the calyx, with limb blue, or less commonly white; the tube paler. Calyx growing in fruit, 5 – 8 mm long, exceeding the nutlets. Nutlets triquetrous-pyramidal, c. 4 mm long, irregularly tubercled and usually puberulent at outer angles, dark brown when young, with maturity becoming glossy grey-olive with a jadelike surface.

Not to be confused with *Ogastemma pusillum* (*q.v.* above), a rather similar-appearing plant that differs in having the corolla without ring-forming appendages in the throat and in having smaller, evenly tuberculate nutlets rounded and not spiny at the corners. The two species may sometimes be found growing together.

Habitat: Silty inland plains and basins. Frequent.

Northern Plains: 567, 658, Khabari Wadha; 688, 20 km WSW Khubayra; 874, 9 km S 'Ayn al-'Abd; 1221(BM), Jabal an-Nu'ayriyah; 1259(BM), 15 km ENE Qaryat al-'Ulya; 1500(BM), 30 km ESE al-Qaysumah; 1533(BM), 36 km NE Hafar al-Batin; 1632(BM), 2 km S Kuwait border at 47 – 06E; 2719, 3 km NW Jabal Dab'; 3074(BM), ad-Dibdibah, 28 – 45N, 47 – 01E; 3080(BM), al-Musannah, 28 – 43N, 46 – 47E.

Northern Summan: 1372(BM), 42 km W Qaryat al-'Ulya; 741, 743, 33 km SW al-Qar'ah wells; 2753, 25 – 43N, 48 – 07E; 8617, near Dahl Umm Hujul, 26 – 35N, 47 – 32E; 8625, Dahl Abu Harmalah, 26 – 29N, 47 – 36E.

Southern Summan: 2139, 24 – 39N, 48 – 51E.

South Coastal Lowlands: 7902, Na'lat Shadqam, 25 – 43N, 49 – 29E.

Vernacular Names: Two northern names reported for this species relate to the idea of 'head' or 'brain' and probably refer to the smooth grey, convoluted, brainlike surface of the nutlets: DIMĀGH AL-JARBŪ' ('jerboa brain', Dickson 1955), 'USHBAT AR-RĀS ('headwort', Musil 1927).

● 6. Arnebia Forssk.

Hispid annual or perennial herbs with red-colored taproot and yellow or violet flowers in a bracteate spicate or racemose inflorescence. Calyx deeply divided, often much enlarging in fruit; corolla tubular-cylindrical without scales at throat. Nutlets 4, variously rugose or sometimes nearly smooth.

Rather common in the desert spring vegetation, sometimes 2 – 3 of the 4 species growing together. Notable for the crimson, water-soluble dye found on the surface of the taproot; this stains the skin as well as herbarium sheets and is sometimes used as a cosmetic rouge by Bedouin girls. This red root is also found in at least one of our other Boraginaceous genera, *Echium*. Forsskal's generic name is apparently based on the Arabic word, *arnab*, 'rabbit, hare'. One or more of the species is known as *shajarat al-arnab*, 'hare's bush', in northern Arab countries.

Key to the Species of Arnebia

1. Corolla pinkish to pale violet (but drying white or pale yellowish), glabrous; nutlets with flat plane at dorsal surface .... 4. *A. tinctoria*
1. Corolla yellow, pubescent; nutlets rounded dorsally
   2. Calyx not or hardly growing in fruit; ¼ to ½ of corolla tube usually exserted beyond calyx .... 2. *A. hispidissima*
   2. Calyx much enlarging in fruit; usually less than ¼ of corolla tube exserted, or only limb exserted
      3. Fruiting calyx lobes 0.7–2 mm broad at middle, 8–12(14) mm long; nutlets 1.5–2 mm long .... 1. *A. decumbens*
      3. Fruiting calyx lobes 3–5 mm broad at middle, 15–30 mm long; nutlets 2.2–3 mm long .... 3. *A. linearifolia*

1. Arnebia decumbens (Vent.) Coss. et Kral.

Ascending annual, usually branched from near base, 5–30 cm high, hispid with appressed hairs on leaf faces and spreading bristles on bract and calyx margins and on the whitish to pale yellowish stems. Lower leaves linear-oblong to linear-lanceolate, 2–7 cm long, 4–10 mm wide, sessile, those above shorter and narrower. Flowers sessile or subsessile, in dense, 1-sided, elongated bracteate spicate inflorescences. Bracts equalling or exceeding the calyx; calyx growing in fruit to 8–12(14) mm long, the lobes abruptly becoming narrowly linear above, 0.5–1.5 mm wide at middle, with indurate, crested, tubercled base. Corolla narrowly cylindrical, pubescent, yellow, 8–15 mm long, the tube exceeding the calyx by ¼–½ its length, or hardly exserted. Nutlets angled ventrally, dorsally rounded, grey to brown, more or less rugose-verrucose, 1.5–2 mm long.

Sometimes difficult to separate from *A. linearifolia* (below) when plants are immature or poorly developed. The two species (if indeed they are specifically distinct) sometimes grow together, and at these sites intermediates and possible hybrids may be found (e.g. No. 7579 – C, below, from Jabal Dab' in the Northern Plains). Other writers, such as Feinbrun-Dothan (1978) have noted this phenomenon in other parts of the plant's range.

Habitat: Sand or silty sands in a wide variety of terrain types. Common.

Northern Plains: 840, 9 km S 'Ayn al-'Abd; 1153(BM), Jabal al-Ba'al; 1003, 12 km N Jabal al-'Amudah; 1237(BM), Jabal an-Nu'ayriyah; 1604(BM), 3 km ESE al-Batin/al-'Awja confluence; 1534(BM), 36 km NE Hafar al-Batin; 1515(BM), 30 km WNW al-Qaysumah; 1477(BM), 50 km WNW an-Nu'ayriyah; 1298(BM), 15 km ENE Qaryat al-'Ulya; 7565, al-Faw al-Janubi; 7964, Jabal al-'Amudah; 7579, Jabal Dab'; 7579-C, from the same site as the last and possibly a hybrid with *A. linearifolia*.

Northern Summan: 1349(BM), 1393(BM), 42 km W Qaryat al-'Ulya; 718, 33 km SW al-Qar'ah wells; 8559, Jibal adh-Dhubabat, 26–45N, 48–01E.

Central Coastal Lowlands: 98(BM), 1820(BM), Dhahran; 7737, 2 km S Ras Tanaqib, 27–49N, 48–53E; 4073, az-Zulayfayn, 26–51N, 49–51E; 3257(BM), 3288(BM), 8 km W Qatif-Qaisumah Pipeline KP–144; 3311(BM), 5 km S Nita'.

Vernacular Names and Uses: KAḤIL (gen.), KAḤAL, KAḤLA' (Al Murrah, Bani Hajir); from verb meaning 'to anoint', as the eyes with collyrium, referring to the red dye obtainable from the surface of the taproot. Bedouin girls at play sometimes rub the fresh root on their face as rouge.

2. Arnebia hispidissima (Lehm.) DC.

Ascending to erect annual herb, sometimes perennating, usually branched from near base, 10 – 30 cm high, appressed-hairy on leaf faces and hispid with rough white spreading bristles on stems, and on margins and nerves of leaves, bracts, and calyx. Lower leaves linear-oblanceolate, to c. 4.5 cm long, 0.7 cm wide, obtusish, those above smaller, more acute. Flowers sessile or subsessile, in crowded-terminal or elongate 1-sided inflorescences. Calyx 5 – 7 mm long, not or hardly growing in fruit, with linear or linear-lanceolate, white-hispid lobes. Corolla yellow, pubescent, with limb spreading to 5 – 6 mm in diameter, the tube clearly exserted from the calyx. Perennating plants in dry season generally with smaller and denser leaves and flowers, and overall greyish with a dense clothing of white bristles. Nutlets angled ventrally, rounded at back, glossy, smooth or verrucose, 1.5 – 2 mm long. *Plate 171.*

Habitat: Sand or silty sand; generally more frequent in the south. Occasional to locally frequent.

Northern Plains: 3073(BM), ad-Dibdibah, 28 – 45N, 47 – 01E; 3079(BM), al-Musannah, 28 – 43N, 46 – 47E.

Central Coastal Lowlands: 129, 164, 418, 426, 1833(BM), 4084, Dhahran; 4058, az-Zulayfayn, 26 – 51N, 49 – 51E; 3213, Jalmudah, N al-Jubayl; 3182(BM), 9 km N Shadqam junction; 1460(BM), 2 km E al-Ajam.

South Coastal Lowlands: 1897(BM), 1 km NE al-Mutayrifi, al-Hasa.

Southern Summan: 2164, 24 – 08N, 48 – 43E; 8064, 24 – 06N, 48 – 48E.

Rub' al-Khali: 7508, 2 km NW al-Hawk, 22 – 07N, 48 – 36E.

Vernacular Names and Uses: FANĪ (Al Rashid), FUNŪN (Qahtan). Names for *A. decumbens* (above) probably also used. Root strongly red and probably used as is *A. decumbens* as a dye or cosmetic.

3. Arnebia linearifolia DC.

Ascending to erect annual, usually branching from near the base and 5 – 25 cm high, with appressed hard hairs on leaf faces and hispid with spreading rough bristles from the stems and from margins and nerves of the bracts and calyx. Lower leaves linear-oblanceolate to spathulate, obtusish, to c. 6 cm long and 0.7 cm wide, sessile or tapering to an indistinct petiole; upper leaves smaller. Flowers in dense, 1-sided bracteate spicate inflorescences, the bracts generally somewhat shorter than the calyx. Flowers nearly sessile, with calyx strongly growing in fruit, 15 – 25(30) mm long, the lobes linear-oblong, tapering very gradually toward the apex and 3 – 4(5) mm wide at middle with distinct mid-vein; corolla yellow, pubescent, 10 – 15 mm long, those toward apex of inflorescence with tubes slightly exserted from the calyx, those below with tubes mostly included. Nutlets angled ventrally, rounded at back, more or less verrucose or pitted, 2 – 3 mm long. *Plate 170.*

See notes with *A. decumbens,* above, for relationship with that species.

Habitat: Sand or silty sand. Frequent.

Northern Plains: 1670(BM), 2 km S Kuwait border in 47 – 14E; 4079, Jabal an-Nu'ayriyah; 7579-A, Jabal Dab'.

Central Coastal Lowlands: 1819(BM), Dhahran; 3258(BM), 8 km W Qatif-Qaisumah Pipeline KP – 144; 3359(BM), Jabal al-Ahass; 7745, 5 km NW Ras Tanaqib, 27 – 53N, 48 – 49E; 7579-C, Jabal Dab', possibly a hybrid with *A. decumbens* (above).

Northern Summan: 8558, Jibal adh-Dhubabat, 26 – 45N, 48 – 01E.

Southern Summan: 8065, 24–06N, 48–48E.
Vernacular Names and Uses: KAḤĪL, KAḤAL; notes on names and uses as for *A. decumbens* (above).

4. Arnebia tinctoria Forssk.

A. tetrastigma Forssk.

Dwarfish ascending annual, usually branched from base and 5–10 cm high, somewhat greyish and densely clothed in silky, mostly appressed, hairs, sometimes with some rougher bristles spreading on the stems and bract margins. Lower leaves linear-oblanceolate or linear-oblong, to c. 3.5 cm long, 0.5 cm wide, obtusish, sessile or tapering below to an indistinct petiole, appressed-pubescent. Flowers in dense, 1-sided spicate inflorescences; calyx growing somewhat in fruit but not crested or indurate, the lobes narrowly linear-oblong, 5–8 mm long, c. 1 mm wide at middle; corolla narrowly cylindrical with minute limb, pale violet or pinkish, sometimes whitish, drying white or very pale yellowish, nearly glabrous, 6–9 mm long, included in the calyx or slightly exserted in the upper flowers. Nutlets distinctly flat at back, convex-rounded ventrally, glossy, nearly smooth, 1.5–2 mm long. *Plate 172*.

Habitat: Silty inland soils and Summan basins; occasional.
Northern Plains: 557(BM), Khabari Wadha; 1635(BM), 2 km S Kuwait border in 47–14E; 1685(BM), al-Hamatiyat, 28–50N, 47–30E; 1707(BM), ash-Shaqq, 28–30N, 47–30E; 1319(BM), 8 km W Jabal Dab'; 7577, Jabal Dab'.
Northern Summan: 2802, 15 km ESE ash-Shumlul; 8577, 14 km S Dahl al-Furayy, 26–39N, 47–05E.
Central Coastal Lowlands: 3313, 5 km S Nita'.
Southern Summan: 8063, 24–06N, 48–48E; 2123, 3 km SW al-Jawamir, Yabrin.
Vernacular Names and Uses: KAḤĪL, KAḤAL; notes on names and uses as for *A. decumbens* (above).

● 7. Echium L.

Annual or perennial hispid herbs with zygomorphic, trumpet-shaped or funnel-shaped corolla with unequally lobed, oblique limb. Filaments long, unequal, inserted below middle of the tube; style exserted.

Echium horridum Batt.

Erect or ascending, usually branched hispid annual herb; taproot reddish. Leaves at base sometimes rosetted, beset with broad-based white bristles, linear-oblong to spathulate, obtuse, tapering at base to a petiole, to c. 15 cm long, 1.5 cm wide, or sometimes larger. Upper leaves smaller, sessile, linear-oblong. Flowers in axillary helicoid cymes; calyx c. 10–18 mm long, the lobes linear-lanceolate; corolla showy, 15–25(30) mm long, trumpet-shaped with oblique open mouth, red to purplish, darkening with age. Stamens reaching the mouth, sometimes slightly exserted; style often exserted. *Plate 173*.

Habitat: Silty soils of the northern plains, extending south through the southern Summan. Occasional to locally frequent.
Northern Plains: 601(BM), al-Batin, 5 km ENE Qulban Ibn Busayyis; 1502(BM), 30 km ESE al-Qaysumah; 1548(BM), 36 km NE Hafar al-Batin.
Northern Summan: 2760, 25–49N, 48–01E; 3889, 14 km ESE Umm 'Ushar; 8579,

8 km S Dahl al-Furayy; 8622, 4 km E 'Irq Hazwa, 26–23N, 47–24E; 8627, Dahl Abu Harmalah, 26–29N, 47–36E.

Southern Summan: 8061, 24–06N, 48–48E.

Vernacular Names: KAḤAL (N, Dickson 1955), a name shared with *Arnebia*, probably because of its similarly red-staining root.

● 8. Gastrocotyle Bge.

Annual herbs with minute flowers solitary in the leaf axils. Corolla funnel-shaped or short-tubular, with obtuse scales in throat. Stamens inserted at or below middle of the corolla tube. Nutlets snail-shaped.

Gastrocotyle hispida (Forssk.) Bge.
Anchusa hispida Forssk.
Annual of variable habit and size, appressed-pubescent to hispidulous, sometimes dwarfed, only c. 5 cm high, but under favorable conditions procumbent-spreading with stems 30 or more cm long. Leaves linear-oblong to spathulate, obtuse, or acute above, tapering at base, more or less repand-wavy, up to c. 8 cm long, 1.4 cm wide. Flowers solitary in the axils; calyx with triangular acute lobes 1.5–3 mm long, spreading in fruit. Corolla about as long as or somewhat exceeding the calyx, pale blue or violet, rarely pink, with papillose scales in the throat. Nutlets ovoid, growing to exceed the calyx lobes slightly, c. 3 mm long, pale to nearly black, snail-shaped with a mouthlike, deeply concave, denticulate-margined basal ring.

Habitat: Shallow sand over silts or limestone. Frequent.

Northern Plains: 897, Jabal al-'Amudah; 1247, Jabal an-Nu'ayriyah; 677, Khabari Wadha; 1273(BM), 15 km ENE Qaryat al-'Ulya; 1499(BM), 30 km ESE al-Qaysumah; 1579(BM), al-Batin/al-'Awja confluence; 3072(BM), ad-Dibdibah, 28–45N, 47–01E.

Northern Summan: 3149(BM), 9 km SSW Hanidh; 3240, Dahl al-Furayy; 3345, 5 km WSW an-Nazim; 7936 (pink flowers), 25–39N, 48–48E; 684, 20 km WSW Khubayra; 705, 33 km SW al-Qar'ah wells; 774, 69 km SW al-Qar'ah wells; 2787, 41 km SE ash-Shumlul; 1380(BM), 42 km W Qaryat al-'Ulya.

Central Coastal Lowlands: 59, 94, 105, Dhahran.

Northern Dahna: 3248, S edge 'Irq Jaham.

Southern Summan: 2077, Jaww ad-Dukhan, 24–48N, 49–05E; 2087, al-Ghawar, 24–27N, 49–05E; 2219, 26 km SSE al-Hunayy; 2248, 26 km ENE al-Hunayy; 2046, 25–00N, 48–30E.

Vernacular Names: KAḤIL.

51. VERBENACEAE

Herbs, shrubs or trees with opposite, usually simple, leaves. Flowers bisexual, mostly zygomorphic with (2–4)5-merous perianth, the corolla tubular to campanulate, often with a 2-lipped limb. Stamens 4 (rarely 2). Ovary superior, 2-carpeled; fruit a 2- or 4-celled drupe or schizocarp.

The mangrove *Avicennia* is included here, although some authorities segregate it in a separate family, the Avicenniaceae.

Key to the Genera of Verbenaceae

1. Creeping herb with serrate leaves .... 1. *Phyla*
1. Erect or climbing-scrambling shrubs; leaves entire
    2. Climbing or scrambling terrestrial shrub; corolla white with tube
      exceeding the limb .... 2. *Clerodendrum*
    2. Erect shrub of seashore mud; corolla yellow, with limb longer than the
      tube .... 3. *Avicennia*

● 1. Phyla Lour.

Trailing perennial herbs with medifixed hairs and opposite simple leaves. Flowers bracteate and sessile in spicate inflorescences; calyx compressed, 2-lobed; corolla funnel-shaped, 4-lobed. Stamens 4, inserted near middle of the corolla tube. Ovary 2-locular; fruit a dry drupe separating into 2 pyrenes.

Phyla nodiflora (L.) Greene
Lippia nodiflora (L.) Michx.

Prostrate perennial herb, finely appressed-pubescent, branched from the base with trailing stems to c. 50 cm long. Leaves cuneate-spathulate to obovate, 10 – 30 mm long, 5 – 12 mm wide, dentate at upper margins, tapering gradually below to a short petiole or subsessile. Peduncles solitary, axillary, exceeding the subtending leaf; spikes very dense, ovoid, at length becoming cylindrical, 7 – 22 mm long, c. 5 mm wide. Bracts imbricate; calyx 2-lobed with lobes acute, keeled; corolla pink, c. 2 – 3 mm long with oblong lobes. Drupe compressed, c. 1.5 mm long.

Habitat: Known only as a weed of wet places in gardens and oases; sometimes in lawns. Frequent.

Central Coastal Lowlands: 1100(BM), 3 km E al-Qatif town; Dhahran in lawns.

South Coastal Lowlands: farms in al-Hasa Oasis.

Vernacular Names: FARFAKH (Hasawi farmer).

● 2. Clerodendrum L.

Shrubs, climbers, or trees with simple, opposite or whorled leaves. Flowers 5-merous; corolla narrowly tubular, with oblique, spreading 5-lobed limb. Stamens 4, exserted. Ovary 4-celled; fruit a drupe separating into 2 or 4 pyrenes.

Clerodendrum inerme (L.) Gaertn.

Climbing or scandent, somewhat rank-smelling shrub, up to 3 m or more high when supported; mostly glabrescent but often with some very fine and short pubescence on stems. Leaves elliptical to ovate, opposite, entire, somewhat coriaceous, acute or obtuse, mostly 2.5 – 4 cm long, 1.5 – 3 cm wide, but rather variable in size depending on local conditions, on petioles ⅙ to ¼ as long as the blades. Flowers mostly near ends of the branches in 3-flowered axillary umbellate cymes with peduncle 2 – 3 cm long and pedicels 5 – 10 mm long. Calyx tubular-campanulate, c. 5 mm long in flower, minutely 5-toothed, growing in fruit; corolla white, 2.5 – 4 cm long, with very narrow tube and spreading limb with 5 unequal lobes c. 5 – 7 mm long. Stamens 4, long-exserted, with purple filaments; style purple, exserted, shorter than the stamens. Drupe subglobose, 8 – 12 mm long (but very rarely seen in our area).

Eastern Province plants flower profusely but usually do not set fruit or seed; the only example of fruiting seen by the author was a specimen collected by Sheila Collenette in al-Qatif Oasis in late March, 1987.

Habitat: A cultivated plant found in towns, gardens and the oases, where it sometimes escapes, or at least spreads vegetatively along roadsides. Locally common.

Central Coastal Lowlands: Many sight records in towns and scandent along roadsides in the oasis south of al-Qatif town.

Vernacular Names and Uses: There is no record of an Arabic name for this plant; in English it is locally called 'false jasmine' or 'Aramco hedge'. It is a native of India and has been long, widely, and very usefully planted as a hedge in the family communities of the Saudi Arabian Oil Company (Aramco) at Dhahran, Abqaiq, and Ras Tanura. It was probably introduced by the oil company from Bahrain in the 1930s but may have existed locally in the oases before then. It is easily propagated by cuttings and forms dense, lush, inpenetrable hedges on post and wire supports. The species has traditional medicinal applications in India, being used to treat fevers and skin diseases.

● 3. Avicennia L.

Shrubs or trees of seashore muds with opposite coriaceous leaves and pneumathodia ('breathing roots') arising vertically from horizontal rhizomes. Flowers cymose with 5-lobed calyx and funnel-shaped, 4-lobed corolla. Stamens 4. Fruit a rather fleshy, somewhat compressed capsule, the embryo often germinating on the mother plant before fall of the fruit.

Avicennia marina (Forssk.) Vierh.

A. officinalis L.

Erect-ascending shrub or small tree, often mealy-tomentose, 1–3 m high. Leaves opposite, lanceolate to elliptical, mostly acute, entire, 3–7 cm long, 1–3 cm wide, tapering at base to a short petiole, coriaceous, darker and somewhat glossy-green above, often dull-greyish below, with numerous excreted salt crystals. Flowers in dense, capitate, short-peduncled cymes, bracteate; calyx lobes 2–4 mm long, obtusish, fine-fimbriate-margined; corolla yellow, exceeding the calyx, with 4 subequal spreading lobes exceeding the tube. Capsule almond-shaped, 1.5–2.5 cm long. *Plate 174.*

● This shrub is usually seen surrounded by pneumathodia (also sometimes called 'pneumatophores'), which are erect, leafless, stem-like growths thought to oxygenate the roots, rising vertically above the grey mud in which the mangrove is rooted. These, arising from horizontal roots, often stand in straight lines radiating from the shrub base. Another adaptation to its severe, saline, oxygen-depleted habitat is its habit of vivipary, in which seeds germinate before dropping from the mother plant and send out a radicle, or embryonic root. This has been interpreted as a mechanism to avoid soil salinity and to provide oxygen at the critical time of seed germination.

This is the only mangrove on the Gulf coast of Saudi Arabia, and its distribution is restricted to a few of the more protected bays and inlets. Its natural northern limit appears to be around latitude 27°12' N, where a few individuals are found in inlets near al-Khursaniyah. It forms dense stands in parts of Tarut Bay, where it is dominant in an ecologically important and highly productive littoral biotope. It has been much destroyed, however, by seafront earth filling, and it was gradually disappearing by the 1980s in many of its former prime habitats around al-Qatif and Tarut Island.

Habitat: Intertidal mud of protected inlets on the Gulf shore. Locally abundant but overall with a rather sparse, interrupted distribution.

Central Coastal Lowlands: 1102(BM), 7811 – A, Tarut Island; al-Qatif (Dickson 1955, 25).

Vernacular Names and Uses: QURM (Bani Khalid, Bani Hajir). Sometimes browsed by camels when other plants are not available but generally not considered a very good grazing plant because of its high salinity.

52. LABIATAE

Herbs or shrubs, often aromatic, with opposite, usually simple leaves. Inflorescence often verticillate; flowers bisexual, zygomorphic with 5(4)-lobed calyx; corolla usually 2-labiate. Stamens 4 or 2. Ovary superior, 4-lobed; fruit of 4, 1-seeded nutlets.

Key to the Genera of Labiatae

1. Stamens 2 .... 3. *Salvia*
1. Stamens 4
    2. Corolla appearing 1-lipped; upper lip absent or very indistinct; desert
       perennials .... 1. *Teucrium*
    2. Corolla 2-lipped or nearly regular; desert annual, or oasis plants
        3. Bracts aristate; dwarf desert annual .... 2. *Lallemantia*
        3. Bracts not aristate; bushy herbs of oases or garden edges
            4. Calyx lobes equal or subequal, all acute; leaves sessile or with petiole
               less than 2 mm long .... 4. *Mentha*
            4. Calyx with upper lobe much broadened, ovate to orbicular, obtuse;
               petioles 5 – 12 mm long .... 5. *Ocimum*

● 1. Teucrium L.

Herbs or shrublets with opposite, usually revolute-margined leaves. Calyx tubular-campanulate, 5-toothed; corolla deciduous, appearing 1-labiate with the lowest lobe much enlarged. Stamens 4. Nutlets obovoid, usually rugose or reticulate.

Key to the Species of Teucrium

1. Flowers in pairs at the nodes, 12 – 20 mm long, on pedicels 10 – 15 mm
   long .... 1. *T. oliverianum*
1. Flowers sessile in dense peduncled heads, 5 – 9 mm long .... 2. *T. polium*

1. Teucrium oliverianum Ging.

Velvety-canescent, non-aromatic shrublet with ascending-erect branches from the base, 20 – 60 cm high. Leaves cuneate, tapering at base and 3 – 5-lobed at apex, those below 1.5 – 3 cm long, 1 – 1.5 cm wide distally, revolute-margined and with veins elevated on lower face, sessile or obscurely petiolate; upper leaves smaller, sessile, subentire, elliptical. Flowers solitary, opposite-paired at the upper nodes, on pedicels 10 – 15 mm long and sometimes reflexed below the calyx; calyx c. 5 mm long, velvety, divided more than half-way to the base with triangular-lanceolate acute lobes; corolla showy, soon deciduous, pale to deep blue-violet, pubescent, much exceeding the calyx, 15 – 18 mm long, the much enlarged central lobe of the limb forming an apparent single lip. Stamens long-exserted. Ovary densely fleecy above. *Plate 175.*

Habitat: Silt soils of inland basins or rocky areas. Occasional to locally frequent.
Northern Plains: 544, 2 km N Qaryat as-Sufla; 568, 10 km NE Khabari Wadha; 1612(BM), 3 km ESE al-Batin/al-'Awja confluence; 3125, al-Batin, 19 km SW Hafar al-Batin.
Northern Dahna: 8729, 1.5 km SE Mashdhubah, 27 – 20N, 45 – 04E.
Northern Summan: 8113, 8 km NNW Hanidh; 8556, W edge Humr Mathluth, 26 – 57N, 47 – 53E; 8587, 4 km W Umm al-Hawshat.

2. Teucrium polium L.

Low, highly aromatic, whitish woolly-canescent dwarf shrublet, branched from the base with numerous erect, ascending or spreading stems 10 – 35(40) cm long. Leaves oblong, sessile, obtuse, crenulate at least in upper half, strongly revolute, 8 – 20 mm long, 3 – 6 mm wide, smaller above. Inflorescence terminal and corymbose, or subterminal and axillary; flowers mildly fragrant, sessile in dense ovoid to subglobose heads c. 12 – 15 mm in diameter on peduncles mostly exceeding or equalling the subtending leaves. Calyx tomentose-woolly, 3 – 5 mm long with triangular teeth immersed in hairs; corolla c. 7 – 8 mm long, white or cream to very pale pinkish, yellowish in throat, exceeding the calyx with one lip prominent, pubescent at back and in lower parts. *Plate 176.*
Habitat: Rocky, well-drained ground around limestone hills or ravines. Infrequent.
Central Coastal Lowlands: 184, 447, 8272, Dhahran.
Vernacular Names and Uses: JA'DAH (Al Murrah, Shammar, Rawalah, Qahtan, gen.). A widely recognized medicinal used as a tea in earlier times to treat malaria and other fevers. Bedouins also report use of the dried leaves smoked in a pipe as a treatment for rheumatism. Musil (1928a, 53) notes that the northern Bedouins used the dried plant as an insect repellent to protect the leather parts of stored armor.

● 2. Lallemantia Fisch. et Mey.

Annual or biennial herbs with lower leaves petiolate, the upper ones subsessile. Inflorescence spicate, with conspicuous dentate bracts. Calyx tubular, 15-veined, with unequal teeth. Corolla bilabiate, the upper lip 2-lobed, the lower one 3-lobed with the central lobe broadened. Stamens 4, didynamous, included in the folds of the upper lip.

Lallemantia royleana Benth.

Pubescent annual, 5 – 20(25) cm high, with stems simple or branched from the base; fresh foliage with a weak but distinct mint odor. Lower leaves ovate, crenate, 10 – 20 mm long, 8 – 15 mm wide, on petioles equalling or exceeding the blade; upper leaves narrower, becoming oblanceolate, serrate, acutish, nearly sessile. Inflorescence dense, spicate, with the upper leaves exceeding the flowers, the flowers subtended by bracts with (mostly) 5 dentate, aristate lobes. Calyx 6 – 7 mm long, 15-nerved, with the 2 lower teeth smaller than the 3 above. Corolla c. 8 mm long with tube mostly included; upper lip weakly 2-lobed with folds including the stamens; lower lip larger, 3-lobed, the central lobe broadened. Nutlets oblong, 2.5 mm long, brown, somewhat compressed and 3-angled, very finely reticulate-punctate. The corollas in our specimens are virtually white, sometimes with the faintest tinge of pinkish; a cluster of pink-mauve flecks on the lower lip near the throat is apparent in fresh material. *Plate 177.*

Habitat: Silty ledges along rocky wadis.
Northern Plains: 8718, tributary to al-Batin, 20 km SW ar-Ruq'i wells, 28–51N, 46–25E; 8722, tributary to al-Batin, 21 km SW ar-Ruq'i wells, 28–51N, 46–25E; 8722, tributary to al-Batin, 21 km SW ar-Ruq'i wells; 8741, tributary to al-Batin, 28–50N, 46–24E.

● 3. Salvia L.

Herbs or shrubs with opposite leaves, aromatic or non-aromatic. Flowers verticillate in spicate or racemose inflorescences. Calyx somewhat 2-labiate with 3 upper and 2 lower lobes. Corolla 2-labiate, the lower lip 3-lobed. Stamens reduced to 2. Nutlets obovoid-oblong, smooth.

Key to the Species of Salvia
1. Calyx 15–25(30) mm long; leaves ovate, subentire, 8–20 cm or more long at base of plant .... 3. *S. spinosa*
1. Calyx 5–9 mm long; leaves elliptical or pinnatisect, less than 8 cm long
 2. Leaves elliptical to oblanceolate, crenulate but undivided .. 1. *S. aegyptiaca*
 2. Leaves pinnatisect with linear lobes .... 2. *S. lanigera*

1. Salvia aegyptiaca L.
Rather diffuse dwarf shrublet, not at all or hardly aromatic, with many ascending branches from the base, more or less tomentose with short pubescence, 10–30 cm high. Leaves in distant pairs, elliptical to linear-oblanceolate, crenate or bluntly dentate, acute, weakly rugose-bullate, 10–20 mm long, 2–6 mm wide, tapering at base to an indistinct petiole, or subsessile. Flowers in rather distant verticillasters toward ends of the branches. Calyx growing in fruit to c. 6–8 mm long, furnished below with spreading, gland-tipped hairs, with acuminate, sometimes purplish teeth. Corolla ground color white to very pale bluish, rather densely spotted blue-violet, about 1.5 times as long as the calyx, with lower lip longer than the upper. Nutlets oblong, smooth black, 2–2.5 mm long.
Habitat: Shallow sand or silt, often in rocky terrain. Occasional.
Northern Plains: 3092(BM), al-Batin, 28–01N, 45–29E; 4014, ad-Dibdibah, 27–56N, 45–35E; 8723, 30 km SW ar-Ruq'i wells; 8745, tributary to al-Batin, 28–50N, 46–24E.
Northern Summan: 720, 33 km SW al-Qar'ah wells; 757(BM), 69 km SW al-Qar'ah wells; 8616, near Dahl Umm Hujul, 26–35N, 47–32E.
Central Coastal Lowlands: 345(BM), Jabal Umm ar-Rus, Dhahran.

2. Salvia lanigera Poir.
Dwarf shrublet with branches ascending-erect from the base, 10–20(30) cm high, the stems and leaves pubescent with rather long crisped white hairs, the flowering parts weakly aromatic with a pungent odor. Leaves oblong-lanceolate in outline, pinnatisect with linear, crenulate, 1–2 mm-wide obtuse lobes, bullate, revolute-margined, 2–8 cm long, 0.5–2 cm wide, smaller above, petiolate or subsessile. Flowers verticillate, subsessile; calyx 5–6 mm long, strongly 2-labiate, the upper lip with teeth obscure, often purplish, densely hirsute with spreading straight whitish eglandular hairs; corolla blue-violet to rather deep purple, 2–2.5 times as long as the calyx, the upper lip exceeding the lower. Nutlets ovoid, 2–2.5 mm long, smoothish, nearly black. *Plate 178.*

265

Habitat: Shallow silts inland, usually along rocky edges of small ravines and wadis. Locally frequent.

Northern Plains: 1616(BM), 3 km ESE al-Batin/al-'Awja confluence; 8710, 20 km SW ar-Ruq'i wells, al-Batin; 8725, al-Batin, 30 km SW ar-Ruq'i wells.

Northern Summan: 810(BM), 37 km NE Jarrarah; 1334(BM), 42 km W Qaryat al-'Ulya; 2864, 26 – 01N, 48 – 29E.

Vernacular Names: JURAYBA' ('mange-weed', likening the rugose-bullate leaves to the thickened rough skin of a camel afflicted with sarcoptic mange, *jarab*, Shammar).

3. Salvia spinosa L.

Perennial herb, 20 – 40 cm high, with upper stems and inflorescence hirsute with spreading crisped white hairs, often somewhat viscid; flowering parts mildly aromatic with a sweetish, lemony scent. Basal leaves petiolate, ovate, obtusish, with puckered bullate surface, rounded or subcordate at base, erose-subentire or obscurely lobed, 8 – 25 cm long, 6 – 15 cm wide, or sometimes larger, the plant sometimes growing strikingly large basal leaves flat on the ground on favorable sandy or silty substrates; upper leaves smaller, those in the inflorescence sessile, cordate-clasping, acuminate. Flowers verticillate, 2 – 6 together, in a rather dense pyramidal inflorescence with ascending stiff branches; calyx 15 – 20 mm long in flower, growing in fruit to c. 25 mm long, with spreading white crisped hairs, dentate with somewhat prickly aristate teeth; corolla white, 20 – 25(30) mm long. *Plate 179.*

Habitat: Silty basins or shallow sand over silts. Infrequent.

Northern Plains: 560(BM), Khabari Wadha; 4045, al-Batin, 27 – 56N, 45 – 23E.

Northern Summan: 623(BM), al-Lisafah; 807-A(BM), 37 km NE Jarrarah; 696, 33 km SW al-Qar'ah; 8618, near Dahl Umm Hujul, 26 – 35N, 47 – 32E.

● 4. Mentha L.

Perennial herbs, mint-scented, with opposite, usually serrate leaves. Inflorescences verticillate, often spicate. Calyx 5-lobed with nearly equal teeth; corolla funnel-shaped with limb nearly equally 4-lobed. Stamens 4. Nutlets obovoid.

Not a native desert genus in eastern Arabia but often cultivated and sometimes seen as an apparent escape in wet habitats of the oases.

Key to the Species of Mentha

1. Leaves minutely but densely short-pubescent, flat; flowers pink-
   violet . . . . 1. *M. longifolia*
1. Leaves glabrous or glabrescent, rugose-bullate with veins elevated below;
   flowers white or nearly white . . . . 2. *M. x villosa*

1. Mentha longifolia (L.) Huds.

Ascending-erect, branched perennial herb, fine tomentose-pubescent, to c. 50 cm high. Leaves elliptical-lanceolate, or subovate, serrate, acute, sessile or subsessile, rounded at base, 2 – 4 cm long, 0.8 – 2 cm wide. Flowers 2 – 3 mm long, verticillate in a dense terminal spike; calyx c. 1.3 mm long, pubescent with acute teeth; corolla c. 2 – 2.5 mm long, pink-lilac, pubescent, with obtuse lobes.

R. M. Harley at the Royal Botanic Gardens, Kew, has provisionally assigned one of our specimens to subsp. *typhoides* (Briq.) Harley.

Habitat: Cultivated in the oases and sometimes seen apparently spontaneous on damp ground around farms, water channels and gardens. Occasional.

South Coastal Lowlands: 2912(BM), al-'Arfaj farm, al-Hasa; 3797(BM), Umm al-Khis, al-Hasa; 7799(BM), 5 km SE al-Hufuf, al-Hasa.

Vernacular Names and Uses: NA'NA' (gen.). Horsemint. Cultivated or collected on a small scale as a condiment. Often used as a flavoring in tea.

2. Mentha x villosa Huds.

Perennial herb with erect stems, to c. 50 cm high. Leaves broadly elliptical-lanceolate to ovate, serrate or crenulate-serrate, acute or obtuse, sessile or subsessile, rugose-bullate with veins elevated below, the lamina (in ours) glabrescent or sparingly tomentose below, 1.5 – 3.5(5) cm long, 1 – 2(3) cm wide. Flowers verticillate in a terminal spike, often branched and interrupted below; calyx 1 – 2 mm long; corolla whitish; anthers pale purple.

● R. M. Harley at the Royal Botanic Gardens, Kew, on the basis of the one, incomplete, specimen available for his review, suggested provisional use of this name for our plant. Ours is a glabrescent form. He notes, however, that it could also prove to be a form of *M. spicata* L. and that additional material is required for a final decision. *M. x villosa* is a hybrid between *M. spicata* L. and *M. suaveolens* Ehr. Our plant has the bullate leaves characteristic of the latter element. It is found along with *M. longifolia* in the oases and is always quite distinct, recognizable by its rugose-bullate, glabrescent, darker leaves and paler flowers. It has a scent distinctly sweeter than that of *M. longifolia*.

Habitat: Cultivated, and apparently occasional as an escape on damp ground in the oases.

South Coastal Lowlands: 2913, al-'Arfaj farm, al-Hasa; 7800, 5 km SE al-Hufuf, farm.

Vernacular Names and Uses: NA'NA' (gen.). Cultivated, like the preceding, as a culinary herb.

● 5. Ocimum L.

Annual or perennial aromatic herbs, or subshrubs, with opposite leaves. Calyx deflexed in fruit, unequally lobed, the upper lobe much broadened and ovate-orbicular; corolla 2-labiate, the upper lip 4-lobed, the lower entire or shortly fimbriate.

Ocimum basilicum L.

Erect, spicy-sweet to peppery aromatic, perennial bushy herb, glabrous or glabrescent, with foliage often purplish but sometimes green at least in white-flowered forms, to c. 50 cm high. Leaves elliptical-lanceolate, entire or obscurely dentate, acute, 2 – 4 cm long, 1 – 1.5 cm wide, tapering at base to a petiole 0.5 – 1.2 cm long. Flowers verticillate in terminal spikes dense toward apex, becoming loose below. Calyx deflexed and growing in fruit to c. 6 mm long, reticulate-veined, with lobes unequal, the upper one broadly obovate, obtuse, free nearly to the base, the others shallowly divided from each other, acute to shortly acuminate. Corolla pink-purplish or white, 6 – 12 mm long, 2-lipped with the upper lip 4-lobed and somewhat spreading, the lower entire or shortly fimbriate. Nutlets obovoid, 2 – 2.5 mm long, minutely punctate and nearly black.

Habitat: Cultivated around gardens and sometimes seen as an apparent escape in the oases.

South Coastal Lowlands: 3796, Umm al-Khis, al-Hasa; 7801, 5 km SE al-Hufuf, farm.

Vernacular Names and Uses: MASHMŪM ('smelled one', gen.), RAYḤAN ('sweet-scent'). Sweet basil. A few plants often cultivated on farms or in gardens as a culinary herb or simply for its aromatic foliage.

53. PLANTAGINACEAE

Herbs or rarely shrublets with leaves rosetted or alternate. Inflorescence spicate; flowers 4-merous with scarious corolla. Stamens 4. Ovary superior, 1 – 4-celled. Fruit a capsule or a nut.

● Plantago L.

Annual or perennial, often stemless herbs with leaves rosulate or alternate and often elongate, appearing parallel-veined. Flowers sessile in a bracteate spike on a scape, bisexual, 4-merous. Calyx persistent, 4-lobed often nearly to base. Corolla scarious-membranous, salverform with 4-lobed limb, persistent. Stamens 4. Ovary superior, usually with 2 locules. Fruit a scarious circumscissile capsule with 2 to several plano-convex seeds.

A very important genus in the desert annual flora with several species, notably *P. boissieri* on stable sands, assuming vernal dominance over extensive areas.

As in some other genera of desert annuals, individual plants exhibit great plasticity in overall size depending on local moisture conditions. On drying ground dwarfed specimens considerably smaller than those of the more usual dimensions listed may be encountered.

Key to the Species of Plantago
1. Leaves pinnate-dentate or pinnatifid with linear acute lobes
   2. Spike 3 – 5 mm in diameter; sepals ciliate .... 4. *P. coronopus*
   2. Spike 7 – 15 mm in diameter; sepals glabrous .... 1. *P. amplexicaulis*
1. Leaves entire
   3. Corolla lobes villous at back
      4. Spikes ovoid, 1 – 1.5 cm long; leaves obovate-spathulate, 2 – 4 cm long .... 3. *P. ciliata*
      4. Spikes cylindrical, 2 – 7 cm long; leaves lanceolate, 5 – 10 cm long .... 8. *P. psammophila*
   3. Corolla lobes glabrous
      5. Calyx lobes 3, the 2 anterior sepals coalescent forming a single 2-striped lobe; weed of gardens, farms .... 5. *P. lanceolata*
      5. Calyx lobes 4; sepals free, all 1-striped; rare weed (*P. major*) or primarily desert plants (others)
         6. Leaves ovate to elliptical, obtuse, not more than twice as long as broad .... 6. *P. major*
         6. Leaves lanceolate to linear, acute, much more than twice as long as broad
            7. Sepals glabrous; spikes 1 – 4 times as long as wide
               8. Leaves glabrous at least above or with a few scattered hairs; seeds 4 – 6 mm long .... 1. *P. amplexicaulis*

8. Leaves soft-villous with fine silky hairs; seeds 2 – 3 mm long .. 7. *P. ovata*
7. Sepals villous; spikes 7 – 25 times as long as wide .... 2. *P. boissieri*

1. Plantago amplexicaulis Cav.

Ascending or decumbent annual herb, 5 – 20(30) cm high, stemless or developing simple
or branching stems, glabrescent above but often pubescent with scattered hairs on young
growth or leaf bases. Leaves linear-lanceolate, acuminate, entire or with a few obscure
fine teeth, 3 – 10 cm long, 0.3 – 1 cm wide, tapering below to an amplexicaul petiole,
glabrous or glabrescent above, sometimes ciliate and with scattered hairs at base. Spikes
ovoid to cylindrical, 1 – 4.5 cm long, 0.8 – 1.5 cm wide, on scapes well exceeding the
leaves or shorter than the leaves on smaller plants. Bracts broadly ovate, glabrous, about
equalling or exceeding the calyx, with herbaceous center and broadly scarious margins;
sepals ovate, glabrous, scarious-margined, keeled. Corolla lobes glabrous, ovate-oblong,
acute. Capsule ellipsoid, 4 – 6 mm long, usually conspicuous and well exceeding the
calyx, glossy, darkening with age. Seeds 2, oblong, boat-shaped, brown, 4 – 5 mm long.

The only *Plantago* of our area that develops stems of significant length. These are
most frequently seen in older plants under favorable moisture conditions; smaller plants
often remain stemless. The species is usually easy to recognize by its large capsules.

Habitat: Silts or shallow sands, usually on or near rocky ground. Frequent.

Northern Plains: 847, 9 km S 'Ayn al-'Abd; 1231(BM), Jabal an-Nu'ayriyah;
1566(BM), al-Batin/Wadi al-'Awja confluence; 1600(BM), 3 km ESE al-Batin/Wadi
al-'Awja confluence; 1743(BM), Jabal al-'Amudah.

Northern Summan: 2774, 25 – 50N, 48 – 00E; 7553, 65 km ESE Umm 'Ushar; 7926,
25 – 35N, 48 – 50E; 7939, 25 – 39N, 48 – 48E; 2798, 15 km ESE ash-Shumlul.

Southern Summan: 2218, 26 km SSE al-Hunayy; 2042, 25 – 00N, 48 – 30E; 2053, Jaww
ad-Dukhan, 24 – 48N, 49 – 05E; 8013, 24 – 00N, 49 – 00E.

Central Coastal Lowlands: 7910, Na'lat Shadqam, 25 – 43N, 49 – 29E.

South Coastal Lowlands: 8313, 5 km E al-Hufuf/Salwah junction, al-Hasa.

2. Plantago boissieri Hausskn. et Bornm.

P. albicans auct.

Ascending annual herb, stemless or with a very short stem crowded with leaf bases,
villous with fine silky hairs, 10 – 30 cm high. Leaves rosulate or crowded on a very
short stem, linear-lanceolate, tapering to an acute apex, 5 – 12(15) cm long, 3 – 7 mm
wide. Spikes cylindrical, 2 – 12(20) cm long, 3 – 5 mm wide, narrowing and becoming
somewhat loose below with age, on scapes shorter or longer than the leaves. Bracts
ovate-oblong, slightly exceeding the calyx, scarious-margined, villous-ciliate at apex
and sometimes pubescent at back. Sepals oblong, scarious-margined, villous at margins
and apex, rounded at back, not or only weakly keeled. Corolla lobes ovate-oblong,
acute, glabrous. Capsule enclosed in the calyx, ovoid-globular, scarious, dehiscing near
the middle, 2 – 2.5 mm long, the lid and seeds falling with the corolla limb. Seeds 2,
oblong, plano-convex, boat-shaped with hollowed flat side, 1.5 – 2 mm long, translucent-
brown. *Plates 180, 181.*

● Perhaps the most abundant and productive annual of sandy habitats in our area
and one of the most characteristic plants of northeastern Arabia in several different
plant communities. It accounts for a major portion of the spring floral biomass. The
seeds are wind-spread at least for a short distance, the corolla limb being adherent to the

dehiscing capsule and, like a parachute, pulling off the lid along with the seeds leaving the empty base of the fruit among the persistent sepals.

Habitat: Shallow or deep sands, or sandy silts, in a wide variety of terrain types. Common and often dominant in the annual flora.

Northern Plains: 1312, 8 km W Jabal Dab'; 689, 20 km WSW Khubayra; 848, 9 km S 'Ayn al-'Abd; 992, 12 km N Jabal al-'Amudah; 1181, Jabal an-Nu'ayriyah; 1560(BM), al-Batin/Wadi al-'Awja confluence; 1646(BM), 2 km S Kuwait border in 47−14E; 1718(BM), 28 km E Abraq al-Kabrit.

Northern Dahna: 799, 15 km SE Hawmat an-Niqyan; 2326, 22 km ENE Rumah; 7532, 40 km NE Umm al-Jamajim; 4035, 27−35N, 44−51E.

Northern Summan: 7928, 25−35N, 48−50E; 2805, 15 km ESE ash-Shumlul; 2833, 26−21N, 47−25E.

Southern Summan: 8365, hills at Khashm az-Zaynah.

Central Coastal Lowlands: 185, Dhahran; 905, Mulayjah; 925, 10 km S al-Khafji-as-Saffaniyah junction; 1416(BM), 7 km NW Abu Hadriyah; 1777(BM), Abu Hadriyah; 2722, 23 km N Jabal Fazran; 3185(BM), 9 km N Shadqam junction; 3293, 2 km W Sabkhat Umm al-Jimal; 3724, 8 km N Abu Hadriyah.

South Coastal Lowlands: 1443(BM), 2 km E Jabal Ghuraymil; 2115, 3 km SW al-Jawamir (Yabrin); 7781, adh-Dhulayqiyah; 7916, 3 km S Jabal Ghuraymil; 8018, 5 km N ad-Dabatiyah; 23−11N, 48−53E.

Rub' al-Khali: 7695, al-Mulayhah al-Gharbiyah, 22−53N, 49−29E.

Vernacular Names and Uses: RIBL, RABL (Mutayr, Shammar, Rawalah, Qahtan, gen.), YANAM (Al Murrah). An important grazing plant. It is reportedly sometimes added to the milk in the preparation of *iqt*, boiled and dried sour milk curd (Bani Hajir).

3. Plantago ciliata Desf.

Villous, stemless or very short-stemmed dwarf annual, 3−5(7) cm high. Leaves rosulate, obovate-spathulate to oblanceolate, often mucronate, obtuse or acute, 1.5−4(6) cm long, 0.3−1 cm wide, tapering at base to a petiole. Spike ovoid to oblong-short-cylindrical, 1−1.5(2) cm long, c. 0.5−0.7 cm wide, on scapes shorter than or exceeding the leaves. Bracts elliptical, villous particularly at margins, mostly slightly shorter than the calyx, white scarious-margined. Sepals elliptical-oblong, villous at margins, rounded at back, not keeled, scarious-margined. Corolla lobes villous at back, lanceolate-triangular, acute, folded upward along the long axis and thus appearing almost linear. Capsule obovoid, c. 2.5 mm long, dehiscing near the middle. Seeds 2, oblong, somewhat compressed, hardly concave, c. 2 mm long, darkish brown. *Plate 182.*

Habitat: Silty soils of inland basins and wadis. Locally frequent to common.

Northern Plains: 656, Khabari Wadha; 1508(BM), 30 km ESE al-Qaysumah; 823, 13 km W Qaryat al-'Ulya; 849, 9 km S 'Ayn al-'Abd; 1198, Jabal an-Nu'ayriyah; 1287(BM), 15 km ENE Qaryat al-'Ulya; 1532(BM), 36 km NE Hafar al-Batin; 1572(BM), al-Batin/Wadi al-'Awja confluence; 1660(BM), 2 km S Kuwait border in 47−14E; 1700(BM), ash-Shaqq, 28−30N, 47−42E; 2711, 14 km ENE Qaryat al-'Ulya; 3066(BM), ash-Shaqq, 28−41N, 47−36E; 3070(BM), ad-Dibdibah, 28−45N, 47−01E; 4055, ad-Dibdibah, 28−06N, 46−25E; 7990, 28−12N, 48−07E.

Northern Summan: 1351(BM), 42 km W Qaryat al-'Ulya; 619, 20 km N al-Lihabah; 740, 33 km SW al-Qar'ah wells; 779, 69 km SW al-Qar'ah wells; 2804, 15 km ESE ash-Shumlul; 2822, 2823, 3 km ESE ash-Shumlul; 2875, 20 km W 'Uray'irah.

Southern Summan: 2049, Jaww ad-Dukhan, 24–48N, 49–05E; 2237, 15 km WSW al-Hunayy; 2093, 8 km N Harad station; 2215, 26 km SSE al-Hunayy; 2242, 26 km ENE al-Hunayy.

South Coastal Lowlands: 374, Jabal Sha'ban, al-Hasa.

Vernacular Names: YANAM (Al Murrah), QURAYṬA' ('lop-ear', N, Dickson 1955).

4. Plantago coronopus L.

Annual stemless herb, often with scapes decumbent, pubescent with rather short, stiff, mostly appressed hairs, 5–15(20) cm high. Leaves rosulate, linear to lanceolate in outline, pinnately dentate or pinnatifid with linear-lanceolate, acute lobes set at right angles to the leaf axis, rarely subentire, 5 – 10(15) cm long, 0.3 – 1 cm wide (including lobes). Spikes narrowly cylindrical, dense, 3 – 10 cm long, c. 3 – 4 mm wide, exceeding or shorter than the leaves, on scapes mostly shorter than the spike. Bracts ovate-triangular, acuminate, acute, ½ to ⅔ as long as the calyx, white-margined, somewhat keeled, finely ciliate. Sepals ovate, white-margined, ciliate, imbricate, the anterior ones gently rounded at back, the posterior ones strongly keeled, winged on keel. Corolla tube densely hirsute, the lobes glabrous, ovate-triangular, acute. Capsule ovoid, 1.5 – 2 mm long, often dehiscing irregularly in the thin-membranous lower portion. Seeds often 4, oblong, brown, somewhat compressed, very minutely punctate, with a shallow fosse on one side. *Plate 183.*

● The species appears to be rather variable in habit, but no attempt has been made to segregate our forms or varieties. In one growth form sometimes seen under good local soil moisture conditions, the scapes are very numerous, radiating almost horizontally from the plant base and recurving upward into erect spikes which surround the plant as a circular curtain or crown. In another form, the scapes are shorter, fewer, ascending irregularly from the base. No. 742 (below), collected on drying ground, was flowering and producing seeds while only 11 mm in total height.

Habitat: Usually on silty inland soils; sometimes on sand near the northern coasts. Occasional.

Northern Plains: 657, Khabari Wadha; 1297(BM), 15 km ENE Qaryat al-'Ulya; 1524(BM), 30 km ESE al-Qaysumah; 1529(BM), 36 km NE Hafar al-Batin; 7562, al-Faw al-Janubi; 1640(BM), 2 km S Kuwait border in 47 – 14E; 8753, ravine to al-Batin, 28 – 50N, 46 – 24E.

Northern Summan: 722, 742, 33 km SW al-Qar'ah wells.

Central Coastal Lowlands: 3268, 3273(BM), 8 km W Qatif-Qaisumah Pipeline KP – 144; 3741, Ras az-Zawr; 3319, 12 km S Nita'; 7727, 2 km S Ras Tanaqib, 27 – 49N, 48 – 53E.

South Coastal Lowlands: 374, Jabal Sha'ban, al-Hasa.

Vernacular Names: QURAYṬA' ('lop-ear', N, Musil 1927, 609).

5. Plantago lanceolata L.

Perennial stemless herb, nearly glabrous to sparsely pubescent, 15 – 60 cm high. Leaves rosulate, long-elliptic to lanceolate, or rarely broadly elliptic, up to c. 25 cm long, 6 cm wide, acute, tapering at base to a petiole, 3 – 7 parallel-veined. Spikes cylindrical, tapering somewhat toward apex, dense, 2 – 6(10) cm long, c. 0.8 cm wide, on sulcate angled scapes usually well exceeding the leaves when mature. Bracts ovate-acuminate, glabrous, scarious-margined, equalling or shorter than the calyx. Sepals sometimes sparsely ciliate at apex, scarious-margined, reduced to 3, the 2 anterior ones coalescent into a single, 2-striped, weakly emarginate unkeeled lobe, the 2 posterior ones separate, keeled. Corolla lobes ovate-triangular, glabrous. Capsule oblong, 2 – 3 mm long. Seeds 2, narrowly oblong, dark brown, c. 2 mm long.

Habitat: Damp ground in the oases or as a weed of irrigated crops. Occasional.

Northern Plains: 7854, farm in al-Batin, 28 – 15N, 45 – 51E.

Central Coastal Lowlands: 7889, al-Khubar.

South Coastal Lowlands: 1903(BM), ash-Shuqayq, al-Hasa; 2903(BM), al-'Arfaj farm, al-Hasa; 3929, near 'Ayn al-Khadud, al-Hasa; 8220, near al-Qurayn, al-Hasa Oasis.

6. Plantago major L. subsp. pleiosperma Pilger

Glabrous to sparingly pubescent perennial herb 25 – 60 cm high. Leaves rosulate, 5 – 30 cm long, 3 – 20 cm broad, ovate to elliptic, obtuse, entire (or sometimes obscurely sinuate), 3 – 7 parallel-nerved, narrowing rather abruptly below to a petiole about equalling or somewhat shorter than the blade. Scapes erect or ascending, usually somewhat exceeding the leaves. Spikes narrowly cylindrical, 5 – 30 cm long, dense or loose, with flowers subtended by ovate, white-margined bracts slightly shorter than the calyx. Calyx lobes equal, obtuse, scarious with green, keeled medial stripe. Corolla lobes acute. Capsule ovoid-ellipsoid.

Habitat: A weed of moist ground in the oases or in irrigated fields. Rare; not nearly as common as *P. lanceolata* in such habitats.

South Coastal Lowlands: The author has seen only one example of this species from the Eastern Province, in the National Herbarium at the Regional Agriculture and Water Research Center, Riyadh. This specimen (RIY 7675) was collected by M. A. Zawawi and labeled 'Eastern Province', and is believed to have come from the al-Hasa Oasis area. It has leaves up to 13 cm long, 8.5 cm wide; the 13 cm-long spike is loose, with rather distant flowers.

7. Plantago ovata Forssk.

Ascending stemless annual, 5 – 15 cm high, moderately villous with fine white silky or cottony hairs. Leaves rosulate, linear-lanceolate, acute, tapering at base to a petiole, 4 – 12(15) cm long, 2 – 6 mm wide. Spikes ovoid to cylindrical, 0.8 – 3 cm long, 0.5 – 1 cm wide, on scapes shorter than or slightly exceeding the leaves. Bracts glabrous, obovate, obtusish, equalling the calyx, broadly white-scarious-margined. Sepals glabrous, elliptic, obtuse, scarious-margined, weakly keeled above. Corolla lobes glabrous, obovate, mucronate. Capsule ellipsoid, dehiscent near the middle, c. 3 mm long. Seeds 2, oblong, plano-convex and boat-shaped with the flattish side hollowed, brown, 2.3 – 3 mm long. *Plate 184.*

Habitat: Silty inland soils. Frequent to common.

Northern Plains: 1515(BM), 30 km ESE al-Qaysumah; 660, 665, Khabari Wadha; 821, 13 km W Qaryat al-'Ulya; 860, 9 km S 'Ayn al-'Abd; 1563(BM), al-Batin/Wadi al-'Awja confluence; 1201(BM), 1242, Jabal an-Nu'ayriyah; 1293(BM), 15 km ENE Qaryat al-'Ulya; 1541(BM), 36 km NE Hafar al-Batin; 1658(BM), 2 km S Kuwait border in 47 – 14E; 3064(BM), ash-Shaqq, 28 – 25N, 47 – 47E; 7564, al-Faw al-Janubi; 7966, Jabal al-'Amudah.

Northern Summan: 7543, 65 km ESE Umm 'Ushar; 3148(BM), 9 km SSW Hanidh; 2791, 41 km SE ash-Shumlul; 721, 33 km SW al-Qar'ah wells; 1338(BM), 42 km W Qaryat al-'Ulya.

Central Coastal Lowlands: 1156, 1163(BM), Jabal al-Ba'al; 3270, 3284, 8 km W Qatif-Qaisumah Pipeline KP – 144.

Southern Summan: 2041, 25 – 00N, 48 – 30E; 2071, Jaww ad-Dukhan, 24 – 48N, 49 – 05E; 2092, 8 km N Harad station; 2185, 24 – 34N, 48 – 45E.

South Coastal Lowlands: 359, 1 km S al-Munayzilah, al-Hasa.

Vernacular Names and Uses: QURAYṬA' (Musil 1927, 609). The seeds reportedly have been used by the Bedouins as a laxative (Carter 1917, 201).

8. Plantago psammophila Agnew et Chalabi-Ka'bi

Decumbent to ascending annual herb, stemless or with short stems branching at base, villous with soft silky hairs, 10 – 15(20) cm high. Leaves lanceolate, acute, tapering at base, sometimes only sparsely pubescent, 5 – 10 cm long, 0.5 – 1.5 cm wide. Spikes cylindrical, elongating and becoming somewhat loose below with age, 2 – 6 cm long, c. 0.7 cm wide, on scapes shorter or longer than the leaves. Bracts ovate, mostly obtuse, equalling the calyx, more or less pubescent at back and strongly ciliate, with herbaceous center often broader than the scarious margins. Sepals villous, ciliate, obtuse, the anterior 2 elliptic-oblong, the posterior 2 ovate, scarious-margined, not or hardly keeled at back. Corolla lobes lanceolate-triangular, acute, villous at back. Capsule ovoid-ellipsoid, dehiscing near the middle, c. 2.5 mm long. Seeds 2, elliptical-oblong, plano-convex, somewhat hollowed on the flattish side, 2 – 2.5 mm long, all grey, grey with brown center stripe, or all brown. *Plate 185.*

This species has the general aspect of a rather coarse, thick-spiked *P. boissieri* but with the villous corolla lobes of *P. ciliata.* The seed color is variable but apparently consistent within a given individual; further study is required to determine whether it correlates with stage of maturity.

Habitat: Sand over silts or silty soils of inland basins or plains. Occasional.

Northern Plains: 1286(BM), 15 km ENE Qaryat al-'Ulya; 1659(BM), 2 km S Kuwait border in 47 – 14E; 3067(BM), ad-Dibdibah, 28 – 45N, 47 – 01E; 8764, ravine to al-Batin, 28 – 49N, 46 – 23E.

Northern Dahna: 7547, 65 km ESE Umm al-Jamajim.

Northern Summan: 1379(BM), 42 km W Qaryat al-'Ulya; 2806, 15 km ESE ash-Shumlul; 2824, 3 km ESE ash-Shumlul; 8571, 8 km S ash-Shumlul.

Southern Summan: 8059, 24 – 06N, 48 – 48E; 7929, 25 – 35N, 48 – 50E.

Vernacular Names: None recorded but probably known as RIBL or YANAM, at least among the central and southern tribes.

54. SCROPHULARIACEAE

Herbs or shrubs with leaves opposite or alternate, simple. Flowers bisexual, usually zygomorphic; calyx 4 – 5-lobed; corolla campanulate, rotate or bilabiate, 5(4)-lobed. Stamens 4 or 2, rarely 5. Ovary superior, bilocular. Fruit a 2-celled capsule.

Key to the Genera of Scrophulariaceae

1. Prostrate creeping herb, rooting at the nodes .... 3. *Bacopa*
1. Erect herbs or shrublets, not rooting at nodes
   2. Stamens 2; corolla 4-lobed .... 4. *Veronica*
   2. Stamens 4; corolla bilabiate, unequally 5-lobed

3. Corolla spurred; annual herbs .... 1. *Linaria*
3. Corolla not spurred; shrublets, or perennials woody at base ... 2. *Scrophularia*

● 1. Linaria Mill.

Annuals or perennials with mostly opposite or verticillate entire leaves. Flowers with linear bracts, in terminal racemes or spikes. Calyx deeply and nearly equally 5-lobed; corolla spurred at base, bilabiate, the upper lip with 2 lobes, the lower 3. Stamens 4, didynamous. Capsule ovoid or globose, dehiscing by apical tooth-like valves. Seeds numerous.

All of our specimens are clearly assignable to *L. tenuis*. Dickson (1938; 1955) reported two other species from Kuwait not yet collected by the author in our area: *L. simplex* Desf. and *L. micrantha* (Cav.) Hoffm. et Link (author given as 'Spreng.' in Dickson 1955). Differentiating characters are provided in the following key:

1. Corolla yellow; spur conspicuous, about as long as the calyx
  2. Seeds transversely wrinkled, unwinged .... *L. tenuis*
  2. Seeds discoid, surrounded by a membranous wing .... *L. simplex*
1. Corolla bluish to whitish; spur minute, much shorter than the
  calyx .... *L. micrantha*

Al-Rawi and Daoud (1985) also report *L. simplex* from Kuwait, and this species might be found to occur in our northern borderlands. *L. micrantha* was not included in their account of the Kuwait flora.

Linaria tenuis (Viv.) Spreng.
L. ascalonica Boiss. et Kotschy
Erect annual herb, simple or branched, mostly glabrous, or with some spreading hairs in the inflorescence, 10–30 cm high. Cauline leaves mostly alternate, linear-filiform, 1–3 cm long, 1–3 mm wide; those on sterile shoots opposite, wider. Flowers racemed, rather distant, with the narrow bract and the pedicel about equalling the calyx. Calyx c. 4 mm long, often with fine spreading hairs, with linear lobes. Corolla yellow, 7–9 mm long including the spur; spur descending, nearly straight, about equalling or somewhat shorter than the corolla. Capsule oblong-cylindrical, equalling or exceeding the calyx. Seeds 0.5–0.7 mm, black, transversely wrinkled. *Plate 186.*

Habitat: Sands and sandy silts. Occasional to locally frequent.

Northern Plains: 1180(BM), Jabal an-Nu'ayriyah.

Northern Dahna: 4033, 27–35N, 44–51E.

Northern Summan: 1366(BM), 42 km W Qaryat al-'Ulya; 8448, 8 km W Nita'; 8551, 10 km W Nita'; 8440, edge 'Irq al-Jathum, 25–54N, 48–36E.

Central Coastal Lowlands: 1773(BM), Abu Hadriyah; 1803(BM), 5 km NE al-Khursaniyah; 3728(BM), 8 km N Abu Hadriyah; 3300(BM), 2 km W Sabkhat Umm al-Jimal.

Vernacular Names: (for similar *Linaria* spp.): SHILWAH (N tribes, Musil 1928b, 363; 1927, 624), UMM AS-SUWAYQ (Kuwait, Dickson 1955).

● 2. Scrophularia L.

Annuals or perennials with leaves opposite or alternate. Flowers small, in a cymose

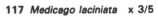

19 *Lotus halophilus* x 3/4

118 *Melilotus indica* x 1/2

120 *Astragalus annularis* x 1/2

122 *Astragalus corrugatus* x 3/5

121 *Astragalus bombycinus* x 1

123 *Astragalus eremophilus* x 2/5

125 *Astragalus kahiricus* x 1/10

124 *Astragalus hauarensis* x 1

130 *Astragalus spinosus* x 2/3

127 *Astragalus schimperi*
x 2/3

126 *Astragalus kahiricus* x 1

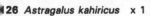

129 *Astragalus sieberi* x 4/5

128 *Astragalus sieberi* x 1/7

131 *Hippocrepis bicontorta* x 1 1/3

132 *Onobrychis ptolemaica* x 1/2

134 *Cynomorium coccineum* x 1/8

135 *Chrozophora tinctoria* x 2/5

133 *Taverniera spartea* x 2

136 *Ziziphus nummularia* x 1 1/4

139 *Seetzenia lanata* x 3/4

37 *Haplophyllum tuberculatum* x 1/6

138 *Fagonia bruguieri* x 4/5

140 *Zygophyllum mandavillei* x 2/5   **141** *Zygophyllum migahidii* x 2/5   **142** *Zygophyllum qatarense* x 1/4

144 *Tribulus arabicus* x 1

143 *Tribulus arabicus* x 1/6

145 *Tribulus pentandrus* agg. x 1 2/5

146 *Nitraria retusa*   x 1/5

147 *Erodium deserti*   x 1

149 *Erodium glaucophyllum*   x 1/3

148 *Erodium deserti*   x 2 1/2

150 *Anisosciadium lanatum* x 1/6

151 *Deverra triradiata* (fruiting umbel) x 1 3/5

152 *Ducrosia anethifolia*
x 1/4

153 *Centaurium pulchellum*
x 1

154 *Rhazya stricta* x 1/30

156 *Calotropis procera* x 1/2

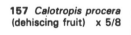

159 *Lycium shawii* x 4/5

158 *Lycium depressum* x 1 1/2

157 *Calotropis procera*
(dehiscing fruit) x 5/8

155 *Glossonema varians* (fruit)
x 3/5

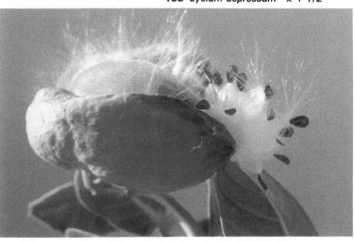

160 *Hyoscyamus pusillus* x 1

162 *Convolvulus cephalopodus* x 3/5

164 *Convolvulus oxyphyllus* x 1/6

161 *Withania somnifera* x 3/10

163 *Convolvulus cephalopodus* x 1/5

166 *Cuscuta planiflora*
(flowers, on *Rhanterium*) x 1
3/4

165 *Cuscuta planiflora* (on
Citrullus) x 1/7

169 *Moltkiopsis ciliata* x 1/5

167 *Heliotropium bacciferum* x 2/5   **168** *Heliotropium digynum* x 3/4

171 *Arnebia hispidissima* x 1/2

172 *Arnebia tinctoria* x 3/5

170 *Arnebia linearifolia* x 1/2

173 *Echium horridum* x 1/4

176 *Teucrium polium* x 1/6

177 *Lallemantia royleana*
x 1/2

175 *Teucrium oliverianum*
x 1 1/5

78 *Salvia lanigera* x 1/4   **179** *Salvia spinosa* x 1/2   **180** *Plantago boissieri* x 1/8

182 *Plantago ciliata* x 1/3

183 (Below left) *Plantago coronopus* x 1/3

184 (Below right) *Plantago ovata* (red spots on one spike are giant velvet mites, *Dinothrombium tinctorium*) x 1/2

1 *Plantago boissieri* x 1/2   **185** *Plantago psammophila* x 1/4

186 *Linaria tenuis*   x 1 1/5

188 *Scrophularia hypericifolia*   x 1

189 *Cistanche tubulosa*   x 1/6

187 *Scrophularia hypericifolia*
x 2/5

190 *Cistanche tubulosa*   x 1/6

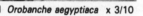

191 *Orobanche aegyptiaca* x 3/10   **192** *Orobanche cernua* x 1/3

193 (Top right) *Orobanche cernua* (on *Lycium shawii*) x 1/8

195 (Above) *Orobanche mutelii* x 3/4

194 *Orobanche crenata* x 1/3

196 *Blepharis ciliaris* x 1/4

197 *Crucianella membranacea* x 2/5

198 *Scabiosa olivieri* x 1/3

199 *Scabiosa palaestina* x 1/2

200 *Ifloga spicata* x 1/3

201 *Gymnarrhena micrantha*
x 1/2

inflorescence. Calyx deeply 5-lobed. Corolla bilabiate, unequally 5-lobed with somewhat inflated short-cylindrical or urn-shaped tube. Stamens 4. Capsule 2-valved, septicidal, with numerous seeds.

Key to the Species of Scrophularia
1. At least some leaves distinctly lobed, the blade oblong to obovate in outline .... 1. *S. deserti*
1. All leaves entire, elliptical-oblong, acute .... 2. *S. hypericifolia*

1. Scrophularia deserti Del.

Perennial branching from a woody base, 20–50 cm high, glabrous except for minute glands in the inflorescence. Leaves opposite and alternate, oblong or obovate in outline, 3–6-lobed about half way to the midrib or some obovate, subentire, denticulate, petiolate to subsessile, the blades mostly 1–3 cm long, 0.4–1 cm wide, sometimes narrowly white-cartilaginous at margins. Flowers short-pedicelled, in elongate terminal inflorescences c. 10–25 cm long with side branches often forked. Calyx 1.5–2 mm long, obtusely lobed; corolla partly dark red, partly paler, 4–6 mm long. Capsule depressed-globose, weakly bilobed, smooth, somewhat woody, c. 2.5 mm long, 3 mm wide. Seeds c. 0.5 mm long, nearly black, transversely wrinkled.

Habitat: Usually on elevated rocky ground, or in shallow sand on hills. Occasional to frequent.

Northern Plains: 3106(BM), al-Batin, 28–00N, 45–28E; 1215, Jabal an-Nu'ayriyah; 878, 9 km S 'Ayn al-'Abd; 1165(BM), Jabal al-Ba'al; 3847, Qatif-Qaisumah Pipeline KP–380.

Northern Summan: 818, near Wabrah.

Central Coastal Lowlands: 108, 1044, Dhahran; 365A, 365B, Jabal Umm ar-Rus, Dhahran.

Southern Summan: 2098, 16 km SSE Harad station.

Vernacular Names: 'ALQA (N tribes, Dickson 1955).

2. Scrophularia hypericifolia Wydl.

Ascending branched shrublet, 30–60 cm high, glabrous except for minute glands in the inflorescence; intermediate branches glossy white. Leaves alternate or subopposite, oblong-elliptical to oblanceolate, entire (or reportedly sometimes denticulate), acute, sessile or tapering to a short petiole, 7–15 mm long, 2–5 mm wide. Flowers short-pedicellate, cymose in a branching terminal inflorescence to c. 30 cm long. Sepals ovate, obtuse, hyaline-margined, 1–1.5 mm long; corolla dark red with lower lobes whitish, 4–5 mm long. Capsule short-pedicellate, depressed-globose, mucronate, weakly bilobed, smooth, becoming woody, c. 3 mm long, 5 mm broad, with valves dehiscing longitudinally. Seeds 1–1.5 mm long, nearly black, rugose. *Plates 187, 188.*

Habitat: Nearly always associated with rather deep sand and dunes. A plant of the northern sands, frequent or even locally dominant in the northern Dahna but apparently rare or absent in that sand body south of approximately the 26th Parallel.

Northern Dahna: 605(BM), Nafud al-Jur'a, 5 km SE ath-Thumami; 801(BM), 15 km SE Hawmat an-Niqyan; 7534, 40 km NE Umm al-Jamajim; 7536, 8 km NE Umm al-Jamajim; 8570, E edge 'Urayq al-Khufaysa, 26–24N, 47–12E.

Vernacular Names: 'ALQA (N tribes, Musil 1928b, 335; 1927, 589).

● 3. **Bacopa** Aublet

Perennial herbs with opposite leaves and flowers solitary, axillary. Calyx lobes 5, unequal; corolla 5-lobed. Stamens 4. Capsule 4-valved with numerous minute seeds.

Bacopa monnieri (L.) Pennell
Herpestis monniera (L.) Humb., Bonpl. et Kunth

Glabrous perennial herb, procumbent with stems repent, up to c. 30 cm long and rooting at the nodes. Leaves opposite, subsessile, oblong, somewhat succulent, obtuse, entire or obsoletely crenate-dentate, 1 – 1.5 cm long, 0.3 – 1 cm wide. Flowers solitary in the axils on pedicels shorter or longer than the subtending leaf; calyx 3 – 6 mm long, lobed very unequally; corolla campanulate, white to pinkish-violet, with ovate obtuse lobes, c. 1.5 times as long as the calyx. Capsule ovoid, acute, 2 – 3 mm long.

Habitat: A weed of wet ground in gardens, farms. Frequent.

Central Coastal Lowlands: Dhahran in lawns; al-Qatif in gardens.

South Coastal Lowlands: 3805, 10 km NW Salwah at a flowing water well; 'Ayn Umm Sab', al-Hasa (Dickson 1955, 50).

● 4. **Veronica** L.

Annuals or perennials with leaves opposite, or alternate above. Calyx deeply 4(5)-lobed. Corolla rotate, unequally 4(5)-lobed. Stamens 2. Capsule 2-loculed.

Veronica polita Fries

Ascending or decumbent, somewhat delicate, glabrescent to sparingly pubescent annual herb, 5 – 25 cm high, branched from base. Leaves mostly alternate, or opposite below, ovate, coarsely dentate or crenate, or shallowly lobed, mostly 1 – 1.8 cm long, 1 – 1.5 cm wide, on petioles 2 – 5 mm long. Flowers solitary in the axils, on pedicels to c. 1.3 cm long. Sepals elliptical to ovate, acute, 3 – 6 mm long, unequal. Corolla blue, 5 – 8 mm in diameter, with 4 unequal lobes. Capsule broader than long, emarginate, equalling or shorter than the calyx, pubescent or glabrous. Seeds 1 – 1.5 mm long.

Habitat: A weed; to be expected only in or around cultivation. Rare.

● Collected by the author in central Arabia (No. 3954, BM, from 'Unayzah) but not yet seen personally in the east. It is reported as a weed of the Eastern Province by Chaudhary, Parker, and Kasasian (1981).

55. OROBANCHACEAE

Annual or perennial herbs without chlorophyll, parasitic on the roots of dicotyledonous plants. Stems erect, simple or branched, arising scapelike from a subterranean tuber having threadlike connections to roots of the host. Leaves reduced to scales. Inflorescence a terminal, usually dense spike or raceme, each flower subtended by a bract and sometimes also by 2 narrower bracteoles. Calyx 4 – 5-lobed. Corolla tubular to funnel-shaped with limb more or less bilabiate, mostly 5-lobed. Stamens 4, didynamous. Ovary superior, 1-loculed. Fruit a loculicidal capsule enclosed in the calyx, with numerous minute seeds.

Members of this family are rather conspicuous in the northeastern Arabian desert vegetation, where their parasitic habit has apparently provided a useful drought-evasion strategy. They require further study, not only in regard to host specificity but also with respect to their general life history, which is poorly understood in many cases.

Key to the Genera of Orobanchaceae
1. Calyx lobes 5, obtuse; corolla hardly bilabiate, with 5 subequal
   lobes .... 1. *Cistanche*
1. Calyx lobes 2 – 4, acute-acuminate; corolla distinctly bilabiate with unequal
   lobes .... 2. *Orobanche*

● 1. **Cistanche** Hoffm. et Link

Annual or perennial herbs consisting of a stout, simple dense spike of flowers emerging from the ground, with leaves reduced to scales on the lower stem. Each flower subtended by a single bract and 2 small lateral bracteoles. Calyx with 5 rounded obtuse lobes; corolla tubular and funnel-shaped, with a limb of 5 nearly equal spreading lobes, the lowest often slightly longer. Stamens 4 with woolly anthers.

Cistanche tubulosa (Schrenk) Wight
C. lutea Wight
Perennial arising from tubers, glabrous with succulent stem, 20 – 80 cm high. Scales lanceolate, 1 – 3(5) cm long. Spike dense, showy, cylindrical. Bracts ovate to lanceolate, acute or obtuse, equalling or often exceeding the calyx; bracteoles oblong-linear, shorter than or about equalling the calyx. Flowers mostly 4 – 5 cm long, with the corolla c. 2.5 – 3 times as long as the calyx; calyx with 5 obtuse, oblong-ovate lobes. Corolla tube ascending, bent near the middle, cylindrical below, expanding and funnel-shaped above, with a spreading or somewhat reflexed limb of 5 obtuse, more or less equal rounded lobes c. 6 – 10 mm long. Corolla color variable, bright pure yellow to pinkish-yellow or violet within, whitish yellow to pinkish-violet outside; buds at apex of spike white to yellow or pinkish violet; flowers at bottom of spike ageing brownish. Stamens 4, mostly 25 – 30 mm long, inserted in the corolla tube about ¼ to ⅓ above its base, more or less woolly-pubescent at their bases; anthers white-woolly, connivent below the down-turned stigma, their wool often enmeshed in a common mass. Capsule ovoid-oblong, c. 1.5 cm long. *Plates 189, 190.*
● This is one of the showiest spring plants in eastern Arabia with its bright yellow, dense column of flowers sometimes approaching a meter high. It is rather variable in our territory, with varying flower color and with the flowers tightly packed in the spike or sometimes rather loose. The connective of the anthers is apiculate to some degree in most of our specimens, although the point is rather variable in size and form. It may be from c. 0.2 to 0.8 mm long, and straight or inflexed above.

Careful deep digging is required to demonstrate the rather fine and fragile connection with the root of the host plant. *Cistanche* produces great quantities of dust-fine seeds, but these do not apparently germinate as surface plants. Further study is required to determine whether germination might occur on or near roots of the host. The author, in midsummer near Dhahran, has found living aestivation tubers attached to roots of *Zygophyllum qatarense* at a depth of 20 – 30 cm.
Habitat: Widespread on sandy or sandy-silty ground; tolerant of somewhat disturbed conditions and often seen along roadsides wherever its hosts are growing.

Host Plants: *Haloxylon salicornicum, Cornulaca arabica,* and other Chenopodiaceae. Also frequently seen on *Zygophyllum,* and sometimes apparently on *Calligonum.*
Northern Dahna: 7540, 8 km NE Umm al-Jamajim.
Central Coastal Lowlands: Dhahran; 1789(BM), 13 km E Abu Hadriyah (apparently on *Calligonum*); 8409 – 13 (detached flowers only), 9 km S Dhahran, on *Zygophyllum qatarense;* 8651 (aestivation tuber in liquid), 9 km S Dhahran (on *Zygophyllum qatarense*).
Rub' al-Khali: 7612, at-Tara'iz, 21 – 09N, 50 – 29E; 7675, as-Sanam, 22 – 27N, 50 – 31E (on *Cornulaca arabica*).
Vernacular Names: DHĀNŪN (gen.), DHUNŪN (Al Murrah), BĀṢŪL (Al Rashid, S Rub' al-Khali).

● 2. Orobanche L.

Annual or perennial herbs consisting of a simple or branched inflorescence supported by a short stem and with leaves reduced to scales appressed to the stem below. Inflorescence a terminal, dense or somewhat loose, spike or raceme. Each flower subtended by a bract, some species having also a pair of narrower lateral bracteoles. Calyx 2 – 4 lobed, often unequally in 2 divided groups, the lobes acute and usually acuminate. Corolla with tubular or funnel-shaped tube and a bilabiate limb of 5 lobes. Stamens 4 with anthers hairy or glabrous. Capsule ovoid or oblong-cylindrical.

For effective use of the key below, flowers should be at least roughly dissected for careful distinction between bracts, bracteoles (which may be minute and adherent to the calyx although clearly outside its whorl), and the calyx lobes.

Key to the Species of Orobanche

1. Each flower subtended by 1 bract and 2 lateral bracteoles
    2. Flowers 25 – 35(40) mm long, strong blue-violet with 2 white folds in the lower lip . . . . 1. *O. aegyptiaca*
    2. Flowers 12 – 15(22) mm long, all dirty white, or sometimes tinged with very pale violet . . . . 4. *O. mutelii*
1. Each flower subtended by 1 bract only; lateral bracteoles absent
    3. Filaments inserted at middle of corolla tube, not hairy at base; corolla limb not or only weakly spreading, with entire oblong lobes . . 2. *O. cernua*
    3. Filaments inserted near base of tube, hairy at base; corolla limb distinctly spreading with broad, crenate-dentate lobes . . . . 3. *O. crenata*

1. Orobanche aegyptiaca Pers.

Ascending to erect, glandular-tomentose annual herb, simple or branched from near base, 10 – 40 cm high. Scales ovate-triangular to lanceolate, acute, mostly 5 – 10 mm long. Flowers mostly sessile in rather loose and broad spikes 5 – 20 cm long, 4 – 7 cm wide. Bracts lanceolate, acuminate, mostly shorter than the calyx; bracteoles linear-lanceolate, long acuminate-subulate, shorter than the calyx. Calyx 8 – 12 mm long, divided about half-way to base with 4 lanceolate to triangular, acuminate lobes. Corolla 25 – 35(40) mm long, blue-violet except at base and lower lip, pubescent, with tube cylindrical in lower third, gradually expanding and funnel-shaped above, gently deflexed at c. ⅓ from base; limb blue-violet except on lobes of lower lip, which has 2 conspicuous white infolds. Anthers connivent, white-woolly at margins. Capsule long-ovoid, tapering to an acute apex, 6 – 8 mm long, longitudinally 4 – 5-ridged, rather woody, sometimes sterile. *Plate 191.*

Habitat: Sands or sandy silts. Occasional to locally frequent.

Host Plants: Often found with *Rhanterium epapposum* but relationship unsure.

Northern Plains: 1474(BM), 50 km WNW an-Nu'ayriyah (in *Rhanterietum* but with annuals); 1717(BM), 28 km E Abraq al-Kabrit (with *Rhanterium*).

Northern Summan: 7930, 25 – 35N, 48 – 50E; 7933, 25 – 31N, 48 – 42E; 8560, Jibal adh-Dhubabat, 26 – 45N, 48 – 01E.

Central Coastal Lowlands: 7711, Ras al-Ghar, 27 – 32N, 49 – 12E, an atypical plant reported by one specialist as having affinities with *O. caucasica* G. Beck. It is smallish, arachnoid-pubescent, with anthers almost glabrous and lobation of calyx differing from that usually seen in *O. aegyptiaca*.

Vernacular Names: DHUNŪN.

2. Orobanche cernua Loefl.

Erect, glandular-puberulent, unbranched annual herb, 10 – 30(40) cm high. Scales triangular to lanceolate, acute, 5 – 10 mm long. Flowers 15 – 20 mm long in a spike 7 – 15(20) cm long, c. 2 – 3.5 cm wide. Bracts lanceolate, acute, mostly exceeding the calyx and about as long as the corolla tube. Bracteoles absent. Calyx about half as long as the corolla, with 2 main lobes, 1 on each side, these entire-acuminate or divided into 2 acute sublobes. Corolla tube cylindrical, hardly widening above, constricted above fruit, yellowish, often glossy, or pale violet in upper part; limb with rather small, weakly spreading lobes, usually pale to dark violet. Stamens inserted near middle of the corolla tube, with filaments glabrous at base; anthers glabrescent. Capsule ovoid to ellipsoid with acute apex, longitudinally ridged, 6 – 10(12) mm long, hard. Seeds dust-fine. *Plates 192, 193.*

Habitat: Sand or sandy silts. Frequent.

Host Plants: *Lycium shawii, Artemisia monosperma, Anvillea garcinii.* Particularly associated with *Lycium*, almost every large stand of which has its population of the parasite.

Northern Plains: 1726(BM), Jabal al-'Amudah; 7969, 1 km NE Jabal al-'Amudah (on *Lycium shawii*).

Northern Dahna: 7539, 8 km NE Umm al-Jamajim; 8621, E edge 'Irq Hazwa, 26 – 24N, 47 – 21E (on *Artemisia monosperma).*

Northern Summan: 7401, Batn Sabsab, 26 – 16N, 48 – 23E; 2807, 12 km ESE ash-Shumlul.

Central Coastal Lowlands: 4068(BM), az-Zulayfayn, 26 – 51N, 49 – 51E (on *Lycium shawii*); 7757, 7893, Dhahran (on *Lycium shawii*); 1975, ad-Dawsariyah; 7921 Batn al-Faruq, 25 – 34N, 49 – 05E (on *Artemisia monosperma*).

South Coastal Lowlands: 7901, Na'lat Shadqam, 25 – 43N, 49 – 29E.

Southern Summan: 8060, 24 – 06N, 48 – 48E; 8002, 15 km N Harad (on *Anvillea garcinii*).

3. Orobanche crenata Forssk.

Erect, more or less glandular-pubescent annual herb, single or aggregate, 15 – 50(70) cm high, often with reddish-purple stems. Scales sometimes absent on above-ground stems, those below mostly oblong-lanceolate, subacute, 1 – 2 cm long. Flowers 15 – 25 mm long, somewhat verticillate, in a spike 10 – 35(50) cm long, 2.5 – 4 cm wide, becoming rather loose with age. Bracts lanceolate, acute, about as long as the calyx.

Bracteoles absent. Main calyx lobes 2, lateral, c. half as long as the corolla, each divided into 2 acuminate sublobes or 1 sometimes entire, acuminate. Corolla white to cream, or yellowish when young, the tube campanulate; the limb widely spreading, up to 20(25) mm broad, with reddish-purple veins and lobes crenulate-denticulate and somewhat undulate at margins; lower lip elongate, tongue-like. Stamens inserted at base of the tube with filaments pubescent at base and with a bright orange gland between their bases and the corolla wall, producing a viscid, clear, non-drying, exudate. Stigma bilobed, brownish-orange or flesh-colored. Capsule elongate-ovoid, acute at apex, hard, 7 – 10 mm long. Seeds brown, dust-like. *Plate 194.*

The largest *Orobanche* of our area and readily recognizable by its wide-spreading, crenulate, veined corolla limb.

Habitat: A weed in gardens, farms. Also collected in desert near a roadside. Rare.

Host plants so far recorded in our area: *Melilotus indica*, *Centaurea pseudosinaica*.

Central Coastal Lowlands: 7887, al-Khubar (on *Melilotus indica*); 8104, roadside 13 km SSE Hanidh (on *Centaurea pseudosinaica*).

4. Orobanche mutelii F. W. Schultz

Erect, dwarfish, glandular-tomentose annual herb, 5 – 10(15) cm high, often consisting of only a dense spike of flowers emerging from the ground with little or no stem visible below, simple or sometimes feebly short-branched at base. Scales triangular-lanceolate, acute, c. 5 – 8 mm long. Flowers 12 – 22 mm long, in a dense terminal spike 5 – 12 cm long, 2 – 3 cm wide. Bracts lanceolate, acute, slightly shorter than the calyx. Bracteoles narrowly lanceolate, acuminate, shorter than the calyx and adherent to it below. Calyx about half as long as the corolla, with 4 lanceolate, acuminate lobes. Corolla dirty white, often tinged with very pale violet, with tube narrowly funnel-shaped, constricted above the ovary; limb c. ⅕ as long as the corolla, weakly spreading with obtuse to acute lobes. Stamens inserted at mid-tube; filaments glabrous at base; anthers glabrous. Capsule ovoid, 6 – 8 mm long, with minute, grey-brown, punctate seeds. *Plate 195.*

Habitat: Sands or sandy silts. Occasional.

Host Plants: Uncertain; associations seldom demonstrated.

Northern Plains: 1738(BM), Jabal al-'Amudah; 7992, 10 km S Abraq al-Kabrit.

Northern Summan: 7924, 25 – 35N, 48 – 50E; 7937, 25 – 39N, 48 – 48E; 8561, al-Malsuniyah, 25 – 31N, 48 – 13E; 8595, 8 km WSW al-Lihy al-'Atshan, 26 – 39N, 47 – 58E.

Central Coastal Lowlands: 1760(BM), Abu Hadriyah; 1975, ad-Dawsariyah; 7899, 5 km S Thaj.

Southern Summan: 8067, Wadi as-Sahba, 24 – 12N, 48 – 38E.

56. ACANTHACEAE

Herbs or shrubs, sometimes spinose, with opposite leaves. Inflorescence cymose or racemose, bracteate. Flowers bisexual, zygomorphic, 5(4)-merous, frequently 2- or 1-labiate. Calyx 4 – 5-lobed; corolla 4 – 5-lobed. Stamens 4 or 2. Ovary superior, 2-locular. Fruit a loculicidal capsule.

● Blepharis Juss.

Spiny perennial herbs or shrublets with leaves opposite, decussate, in 4 ranks.

Inflorescence a spike, with flowers subtended by a spiny bract and smaller bracteoles. Calyx unequally 4-lobed. Corolla 1-labiate with 5- or 3-lobed limb. Stamens 4, inserted at base of the corolla tube. Anthers pubescent. Capsule 2(4)-seeded.

Blepharis ciliaris (L.) B. L. Burtt
B. edulis (Forssk.) Pers.

Ascending, rigid, spiny-prickly perennial herb, glabrescent to sparsely pubescent, much branched from the base, 10 – 30 cm high. Stems leaves oblong or lanceolate, tapering at base to a petiole, entire or with small remote spinules at margin. Spikes dense with flowers much exceeded by spreading, channeled, 3 – 5 cm-long, lanceolate-acuminate, strongly veined bracts with several pairs of marginal spines. Bracteoles 10 – 15 mm long. Calyx scarious, with posterior lobe larger, denticulate. Corolla blue, veined darker, 2 – 2.5 cm long with short tube and a single spreading, obtusely 3-lobed lip. *Plate 196.*

Habitat: Silty bottoms of rocky wadis or basins. So far found only in the southern parts of our area, and then rarely.

Southern Summan: 1028(BM), 28 km NE Harad; also seen in 25 – 05N, 48 – 30E.

Vernacular Names: NAQΓ', NAQQΓ' (gen.).

57. RUBIACEAE

Woody plants or herbs with opposite branches and opposite or whorled simple leaves. Inflorescence basically cymose. Flowers bisexual, mostly regular and 4 – 5-merous. Limb of calyx often obsolete; corolla 4(5)-lobed. Stamens 4 – 5, inserted on the corolla and alternate with the lobes. Ovary inferior. Fruit a capsule, a berry or paired mericarps.

Key to the Genera of Rubiaceae
1. Corolla 5-lobed; flowers sessile in dense spikes with imbricate leaflike
   bracts .... 1. *Crucianella*
1. Corolla 4-lobed; flowers axillary in ebracteate peduncled cymes .... 2. *Galium*

● 1. Crucianella L.

Annual or perennial herbs with leaves 4 – 6 whorled. Flowers sessile in terminal spikes, each subtended by a white-margined bract and 2 bracteoles. Calyx with limb obsolete; corolla salverform or funnel-shaped with 4 – 5-lobed limb. Fruit of 2 oblong mericarps.

Crucianella membranacea Boiss.
Scabridulous, often dwarf, annual herb, ascending or somewhat spreading-branched from base, with tetragonal stems 5 – 15(20) cm long. Leaves in whorls of 4 or 6, linear-lanceolate, acute, sessile, 5 – 10 mm long, c. 1 – 1.5 mm wide. Spikes of variable length, elongating to 4(8) cm long, 1 – 1.5 cm wide. Bracts lanceolate-acuminate, white-scabridulous-margined, the outer and longer one recurved, to c. 10 mm long, exceeding the straighter inner ones. Corolla c. 5 – 7 mm long, shorter than the outer bract, with narrow cylindrical tube and short minute limb of 5 oblong-linear obtuse lobes. Mericarps 2, oblong, c. 2 mm long, marked with short longitudinal white lines. *Plate 197.*

Habitat: Sandy-silt soils. Occasional to locally frequent.

Northern Plains: 875, 9 km S 'Ayn al-'Abd; 1009(BM), 12 km N Jabal al-'Amudah; 1744(BM), Jabal al-'Amudah; 7578, Jabal Dab'; 7988, 28 – 12N, 48 – 07E.

Central Coastal Lowlands: 1792(BM), 5 km NE al-Khursaniyah; 7721, 2 km S Ras Tanaqib, 27–49N, 48–53E; 7744, 5 km NW Ras Tanaqib, 27–53N, 48–49E.

Southern Summan: 2079, Jaww ad-Dukhan, 24–48N, 49–05E.

● 2. Galium L.

Annual or perennial herbs or shrublets with tetragonal stems and leaves opposite or whorled. Flowers 4-merous in ebracteate cymes. Calyx limb obsolete. Corolla rotate with 4 lobes often appendaged at apex. Fruit mostly of paired globose to ellipsoid mericarps, or mericarps single by abortion.

Key to the Species of Galium

1. Flowers white; fruiting pedicels strongly curved; mericarps c. 1.5–4(5) mm in diameter
   2. Peduncles 5 mm or less long .... 1. *G. ceratopodum*
   2. Peduncles 8–15(20) mm long .... 3. *G. tricornutum*
1. Flowers purple; fruiting pedicels straight; mericarps 0.5–0.7 mm in diameter .... 2. *G. setaceum*

1. Galium ceratopodum Boiss.

Ascending or climbing, somewhat delicate annual herb, 10–30 cm high with stems retrorsely rough scabridulous. Leaves in whorls of 5–7, oblanceolate, mostly 5–15 mm long, 1–3 mm wide, sometimes larger, acute-acuminate, retrorsely scabridulous at margins. Peduncles axillary, mostly 2–3 mm long. Flowers minute; corolla with 4 white lobes; stamens 4, alternating with the corolla lobes. Fruit pedicels strongly recurved, about as long as the peduncles. Mericarps twin, or often single by abortion, globose, dark brown, tuberculate, c. 1.5–1.8 mm in diameter.

Habitat: Silty bottoms of larger wadis or basins; sometimes in shelter of large shrubs such as *Ziziphus nummularia*. Locally frequent.

Northern Plains: 613(K), al-Batin, 30 km NE Umm 'Ushar; 3895, 14 km ESE Umm 'Ushar.

Vernacular Names: ABU NASHR ('saw-plant', in reference to its scabrid stems and leaves, N tribes, Musil 1927, 587).

2. Galium setaceum Lam. subsp. decaisnei (Boiss.) Ehrendf.
   G. decaisnei Boiss.

Delicate, filiform, branched ascending annual herb, 5–20 cm high, minutely scabridulous or smooth-glabrous. Leaves in whorls of 5–8, linear-lanceolate to subulate-filiform and revolute, to c. 10 mm long, petiolate or sessile. Cymes mostly near ends of stems, with straight diverging filiform peduncles and pedicels. Corolla rotate, purple, minute, with lobes short-appendiculate at apex. Mericarps minute, globose-ovoid, 0.5–0.7 mm long, pubescent or glabrous.

Habitat: In shade sheltered by boulders or large shrubs.

● Possible sight records in al-Batin, in the northern plains, but not yet confirmed by specimens. Recorded from Kuwait and Iraq and most probably present in rocky parts of our northern plains and the Summan.

3. Galium tricornutum Dandy

Climbing or scrambling, retrorsely scabridulous annual herb with flaccid quadrangular stems 10 – 60 cm long. Leaves in whorls of 6 – 8, linear-oblanceolate, acute-acuminate, 10 – 35 mm long, mostly 1.5 – 4 mm wide, retrorsely scabrid at margins. Cymes axillary, 1 – 3(5)-flowered, on peduncles mostly 8 – 15 mm long and equalling or somewhat exceeding the subtending leaves. Pedicels recurved in fruit, 3 – 8 mm long. Corolla white with acute lobes; mericarps subglobose, 2.5 – 5 mm in diameter, finely tuberculate.

Habitat: A weed of gardens or field crops. Infrequent to rare overall but possibly abundant in individual fields.

Central Coastal Lowlands: 8246, Dhahran, with other Mediterranean weeds in newly laid-out building grounds.

58. DIPSACACEAE

Herbs, rarely shrublets, with opposite leaves. Flowers in a capitulum surrounded by involucral bracts, the marginal ones often radiate, strongly zygomorphic. Calyx cupuliform or 5 – 10-divided into awns. Corolla 4 – 5-lobed, mostly funnel-shaped. Stamens usually 4. Ovary inferior, 1-locular, surrounded by an involucel or 'outer calyx'. Fruit an achene enclosed in the involucel.

● Scabiosa L.

Annual or perennial herbs, rarely shrubs, with leaves entire or pinnatisect. Heads long-peduncled, often radiate, with herbaceous involucral bracts. Corolla unequally 5-lobed. Fruiting involucel with a membranous, often conspicuous, campanulate to rotate corona. Achenes topped by the 5 awns of the 'inner calyx', within the corona.

Key to the Species of Scabiosa

1. Flowers pink-lilac, in heads 4 – 10 mm wide .... 1. *S. olivieri*
1. Flowers pale yellowish-white in heads 15 – 35 mm wide ...... 2. *S. palaestina*

1. Scabiosa olivieri Coult.

Erect or ascending annual herb with both appressed and spreading hairs; stems often reddish above and leaves somewhat greyish-canescent. Leaves linear-oblong or oblanceolate, acute or somewhat obtuse, entire or the basal ones rarely dentate-lobed, 1 – 4(6) cm long, 0.3 – 1.2 cm wide. Heads 4 – 10 mm in diameter on fine peduncles 1 – 8(15) cm long. Involucral bracts elliptical, acute, densely appressed hairy, 3 – 5 mm long. Flowers pink-lilac, pubescent, 5 – 10 mm long. Corona of involucel membranous, c. 2 mm long, 20 – 40-nerved. Awns of inner calyx fine, 4 – 7 mm long, reddish-purple, minutely antrorsely scabridulous, conspicuous and exceeding the head at all stages of flowering and fruiting. *Plate 198.*

Habitat: Silty soils of the northern plains and Summan. Locally frequent.

Northern Plains: 1613(BM), 3 km ESE al-Batin/Wadi al-'Awja confluence; 564(BM), Khabari Wadha; 571(BM), 105 km E Hafar al-Batin; 687, 20 km WSW Khubayra; 1489(BM), 50 km WNW an-Nu'ayriyah; 1521(BM), 30 km ESE al-Qaysumah; 1556(BM), al-Batin/Wadi al-'Awja confluence; 1672(BM), 2 km S Kuwait border in 47 – 14E; 2718, 3 km NW Jabal Dab'; 3848, Qatif-Qaisumah Pipeline KP – 380; 8795, 30 km SW ar-Ruq'i Post, 28 – 54N, 46 – 27E.

Northern Summan: 773(BM), 69 km SW al-Qar'ah wells; 739, 33 km SW al-Qar'ah wells; 1347(BM), 42 km W Qaryat al-'Ulya; 2767, 25 – 49N, 48 – 01E.

Vernacular Names: WUBAYRAH ('fur-weed', N tribes, Musil 1927, 629).

2. Scabiosa palaestina L.

Erect annual herb, 5 – 40(50) cm high, with main stem branched above, more or less pubescent with both shorter, mostly appressed, hairs and longer spreading hairs. Leaves linear-oblong to lanceolate, acute or obtuse, 2 – 6(8) cm long 0.1 – 1 cm wide, entire or faintly serrate near apex, the lowest sometimes somewhat divided, those above linear. Heads 1.5 – 3 cm in diameter, somewhat radiate, on peduncles 3 – 10 cm long. Involucral bracts lanceolate, acute, 5 – 12 mm long, 1 – 3 mm wide. Flowers pale yellowish-white (or reportedly sometimes bluish), fragrant, 12 – 15 mm long, the lobes with linear-oblong, mostly obtuse lobules. Fruiting heads globose or subglobose, membranous with the prominent coronas of the involucels. Corona 8 – 10 mm long, 30 – 35-nerved. Awns of inner calyx c. 5 – 7 mm long, included or slightly exserted from the fruiting corona. Achenes densely white-villous at base. *Plate 199.*

Habitat: Sandy plains or sand along rocky wadi edges; also collected on silt. Occasional to locally frequent.

Northern Plains: 543, 1 km NW Qaryat as-Sufla; 603(BM), 5 km ENE Qulban Ibn Busayyis; 1228(BM), Jabal an-Nu'ayriyah; 1304(BM), 10 km W Qaryat al-'Ulya; 1602(BM), 3 km ESE al-Batin/Wadi al-'Awja confluence.

Northern Dahna: reported by Dickson (1955, 83).

Northern Summan: 3870, 15 km WSW Umm 'Ushar; 2759, 25 – 49N, 48 – 01E; 7927, 25 – 35N, 48 – 50E.

Vernacular Names: UMM AR-RUWAYS ('headlet-wort', N tribes, Musil 1927, 628; 1928a, 710), FANĪ (Al Murrah).

59. COMPOSITAE

Annual or perennial herbs or shrubs, rarely trees, with leaves alternate, rarely opposite. Flowers *(florets)* sessile, crowded on a common receptacle, and together forming a head *(capitulum)* surrounded at base by an involucre of bracts. Receptacle with or without bracts. Florets mostly 5-merous, and may be bisexual, staminate, pistillate, or neuter. Calyx usually replaced by a *pappus* from base of the corolla. Corolla of 2 basic types: *tubular*, mostly regular; and *ligulate*, with one side much extended in a strap-shaped lobe. Stamens 4 – 5, inserted on the corolla tube. Ovary inferior, 1-locular and 1-ovuled. Fruit an achene, with or without an apical pappus consisting of hairs, bristles, pales, a crown, or an auricle.

The basic types of the head are: *discoid*, with all florets tubular; *ligulate*, with all florets ligulate, bisexual, with the ligules usually 5-dentate; and *radiate*, with florets in the middle of the head tubular and those around the periphery ligulate, the ligules generally 3-dentate. Care should be taken in examining some ligulate heads, such as those of *Picris*, not to mistake the immature, still non-ligulate (and sometimes different-colored) central florets for tubular florets and thus misinterpret the head as radiate.

A very important plant family in eastern Arabia, second only to the Gramineae in number of species. Classification is based partly on the form of the achene and its appendages, and some dissection may be required to separate some genera and species although gross features are employed where possible in the following artificial key.

Key to the Genera of Compositae
1. All florets, or at least the central ones, tubular (Tubuliflorae)
   2. Leaves spiny or prickly-margined
      3. Heads compound, composed of 1-flowered headlets each surrounded by an involucre of spine-tipped or awn-tipped bracts
         4. Heads globular, not fleecy, without involucre at base ... 23. *Echinops*
         4. Heads hemispherical, densely fleecy, with an involucre of leaf-like bracts at base .... 24. *Acantholepis*
      3. Heads not compound; florets non-bracteate
         5. Stem spiny- or prickly-winged
            6. Pappus of simple soft hairs; green annual .... 26. *Carduus*
            6. Pappus short-plumose; grey tomentose biennial ... 28. *Onopordum*
         5. Stem not spiny- or prickly-winged
            7. Leaves green, blotched with white .... 27. *Silybum*
            7. Leaves, if with white, then only along veins, or all grey-tomentose
               8. Achene glabrous .... 32. *Carthamus*
               8. Achene hairy .... 25. *Atractylis*
   2. Leaves unarmed, or only minutely prickly-dentate
      9. Outer involucral bracts spiny- or prickly-tipped .... 30. *Centaurea*
      9. Outer involucral bracts unarmed at apex, or tipped with fine red awns only
         10. Head with radiating, tubular, marginal florets
            11. Involucral bracts fine pectinate-margined, tipped with a reddish awn exceeding the head .... 29. *Zoegea*
            11. Involucral bracts entire, scarious at apex, without awn .... 31. *Amberboa*
         10. Head with radiating ligulate marginal florets, or head discoid
            12. Leaves opposite
               13. Heads sessile, yellow-flowered; plant glabrous . 13. *Flaveria*
               13. Heads peduncled, white-flowered; plant appressed-pubescent .... 12. *Eclipta*
            12. Leaves alternate
               14. Pappus of hairs or bristles
                  15. Plants less than 15 cm high, stemless or dwarf
                     16. Involucre wholly scarious, translucent; leaves thread-like, dense .... 4. *Ifloga*
                     16. Involucre not wholly scarious
                        17. Glabrous, stemless herb .... 6. *Gymnarrhena*
                        17. Grey-woolly dwarf herb .... 5. *Filago*
                  15. Stemmed herbs or shrubs over 15 cm high
                     18. Involucral bracts in 1 row; glabrous desert annual .... 21. *Senecio*
                     18. Bracts imbricate in several rows; shrubs or pubescent annuals, or glabrous weed or ruderal
                        19. Pappus surrounded or joined at base with a scarious cup .... 8. *Pulicaria*

285

19. Pappus sessile and without surrounding cup
   20. Involucral bracts densely white-villous with hairs exceeding the width of the bract .... 7. *Vicoa*
   20. Bracts glabrous, or short-pubescent, or ciliate, with hairs shorter than width of the bract
      21. Involucral bracts glabrous, not ciliate; plant glabrous .... 1. *Aster*
      21. Bracts short-pubescent or with margins minutely ciliate; plant short-pubescent at least in the inflorescence
         22. Herb with narrowly linear stem leaves; flowers yellowish .... 2. *Conyza*
         22. Shrub with ovate to elliptical leaves; flowers pinkish .... 3. *Pluchea*
14. Pappus scarious or absent
   23. Ray flowers present
      24. Ray flowers white
         25. Receptacle with chaffy bracts at least in upper part .... 14. *Anthemis*
         25. Receptacle entirely naked .... 16. *Matricaria*
      24. Ray flowers yellow or orange
         26. Much-branched woody shrublet .... 9. *Rhanterium*
         26. Annual herbs
            27. Outer involucral bracts long, leaflike; stemless plant at ground surface .... 11. *Asteriscus*
            27. Outer bracts not long, leaflike; stemmed plants
               28. Leaves entire; flowering heads 5 – 15 mm wide .... 22. *Calendula*
               28. Leaves lobed or pinnatisect; flowering heads 20 – 60 mm wide .... 19. *Chrysanthemum*
   23. Ray flowers absent
      29. Perennials, woody at least at base
         30. Heads solitary, subtended by large, leaflike, outer involucral bracts .... 10. *Anvillea*
         30. Heads grouped in inflorescences, not subtended by leaflike bracts
            31. Heads corymbed; leaves entire, serrulate ..... 15. *Achillea*
            31. Heads panicled; leaves pinnatisect, lobed, or entire and non-serrulate .... 20. *Artemisia*
      29. Annual herbs
         32. Achene entirely without pappus or auricle, 3-ribbed on ventral side, 0.7 – 0.9 mm long .... 16. *Matricaria*
         32. Achene with white-scarious auricle at apex, 1 – 1.5 mm long (not including auricle)
            33. Achene smooth at back, 3-ribbed ventrally, sometimes 2-glandular at dorsal apex .... 17. *Tripleurospermum*
            33. Achenes finely c. 15-striate at back; rounded or obscurely prismatic ventrally .... 18. *Aaronsohnia*

1. All florets ligulate (Liguliflorae)
   34. Pappus absent .... 33. *Koelpinia*
   34. Pappus of hairs present
      35. Pappus plumose
         36. Involucral bracts imbricated, unequal, in several rows
            37. Leaves pinnatifid into narrowly linear, acuminate
               lobes .... 35. *Leontodon*
            37. Leaves narrowly linear or erose-subentire .. 37. *Scorzonera*
         36. Involucral bracts in a single row, equal or with a few much
           smaller ones at base
            38. Head without smaller involucral bracts at base; achenes
               (with beak) over 10 mm long; weed ..... 34. *Urospermum*
            38. Head with smaller bracts at base of involucre; achenes
               (with beak) c. 5 mm long; desert plant .... 36. *Picris*
      35. Pappus of simple hairs
         39. Pappus of the central or all achenes beaked, at least shortly
            40. Involucral bracts glabrous to fine-glandular; coarse weed
               1–2.5 m high .... 41. *Lactuca*
            40. Involucral bracts pilose or bristly-hispid; herbs under
               0.5 m high
               41. Achenes minutely echinate-tuberculate around the
                  apex; involucre glabrous or fine-glandular beneath
                  the longer bristles .... 38. *Heteroderis*
               41. Achenes smooth around apex; involucral bracts
                  at least very sparsely tomentose beneath the
                  bristles .... 43. *Crepis*
         39. Pappus sessile
             42. Achenes all of the same kind, strongly compressed,
               brown .... 40. *Sonchus*
            42. Achenes of 2 kinds in the same head, terete or prismatic,
               pale
               43. Pappus caducous with hairs connate in a hollow ring
                  at base; ligules reddish at back .... 42. *Reichardia*
               43. Pappus persistent, or deciduous and connate in a
                  solid disc at base; ligules yellow, whitish, or silvery
                  at back .... 39. *Launaea*

TUBULIFLORAE

● 1. Aster L.

Perennial or annual herbs with entire or dentate leaves. Heads with both tubular and ligulate florets, radiate or not. Involucre imbricate; receptacle ebracteate. Achenes compressed, with a pappus of hairs or bristles.

Aster squamatus (Spr.) Hieron.
Erect glabrous annual, sometimes possibly perennating, rigid and sometimes somewhat

woody below, branched above, 20 – 120 cm high. Lower leaves oblanceolate, acute, entire or very faintly and remotely serrulate, to c. 12 cm long, 1.3 cm wide, tapering at base to a petiole; those above shorter, sessile, narrowly linear. Flowering heads 4 – 5 mm in diameter with yellow disc florets, the rays white or sometimes pale pink-violet, weakly spreading and only slightly exceeding the involucre; heads in a rather rich, rounded paniculate or somewhat corymbose terminal inflorescence. Fruiting heads obconical, c. 7 – 9 mm long (including the slightly exserted pappi). Achenes linear-lanceolate, glabrous, somewhat compressed and faintly striate, c. 2 – 2.3 mm long, with pappus of whitish hairs 5 – 6 mm long and sometimes kinky near base.

Habitat: A weed found only around farms or waste places, usually near irrigation run-off. Locally frequent.

Central Coastal Lowlands: 1097, al-Qatif; 1917, 1921, al-Jarudiyah, al-Qatif Oasis; 7439(BM), Dhahran; 7769, 4 km S al-Qatif town.

A plant that has been spreading as a weed in the Middle East, including the Gulf area, in recent years. Its range is apparently expanding in the Eastern Province, where it has existed in the oases at least since the mid-1960s.

● 2. Conyza Less.

Annual or perennial herbs or shrubs with discoid or sometimes somewhat radiate heads. Central florets bisexual, tubular, the marginal ones filiform, staminate. Involucral bracts imbricate; receptacle ebracteate. Anthers without tails at base; style arms appendaged. Achenes oblong, compressed, often somewhat pubescent, with a pappus of scabrid hairs or bristles.

Conyza linifolia (Willd.) Täckh.
Erigeron linifolium Willd.
Conyza bonariensis (L.) Cronquist
Erect or ascending appressed-pubescent annual herb, 20 – 80(100) cm high. Leaves entire or with some irregular acute lobes, those at base oblong-lanceolate and tapering to a petiole, those on the stem sessile, narrowly linear, acute. Heads numerous, solitary at branch apices, in a raceme or a somewhat corymbose panicle. Involucre c. 5 mm long, with short-pubescent linear-lanceolate, acute bracts. Florets pale yellowish. Achenes 1.5 – 2 mm long with yellowish to brownish pappus well exceeding the achene.

Habitat: Weed in and around gardens and farms. Infrequent.

Northern Plains: 7449, Qaisumah Pump Station.

South Coastal Lowlands: 8309, al-Hufuf.

● *Conyza triloba* Decne. has been collected by the author as a weed at al-Majma'ah, central Arabia, and it might also occur sporadically in the east. It may be recognized by its leaf form: spathulate in outline with 3 – 5 obtusish distal lobes. The heads are only c. 3 mm long, in rather dense terminal corymbs.

● 3. Pluchea Cass.

Shrubs or perennial herbs with discoid heads in corymbs. Involucral bracts imbricate, scarious or membranous; receptacle flat, ebracteate. Outer florets pistillate, filiform; central florets bisexual, tubular. Style arms without appendages; anthers tailed at base. Achenes terete or angular with a pappus of hairs or bristles in a single row.

288

Pluchea dioscoridis (L.) DC.

Conyza dioscoridis (L.) Desf.

Ascending, much-branched, rather dense, short-pubescent woody shrub, 1 – 3 m high with aromatic foliage. Leaves ovate to elliptical-oblong, serrate, sessile or subsessile, often shortly auriculate at base and the younger, upper ones sometimes decurrent, 2 – 6(8.5) cm long, 1 – 3(5) cm wide, smaller in the inflorescence. Heads numerous in dense terminal corymbs. Involucre c. 5 mm long, 4 mm wide, with imbricate, lanceolate, acute, ciliate, sometimes purplish bracts. Florets pink to purplish. Achenes linear, faintly ribbed, c. 1 – 2 mm long with a pappus of c. 15 – 18 hairs c. 5 mm long.

Habitat: Canal banks or field edges in the oases. Occasional to frequent.

Central Coastal Lowlands: 527(BM), al-Qatif; 8656, Dhahran; 8658, Darin, Tarut Island.

South Coastal Lowlands: 8662, 2 km E al-Hufuf.

Vernacular Names: SĀHIKĪ, often pronounced *'sāhichī'* (farmers in al-Qatif Oasis).

● Some modern authors follow Desfontaines in placing this plant in *Conyza* (Tribe Astereae). The assignment to *Pluchea* appears to be preferred, based on the clearly tailed anthers and non-appendaged style branches of our specimens. A question remains about the relationship of our plant to *P. ovalis* (Pers.) DC., described from Africa and usually separated on the basis of its decurrent leaves. Dorothy Hillcoat and Sylvia Gould at the British Museum (Natural History) herbarium report that our specimens are a good match for Bornmüller 400, collected in Bahrain and called *P.* (or *Conyza*) *dioscoridis* var. *ovalifolia* Haussk. et Bornm., and probably the type for that variety. They also note that this specimen has some decurrent leaves and that it appears to be indistinguishable from *P. ovalis*. Some other specimens from Arabia and Mesopotamia have been named as the latter. The character of decurrent leaves, however, is not constant in our plants, which appear to be all of one species. Many shrubs have none at all, while some carry them on younger shoots. The character seems to appear mainly on new branchlets with lush foliage and to be associated with well-watered or seasonal (spring) conditions. It seems preferable, therefore, to use the name *dioscoridis* provisionally until other differences between the two taxa are studied.

● 4. Ifloga Cass.

Dwarf annual herbs with linear to subulate leaves. Heads minute, discoid, clustered 1 to several in the leaf axils. Involucral bracts glabrous, imbricate, grading into pales. Achenes compressed, glabrous, both epappose and with a pappus of apically plumose bristles.

Ifloga spicata (Forssk.) Sch.-Bip.

Erect dwarf annual herb, simple or branched at base and 3 – 12 cm high, glabrescent, often viscid and with adhering sand grains. Leaves dense, sessile, very fine, linear-subulate, mostly 10 – 15 mm long, less than 1 mm wide, often tomentellous on 1 side. Heads ovoid, acute at apex, c. 3 mm long, sessile in clusters of 1 – 3 in the axils, few-flowered. Involucral bracts wholly membranous, golden-yellow, ovate, the outer ones acuminate. Achenes minute, less than 1 mm long. *Plate 200.*

A plant of peculiar appearance, the stems with their dense subulate foliage resembling small cylindrical brushes growing out of the ground.

Habitat: Stable sands or gritty plains. Common and widespread.

Northern Plains: 1322(BM), 8 km W Jabal Dab'; 670, Khabari Wadha; 830, 13 km W Qaryat al-'Ulya; 858, 9 km S 'Ayn al-'Abd; 995, 12 km N Jabal al-'Amudah; 1152(BM), Jabal al-Ba'al; 1191, Jabal an-Nu'ayriyah; 1591(BM), al-Batin/Wadi al-'Awja junction; 1638(BM), 2 km S Kuwait border in 47–14E.

Northern Summan: 1384(BM), 42 km W Qaryat al-'Ulya; 738, 33 km SW al-Qar'ah wells; 777, 69 km SW al-Qar'ah wells; 3329, 12 km WSW an-Nazim.

Central Coastal Lowlands: Dhahran; 1766(BM), Abu Hadriyah; 1791(BM), 5 km NE al-Khursaniyah; 1986, ad-Dawsariyah; 3782, Ras al-Ghar; 4062, az-Zulayfayn, 26–51N, 49–51E.

Southern Summan: 2051, Jaww ad-Dukhan, 24–48N, 49–05E; 2149, 23–15N, 48–33E; 2177, 24–25N, 48–41E; 2233, 15 km WSW al-Hunayy; 2125, 4 km SW al-Jawamir (Yabrin).

South Coastal Lowlands: 1426(BM), 2 km E Jabal Ghuraymil.

Vernacular Names: TARABAH ('earth-wort', Bani Hajir, a name more commonly used for *Silene villosa* but also applied to this and other viscid annuals that tend to collect adherent sand on their surfaces), ZUNAYMAH ('slit-ear', Rawalah, Musil 1927, 631).

● 5. Filago L.

Annuals, often grey-tomentose, with entire leaves. Heads very small, discoid, often woolly, clustered into glomerules subtended by floral leaves. Florets yellowish, inconspicuous. Achenes compressed, often pappose and epappose in the same head.

Filago desertorum Pomel
F. spathulata auct. non C. Presl

Prostrate to ascending dwarf annual, grey woolly-tomentose, branched from base with stems usually 5–8 cm long. Leaves oblong-spathulate, entire, tapering at base, often only 5–12 mm long, 2–5 mm wide. Glomerules densely white-woolly or cottony, c. 10 mm in diameter, with 6–12 headlets. Involucral bracts of headlets in several whorls, densely fleecy, green medially with white-scarious margins and with white, awnlike, acuminate apex. Florets c. 1.5 mm long. Achenes 0.5–0.8 mm long, the inner ones with a pappus.

Habitat: Silty desert wadi beds and basins. Frequent.

Northern Plains: 1636(BM), 2 km S Kuwait border in 47–14E; 1585(BM), al-Batin/Wadi al-'Awja junction; 597(BM,K), al-Batin, 16 km ENE Qulban Ibn Busayyis; 669, Khabari Wadha; 1236, Jabal an-Nu'ayriyah; 1285(BM), 15 km ENE Qaryat al-'Ulya; 1523(BM), 30 km ESE al-Qaysumah; 1540(BM), 36 km NE Hafar al-Batin; 7561, al-Faw al-Janubi.

Northern Summan: 2821, 3 km ESE ash-Shumlul; 730, 33 km SW al-Qar'ah wells; 1376(BM), 42 km W Qaryat al-'Ulya.

Central Coastal Lowlands: 1762(BM), Abu Hadriyah; 3262, 8 km W Qatif-Qaisumah Pipeline KP–144.

Southern Summan: 2048, Jaww ad-Dukhan, 24–48N, 49–05E.

Vernacular Names: QUṬṬAYNAH ('cotton-weed', Mutayr).

● 6. Gymnarrhena Desf.

Stemless annuals with rosulate leaves. Heads discoid, crowded in dense sessile clusters in center of the plant at ground level. Involucral bracts short-ovate; receptacular bracts boat-shaped, each enclosing a floret, bristly at center of the receptacle. Fertile achenes obconical, villous; pappus of an outer row of scabrous bristles and an inner row of 7 – 9 pales. Subterranean heads sometimes present with fewer florets and epappose achenes. A monotypic genus.

Gymnarrhena micrantha Desf.

Dwarf, glabrous, virtually stemless annual consisting of a flat rosette of leaves and flattish clusters of sessile heads at ground level in center. Leaves sessile, rosulate, lanceolate, acute, sometimes acuminate, entire or (less commonly) dentate, mostly 3 – 10 cm long, 0.5 – 1(1.5) cm wide; inner leaves often smaller. Heads tightly aggregated in basal, central, flattish green to brownish clusters. Achenes c. 2 mm long with a pappus c. 3 times their length. *Plate 201.*

● The plant is known to produce subterranean as well as surface heads. These are few-flowered, with achenes often without pappus. They may germinate within the mother plant and serve to maintain local propagation, a mechanism believed to be an adaption to extreme desert environments where conditions may be inhospitable to germination of the usual wind-borne, pappose achenes (Zohary and Feinbrun-Dothan 1978).

Habitat: Gritty soils or limestone crevices, often associated with stony ground. Infrequent.

Northern Plains: 1226, Jabal an-Nu'ayriyah; 1279(BM), 15 km ENE Qaryat al-'Ulya; 1519(BM), 30 km ESE al-Qaysumah; 1631(BM), 2 km S Kuwait border in 47 – 06E; 1732(BM), Jabal al-'Amudah.

Northern Summan: 1382(BM), 42 km W Qaryat al-'Ulya.

Southern Summan: 2107, 49 km S Harad.

Vernacular Names: KAFF AL-KALB ('dog's paw', Shammar; Musil 1927, 594).

● 7. Vicoa Cass.

Annual herbs with simple, entire, alternate leaves and radiate heads. Involucral bracts imbricate in several rows. Achenes subterete; pappus of unequal bristles.

Vicoa pentanema Aitch. et Hemsl.

Erect or ascending, branched annual with white, spreading, soft hairs and purplish-red stems. Basal leaves oblong-spathulate, sessile, obtuse, entire or dentate, mostly 2 – 4 cm long, 0.7 – 1 cm wide; the stem leaves oblong to broad-lanceolate or ovate, sessile, broad-rounded at base, mostly 1 – 2 cm long but smaller above. Heads numerous, mostly terminal, villous, c. 6 – 7 mm long and equally broad, on fine reddish peduncles 1 – 3 times as long as the head. Involucral bracts narrowly linear-lanceolate, acute, with dense white spreading hairs. Flowers yellow, the rays hardly exceeding the disc florets. Achene minute, less than 0.5 mm long, with pappus or apparently sometimes without.

Habitat: A weed of disturbed ground around farms or roadsides. Infrequent.

Northern Summan: 624(K), 20 km NE al-Lihabah, near farm.

Central Coastal Lowlands: 8115, 15 km SSE al-Wannan.

Vernacular Names: ʿUFAYNAH (Mutayr).

● 8. Pulicaria Gaertn.

Annual or perennial herbs, sometimes somewhat shrubby, with radiate or discoid heads and upper leaves often semi-amplexicaul. Involucral bracts imbricate in several rows; receptacle ebracteate. Florets yellow, the rays in 1 row. Achenes more or less cylindrical, often pubescent, with a biseriate pappus, the outer row consisting of a very short dentate scarious cup or scarious teeth, these persistent or falling together with the much longer inner bristles.

Key to the Species of Pulicaria

1. Flowers in corymbose terminal groups; pedicels mostly under 2.5 cm long, with white spreading hairs .... 1. *P. arabica*
1. Flowers nearly all solitary at the branch tips, or if corymbose then pedicels mostly over 3 cm long; pedicels close woolly-tomentose
    2. Pappus of 14 – 18 bristles .... 3. *P. incisa*
    2. Pappus of 6 – 12 bristles
        3. Leaves dentate and crisp-wavy at margins, mostly under 3 mm wide; heads 5 – 10 mm wide .... 4. *P. undulata*
        3. Leaves somewhat wavy but not dentate, over 3 mm wide; heads 15 – 25 mm wide .... 2. *P. guestii*

1. Pulicaria arabica (L.) Cass.

Erect or ascending dichotomously branched annual, 10 – 60 cm high, leafy throughout, glabrescent to rather densely pubescent, especially in younger parts, but older stems often glossy yellowish or red, glabrescent or with sparse hairs. Leaves oblong to lanceolate or oblanceolate, mostly 2 – 4 cm long, acute or obtuse, the lower ones tapering at base and subpetiolate, those above sessile, somewhat half-clasping at base. Heads c. 7 – 15 mm in diameter, terminal and axillary on rather fine, c. 5 – 20 mm-long peduncles, yellow-flowered with rays slightly exceeding the involucre. Involucre c. 5 mm long, with lowest bracts lanceolate, the inner ones narrowly linear. Achenes oblong, brown, weakly compressed, sparsely appressed-hairy, 1 – 2 mm long (in ours mostly 1 mm), with a pappus of 7 – 12 bristles (in ours mostly 7 – 10), 3 times as long as the achene.

Habitat: Found only as a weed, often on rather wet ground, in the oases. Infrequent.
South Coastal Lowlands: 1905(BM), ash-Shuqayq, al-Hasa Oasis; 'Ayn Umm Sab', al-Hasa Oasis (Dickson 1955, 77); 8224, near al-Qurayn, al-Hasa Oasis.

2. Pulicaria guestii Rech. f. et Rawi

Ascending, numerous-stemmed, branched perennial, somewhat aromatic, grey-green woolly-tomentose, 15 – 35 cm high. Leaves obovate-spathulate, obtuse, mostly 1.5 – 2.5 cm long, those below tapering at base to a petiole, those above minute; upper stems leafless or nearly so beneath the heads, scape-like. Heads solitary, terminal, in flower 15 – 25 mm in diameter, with yellow florets, the rays short-liguled. Involucral bracts narrowly linear even below, appressed silky-pubescent, the inner bracts with a strongly acuminate scarious apex. Achenes appressed silky, c. 1.5 mm long, with a pappus of c. 8 bristles 2.5 times as long as the achene.

Habitat: Small silty basins on stony ground or wadi bottoms. Infrequent to rare.
Northern Plains: 576(K), 76 km E Hafar al-Batin; see also below.

Northern Summan: 2871(BM), 16 km W an-Najabiyah wells, 25 – 59N, 48 – 38E.

● Our specimens appear to be the southernmost records for this species described from Iraq. It is almost certainly the *Pulicaria* noted as 'may be undescribed' (as it indeed was at the time) by Violet Dickson (1955, 77). She collected it northeast of Hafar al-Batin in April 1949 and described it as a 'bush about 1 ft. high; large yellow flowers with small turned-under petals; soft grey leaves, aromatic'.

3. Pulicaria incisa (Lam.) DC.

P. undulata auct. non L.

Ascending, many-stemmed, branched, more or less crisped-tomentose perennial, 20 – 40 cm high, highly aromatic with a sweet-minty fragrance. Leaves oblong to linear, entire or with shallow serrate lobes, often wavy-margined, mostly obtusish, those at base tapering to a petiole, those on stems 1 – 3(4) cm long, 0.2 – 0.6 cm wide, sessile, clasping-auriculate at base, becoming smaller above. Heads terminal, solitary, 1.5 – 2 cm wide in flower with prominent spreading ray florets, the ligules often ⅓ or even ½ as long as the diameter of the disc, paler yellow than the disc florets. Involucre c. 5 mm long with outer bracts oblong, green-herbaceous, tomentose, the innermost linear with scarious acuminate apex. Achenes cylindrical, bullet-shaped with truncate apex, pale, c. 1.5 mm long with sparse appressed hairs; pappus of 14 – 18 bristles c. 2.5 times as long as the achene. *Plate 202.*

● This species has long been known in most references as *P. undulata* L., but this usage has been shown to be based on an erroneous attribution of the type (see *Taxon 29*: 694 – 95, and the report of the IAPT Committee for Spermatophyta: 25, *Taxon 32*: 282 – 83). The prominent ray florets and strong sweet aroma of this plant are distinctive.

Habitat: Silt-floored basins or wadi beds. So far recorded only from the south in and around Wadi as-Sahba; probably intrusive from central Arabia along that channel.

Southern Summan: 8006, 24 – 00N, 49 – 00E; 8058, 24 – 06N, 48 – 48E.

4. Pulicaria undulata (L.) C. A. Meyer

Pulicaria crispa (Forssk.) Benth.

Francoeuria crispa (Forssk.) Cass.

Ascending, often hemispherical, suffrutescent perennial, 12 – 75 cm high, intricately and densely branched from the base with stems closely white-woolly tomentose. Leaves narrowly linear or broadening somewhat distally, sessile and semi-amplexicaul at base, acute to obtusish, strongly repand-undulate, and appearing somewhat toothed, mostly 5 – 20 mm long, 1.5 – 3 mm wide (but see below), more or less woolly-tomentose, much smaller on the upper stems. Heads solitary-terminal, hemispherical, 5 – 10 mm in diameter, the disc convex with golden-yellow to orangish florets, the rays in a single marginal row, very short and somewhat deflexed, slightly paler yellow, sometimes hardly visible. Involucral bracts numerous, imbricate, narrowly linear-lanceolate, acuminate, with green medial nerve and yellowish-scarious margins. Achene glabrous, 0.5 – 0.9 mm long, with a pappus of 7 – 10, 3 mm-long bristles adnate at base to an outer series of very short scarious teeth.

● No. 8116, from the edge of the northern Summan, was atypical in having very large leaves below, up to 4.5 cm long and 0.7 cm wide. The (immature) florets were close to the typical form, although smallish. It was a young plant on a well watered silt basin and may represent a spring juvenile form.

The long-used name *P. crispa* has to be dropped for this plant because of the discovery that a specimen of it was the type for *Inula undulata* L. (see *Taxon* 29: 694–95, also the decision of the IAPT nomenclature committee not to reject the name *P. undulata, Taxon* 32: 282–83).

Habitat: Silt-floored basins or wadi bottoms; sometimes as a ruderal. Occasional to locally frequent.

Northern Plains: 239, 27–58N, 47–07E; 315, Qaisumah Pump Station; 853, 9 km S 'Ayn al-'Abd.

Northern Summan: 2745, 25–20N, 48–14E; 8116, 15 km SSE al-Wannan.

Central Coastal Lowlands: 1974, 8 km NE Dhahran.

Southern Summan: 1870, 25 km S Jaww al-Hawiyah; 2204, 26 km SSE al-Hunayy; 8076, Mishash al-'Ashawi, 24–17N, 48–53E; 8081, 24–33N, 49–06E.

South Coastal Lowlands: 7908, Na'lat Shadqam, 25–43N, 49–29E; 8290, 6 km N al-Mutayrifi; 8295, al-Hufuf, waste ground.

Vernacular Names: JATHJĀTH (gen.), JATHYĀTH (some N tribes), 'URAYFIJĀN (Al Murrah, a diminutive form of the *'arfaj (Rhanterium)*, which it somewhat resembles.

● 9. Rhanterium L.

Shrublets with white-tomentose young stems, narrow leaves, and numerous radiate, yellow-flowered heads. Involucral bracts imbricate, several-rowed, glabrous; receptacle chaffy-bracteate. Achenes narrow, glabrous, with or without pappus.

Rhanterium epapposum Oliv.

Rounded, often hemispherical, intricately branched shrublet, 30–70(100) cm high, somewhat aromatic, with closely white-tomentose young stems. Leaves rather distant, sessile and linear, 1–2(4) cm long, 0.2–0.3 cm wide, entire or remotely dentate, glabrous or sparingly pubescent. Heads numerous, solitary, terminal, 0.7–1.5 cm broad, yellow-flowered with c. 7–15 spreading, sometimes inflexed ray-ligules. Involucral bracts coriaceous, glabrous, long-triangular to lanceolate, all acute, often somewhat reflexed, imbricate in many rows and shortest near base. Receptacle with chaffy bracts. Achene elliptical-oblanceolate in outline, somewhat curved and compressed, weakly angled longitudinally, glabrous, c. 2.5 mm long, clasped by a folded bract, without pappus. *Plates 203, 204.*

Wiklund (1986) provides a full technical description of the plant as well as dissection figures of our specimens 1756 and 2709 (listed below).

● A species very important in the east Arabian vegetation, dominant in a distinctive shrublet community over thousands of square kilometers in the Eastern Province and of major ecological significance (see chapter 6). Thalen, as part of rangeland studies in Iraq (1979), provides valuable information on the autecology and synecology of this shrub. He shows that the dispersal unit of the plant is the entire, stiff-bracted head, which is deciduous as a unit and is spread by wind and possibly in animal hair. The achenes remain in the fallen head and germinate within it. The present writer has observed that the fruits are clasped by folded bracts, a habit which no doubt contributes to this mechanism. The shrub drops its leaves at the beginning of the hot season and remains dormant through summer. New foliage is produced after winter rains and after the coldest part of the season has passed.

Habitat: Shallow sands, often somewhat loamy in the best stands, usually on somewhat elevated, well-drained ground rather closely underlain by limestone.

Northern Plains: Dominant over many parts of this region; particularly well developed in the country around Qaryat al-'Ulya.

Northern Summan: 2741, 25 – 18N, 48 – 20E.

Southern Summan: 2155, 23 – 27N, 48 – 34E.

Central Coastal Lowlands: 140, 180, 1046, 1057, Dhahran; 2709(BM), al-'Alah; 1756(BM), Abu Hadriyah.

South Coastal Lowlands: 421, Na'lat Shadqam.

Vernacular Names and Uses: ʿARFAJ (gen.). A plant of great economic importance, being the dominant constituent of valuable rangelands. It is itself grazed by livestock; perhaps equally important, it stabilizes the soil and provides protection for dense growths of associated annuals. It is seldom used by Bedouins as firewood if other woody species such as *Calligonum* or *Haloxylon* are available, as it burns too quickly when dry to be very useful and is rather smoky when green. In some areas where other woody plants are not found close at hand, as in parts of Kuwait and around some desert villages in Saudi Arabia, broad stands of *Rhanterium* have nevertheless been destroyed by uprooting for fuel. Such areas, after wind erosion has begun, lose most of their value as a grazing resource.

• 10. **Anvillea** DC.

Somewhat woody perennials with flattish discoid heads. Involucre hardening after flowering, the inner bracts spinescent, the outer ones leaflike, herbaceous, spreading. Receptacle bracteate. Achenes compressed, tetragonal, without pappus.

Anvillea garcinii (Burm. f.) DC.

Ascending greyish-green, woolly-canescent, rigidly branched perennial, rounded, often broader than high, more or less woody, c. 20 – 50 cm high. Leaves obovate to spathulate, long-tapering to the base, repand, entire or irregularly dentate or lobed. Heads discoid, rather flat, golden yellow-flowered, on short thick peduncles near the branch ends, 2 – 3 cm in diameter. Inner involucral bracts about equalling the florets, acute, somewhat spinescent, the outer ones resembling the leaves, obtuse, spreading, shorter or longer than width of the disc. Bracts of the receptacle nearly linear, keeled, broadening somewhat distally then narrowing abruptly to an acute apex. Achenes 3 – 4 mm long, 4-angled, somewhat compressed, sometimes ciliate at margins, without pappus. *Plate 205.*

Habitat: Silt floors of basins and wadis in the northern plains and throughout the northern and southern Summan. Frequent to locally common.

Northern Plains: 1505(BM), 30 km ESE al-Qaysumah. Numerous sight records throughout the silty terrain types of the Province.

Vernacular Names: NUQD (gen.).

• 11. **Asteriscus** Mill.

Annuals or perennials with entire or dentate leaves. Heads radiate. Involucre imbricate, hardening in fruit, the outer bracts herbaceous, spreading leaflike, exceeding the rays. Receptacle with oblong bracts; florets yellow. Achenes with a pappus of lanceolate pales.

Asteriscus pygmaeus (DC.) Coss. et Dur.
Odontospermum pygmaeum Benth. et Hook.
Dwarf annual herb, stemless at ground level or branched with very short stems, 1- or several-headed, appressed-pubescent on leaf faces and spreading-villous-ciliate at leaf margins. Leaves entire, oblong to obovate-spathulate, more or less obtuse, tapering at base to a petiole. Heads solitary, sessile, c. 1.5 – 2 cm in diameter, with involucre merging with the spreading subtending leaves. Inner bracts indurate in fruit, closing over and enveloping the achenes in the woody dry plant. Bracts of the receptacle oblong, about equalling the tubular florets. Ray florets pale yellow, with short, 3-toothed ligules in 1 marginal row; disc florets shorter, darker yellow. Achenes silky with subulate-tipped pales. *Plate 206.*
Habitat: Silty floors of basins, wadis. Occasional.
Northern Plains: 574, 95 km E Hafar al-Batin; 1274, 15 km ENE Qaryat al-'Ulya; 1520(BM), 30 km ESE al-Qaysumah; 1731(BM), Jabal al-'Amudah.
Northern Summan: 8628, Dahl Abu Harmalah, 26 – 29N, 47 – 36E.

● 12. Eclipta L.

Annual herbs with opposite leaves. Heads radiate with herbaceous, nearly equal, involucral bracts and receptacle with narrow pales. Florets white, the rays short-liguled. Achene compressed, more or less angular, truncate at apex, mostly epappose.

Eclipta alba (L.) Hassk.
E. prostrata (L.) L.
Erect or prostrate annual herb, 20 – 50 cm high, shortly appressed-pubescent with straight, rigid-whitish, often bulbous-based hairs. Leaves opposite, elliptic-lanceolate, acute, subsessile, entire to obscurely denticulate, 2 – 6 cm long, 0.5 – 1.7 cm wide. Heads solitary or paired on elongate axillary peduncles, 10 – 12(15) mm in diameter, hemispherical, with white florets. Involucral bracts thin, green-herbaceous, ovate, mostly acute. Achenes somewhat compressed and winged at margins, or somewhat 3-angled, medially tuberculate, with apex truncate, blackish, and somewhat pubescent, without pappus, 2.5 – 3 mm long, c. 1.5 mm wide. *Plate 207.*
Habitat: A weed of wet ground in oasis gardens. Frequent.
Central Coastal Lowlands: 528(BM), 8263, al-Qatif; C. Parker 64(BM), al-Qatif Oasis.
South Coastal Lowlands: 1906, ash-Shuqayq, al-Hasa Oasis.

● 13. Flaveria Juss.

Annuals with opposite leaves. Heads radiate, few-flowered, with an involucre of 2 – 5 keeled concave bracts. Achenes epappose.

Flaveria trinervia (Spreng.) Mohr
Erect, glabrous, somewhat diffusely branched annual herb, c. 20 – 100 cm high. Leaves opposite, elliptical-oblanceolate, acute, tapering at base to an indistinct petiole or sessile, more or less serrate, with 3 parallel nerves beneath, mostly 2 – 6 cm long, 0.4 – 1.5 cm wide. Heads clustered, sessile or subsessile in the stem forks, the clusters 1 – 2(3) cm in diameter, with yellow florets. Achenes c. 2 mm long, 10 – 12-striate, without pappus.

Habitat: A weed of farm or gardens. Locally frequent. An American native now widely naturalized in tropical countries.

Central Coastal Lowlands: 7766, between Sayhat and 'Anik, al-Qatif Oasis.

South Coastal Lowlands: 1891(BM), al-Mutayrifi, al-Hasa; 1900(BM), ash-Shuqayq, al-Hasa Oasis; C. Parker 5(BM), vicinity of al-Hufuf; 8192, near ash-Shuqayq, al-Hasa Oasis.

● 14. Anthemis L.

Annuals or perennials, usually with 1–2-pinnatisect leaves. Heads radiate, rarely discoid, with involucral bracts imbricate, scarious-margined, in several rows. Receptacle hemispherical to conical, bracteate. Ray florets mostly white, disc florets yellow. Fertile achenes more or less obconical, ribbed or tuberculate, naked at apex or with a crown or auricle.

Key to the Species of Anthemis

1. Achenes all deciduous at maturity; leaves 1-pinnatisect or partially and obtusely 2-pinnatisect
    2. Achenes broadly obpyramidal or obconical, ripening dark brown and coarsely tuberculate-rugose, with apex truncate or (more rarely) furnished with an auricle ⅙ to as long as the achene body; tube of floret darkened, indurated . . . . 1. *A. melampodina*
    2. Achenes obconical-clavate, with rounded nonauriculate apex, whitish, longitudinally ribbed, not or only weakly tuberculate; floret tube pale, not indurated . . . . 2. *A. scrobicularis*
1. Achenes (at least the broader marginal ones) persistent at maturity; leaves finely 2-pinnatisect . . . . 3. *A. pseudocotula*

1. **Anthemis melampodina** Del.

Ascending annual herb, branched from near the base and appressed-pubescent to somewhat sparingly tomentose-canescent, 5–25(30) cm high. Leaves oblong in outline, sessile or petiolate, c. 1–3 cm long, pinnatisect into linear, mucronulate lobes. Heads 2–3 cm broad with white ray petals and yellow disc, on peduncles overtopping the leafy parts of the plant. Involucral bracts lanceolate-oblong, scarious-margined, acute or obtusish, sometimes somewhat fimbriate around apex. Receptacular bracts oblong-oblanceolate, often attenuate or mucronate, scarious but medially green-nerved, keeled-folded or not, c. 3 mm long and 0.7–1.2 mm wide. Tube of florets inflated and indurate, darkened, sometimes persistent on the achene and appearing jointed with it. Achenes obpyramidal or obconical, ripening dark brown, rather coarsely tuberculate-rugose, c. 1.4–1.7 mm long (not including any auricle), the apex truncate or weakly rimmed or (less commonly in our area) surmounted at one side with a rounded scarious auricle ⅙ as long to longer than the achene body. *Plates 208, 209.*

● Earlier convention in the Arabian botanical literature has been to maintain subsp. *deserti* (below) as a distinct species. In the present author's view that taxon, seeming to differ virtually only by the absence of auricles on the achenes, is too close not to be accepted as conspecific. This impression is reinforced by the presence in our area of some apparently intermediate plants with auricles short, sometimes only ⅙ as long as the achene body. Eig's treatment as two subspecies is followed here, although

297

arguments could be made for Ascherson's view that these are simple varieties. We should note that our interpretation of subsp. *deserti* does not correspond exactly with that of Eig, who separated it not by the exauriculate state of its achenes but by their being less tuberculate, more striate, and more rounded at apex (Eig 1938). Subsp. *deserti* appears to be by far the more common form in our territory. Further work on these two taxa is needed. The specimens cited below represent only a small fraction of our collections attributed to this species but are selected as sheets with mature fruits that have been studied critically.

● subsp. **melampodina,** with achenes furnished at apex with a rounded scarious auricle ⅙ to as long as the achene body.

Habitat: Sandy silts along small wadis, often in rocky terrain but apparently sometimes also on open plains. Infrequent to rare by present records but perhaps more frequent than currently appreciated.

Northern Plains: 3744(BM), 24 km W as-Saffaniyah.

Northern Summan: 8573, 12 km WNW ash-Shumlul; 7925, 25 – 35N, 48 – 50E.

● subsp. **deserti** (Boiss.) Eig (*A. deserti* Boiss.), with achenes nonauriculate, truncate or weakly rimmed at apex.

Habitat: Sandy silts in ravines or on open ground. Frequent to common.

Northern Plains: 1492(BM), 30 km ESE al-Qaysumah; 3846(BM), Qatif-Qaisumah Pipeline KM Post 380.

Northern Summan: 8580, 8 km S Dahl al-Furayy; 8586, Dahl al-Furayy; 8599, 17 km ESE Jabal Burmah, 26 – 41N, 47 – 45E.

Central Coastal Lowlands: 3776(BM), 3 km E Ras al-Ghar; 1765(BM), Abu Hadriyah.

Vernacular Names: QAHWIYĀN (Al Murrah, Bani Hajir, Mutayr, Harb, gen.), ZAMLŪQ (az-Zafir).

2. Anthemis scrobicularis Yavin

Ascending greyish-green tomentose annual with 1 to several branches arising from the base, 5 – 25 cm high. Leaves oblong in outline, at least the lower ones petiolate, mostly 1 – 2 cm long, 0.3 – 0.8 cm wide, 1 – 2 pinnatisect with 5 – 7 main lobes. Peduncles overtopping the leafy part of the plant, not or hardly thickened above; heads c. 1 – 2.5 cm across (including the rays). Involucral bracts lanceolate to oblong, mostly acute, green-herbaceous medially with scarious margins, sometimes fimbriate at apex. Receptacular bracts narrowly linear-oblanceolate, acute, 0.3 – 0.5 mm wide, scarious with a pale, herbaceous medial stripe. Floret tube pale, glabrous, not inflated. Achenes obconical-clavate, often slightly curved, nonauriculate, mostly rounded at apex, (1.7)2 – 2.3 mm long, very pale to whitish with longitudinal ribs, sometimes obscurely scrobiculate or when fully mature bearing minute, darker, gland-like tubercles along the ribs.

● Our records, confirmed by comparison with type material at the herbarium of the Royal Botanic Garden, Edinburgh, provide an interesting eastern extension of the range of this species, which was described from specimens collected in Jordan and Sinai. It appears to favor the deeper sands in contrast to *A. melampodina,* which is usually found on silty soils. Its distribution is probably centered in Arabia, and it will probably be found to be common in deeper sand terrain across the northern part of the Peninsula.

Habitat: So far collected only on fairly deep sand. Future field work may prove it to be widespread in such habitats.

Northern Plains: 4080(BM), Jibal an-Nu'ayriyah.
Northern Dahna: 8565, east edge of 'Urayq al-Uqayhab, 25 – 47N, 47 – 45E; 2831(BM), S tip of 'Irq al-Hazwa, 26 – 21N, 47 – 25E.
Southern Summan: 7917, Jabal ar-Ruhayyah, 25 – 34N, 49 – 05E.
South Coastal Lowlands: 1446(BM), 2 km E Jabal Ghuraymil.
Vernacular Names: QAḤWĪYĀN (Al Murrah, Bani Hajir, gen.).

3. Anthemis pseudocotula Boiss.

Annual herb with stems erect or ascending and branched above the base, glabrescent (in weed form) or moderately appressed-pubescent (in some desert plants), 10 – 30(50) cm high and leafy nearly to the top. Leaves finely 2-pinnatisect (notably finer than in the preceding 2 species) with segments crowded. Heads 1.5 – 3 cm wide with white spreading rays and yellow disc, on rather short peduncles more or less thickened beneath the fruiting heads. Involucral bracts scarious-margined, obtusish. Receptacular bracts linear-subulate to narrowly lanceolate, aristate. Achenes truncate-obpyramidal or obconical, c. 1.2 – 1.5 mm long, pale brown, usually strongly grooved on the faces and not or hardly tuberculate, concave and flat-rimmed at apex, the rim sometimes produced at one side as a small, opaque, entire or lobed auricle; achenes persistent at maturity, particularly the broader ones near the margins which are often strongly attached.

● Our desert specimens are somewhat atypical in bearing bracts on the receptacle well down to near the margins. The thickening of the peduncles below the heads may not be very prominent but is usually seen to some extent under the heads closest to the main stems.

Habitat: Found both as a weed and as a desert plant, but as the latter apparently tending to occur on disturbed ground such as Bedouin camp sites. Infrequent to rare.
Northern Plains: 3863(BM), 15 km SSE Samudah, 27 – 44N, 45 – 00E (barely outside the formal limits of our area).
Northern Summan: 8607, Wadi Mibhil, 26 – 40N, 47 – 34E.
Central Coastal Lowlands: 7425(BM), 7419(BM), 8247, Dhahran, garden weeds.

● 15. Achillea L.

Perennial herbs or shrublets with leaves finely dissected or rarely, (as in ours) entire. Heads radiate or discoid, solitary or in corymbose inflorescences. Involucre imbricate and receptacle bracteate. Disc flowers usually compressed. Achenes compressed, without pappus.

Achillea fragrantissima (Forssk.) Sch.-Bip.

Closely woolly-tomentose, very strongly fragrant perennial with stems erect, virgate, branched from a woody base, 30 – 75(100) cm high. Leaves ovate-triangular, sessile, rather thick, subacute or obtuse, sometimes somewhat appressed to stem, callose and bluntly serrulate at margins, c. 4 – 6 mm long, 2.5 mm wide. Heads oblong-ovoid, c. 3 – 4 mm wide, discoid, short-peduncled, grouped 4 – 10 in short, rather dense terminal corymbs. Involucral bracts oblong, obtuse, imbricate, closely woolly. Florets yellow.

● This is easily the most powerfully fragrant plant in the Eastern Province; walking over a desert silt basin and thus crushing some of the plants produces a unique heady scent that fills the surrounding air. The species is much more frequent to the west-northwest of our territory, in the rocky country known as al-Hajarah. It is there not only frequent but often dominant in shallow silt-floored basins. Its range extends down into the northern Summan, perhaps as far south as about the 26th Parallel.

Habitat: Silt-floored basins, usually in rocky country. Locally common but not widespread.

Northern Plains: 3197, al-Batin, 5 km NE Dhabhah; Rawdat al-Qaysumah; 8798, 18 km WSW ar-Ruq'i Post, 28 – 59N, 46 – 31E.

Northern Summan: 735, 33 km SW al-Qar'ah wells.

Vernacular Names: QAYṢŪM (gen.). The pump station and town on the Trans-Arabian Pipeline, Qaisumah (al-Qaysumah), takes its name from this plant. The species is not found at the town itself, but at a basin some distance to the southwest known as Rawdat al-Qaysumah.

● 16. Matricaria L.

Annual aromatic herbs with radiate or discoid heads and finely dissected leaves. Involucral bracts imbricate, scarious-margined; receptacle conical and ebracteate. Achenes oblong in outline, finely 3-ribbed ventrally, smooth, tuberculate or minutely striate dorsally; pappus of a crown or auricle present or absent.

Key to the Species of Matricaria
1. Heads discoid, without radiating florets .... 1. *M. aurea*
1. Heads with white radiating florets .... 2. *M. parviflora*

1. Matricaria aurea (Loefl.) Sch.-Bip.

Glabrous, sweet-aromatic annual, 5 – 25 cm high, branched from the base with several to numerous stems ascending or erect. Leaves low on the stems and often present into their upper halves, very finely dissected, 1 – 2-pinnatisect into linear segments under 1 mm wide. Heads discoid, mostly terminal, also sometimes axillary, numerous, dome-shaped to rather high-conical, yellow-flowered, c. 5 mm in diameter. Involucral bracts oblong, c. 2 mm long, obtuse, green-herbaceous with brownish or translucent scarious margins. Receptacle ovoid-conoid. Achenes minute, 0.7 – 0.9(1) mm long, subterete, gently curved, whitish to brown, with 3 ventral ribs, smooth to minutely striate dorsally, with oblique truncate apex devoid of any pappus (reportedly sometimes with some achenes bearing an apical auricle, but these not seen in our specimens).

● A plant rather closely resembling *Aaronsohnia factorovskyi* and *Tripleurospermum auriculatum*. The achenes should be examined with good magnification to confirm identifications. Two, or even 3, of these species may be found in the same habitat.

Habitat: Silt-floored basins of the Summan and northern plains. Apparently rare but may be more common that realized due to confusion of sight records with *Aaronsohnia* and *Tripleurospermum*.

Northern Summan: 8118, 15 km SSE al-Wannan; 8606, Wadi Mibhil, 26–40N, 47–34E.

Vernacular Names and Uses: BĀBŪNAJ (gen.). A medicinal camomile tea is sometimes brewed from the flower heads; known to both Bedouins and villagers.

2. Matricaria parviflora (Willd.) Poir. ex Lam.

Many-branched herb with decumbent to ascending stems, c. 30 cm high. Leaves mostly 3–5 cm long, oblong in outline, several-pinnatisect into fine, filiform lobes. Heads solitary, terminal, 2–3 cm in diameter. Involucre c. 4–5 mm long, with oblong to long-triangular bracts herbaceous medially, scarious at margins; disc hemispherical, yellow, 8–13 mm broad; radiating ligules in a single row, white, linear-oblong, 8–10 mm long, 2–3 mm wide. Achenes c. 2 mm long, truncate, concave and whitish-rimmed at apex, tapering somewhat below, flattish, blackish and finely tuberculate-wrinkled at back; margins with whitish ribs or wings; face with a medial, white, longitudinal rib or keel (description entirely from our single specimen).

Habitat: A central and SW Asian weed, rare in Arabia around cultivation.

Central Coastal Lowlands: 8270(BM), Dhahran, with Mediterranean weeds near garden area.

● 17. Tripleurospermum Sch.-Bip.

Annual or perennial herbs with discoid or radiate heads and finely dissected leaves. Involucral bracts in few rows, membranous-margined; receptacle ebracteate. Achenes subtriquetrous, slightly compressed, 3-ribbed and 2-furrowed ventrally, usually smooth-rounded at back, sometimes with 2 reddish glands near apex.

Tripleurospermum auriculatum (Boiss.) Rech. f.

Nearly glabrous annual, branching from the base with erect stems 10–20 cm high. Leaves more or less oblong in outline, finely 1–2-pinnatisect with linear lobes. Heads solitary, terminal on scapelike stems, hemispherical, discoid, yellow-flowered, 7–10(12) mm in diameter. Involucral bracts ovate-oblong, obtuse, broadly scarious-margined. Achenes c. 1.5 mm long, slightly incurved, 3-ribbed and 2-furrowed ventrally, smooth at back, sometimes with 2 glands at the dorsal apex; pappus present in the form of an oblong auricle equalling or exceeding the achene.

● Rather easily confused with *Aaronsohnia factorovskyi*, which is generally more common and has somewhat coarser leaf lobules. Careful examination of achenes is required for identifications.

Habitat: Silt flats in rocky Summan country. Apparently rare, but perhaps more common than realized due to confusion with *Aaronsohnia*.

Northern Summan: 3226(BM), 74 km S Qaryat al-'Ulya on old Riyadh track.

● 18. Aaronsohnia Warbg. et Eig

Annual herb with dissected leaves. Heads discoid, hemispherical, with imbricate, scarious-margined involucral bracts and hemispherical to conical receptacle. Achenes dimorphic, striate, with a scarious auricle.

Aaronsohnia factorovskyi Warbg. et Eig

Annual herb, 8–25 cm high, branched at base with stems erect or decumbent at base, glabrous or sparingly hairy at leaf bases. Leaves much dissected, slightly succulent, bipinnatisect into linear lobes c. 1 mm wide. Heads discoid, hemispherical, (5)6–8(10) mm in diameter, solitary on numerous leafless peduncles exceeding the leafy parts of the stems. Involucral bracts mostly oblong, obtuse, white-scarious-margined. Achenes

15–20 fine-striate at back, rounded or faintly prismatic ventrally, of 2 types: one, at the margin of the head, broader, rounder, with a scarious auricle somewhat shorter than, to about equalling, the achene; the other (usually more numerous) thinner, with auricle well exceeding the achene. *Plates 210, 211.*

● The plant closely resembles two other species that may share the same habitats, *Tripleurospermum auriculatum* and *Matricaria aurea*, and achenes should be examined with good magnification to confirm identifications. *Aaronsohnia* has somewhat coarser and more succulent leaves than the other two and is the most commonly encountered of the three.

Habitat: Silt-floored basins and wadi bottoms of the northern plains and Summan. Also sometimes on shallow sands. Frequent to locally common.

Northern Plains: 1239(BM), Jabal an-Nu'ayriyah; 1314(BM), 8 km W Jabal Dab'; 1567(BM), al-Batin/Wadi al-'Awja confluence; 1667(BM), 2 km S Kuwait border in 47–14E; 7996, near Sabkhat an-Nu'ayriyah; 7982, 24 km W al-Khafji.

Northern Summan: 761, 69 km SW al-Qar'ah wells; 1397(BM), 42 km W Qaryat al-'Ulya; 8110, 2 km S Hanidh; 8584, Dahl al-Furayy, 26–47N, 47–05E.

Central Coastal Lowlands: 8117, 15 km SSE al-Wannan.

Southern Summan: Extends through the Summan and known as far southeast as Qatar.

Vernacular Names and Uses: QURRAYṢ (az-Zafir), QARRĀṢ (Qahtan). Dickson (1955, 11) notes that the plant is sometimes eaten raw by Bedouins, who also may use it in the preparation of *iqt* (dried sour milk cakes). The name is probably derived from Arabic *qurṣ*, meaning 'disk', as in a disk of bread or the disk of the sun, a reference to the shape of the heads.

● 19. Chrysanthemum L.

Annual herbs with leaves dentate to much-divided. Heads radiate, solitary-terminal or in corymbs. Involucral bracts imbricate, membranous-margined; receptacle ebracteate. Marginal achenes somewhat triquetrous, the inner ones obpyramidal or compressed, without pappus or crowned.

A genus represented in our territory only by very rare weeds or by garden escapes not far from cultivation.

Key to the Species of Chrysanthemum
1. Leaves irregularly lobed and dentate, the lower often with 3(5) main lobes at apex, tapering-dentate below; achenes 10-ribbed .... 2. *C. segetum*
1. Leaves bipinnatisect with oblong acute lobules; achenes 4 or 6-angled .... 1. *C. coronarium*

1. Chrysanthemum coronarium L.

Erect glabrous branched annual to c. 80 cm high with rather highly dissected 2-pinnatisect leaves. Heads terminal, showy-radiate, 5–6 cm or more broad with ligules and disc both yellow or the rays cream to white. Involucral bracts oblong-ovate, obtuse, membranous-margined. Achenes 4–6-angled and more or less winged at margins. Pappus absent.

Habitat: This is the *Chrysanthemum* of garden cultivation; found along roadsides or on waste ground as an escape, seldom at any distance from its point of cultivation.

Its ability to reproduce outside cultivation in our conditions is not confirmed, but it does tend to appear regularly in the same ruderal locations. Rare.
Central Coastal Lowlands: 3757(BM), Dhahran.

2. Chrysanthemum segetum L.
Glabrous annual herb with simple or branched stems, 20 – 50 cm high. Lower leaves as described in key, those above sessile, less lobed-dentate, oblong to lanceolate, the uppermost often entire. Heads radiate, 2 – 4 cm in diameter, with yellow florets. Involucral bracts ovate, the inner ones with membranous margins. Achenes 10-ribbed, the inner ones equally, the marginal ones with marginal ribs wider, all without pappus.
Habitat: Only known in our territory as a very rare weed.
Northern Summan: 8099, Batn al-Faruq, 25 – 43N, 48 – 53E. This only specimen was collected on rather deep but stable sand in remote desert at a former Bedouin campsite. It was associated here with a number of otherwise rare or locally unrecorded Mediterranean weeds that had grown with spilled grain. The head was immature, but its identification is reasonably certain, with the leaves typical of this species' rather characteristic form.

● 20. Artemisia L. incl. *Seriphidium* (Besser) Poljak

Herbs or shrubs, often with divided aromatic leaves and small discoid, often panicled, heads. Involucre with imbricate bracts; receptacle glabrous or somewhat hairy. Florets usually few, yellow, inconspicuous. Achenes cylindrical to obovoid, without pappus.

Key to the Species of Artemisia
1. Heads hemispherical, about as broad as long; involucre hairy ... 2. *A. judaica*
1. Heads ovoid to oblong; involucre glabrous
   2. Leaves pinnatisect with 1 – 4 mm-long obtuse lobes,
     grey-woolly-tomentose; plant under 50 cm high .... 1. *A. sieberi*
   2. Leaves entire, linear or lanceolate, or with linear acute lobes 1 – 10 cm long,
     glabrous to very finely appressed-silky; plant usually 50 – 100 cm high
      3. Middle stem leaves with lobes less than 1 mm wide;
        annual or biennial of silty basins .... 4. *A. scoparia*
      3. Stem leaves, or lobes, 2 – 4(8) mm wide; perennial
        of deep sands .... 3. *A. monosperma*

1. Artemisia sieberi Besser
A. herba-alba auct. non Asso
Ascending, strongly aromatic shrublet, sometimes wider than high, greyish-tomentose with stems becoming glabrescent below, 20 – 50 cm high. Lower leaves ovate to oblong in outline, 1 – 2-pinnatisect into oblong, obtuse lobes mostly 1 – 4 mm long; leaves above smaller, less lobed. Heads sessile in a very rich, dense paniculate inflorescence often broader than long; heads oblong, nearly cylindrical, slightly broadening distally, c. 3 – 4 mm long, 1 – 2 mm wide, with short herbaceous bracts at base and longer, broader, scarious-margined and glossy bracts above, 2 – 4 flowered.
● A species placed by some workers in the segregate genus, *Seriphidium* (Besser) Poljak, as *S. sieberi* (Besser) Bremer et Humphries. It is often referred to in the literature as *A. herba-alba* Asso, a taxon now considered by authorities to be restricted to the western

Mediterranean. Next to *Achillea fragrantissima* this is the most strongly aromatic plant of the northern plains and Summan, the scent of its crushed foliage being lemony-sweet and more intense than that of the two following species.

Habitat: Silt-floors of wadis and basins, usually in rocky country. Occasional to locally frequent.

Northern Plains: 1582, al-Batin/Wadi al-'Awja junction. Also seen 15 km SSW Hafar al-Batin.

Northern Summan: 2862, 26–00N, 48–29E.

The plant is more common outside the edge of the Eastern Province, to the northwest in al-Hajarah district, but may be found down through the Summan at least as far south at the 26th Parallel.

Vernacular Names and Uses: SHĪ\d{H} (gen.). Used by the Bedouins as a medicinal, often by inhaling the smoke. Musil (1928a, 383, 408) reports it was used as a smoke inhalant to treat glanders in horses and to cure 'bewitched' animals. It was also used dried and powdered to provide a tinder for fire-making with flint and steel (Musil 1928a, 128).

2. Artemisia judaica L.

Densely tomentose aromatic shrublet, 30–70 cm high. Lower leaves crowded, rounded in outline, 1–2 pinnatifid with oblong lobes; leaves on flowering branches smaller, clustered. Inflorescence terminal, paniculate, with short-peduncled, sometimes nodding heads. Heads hemispherical, 3–4 mm in diameter, with involucral bracts broadly ovate to orbicular with broadly scarious margins; florets numerous, yellow.

Habitat: Alluvial silts or sands in wadi beds or basins. Rare.

Northern Plains: Vesey-Fitzgerald 15386(BM), Hafar al-Batin; 3207(BM), 28 km SW Hafar al-Batin.

● Vesey-Fitzgerald's 1946 specimen (above, seen by the author at BM) bears flowers and confirms the presence of this northwestern Arabian species in our territory. Our No. 3207, collected in late October 1971, was sterile but closely matched the earlier specimen in foliage characteristics. Natural habitats around Hafar al-Batin have since been greatly affected by urban growth and farm operations; a 1989 search for the plant in al-Batin failed to turn up any specimens. One earlier sight record of a sterile plant in the central Summan may also have been of this species.

Vernacular Name and Uses: BU'AYTHIRĀN (Mutayr, gen.). Musil (1928b, 6) reports that Shammar tribesmen used the cut foliage to flavor and preserve dates.

3. Artemisia monosperma Del.

Ascending aromatic shrub, green to silvery-green, glabrous to very finely appressed-silky, 50–100 cm high. Leaves linear-oblanceolate, solitary or clustered, entire or with linear lobes mostly 3–7(8) cm long, 0.3–0.5(0.8) cm wide, sessile, tapering to base, acute. Heads very short-peduncled to subsessile in somewhat 1-sided numerous racemes emerging laterally from the stem apices, together forming an elongate, compound racemose to paniculate inflorescence up to c. 40 cm long. Heads ovoid, 3–4 mm long, 1.5–2 mm wide, with bracts ovate, longer and more broadly scarious-margined above, 3–6-flowered.

Habitat: Deep sands; characteristic of the northern and central Dahna and the red sand bodies of central Arabia, where it is codominant with *Calligonum*. Locally common.

Northern Plains: 3205, al-Batin, 28 km SW Hafar al-Batin.

Northern Dahna: 341, 38 km W Khurays; 802, 15 km SE Hawmat an-Niqyan.

Central Coastal Lowlands: 7608, 15 km W al-Jubayl; 1111, 20 km N 'Ayn Dar GOSP-2.

Southern Summan: 7918, near Jabal ar-Ruhayyah, 25-34N, 49-05E.

Vernacular Names: 'ADHIR (gen.).

4. Artemisia scoparia Waldst. et Kit.

Erect annual or biennial, glabrous except for some hairs on the lowest leaves, not or hardly aromatic, often 50-100 cm high, rather strict, with stems sometimes reddish. Basal leaves ovate in outline, petiolate, 2-pinnatisect into linear-oblanceolate acute lobes, thinly and finely appressed hairy; stem leaves glabrous, sessile with very narrow, almost filiform lobes less than 1 mm wide. Heads short-peduncled to subsessile in numerous 1-sided racemes all forming an elongate, terminal, paniculate inflorescence to c. 40 cm long. Heads obovoid, 1.5-3 mm long, 1.5-2 mm wide, with ovate bracts, the lower herbaceous, those above with broad, glossy, scarious margins. Heads mostly 4-7-flowered.

Habitat: Silt-floored basins of the northern plains and Summan; sometimes at roadsides on disturbed ground. Infrequent.

Northern Plains: 575, 83 km E Hafar al-Batin; 3188, al-Batin, 35 km SW Hafar al-Batin; 3203, al-Batin at Burq ash-Sharif.

Northern Summan: 806, 37 km NE Jarrarah (southernmost record to date).

Vernacular Names: 'UWAYDHIRAN (Mutayr, a diminutive form of the name of *A. monosperma, 'ādhir*).

● 21. Senecio L.

Annual or perennial herbs, rarely shrubs, with alternate leaves and heads usually radiate. Involucral bracts mostly linear, equal, in a single inner row with a few very short squamulose bracts at base; receptacle ebracteate. Achenes cylindrical, ribbed, with a pappus of simple hairs.

Key to the Species of Senecio

1. Leaves pinnatisect; ligules long-exserted, spreading .... 1. *S. glaucus*
2. Leaves undivided; ligules hardly exserted .... 2. *S. flavus*

1. Senecio glaucus L. subsp. coronopifolius (Maire) Alexander
S. desfontainei Druce

Ascending, branched annual herb, 10-25 cm high, glabrous or with a few crisped hairs. Leaves somewhat succulent, revolute-margined, pinnatisect into linear lobes, or rarely apparently narrowly linear, entire; those on the stems partly clasping-auriculate at base. Heads c. 1.5-2 cm in diameter in flower, in terminal corymbose inflorescences, yellow to rich gold-flowered with conspicuous spreading, later reflexed, ray florets. Involucre broadly cylindrical or constricting above in fruit, c. 5-8 mm long, with bracts mostly equal in a single row, linear, acute, with a few much shorter, often ciliate bractlets at base. Achenes cylindrical, somewhat tapering at base, indistinctly ribbed, dark brown with short whitish hairs appressed toward the apex, c. 1.5-2 mm long, with a very caducous pappus of simple white hairs c. 4 mm long. *Plates 212, 213.*

Habitat: In a variety of terrain types; often in sands near the coast and on Gulf islands, where it may be the most conspicuous flowering annual. Common and widespread.
Northern Plains: 1172, Jabal al-Ba'al; 1210(BM), Jabal an-Nu'ayriyah; 1506(BM), 30 km ESE al-Qaysumah; 1545(BM), 36 km NE Hafar al-Batin.
Northern Summan: 700, 33 km SW al-Qar'ah wells; 1396(BM); 3333, 12 km WSW an-Nazim; 8582, Dahl al-Furayy, 26–47N, 47–05E.
Central Coastal Lowlands: 71, Thaj; 1984, ad-Dawsariyah; 4071, az-Zulayfayn; 7718, Ras az-Zawr, 27–29N, 49–16E; 7037, Kurayn Island; 7043, Karan Island.
Vernacular Names: ZUMLŪQ (Bani Hajir), KURAʿ AL-GHURAB ('crow's shank', Bani Hajir), RIJLAT AL-GHURAB ('crow's foot', Rawalah, Musil), SHAKHIṢ (Shammar), JIRJĪR (Shammar, Musil 1928b, 343; 1927, 600; 1928a, 696). Musil (1927, 220–21; 1928a, 95) reported that the young shoots of the plant were sometimes eaten raw by Bedouins.

2. Senecio flavus (Decne.) Sch.-Bip.

Glabrous branched annual, 15-40 cm high, often dark reddish. Stem leaves oblong-ovate, dentate, cordate-clasping at base. Florets yellow, hardly exceeding the involucre.
Habitat: Steep rocky slopes on limestone hills.
Northern Summan: 8861, Dawmat al-'Awdah, 25 km W Nita'.

● 22. Calendula L.

Annual or perennial herbs or shrublets with alternate undivided leaves. Heads radiate with involucral bracts in 1–2 rows and florets yellow to orange, or sometimes the disc florets reddish. Achenes in several rows, often of 2–4 different forms in the same head.

Our treatment of the genus follows the revision of annual *Calendulas* by Heyn, Dagan, and Nachman (1974). This places the great majority of our desert specimens, formerly called *C. persica* C. A. Meyer by some authors, in *C. tripterocarpa*. *C. arvensis* probably occurs in some disturbed habitats. The two species differ in chromosome number.

Key to the Species of Calendula

1. Heads with at least a few broadly 3-winged achenes without spines
   (but sometimes denticulate) at back; beaked achenes never
   present .... 2. *C. tripterocarpa*
1. Heads without broadly 3-winged achenes; sometimes with boat-shaped
   dorsally spinulose achenes present; beaked achenes absent or
   present .... 1. *C. arvensis*

1. Calendula arvensis L.

Ascending or decumbent, glandular-pubescent, sometimes somewhat viscid, annual herb with stems 5–30 cm long. Lower leaves 2–6(10) cm long, 0.5–1.5 cm wide, oblong to lanceolate, acute or obtuse, entire or obscurely remote-denticulate, sometimes repand. Flowering heads usually 0.5–1.5 cm wide, with yellow to orange ray florets and disc florets of the same color or (but apparently not in our area) reddish-purple. Involucral bracts in ours lanceolate, acute, green with white margins, sometimes reddish-tipped. Heads usually each with several different achene types, beaked or unbeaked, in ours with the inner ones annulate with sculptured or prickly back, the outer ones boat- or helmet-shaped with wings inflexed at margins, spinulose at back. Chromosome number: 2n = 44 (Heyn, Dagan and Nachman 1974).

306

Habitat: Disturbed ground around villages; possibly also around much-used Bedouin camp sites in desert.

● The single record below, assigned provisionally to this species, was collected at the edge of a desert village on much-grazed ground. It appears closest to the *persica* form of Heyn, Dagan and Nachman (1974). The corolla is concolorous, and the achenes are annulate and cymbiform, all rather prominently echinate at back. The plant is somewhat more densely glandular and viscid than our more common *C. tripterocarpa* of the desert.

Central Coastal Lowlands: 7898, Thaj.

Vernacular Names: Most probably known by the same names as the following species (q.v.).

2. **Calendula tripterocarpa** Rupr.

Spreading decumbent to ascending annual herb, glandular-pubescent, rarely glabrescent, branched with stems 5 – 25(40) cm long. Lower leaves oblanceolate to linear-oblong, usually 2 – 5 cm long, 0.4 – 1.3 cm wide, obscurely serrate to entire, often somewhat repand, mostly acute. Flowering heads 0.5 – 1.5 cm wide, with florets palish yellow to rich orange, the disc and ray florets concolorous. Involucral bracts lanceolate to linear, all acute, green-herbaceous with white-scarious and more or less short-ciliate margins, sometimes reddish tipped. Fruiting heads with at least a few 3-winged achenes (2 wings dorsolateral and 1 ventral) smooth or only denticulate at back, the lateral wings not or hardly inflexed at margins; smaller annulate achenes usually present at center of the head, with intermediate forms also often occurring; beaked achenes never present. *Plate 214.*

● The achenes in our specimens have quite pronounced wings and tend to be more or less denticulate, but not echinate, at back. Under very favorable conditions our plants sometimes grow stems up to c. 40 cm long, large for this species as described. Severely dwarfed individuals may be found on dry ground.

Habitat: Sandy or silty ground in desert. Frequent but never very abundant. Also apparently appears very rarely as a weed (No. 8209).

In the following list specimens followed by (!) indicate those reexamined by the author and assigned to this species. The others probably also belong here but do not have mature fruits.

Northern Plains: 1509(BM!), 30 km ESE al-Qaysumah; 1589(BM!), al-Batin/Wadi al-'Awja junction; 686, 20 km WSW Khubayra; 862, 9 km S 'Ayn al-'Abd; 1175, Jabal al-Ba'al; 1246(BM), Jabal an-Nu'ayriyah; 7965(!), Jabal al-'Amudah.

Northern Summan: 1403(BM), 42 km W Qaryat al-'Ulya; 616(BM), 30 km NE Umm 'Ushar; 703(!), 33 km SW al-Qar'ah wells; 3169(BM), 11 km SSW Hanidh; 2828(BM!), 30 km N ash-Shumlul; 3221, 74 km S Qaryat al-'Ulya; 3326, 12 km WSW an-Nazim; 3835, ash-Shayyit, 27 – 29N, 47 – 22E; 7477(!), Batn Sumlul, 26 – 26N, 48 – 15E.

Central Coastal Lowlands: 367, Dhahran; 3732, Ras az-Zawr; 7733(!), 2 km S Ras Tanaqib, 27 – 49N, 48 – 53E; 8114(!), 15 km SSE al-Wannan; 8124(!), 10 km SSE al-Wannan; 8209, Dhahran, as a weed in a lawn.

Southern Summan: 2070, Jaww ad-Dukhan, 24 – 48N, 49 – 05E; 2195, 24 – 34N, 48 – 45E.

Vernacular Names: ḤANWAH (Shammar, Bani Hajir, 'crook-weed', referring to the curved achenes), 'USHBAT AL-GHURAB ('crow-wort', N, Musil 1927, 598), ṢUFAYRA'

307

('yellow-weed', Musil 1928a, 698). Bedouins of Bani Hajir and other tribes believe the plant is injurious to livestock when grazed in any quantity, leading to bloat and other disorders.

● 23. Echinops L.

Spinescent perennial herbs or shrublets with leaves dentate to pinnatifid and spiny-lobed. Heads discoid, terminal-solitary, globose, without apparent common involucre, composed of numerous sessile headlets, each with its own involucre. Involucre of the headlet often with a series of outer bristles or pales forming a pappus-like penicil, the inner bracts numerous, imbricate, sometimes with spines exceeding the globose primary head, which is then termed cornigerous. Florets tubular, regular, with a 5-lobed limb, whitish or bluish. Achenes often densely villous, with a pappus of bristles.

A genus much in need of detailed study and revision for the Arabian Peninsula. There are two species in our area, one of somewhat uncertain nomenclature, the other apparently undescribed.

Key to the Species of Echinops
1. Heads cornigerous with yellowish spines exceeding the headlets; leaves
   divided ½-way to entirely to the midrib .... 2. *E.* sp. B
1. Heads without radiating spines; leaves divided mostly less than ½-way to
   midrib .... 1. *E.* sp. A

1. Echinops sp. A
Greyish, close-felty-tomentose, spiny perennial herb or shrublet, often densely branched and wider than tall, 20–60 cm high. Stems white-felty-tomentose. Stem leaves lanceolate, long-tapering to a yellow-spined apex, mostly 8–17 cm long, 2–4 cm wide (excluding lateral spines), sessile, amplexicaul and auriculate at base, pinnately lobed ¼ to ½-way to the midrib with acute-triangular segments terminating in yellow spines, closely greyish-felty-tomentose on both surfaces, strongly nerved below. Upper leaves shorter, short-lanceolate to triangular. Heads terminal, solitary, non-cornigerous, c. 5–6 cm in diameter, with whitish, rarely very pale bluish, florets. Headlets c. 17–20 mm long, with a penicil of whitish scabridulous bristles about half their length. Involucre c. 9–10 mm long, with c. 14–17 bracts, the outer ones spathulate, acute-deltoid and fimbriate at apex, those at middle tapering to a fine spinule, the innermost 5–6 spineless, connate in their lower two-thirds. Corolla tube glandular. Achene obpyramidal, c. 4–5 mm long, angled at margins, appressed hairy, fragile and subdehiscent basilaterally, with a pappus of very short yellowish bristles connate at base. *Plates 215, 216.*

As suggested by Mr Ian Hedge, Royal Botanic Garden, Edinburgh, this species is probably undescribed. Naming would best await a badly needed regional revision of the genus.

Habitat: A psammophil usually found in stabilized to somewhat wind-drifted sand. Overall rare but sometimes locally frequent. Sometimes seen near roadsides and apparently tolerant of somewhat disturbed ground.

Central Coastal Lowlands: 446(K), Dhahran; 1811(BM), 38 km SE Abu Hadriyah; 4076, Qatif-Qaisumah Pipeline KP–94; sight records at al-'Alah, 25 km NW Dhahran, and also 12 km W Dhahran; 7760(E), 10 km WSW Dhahran; 8642, 10 km W Dhahran.

Northern Dahna: 1883, 24 km E Khurays; sight record in mid-Dahna near 27° N.

Northern Summan: 8437, near edge 'Irq al-Jathum, 25 – 54N, 48 – 36E; 8444, 5 km SW Judah on Riyadh highway, 25 – 48N, 48 – 48E.

Vernacular Names: SHAYYŪKH (Al Murrah), SHUWWAYKH (Bani Hajir).

2. Echinops sp. B

Erect spiny perennial up to 150 cm high; stems with glandular hairs or glabrescent, reddish when young. Lower leaves lanceolate in outline, half-clasping at base, to c. 40 cm long, 15 cm wide, pinnatisect or bipinnatisect with lanceolate lobes or triangular lobules, all terminating in yellow spines; upper surface green, glandular, grey-tomentose along the veins; lower surface grey arachnoid-tomentose with intermixed glandular hairs, whitish on the raised veins; margins revolute. Heads to c. 8 cm in diameter (not including the exserted spines), strongly cornigerous. Corollas pale violet when young, becoming whitish or cream. Non-cornigerous headlets 20 – 25 mm long with a penicil of white hairs c. 12 mm long. Bracts of partial involucre (15)16 – 18 (including the inner 5 connate ones), glabrous, the outer ones obdeltoid-oblanceolate, long attenuate to the base, fimbriate near the apex with an acute point; middle ones similar but larger and terminating in a 4 – 6 mm-long hardened spinule; the inner 5 connate for ⅔ their length in a cylindrical tube, bifid and fimbriate-chartaceous at apex. Spines of cornigerous headlets to c. 3.5 cm long, yellow, slightly curved, exceeding the corollas. Achene densely covered in appressed, yellowish hairs, surmounted by a hollow, crown-like pappus c. 1 mm long composed of connivent barbellate bristles concrete below in a hardened, smooth ring. *Plates 217, 218.*

This is probably the same plant as that listed from Kuwait by Al-Rawi as *E. blancheanus* Boiss. (Al-Rawi and Daoud 1987, 2:250). We defer assignment of a name pending a regional revision of the genus.

Habitat: Wadi, ravine bottoms in rocky country. Abundant very locally.

Northern Plains: 585(K), 60 km SW Hafar al-Batin; 8757, tributary to al-Batin, 28 – 50N, 46 – 24E; 8801, 8782, 18 km WSW ar-Ruq'i Post, 28 – 59N, 46 – 31E; 8768, 35 km NE Hafar al-Batin; 8797, 30 km SW ar-Ruq'i Post, 28 – 54N, 46 – 27E.

Vernacular Names: SHAYYŪKH (Al Murrah), SHUWWAYKH (Bani Hajir, Rawalah), KHARSHAF (Shammar).

● 24. Acantholepis Less.

Small annual with compound heads, the 1-flowered headlets on a common receptacle subtended by a common involucre of leaf-like bracts. Bracts of each headlet imbricate, very densely fleecy, with a stiff awn or slender spinule at apex. Achenes villous with a pappus of short pales coalescent at base.

Acantholepis orientalis Less.

Ascending grey-green, woolly-tomentose annual herb, sometimes dwarf, simple or several-branched and rather flat-topped, 5 – 20(25) cm high. Leaves lanceolate, acute and spinulose at apex, sessile, 2 – 4(6) cm long, 0.2 – 0.6 cm wide, entire or with fine spinules at margins. Heads hemispherical, terminal, peduncled, 1.5 – 2 cm in diameter, very densely fleecy, subtended by leaflike lanceolate spinulose bracts exceeding the head. Bracts of headlets with a stiff awn or fine spinule slightly exceeding the head. Achenes with bearded pales. *Plate 219.*

Habitat: Shallow sand on wadi bottoms or hillslopes; also on silt. Apparently rare.

Northern Summan: 8105, 13 km SSE Hanidh; 8588, 2.5 km NE Umm al-Hawshat, 27–03N, 47–17E.

● **25. Atractylis** L.

Annual or perennial thistles with spinescent leaves. Heads discoid, rarely radiate, with outer involucral bracts leaflike, mostly spiny-dentate; inner bracts imbricate. Receptacle chaffy with scarious bracts. Achenes terete, silky-pubescent, with a persistent pappus of plumose bristles.

Key to the Species of Atractylis

1. Outer involucral bracts divided much more deeply and finely than the stem leaves, c. 1.5 cm long, scarcely longer than, and enclosing the head .... 1. *A. cancellata*
1. Outer bracts spiny-dentate, similar to the stem leaves, 2–5 cm long, well exceeding the head and spreading .... 2. *A. carduus*

1. Atractylis cancellata L.

Ascending annual, glabrescent to more or less arachnoid-tomentose, usually branched from the base, occasionally simple, dwarfed, 4–25 cm high. Basal leaves (when present) spathulate, prickly-dentate, tapering to a petiole; stem leaves sessile, linear-oblanceolate, prickly-dentate, mostly 1.5–2.5 cm long. Heads discoid, terminal, in well developed plants grouped 3–4 in corymbose inflorescences. Outer involucral bracts pectinate-pinnatisect, prickly-spinescent, with c. 8 pairs of primary lobes, erect, enclosing the head, slightly exceeding the head, c. 1.5 cm long. Inner involucral bracts linear-oblanceolate, entire, woolly-herbaceous medially, scarious-margined, with a scarious apical awn or acute scarious apex. Florets pink or purple. Achenes silky or fleecy, with a pappus of plumose bristles c. 7 mm long.

Habitat: Shallow silty sands, often along rocky knolls or ravines. Infrequent.

Northern Plains: 1617(BM), 3 km ESE al-Batin/Wadi al-'Awja junction.

Northern Summan: 2784, 25–50N, 48–00E; 2827, 30 km N ash-Shumlul; 8123, 10 km SSE al-Wannan.

Vernacular Names: JURRAYS (N tribes, Musil 1927, 601). The name is derived from *jurays*, 'little bell', and probably refers to the bract-enveloped head resembling a rounded bell of the kind made in India.

2. Atractylis carduus (Forssk.) C. Christ.

A. flava Desf.

Ascending to decumbent, rarely procumbent, perennial herb, branching from the base with stems c. 10–30 cm long, usually more or less densely grey cobwebby-tomentose, sometimes glabrescent or even glabrous. Leaves coriaceous, linear-lanceolate, acute, 2–6 cm long, 1–2 cm wide, with pinnate, acute, spinescent lobes. Heads solitary, terminal, c. 1.5–3 cm long with involucre 1–2 cm wide, usually radiate (at least obscurely) with some ligulate margin florets, yellow-flowered. Outer involucral bracts spreading, resembling the stem leaves, exceeding the head, the inner ones oblong to obovate, entire, scarious-margined, with a spiny tip. Achene with pappus about twice its length.

The pubescence in this species appears to vary rather widely. One specimen from the south, in Wadi as-Sahba, was completely glabrous in all parts although otherwise closely resembling a moderately tomentose individual growing nearby.

Habitat: Shallow sands, often around rocky areas. Frequent.

Northern Plains: 1619(BM), 3 km ESE al-Batin/Wadi al-'Awja junction; 1734(BM), Jabal al-'Amudah; 1011(BM), 12 km N Jabal al-'Amudah.

Northern Summan: 8076(BM), 37 km NE Jarrarah; 2877, 20 km W 'Uray'irah; 8609, near Dahl Umm Hujul, 26 – 35N, 47 – 32E.

Central Coastal Lowlands: Dhahran.

Vernacular Names: SHUWWAYKH (Bani Hajir), KALBAH ('bitch-bush', Qahtan), JALWAH (Qahtan).

● 26. Carduus L.

Annual or biennial thistles with spiny leaves and stems prickly-winged with decurrent leaves. Heads discoid with imbricate, spiny-tipped involucral bracts and bristly receptacle. Florets mostly pink to purple. Achenes compressed, glabrous, with a raised rim and a central boss at apex, and with a deciduous pappus of hairs connate at base in a ring.

Carduus pycnocephalus L.

Glabrescent to moderately cobwebby annual herb, usually branched at base with several erect stems with long, narrow, prickly-lobed wings, sometimes short-stemmed, 10 – 40 cm high. Leaves oblong to lanceolate, shallowly or deeply lobed pinnately, with prickly margins, those at base rosulate, the stem leaves decurrent. Heads c. 0.5 – 1 cm wide, or wider in fruit, c. 1.5 cm long, with pink to reddish florets, subsessile, clustered 2 – 5 at ends of the stems or sometimes solitary. Involucral bracts imbricate, long-triangular-lanceolate, acute, scarious-margined, those at base shorter, the intermediate ones ending in a spinule c. 2 mm long and somewhat ribbed or keeled medially just below the apex. Achenes oblong, somewhat compressed, smooth and shiny, c. 3 – 4 mm long, 1.5 mm wide, with boss at apex; pappus of numerous equal scabrous hairs, brownish-grey, deciduous. *Plate 220.*

Habitat: Sand or silty soil in various terrain types. Somewhat frequent.

Northern Plains: 581(K), 15 km SSW Hafar al-Batin; 1183(BM), Jabal an-Nu'ayriyah; 1629(BM), 3 km E al-Batin/Wadi al-'Awja junction.

Central Coastal Lowlands: 3290, 8 km W Qatif-Qaisumah Pipeline KP – 144; 7739, Ras Tanaqib, 27 – 50N, 48 – 53E; 7749, 5 km NW Ras Tanaqib, 27 – 53N, 48 – 49E.

Vernacular Names: SHADQ AL-JAMAL ('camel-jaw', 'camel-gape'?, Dickson 1955).

● 27. Silybum Adans.

Stout annual or biennial thistles with large discoid heads. Involucre of imbricate, spiny-tipped bracts; receptacle with bristly bracts. Achenes obovate, compressed, glabrous, smooth, with a boss at apex and a pappus of bristles connate in a ring at base.

Silybum marianum (L.) Gaertn.

Stout, nearly glabrous annual with simple or branched striate stems 25 – 100(200) cm high. Leaves large, often over 20 cm long, white-veined and whitish-blotched around the veins, sinuate or pinnately lobed with roughly triangular lobes, dentate-spinulose-margined, the stem leaves amplexicaul with spiny-dentate auricles. Heads 3 – 4 cm long, c. 3 cm in diameter (not including the involucral spines), solitary, terminal, with exserted purplish florets. Involucral bracts ovate-oblong basally with an appendage margined with spinelets, tapering above into a flat, more or less recurved spine c. 2 – 3 cm long; inner bracts lanceolate, entire, not spinescent. Achene c. 6 mm long, grey-flecked blackish, with a yellow ring and a boss at apex; pappus white to yellowish, c. twice as long as the achene. *Plate 221.*

Habitat: To be expected mainly on disturbed ground and then rarely. Rare; only collected once.

Northern Plains: 3752, 24 km W as-Saffaniyah. This single record was from an abandoned Bedouin camp site, and the plant was probably introduced with sacked grain used for fodder.

● 28. Onopordum L.

Stout perennial thistles with discoid heads, spiny decurrent leaves, and spiny-winged stems. Involucral bracts coriaceous, imbricate, terminating in rigid spines; receptacle alveolate. Florets purple to white. Achenes obovate-oblong, tetragonal, sometimes compressed, with a pappus of scabrous or plumose bristles.

Onopordum ambiguum Fresen.

Stout, erect grey-felty tomentose perennial with stems simple or branched, spiny-winged with decurrent leaves, 50 – 150 cm high. Lower stem leaves sessile, acute, lanceolate to elliptical in outline, 10–20(30) cm long, 6–10 cm wide, or sometimes larger, with short triangular lobes bearing rigid spines at apices and smaller spines at margins; leaves becoming gradually smaller and nearly entire above, decurrent and forming spiny-margined wings on the stems, all grey-felty tomentose on both surfaces. Heads terminal, solitary, the involucre 3 – 5 cm in diameter (excluding the spines), with purple florets exserted c. ½ the length of the involucre. Involucral bracts coriaceous, pale greenish, sometimes purple-blotched, cobwebby-tomentose at least below, the lower ones with recurved spines, those at middle with erect-spreading spines c. 3 – 4 cm long, the innermost erect beneath the exserted florets. Achenes tetragonal, compressed. *Plates 222, 223.*

Habitat: Silt floors of basins and wadi beds. Rare except very locally.

Northern Plains: 561(K), Khabari Wadha.

Northern Summan: 2859, 26 – 01N, 48 – 27E; 7394, Wadi an-Najabiyah; 25 – 59N, 48 – 30E.

● 29. Zoegea L.

Annual herbs with radiate heads. Involucral bracts imbricate, several-rowed, scarious, the outer ones with a pectinate-fringed appendage; receptacle bristly. Achenes obovate, compressed, with concentric furrows at apex; pappus persistent, with an inner series of bristles and an outer series of short pales.

Zoegea purpurea Fresen.

Erect, puberulent to weakly scabridulous annual, 15 – 35 cm high with stems often branched above. Basal leaves short-petiolate, narrowly oblong and entire or pinnatisect with lanceolate to linear lobes; stem leaves narrow-linear or narrowly lanceolate, acute, mostly 1 – 3 cm long, 1 – 2 mm wide. Heads solitary, terminal, subglobose, 5 – 10 mm long, overtopped by fine reddish awns from the involucre. Involucral bracts oblong-lanceolate with pectinate-fringed margins, terminating in a reddish erect awn about as long as the bract. Florets white or purplish, inconspicuous, hardly exceeding the involucre. Achenes c. 1.5 – 2 mm long, smooth, glossy, truncate at apex, with a pappus of unequal fine white bristles.

Habitat: Silty soils of the northern plains or Summan. Apparently rare, but it is an inconspicuous plant easily overlooked.

Northern Plains: 1593(BM), al-Batin/Wadi al-'Awja junction.

● 30. Centaurea L.

Annual, biennial, or perennial herbs with discoid heads. Involucral bracts imbricate in several rows, spiny or unarmed. Receptacle with bristly bracts. Florets all tubular, sometimes spreading at margins of the head. Achenes smooth, glabrous or pubescent, with lateral hilum. Pappus persistent, several-rowed, of bristles or pales, or both.

Key to the Species of Centaurea
1. Stems narrowly winged with decurrent leaves; flowers
   yellow to flesh-colored .... 2. *C. pseudosinaica*
1. Stems wingless; leaves not decurrent; flowers pink .... 1. *C. bruguierana*

Another *Centaurea*, reported from neighbouring parts of southern Iraq and northern Saudi Arabia, may rarely occur in our northernmost districts, although not yet so recorded: *Centaurea ammocyanus* Boiss., a pink-flowered plant with non-winged stems and involucral bracts green-glabrous with a fimbriate appendage and a softish terminal spinule less than 4 mm long. Dickson's desert record in Kuwait for *C. solstitialis* L. may have been a rare accidental; Wagenitz (1984) lists no specimens of that species from Arabia. *C. mesopotamica* Bornm., reported by Dickson as 'common' in Kuwait, has not apparently been collected in our territory. Her characterization of the plant as prostrate with mauve flowers suggests a confusion of notes, as the species is usually described as erect with yellow flowers. Wagenitz (1984) confirms some records of *C. mesopotamica* from northern Arabia and Kuwait.

1. **Centaurea bruguierana** (DC.) Hand.-Mazz.

C. phyllocephala Boiss.
Erect, branched, glandular-papillose annual herb, 15 – 35 cm high, with white stems both glandular and with spreading crisped hairs. Lowest leaves entire or more or less pinnately lobed, up to 13 cm long, tapering toward the base; cauline leaves oblong-linear, subentire, finely sharp-denticulate, mostly obtusish, sessile and somewhat clasping at base, 1.5 – 9 cm long, 0.3 – 2 cm wide. Heads terminal, mostly solitary, 2 – 4 cm in diameter including the prominent white spines of the involucre, subtended by several leaves grading into glossy bracts. Head small in relation to size of the bone-white, finely glandular, straight-spreading, 1.3 – 2 cm-long spines of the involucre. Florets pinkish. Achenes minute with a white pappus of few bristles.

A plant easily recognized by the conspicuous straight white spines radiating from the smallish head. Our plants are assignable to subsp. *bruguierana*.

Habitat: Silty plains or basins. Infrequent.

Northern Plains: 1620(BM), 3 km SSE al-Batin/Wadi al-'Awja junction.
Northern Summan: 2762(BM), 25 – 49N, 48 – 01E; 7554, 65 km ESE Umm 'Ushar.

2. Centaurea pseudosinaica Czerep.
C. sinaica auct. non DC.

Ascending, more or less densely but very shortly pubescent annual, with stems and branches narrowly winged with decurrent leaves, 10 – 35 cm high. Leaves oblong-linear in outline, the lowest more or less pinnatifid with oblong segments, dentate, tapering at base to a petiole, those above linear-oblong to lanceolate, lobed or subentire-denticulate, decurrent. Heads terminal, with ovoid involucre 0.5 – 1.5 cm wide (not including spines), c. 0.5 – 2 cm long, nearly glabrous, or cobwebby above. Involucral bracts ovate below, narrower and longer within, furnished at apex with a yellow, spreading, somewhat recurved spine with pinnately branching spinelets at base; inner bracts nearly or entirely spineless with membranous, somewhat fringed apex. Florets (in ours) yellow. Achene smooth, light-brownish, truncate at apex, 2.5 – 3 mm long, with a pappus of white to pale brownish fine bristles in several rows, about equalling or slightly longer than the achenes. *Plates 224, 225.*

Our Arabian plants, formerly assigned to *C. sinaica* DC. by many authors, have been renamed as this closely related species.

Habitat: Shallow sands around rocky areas, wadi banks, or in desert basins. Frequent.
Northern Plains: Abraq al-Khaliqah (as *C. sinaica*, Dickson 1955, 31).
Northern Summan: 3912(BM), 27 – 41N, 46 – 02E; 2749(BM), 25 – 34N, 48 – 08E.
Central Coastal Lowlands: 554(K), 1834(BM), Dhahran; 1775(BM), Abu Hadriyah.
Southern Summan: 8066, Wadi as-Sahba, 24 – 12N, 48 – 41E; 8074, Wadi as-Sahba, 24 – 12N, 48 – 50E.
Vernacular Names: MARĀR (Al Murrah, Bani Hajir), MURĀR (Qahtan). The name is derived from *murr*, 'bitter'; a Bedouin of Bani Hajir explained that the plant is much liked by camels but that it reputedly taints the milk with a bitter taste.

● 31. Amberboa (Pers.) Less.

Annual herbs with discoid heads and radiating tubular florets. Involucral bracts imbricate, spinose or not; receptacle bristly. Achenes hairy, ribbed and pitted, with a persistent pappus of many-rowed pales.

Amberboa lippii (L.) DC.
Centaurea lippii L.

Scabrous annual with erect branched stems 15 – 40 cm high, with scattered broad white hairs. Leaves mostly rosulate, pinnately divided nearly or entirely to the midrib into 3 – 5 pairs of main obtuse, sometimes serrulate, lobes, the basal leaves petiolate, c. 3 – 10 cm long, 2 – 3 cm wide, those above smaller, sessile and decurrent. Heads ovoid, with involucre c. 5 – 10 mm broad, the florets pink, or blue drying pink, somewhat spreading at margins of the head. Involucral bracts elliptical-lanceolate, entire, herbaceous at center, scarious-margined, scarious and long-acute in upper quarter or third, with spreading hairs at middle, not or hardly spinescent at apex. Achenes c. 3 mm long, finely pitted, with pappus about ⅔ as long as the achene.
Habitat: Sand in limestone ravine. Rare; collected only once.
Northern Plains: 630(K), 15 km SSW Hafar al-Batin in ravine tributary to al-Batin.

● 32. Carthamus L.

Annual, sometimes perennial, herbs with spiny-dentate, coriaceous leaves. Heads discoid with spiny involucral bracts. Receptacle with bristlelike bracts. Achenes glabrous with oblique-lateral hilum, somewhat tetragonal, with or without a pappus.

Carthamus oxyacantha M.B.

Ascending, much-branched, coarse prickly herb with whitish stems, glabrescent to pubescent-cobwebby in younger parts, 30 – 100 cm high. Leaves coriaceous, oblong to lanceolate, acute, sessile and partly clasping at base at least above, dentate, with yellow spines at margins and apex, 2.5 – 5.5 cm long, 0.7 – 2 cm wide. Heads solitary, terminal on the stems, c. 2.5 cm long, exceeded by the leaflike spiny outer involucral bracts; inner bracts less spiny laterally, entire near base. Florets yellow. Achenes glabrous, smooth, rather shiny, obovoid and somewhat compressed, completely without pappus, 4 – 5 mm long, pale brown with blackish marks.

Habitat: Nearly always on disturbed or waste ground. Occasional.

Northern Plains: 7847, Qaisumah Pump Station, 28 – 19N, 46 – 07E.

Central Coastal Lowlands: 644(BM), Dhahran.

South Coastal Lowlands: 2727, ar-Ruqayyiqah, al-Hasa; 8341, 5 km SE al-Hufuf; 8304, al-Hufuf.

Vernacular Names: 'UṢFUR ('yellow-tint', al-Hasa gardener, a name also applied to the dye plant much cultivated in southern Arabia, *C. tinctorius*), SAMNAH ('butterweed', Shammar).

LIGULIFLORAE

● 33. Koelpinia Pall.

Annual herbs with linear leaves and ligulate heads. Involucre in 2 rows, the outer of few very short bracts, the inner much longer of 5 – 7 bracts. Florets yellow. Achenes narrow-cylindrical, curved, long-spreading, striate with prickles at back, epappose.

Koelpinia linearis Pall.

Ascending or decumbent diffuse annual herb branching from the base, finely puberulent to subtomentose in parts, 10 – 30 cm high. Leaves sessile, very narrowly linear, 5 – 15 cm long, 0.5 – 1.5 mm wide, flat or channeled, long-tapering to apex. Heads peduncled in the axils, both at base and on the stems. Involucre puberulent and tomentellous, c. 7 – 10 mm long, the basal bracts c. 3, linear-lanceolate, the inner row of c. 6 – 9 bracts, linear-lanceolate, acute, spreading at maturity. Florets rather few, yellow, with 4 – 5-toothed ligules. Achenes c. 10 – 15, stellate-spreading, very conspicuous and well exceeding the head, c. 15 – 20 mm long, narrowly cylindrical, incurved, striate and tomentellous, furnished at back with rows of somewhat hooked prickles; pappus absent.

● The conspicuous long, curved, prickly achenes of this species are very distinctive. With their hooked prickles they are very tenacious in hair and probably serve as a dispersal mechanism by leading to transport of the achenes by grazing animals.

Habitat: Sands or sandy-silts. Frequent.

Northern Plains: 1510(BM), 30 km ESE al-Qaysumah; 1654(BM), 2 km S Kuwait border in 47 – 14E; 861, 9 km S 'Ayn al-'Abd; 901, Jabal al-'Amudah; 1248, Jabal an-Nu'ayriyah; 1290(BM), 15 km ENE Qaryat al-'Ulya.

Northern Summan: 707, 33 km SW al-Qar'ah wells; 767, 69 km SW al-Qar'ah wells; 1365(BM), 42 km W Qaryat al-'Ulya.

Southern Summan: 2265, 29 km ENE al-Hunayy; 2062, Jaww ad-Dukhan, 24 – 48N, 49 – 05E; 2252, 26 km ENE al-Hunayy.

Central Coastal Lowlands: 7464, 7 km SW Qannur.

Vernacular Names and Uses: LIḤYAT AT-TAYS ('goat's beard', Qahtan, Rawalah, gen.), LIḤYAT ASH-SHAYBAH ('old man's beard', Al Murrah), DHIQNŪN ('beards', Shammar), DHU'LŪQ (Philby 1922). Bedouins sometimes eat this plant raw (Bani Hajir tribesmen, also Carter (1917, 204) and Musil (1928a, 95).

● 34. Urospermum Scop.

Annual or perennial herbs with ligulate heads. Involucral bracts herbaceous in a single row; receptacle ebracteate. Florets yellow. Achenes long-beaked with a pappus of plumose bristles.

Urospermum picroides (L.) Schmidt

Hispidulous to glabrescent annual with erect, often branched stem, 15 – 50 cm high. Leaves at base oblong-spathulate, lyrate, tapering at base; stem leaves oblong, denticulate, sessile, clasping-auriculate at base, the uppermost lanceolate to linear. Heads terminal with ovoid-conical involucre c. 1.5 – 2 cm long; involucral bracts c. 9, broadly lanceolate, acute, connate near base, usually setose. Achenes often over 10 mm long (including beak), with long narrow beak dilated at base; pappus of plumose white bristles shorter than the achene and joined at base in a ring.

Habitat: A weed of farms or waste ground. Apparently rare in the Eastern Province but perhaps overlooked because of its superficial resemblance to *Sonchus*.

Central or South Coastal Lowlands: Not seen by the author but listed as a common weed (presumably in central Arabia) by Chaudhary and Zawawi (1983) and likely to be found also in the east.

● 35. Leontodon L.

Annual or perennial herbs with ligulate heads and rosulate leaves. Involucre imbricate; receptacle ebracteate. Achenes tapering at apex or beaked, striate, transversely wrinkled, with a persistent pappus of plumose bristles.

Leontodon laciniatus (Bertol.) Widder

L. hispidulus (Del.) Boiss. var. *tenuiloba*

Annual herb with ascending, simple or branched scapes, somewhat hispid, usually with some short pubescence on the involucral bracts and at stem bases, or glabrescent. Leaves rosulate, oblong-oblanceolate in outline, 4 – 12 cm long, mostly divided nearly to midrib into narrow triangular to linear acute lobes. Heads yellow-flowered, solitary-terminal on scapes thickened below the head. Involucre 6 – 11 mm long in flower, with narrowly triangular to lanceolate, acute bracts. Achenes tapering at apex, more or less beaked, striate and transversely rugose, with a pappus of about 10 plumose bristles about as long as the achene.

Habitat: Silty soils of the northern plains and Summan. Occasional.

Northern Plains: 596, al-Batin, 16 km ENE Qulban Ibn Busayyis; 1549(BM), 36 km NE Hafar al-Batin; 1581(BM), al-Batin/Wadi al-'Awja junction; 1669(BM), 2 km S Kuwait border in 47 – 14E.

Northern Summan: 4019, 27 – 45N, 45 – 32E; 725, 33 km SW al-Qar'ah wells; 1357(BM), 1387(BM), 42 km W Qaryat al-'Ulya; 2810, 12 km ESE ash-Shumlul.

Vernacular Names: MURAR (Dickson 1955).

● 36. Picris L.

Annual, biennial, or perennial herbs, often hispid, with ligulate heads and leaves mostly rosulate. Involucre usually constricted near the middle, with inner bracts erect, nearly equal, indurate and channeled below. Receptacle ebracteate. Florets yellow. Achenes usually dimorphic, striate and transversely rugulose, tapering at apex or beaked, and with a pappus of plumose bristles.

Picris babylonica Hand.-Mazz.

P. saharae auct. non Coss.

Ascending or decumbent annual with simple or branched stems 5 – 30 cm long, hispid with short, spreading, often terminally barbed bristles. Basal leaves rosulate, oblong-oblanceolate in outline, runcinate-pinnatifid, 3 – 8 cm long, 0.6 – 1.5 cm wide; stem leaves few, smaller, sessile, less strongly lobed. Heads solitary, terminal, with bright yellow, broadly spreading florets, sometimes black-centered with immature florets. Involucre more or less campanulate in flower, cylindrical and often somewhat constricted near middle in fruit, with lanceolate-linear, acute bracts mostly equal in a single row with a few shorter ones at base, at back both closely crisp-tomentose and furnished with longer bristles; fruiting head c. 1 – 1.3 cm long, 0.6 – 1.3 cm in diameter. In homocarpous heads: all achenes subellipsoid, somewhat curved, tapering at base and apex, longitudinally striate and distinctly transversely wrinkled, c. 5 mm long (including the beak), tapering above to a beak ½ to about as long as the achene body, the entire achene ripening to a grey-violet or grey-brown color, with pappus c. 7 mm long, mostly with 10 – 15 scabrous, plumose bristles. In heterocarpous heads: marginal achenes clasped by the inner involucral bracts, terete, gently curved, tapering very gradually to apex and not beaked, more or less red-brown pubescent, and with a very short (c. 1 – 3 mm-long) pappus consisting of a crown of hairs; central achenes (more numerous and over most of the receptacle) subellipsoid, tapering at base and with an apical beak about ½ to ¾ as long as the achene body, finely striate longitudinally and transversely wrinkled, with a pappus of c. 10 – 15 plumose bristles 5 – 7 mm long; both types of achenes brown or grey-brown, or the central beaked ones grey-violet. *Plate 226.*

● Plants with homocarpous heads appear to be more common in our northern and central districts. Specimens with virtually all heads heterocarpous were noted first around Harad and Wadi as-Sahba; they were then found to extend northward into the central Ghawar area. Plants with some heads heterocarpous, however, also appear sporadically as far north as al-Batin, on the northern boundary of our territory. Lack (1973) referred to facultative heterocarpy in *P. babylonica* and expressed interest in the possibility that populations in the southern part of the Arabian Peninsula might be found to exhibit this character more often than those in the north. This seems to be borne out by our experience. Dr Lack has examined examples of our homocarpous and heterocarpous specimens and concurs with the view that both should be treated as forms of

P. babylonica. It might be noted also, however, that the longer-beaked achene forms in some of our northern plants appear to approach those ascribed to *P. cyanocarpa* Boiss. The relationship between these two species would appear to merit further scrutiny.

Facultative heterocarpy might conceivably have some selective advantage with regard to fruit dispersal under extreme desert conditions. The central, free, pappose achenes, through wind dispersal, would provide a mechanism for chance exploitation of new areas. The marginal, nearly epappose achenes, firmly clasped by the persistent bracts and falling with the mother plant, would ensure that some units would at all times remain in a habitat of proved fitness. The soil type there would be of assured suitability; and if this were also in a runnel or other low spot, there would also be increased probability of receiving adequate moisture.

Habitat: Sands or silts in a wide variety of habitats; one of the most abundant annuals in the northern districts and sometimes grows in great patches over flat areas. Very common.

Northern Plains: 1151(BM), Jabal al-Ba'al; 674 Khabari Wadha; 827(BM), 13 km W Qaryat al-'Ulya; 1513(BM), 30 km ESE al-Qaysumah; 843, 9 km S 'Ayn al-'Abd; 1206, Jabal an-Nu'ayriyah; 1295(BM), 15 km ENE Qaryat al-'Ulya; 1326(BM), 8 km W Jabal Dab'; 1405(BM), 15 km W Qaryat al-'Ulya; 1551(BM), 36 km NE Hafar al-Batin; 1561(BM), al-Batin/Wadi al-'Awja junction; 1601(BM), 3 km ESE al-Batin/Wadi al-'Awja junction; 1633(BM), 2 km S Kuwait border in 47–06E; 1671(BM), 2 km S Kuwait border in 47–14E; 1697(BM), 28–41N, 47–35E; 1720(BM), 28 km E Abraq al-Kabrit; 8827, al-Batin in 45–51E; 8828, al-Batin in 45–55E.

Northern Summan: 731, 33 km SW al-Qar'ah wells; 775, 69 km SW al-Qar'ah wells; 2769, 25–49N, 48–01E.

Central Coastal Lowlands: 56, 112, 1032(BM), Dhahran; 1754(BM), Abu Hadriyah; 3307(BM), 5 km S Nita'.

South Coastal Lowlands: 375, Jabal Sha'ban, al-Hasa; 8047, Wadi as-Sahba, 24–00N, 49–08E; 8049, Wadi as-Sahba, 24–02N, 49–03E.

Southern Summan: 2066(BM), Jaww ad-Dukhan, 24–48N, 49–05E; 8346, al-Ghawar, 18 km N Harad; 8370, al-Ghawar, 28 km N Harad; 8373, al-Ghawar, 38 km S 'Uthmaniyah GOSP 13.

Vernacular Names: ḤAWDHĀN (Bani Hajir, Al Murrah, Mutayr, Qahtan, Rawalah), ḤUWWA' (Al Murrah).

● 37. Scorzonera L.

Herbs, usually perennial, or subshrubs, with ligulate heads and sometimes tuberous roots. Involucral bracts herbaceous, imbricate; receptacle ebracteate. Florets yellow, pink, or purplish. Achenes generally columnar, ribbed or striate, sometimes with a hollow stalk at base. Pappus of plumose, scabrous bristles in several rows.

Key to the Species of Scorzonera
1. Flowers yellow; leaves narrow-linear to subulate . . . . 1. *S. tortuosissima*
1. Flowers pink-purplish; leaves elliptical-lanceolate . . . . 2. *S. papposa*

1. Scorzonera tortuosissima Boiss.
Ascending, silvery-greyish, dense or rather diffuse branched perennial, 20–50 cm high, with stems closely whitish-canescent and finely lined or ribbed. Leaves finely linear

to subulate, somewhat triangular in cross section, channeled and white-canescent on one side, 5 – 15 cm long near base and on lower stems, becoming much shorter above. Heads numerous, solitary-terminal at the apices of the corymbosely branched stems, mostly with 5 – 7 yellow florets. Fruiting involucre cylindrical, 13 – 15 mm long, 4 – 5 mm wide, with inner bracts linear-lanceolate, acute, more or less canescent at back, those at base much shorter, triangular. Achenes narrowly columnar-prismatic, 8 – 10(13) mm long, longitudinally grooved, glabrous, with a persistent brownish-white pappus of bristles about equalling or somewhat exceeding the achene, very finely plumose in the lower ⅔, scabrous above. *Plate 227.*

Our plants appear to match this species fairly well. Further research is warranted, however, into the relationships of this taxon, our specimens, and *S. musilii* Vel.

Habitat: Sands, often around rocky ground. Sometimes seen along desert tracks or roadsides. Infrequent but apparently with a wide distribution.

Northern Plains: 3114(BM), al-Batin, 28 – 00N, 45 – 28E; 8788, 30 km SW ar-Ruq'i Post, 28 – 54N, 46 – 27E.

Northern Dahna: 3871, 15 km WSW Umm 'Ushar.

Central Coastal Lowlands: 2888, 8 km WSW 'Ain Dar GOSP – 1, 25 – 51N, 49 – 10E; 7959, 15 km WSW 'Ain Dar GOSP – 1.

Vernacular Names: DHU'LŪQ (N tribes, Musil 1928a, 95), DHU'LŪQ AL-JAMAL (N tribes, Musil 1927, 595). Musil notes that the plant is sometimes eaten raw by Bedouins.

2. Scorzonera papposa DC.

Ascending, rarely decumbent, showy-flowered perennial herb with branched stems, 15 – 50 cm high, sparsely cobwebby-pubescent in parts, or glabrescent, often with a dark brown tuber on the root. Leaves elliptical-oblanceolate, entire or erose, acute, often wavy-margined, tapering at base to a petiole or sessile, 5 – 8(10) cm long, 0.5 – 1.5(2) cm wide, smaller and narrower above. Heads solitary, terminal, 4 – 5 cm in diameter in flower, showy, with pink to purplish florets. Involucre cylindrical, c. 20 mm long, 6 – 8 mm wide, with herbaceous, white-margined, glabrescent bracts broadly ovate at base to oblong-lanceolate within and above. Achenes 8 – 10 mm long, glabrous, grooved-muricate, with a white pappus c. 10 – 13 mm long, finely plumose below, scabrous to apex. *Plates 228, 229.*

Habitat: Shallow silts and sands, usually in crevices on rocky ground. Infrequent.

Northern Plains: 1614(BM), 3 km ESE al-Batin/Wadi al-'Awja junction; 1276(BM), 15 km ENE Qaryat al-'Ulya.

Northern Summan: 1400(BM), 42 km W Qaryat al-'Ulya; 3331, 12 km WSW an-Nazim; 4140, W as-Sarrar, 26 – 58N, 48 – 18E; 8108, 9 km SSE Hanidh; 3318, 12 km S Nita'.

Vernacular Names and Uses: RUBAḤLAH (gen.). Well known to all Bedouins for the edible nutlike tuber found on the root. This is usually extracted only with difficulty as it is often deeply wedged in rock crevices beneath the sand, with the tender stem breaking off above.

● 38. **Heteroderis** Boiss.

Annual herbs with rosulate leaves and ligulate heads. Involucral bracts calyculate; receptacle ebracteate. Achenes dimorphic, the marginal ones beakless, the central ones with a fine beak longer than the achenes.

Heteroderis pusilla Boiss.

Ascending or decumbent branched annual herb, often sparsely pubescent with rather soft bristles, 15 – 30 cm high. Basal leaves rosulate, sessile, oblong-oblanceolate, coarsely dentate or pinnately lobed with acute triangular segments. Stem leaves at the forks, much smaller, narrowly lanceolate, acuminate, with a clasping auriculate, laciniate or dentate base. Heads rather short-peduncled, mostly grouped 1 – 3(4) at the stem apices, nearly cylindrical, c. 10 mm long (including the somewhat exserted pappi), 3 – 5 mm wide. Outer involucral bracts at base very short, ovate, glabrous, the inner ones equal, in 1 row, lanceolate, furnished along the medial herbaceous sector with somewhat softish bristles but glabrous or minutely glandular beneath them. Achenes oblanceolate in outline, ellipsoidal to somewhat prismatic, pale brown, c. 2 mm long, furnished in upper half or third with minute echinate tubercles and with a fine beak c. 5 mm long terminating in a pappus of 2 – 3 mm-long, fine white hairs.

● A plant rather closely resembling *Crepis aspera* in many respects but overall smaller, more delicate, with softish rather than rough bristles, and with the involucral bracts glabrous or only fine-glandular — rather than tomentose-pubescent — beneath the bristles.

Habitat: Our only specimen was from a silty basin on a gravel plain.

Northern Plains: 1550(BM), 36 km NE Hafar al-Batin. Only collected once and apparently rare.

● 39. Launaea Cass.

Annual or perennial, often lactiferous herbs, rarely shrubs, with ligulate heads. Involucre imbricate; receptacle ebracteate. Achenes dimorphic, not beaked, sometimes winged, with a pappus usually of dense soft hairs, sometimes with longer bristles interspersed.

A key to species based largely on achene characteristics is probably most reliable. Our species, however, fall into two basic habit groups: those with thickish, green leafy stems and broader, herbaceous involucral bracts (*L. angustifolia* and *L. mucronata*); and those with runcinate leaves mostly at base, stems thin, scape-like, and narrower, strongly white-margined involucral bracts (the remaining species).

The achenes in a single head are often dimorphic or polymorphic, and they should be examined as a group for key characters. Persistence of the pappus in our species is a generally reliable and useful character.

Key to the Species of Launaea

1. Achenes 2 – 4-winged along margins
    2. Pappus persistent; outer involucral bracts green-herbaceous, not or hardly white-scarious margined; stems erect, green, leafy .... 1. *L. angustifolia*
    2. Pappus deciduous; outer involucral bracts with pronounced white-scarious margins; stems absent, or scapelike, procumbent, not leafy .. 2. *L. capitata*
1. Achenes subquadrangular, ribbed, striate or compressed, but not winged
    3. Achene 4 – 7 mm long; pappus dimorphic, of fine white hairs mixed with fewer, longer bristles .... 3. *L. mucronata*
    3. Achene 2 – 3.5 mm long; pappus with all hairs similar
        4. Achenes truncate at apex; pappus persistent .... 4. *L. nudicaulis*
        4. Achenes short-tapering, acutish at apex; pappus very caducous .... 5. *L. procumbens*

1. Launaea angustifolia (Desf.) O. Kuntze

L. arabica (Boiss.) H. Lindb.

Ascending, glabrous, glaucous annual branched at base and above, 10 – 30 cm high. Leaves somewhat fleshy, oblong to lanceolate in outline, mostly 3 – 6 cm long, variably incised-dentate, those at base tapering to a petiole, those on stems sessile with clasping auricles at base. Heads mostly solitary at the branch ends, to c. 4 cm in diameter in flower, with yellow florets and ovoid involucre c. 10 – 14 mm long, 10 – 15 mm broad, pendulous-nodding in fruit. Outer involucral bracts ovate to oblong, green-herbaceous, not or only very faintly and narrowly white-margined; inner bracts lanceolate, acute, white-scarious margined. Achenes quadrangular, c. 3 mm long, longitudinally winged at angles with coherent hairs, with narrower wings or ribs between, truncate at apex with a very dense, snow-white persistent, somewhat wavy pappus of hairs 4 – 5 mm long. *Plates 230, 231.*

A species generally easily recognized by its incised leaf margins and its nodding fruit heads with the dense white pappi somewhat exserted like a boll of cotton.

Habitat: Generally an inland plant of wadi beds and basins in the Summan; usually in rocky country. Locally frequent.

Northern Summan: 1398(BM), ash-Shayyit al-'Atshan, 42 km W Qaryat al-'Ulya; 3338(BM), 5 km WSW an-Nazim; 4141, W as-Sarrar, 26 – 58N, 48 – 18E.

South Coastal Lowlands: 7911, Na'lat Shadqam, 25 – 43N, 49 – 29E.

Vernacular Names: MURĀR (Dickson 1955).

2. Launaea capitata (Spreng.) Dandy

L. glomerata (Cass.) Hook. f.

Glabrous annual or perennial herb, sometimes dwarf, stemless or with scapelike, mostly procumbent stems 5 – 15 cm long. Leaves rosulate, oblong to spathulate in outline, tapering somewhat toward the base, runcinate-pinnatifid with cartilaginous, denticulate margins, mostly 2 – 6(8) cm long, 0.5 – 2(3) cm wide. Heads yellow-flowered, nearly sessile, clustered, basal at center in non-stemmed plants, or near the ends of scapelike, nearly leafless stems. Involucre oblong-subovoid, c. 8 – 10 mm long, with imbricate, distinctly white-margined bracts. Achenes whitish, compressed, c. 3.5 mm long, finely striate and tuberculate-muricate, broadly 2-winged at opposite margins, with or without narrower winglets between, truncate at apex with a caducous pappus of equal white hairs slightly longer than the achene.

Generally recognizable by its flat, procumbent, habit.

Habitat: Sands and silts; widespread and common in many terrain types.

Northern Plains: 1150(BM), Jabal al-Ba'al; 822, 829, 13 km W Qaryat al-'Ulya; 865, 9 km S 'Ayn al-'Abd; 896, Jabal al-'Amudah; 926, al-Khafji/as-Saffaniyah junction; 1006, 12 km N Jabal al-'Amudah; 1188, 1241(BM), Jabal an-Nu'ayriyah; 1571(BM), al-Batin/Wadi al-'Awja junction; 1679(BM), 2 km S Kuwait border in 47 – 14E; 1266(BM), 15 km ENE Qaryat al-'Ulya.

Northern Summan: 227(BM), 10 km SSW al-'Uwaynah; 1406(BM), 42 km W Qaryat al-'Ulya; 3155(BM), 9 km SSW Hanidh; 3235, Dahl al-Furayy.

Southern Summan: 2122, 3 km SW al-Jawamir (Yabrin); 2216, 26 km SSE al-Hunayy; 2229, 15 km WSW al-Hunayy; 2250, 26 km ENE al-Hunayy; 2260, 29 km ENE al-Hunayy; 8020, 23 – 11N, 48 – 53E.

Central Coastal Lowlands: 395, 420, Dhahran; 405, al-Midra ash-Shimali; 1473(BM), 2 km E al-Ajam; 1794(BM), 5 km NE Abu Hadriyah; 1978, ad-Dawsariyah; 3726, 8 km N Abu Hadriyah; 4061, az-Zulayfayn.

South Coastal Lowlands: 7911, edge Na'lat Shadqam, 25–43N, 49–29E.

Vernacular Names: ḤUWWA' (gen.).

3. Launaea mucronata (Forssk.) Muschl.

Erect, much-branched, rather stout glabrous, glaucous perennial herb, 30–80 cm high, with rather copious latex. Basal leaves petiolate, to c. 15 cm long, oblong-lanceolate in outline, pinnatifid with oblong to linear, dentate to partite lobes; cauline leaves shorter, sessile, with dentate to divided auricles at base. Heads terminal, yellow-flowered, with involucre 1–1.5 cm long. Involucral bracts white-margined, ovate to oblong to narrowly lanceolate above. Achenes (3)4–7 mm long, columnar-prismatic, subtetragonal, 4-striate, smoothish or finely wrinkled, glabrous or (particularly the outer achenes) short-velvety pubescent, often with 4 obscure to distinct hornlets at basal corners, with a persistent pappus 10–15 mm long consisting of soft white hairs and some longer inner bristles somewhat exceeding the copious outer hairs. *Plate 232.*

Habitat: Sands or silts. Widespread and frequent to common in the desert; also not infrequently seen as a ruderal or on disturbed ground around inhabited areas.

Northern Plains: 871(BM), 9 km S 'Ayn al-'Abd; 933, 10 km S al-Khafji/as-Saffaniyah junction; 990(BM), 12 km N Jabal al-'Amudah; 1213(BM), Jabal an-Nu'ayriyah; 1609(BM), 3 km ESE al-Batin/Wadi al-'Awja junction.

Central Coastal Lowlands: 1805(BM), 5 km NE al-Khursaniyah; 2723(BM), 23 km N Jabal Fazran.

Southern Summan: 2258(BM), 29 km ENE al-Hunayy.

South Coastal Lowlands: 1445(BM), 2 km E Jabal Ghuraymil.

Vernacular Names: 'AḌĪD (gen.).

4. Launaea nudicaulis (L.) Hook. f.

Glabrous perennial herb with stems diffusely branched, ascending and 20–50 cm high or sometimes scrambling on rocks and longer. Leaves mostly at base, rosulate, oblong in outline, runcinate-pinnatifid with triangular, dentate lobes, sessile or petiolate, 4–10(15) cm long, to c. 3 cm wide; stem leaves much smaller, often minute above. Heads numerous on rather short peduncles on the upper branches, the ligules yellow above, whitish at back. Involucre cylindrical, 12–15 mm long in fruit, with basal bracts ovate, those above narrowly lanceolate-linear, medially brownish with white scarious margins, often with a small black dorsal keel near the apex. Achenes c. 3.5 mm long, of two basic types: either columnar-subtetragonal as if formed of 4 rounded obtuse-margined ribs, pale, smoothly wrinkled, or: compressed, darker and rougher, tuberculate-hispidulous. Pappus of subequal white hairs, persistent, 8–9 mm long. *Plate 233.*

Plants collected near the seashore at Ras al-Mish'ab were of atypically dense habit, with flowering stems hardly exceeding the leafy parts *(Plate 233)*.

Habitat: Particularly characteristic of small wadis and ravines in the rocky Summan, where it may scramble along rocky banks; also occasionally found on coastal sands in the north. Frequent.

322

Northern Plains: 4013, ad-Dibdibah, 27 – 56N, 45 – 35E; 1607(BM), 3 km ESE al-Batin/Wadi al-'Awja junction; 3111(BM), al-Batin, 28 – 00N, 45 – 28E; 7970, base of Ras al-Mish'ab.

Northern Summan: 7484, Batn Sumlul, 26 – 26N, 48 – 15E; 813(BM), 37 km NE Jarrarah; 2866, 26 – 01N, 48 – 29E; 3868, 15 km WSW Umm 'Ushar; 3133(BM), 6 km SE ad-Dabtiyah; 3176(BM), 13 km SSW Hanidh; 7943, 25 – 39N, 48 – 48E; 8612, near Dahl Umm Hujul, 26 – 35N, 47 – 32E.

Vernacular Names: ḤUWWA' (gen.).

5. **Launaea procumbens** (Roxb.) Ramayya et Rajagopal
L. fallax (Jaub. et Spach) O. Kuntze

Glabrous perennial herb with stems procumbent, decumbent, or ascending, 10 – 30 cm high, sometimes tangled and shrubby at base. Leaves mostly basal, rosulate, oblong to spathulate in outline, sinuate-dentate to runcinate-pinnatified with denticulate, callose margins, 4 – 12(15) cm long, 0.5 – 3 cm wide. Stem leaves nearly obsolete. Heads very short-peduncled, solitary or more commonly clustered 3 – 10, yellow-flowered. Involucre oblong-cylindrical, 8 – 10 mm long in fruit, with basal bracts ovate, those above lanceolate, all broadly white-scarious margined. Achenes brown, c. 2.5 mm long, oblong, weakly compressed and faintly tetragonal, minutely ribbed, tapering near top to an acutish apex, rarely finely denticulate on margins and ribs. Pappus very caducous, of fine white hairs 4 – 5 mm long.

This plant is rather variable in habit. One unusual specimen from the coast (No. 7726) was strikingly similar to *L. capitata* except for the achenes, which were very near typical *procumbens* but with a faint suggestion of marginal wings; it was suggestive of a hybrid of the two.

Habitat: Much more common as a ruderal and a weed around settled areas and farms than as a desert plant; it is the common weed *Launaea* along walk edges and gardens both in towns and the oases. Locally common.

Central Coastal Lowlands: 637(K), Dhahran; 1456(BM), 1472(BM), 2 km E al-Ajam; 7726, 2 km S Ras Tanaqib, 27 – 49N, 48 – 53E; 3775, Ras az-Zawr.

South Coastal Lowlands: 1909(BM), al-Mutayrifi, al-Hasa Oasis.

Vernacular Names: ḤUWWA' (gen.).

● 40. Sonchus L.

Annual, biennial, or perennial herbs with ligulate, yellow-flowered heads. Involucral bracts imbricate in several rows; receptacle ebracteate. Achenes compressed, longitudinally ribbed or striate, unbeaked, with a pappus of simple white hairs. Weeds of farms, gardens, or waste ground.

Key to the Species of Sonchus

1. Leaves linear or linear-lanceolate, mostly 1 – 2 cm wide, virtually entire or obscurely undulate .... 2. *S maritimus*
1. Leaves runcinate- or lyrate-pinnatifid; mostly over 2 cm wide
   2. Achenes papery-thin with winglike margins; auricles of stem leaves rounded-dentate .... 1. *S. asper*
   2. Achenes compressed but not papery, unwinged; auricles of stem leaves pointed, acute .... 3. *S. oleraceus*

1. Sonchus asper (L.) Hill

Glabrous, erect, few-branched annual, 30 – 75(100) cm high, sometimes purple-tinged in some parts. Leaves at base runcinate- or lyrate-pinnatifid with sharply dentate margins, tapering at base to a petiole, acute or rounded at apex; cauline leaves progressively less deeply lobed, amplexicaul, with rounded, dentate, usually appressed, auricles. Heads in a terminal subumbellate inflorescence and short-peduncled in upper axils. Involucre c. 10 mm long, with long-triangular to lanceolate bracts. Achenes oblong, 2.5 – 3 mm long, strongly compressed and almost paper-thin, brown, with 3 main riblets on each face and thin, winglike, entire or denticulate margins. Pappus of fine white hairs 1.5 – 2 times as long as the achene.

The form of the middle to upper cauline leaves of this species, at least in our area, is usually quite different from that of *S. oleraceus*, being oblong or spathulate, not runcinate, virtually entire except for the distinct marginal dentation.

Habitat: A weed of gardens, farms and walk edges. Frequent.

Central Coastal Lowlands: 1073(BM), 3759(BM), Dhahran stables farm; 919, al-Qatif Oasis.

South Coastal Lowlands: 3924, near 'Ayn al-Khadud, al-Hasa Oasis; C. Parker 14(BM), al-Hufuf area; 8228, near al-Qurayn, al-Hasa.

Southern Summan: 8090, Batn al-Faruq, 25 – 43N, 48 – 53E, an abandoned Bedouin camp site.

Vernacular Names: KHUWWAYSH (al-Hasa farmers).

2. Sonchus maritimus L.

Glabrous, perennial rhizomatous herb with erect stems, 20 – 50 cm high. Leaves linear or linear-lanceolate, mostly 10 – 25 cm long, 1 – 2 cm wide, entire or faintly sinuate and remotely denticulate, those below tapering at base, those above sessile, amplexicaul with short auricles. Heads few, yellow-flowered, with involucre c. 10 mm long. Achenes elliptical, compressed, c. 2.5 – 3 mm long, with broad, thickened margins and 2 – 3 riblets on each face. Pappus c. twice as long as the achene.

Habitat: Wet ground along edges of shaded drainage and irrigation ditches in the oases. Locally frequent.

Central Coastal Lowlands: 920, W edge of al-Qatif Oasis opposite al-Qatif town; C. Parker 76(BM), al-Qatif Oasis.

South Coastal Lowlands: 357(BM), 1 km S ad-Dalwah, al-Hasa Oasis; 8222, near al-Qurayn, al-Hasa Oasis.

Vernacular Names: FARAS (gen.).

3. Sonchus oleraceus L.

Erect, glabrous annual herb, usually few-branched above, 15 – 75(100) cm high. Leaves oblong in outline, runcinate-pinnatifid with acute or rounded apex and more or less triangular lobes, sharp-dentate at margins, those below tapering at base, the stem leaves sessile, amplexicaul, with pointed, dentate auricles. Heads mostly on terminal branchlets with a few in upper axils, yellow-flowered, with involucre c. 10 mm long. Involucral bracts long-triangular to linear-lanceolate. Achenes compressed but not papery-thin, oblanceolate to linear-elliptical, sometimes subprismatic with 3 – 4 angles, finely rugulose, brown, c. 2.5 – 3 mm long, with a pappus c. twice as long as the achene.

The achenes of this species differ from those of *A. asper* mainly in being narrower, thicker, and overall more rugulose.

Habitat: A weed of gardens and farms. Frequent. Sometimes also in desert.

South Coastal Lowlands: 7798, 5 km SE al-Hufuf, on farm; 8227, near al-Qurayn, al-Hasa Oasis.

Vernacular Names: KHUWWAYSH (al-Hasa farmers).

● 41. Lactuca L.

Annual, biennial, or perennial lactiferous herbs or shrubs with ligulate heads. Involucre cylindrical or campanulate with imbricate bracts in several rows; receptacle ebracteate. Achenes compressed, beaked, with a pappus of many soft hairs.

Key to the Species of Lactuca

1. Leaf margins continuously fine-spinulose-denticulate; inflorescence a broad panicle .... 2. *L. serriola*
1. Leaf margins not continuously spinulose-denticulate; inflorescence narrowly racemose, or spicate .... 1. *L. saligna*

1. Lactuca saligna L.

Erect glabrous annual or biennial, usually 60 – 100 cm high, with smooth, whitish stem. Leaves prickly at lobe apices or unarmed, sagittate at base, the lower ones runcinate-pinnatifid with oblong, dentate lobes, often spinulose on the midrib, the upper ones usually entire, linear-lanceolate, with sagittate base. Heads short-peduncled in an elongate, narrowly racemose inflorescence. Involucre c. 7 – 15 mm long; florets yellow, drying bluish. Achenes 5 – 9-striate on each side, compressed, obovate-oblong, with a beak 1 – 2 times as long as the achene; pappus deciduous.

Habitat: A weed of farm fields or waste ground in the oases. Rare.

Central Coastal Lowlands: 917, W edge al-Qatif Oasis. A questionable specimen, not yet flowering, and assigned to this species on the basis of leaf characters. The species is listed by Chaudhary, Parker and Kasasian (1981) as an Eastern Province weed, but their specimens have not been seen by the present author. Further collections are required to establish the status of the species in our area. It would not be unexpected as a weed around agricultural areas, but it certainly is not as frequent as *L. serriola* (below).

2. Lactuca serriola L.

Erect, glabrous annual or biennial, 30 – 150(200) cm high, with a glossy, whitish to yellowish stem. Leaves rigid, coriaceous, oblong in outline, mostly 6 – 14 cm long, 1.5 – 3.5 cm wide, either entire (as in many of our specimens) or somewhat lobed, or runcinate-pinnatifid, sessile, sagittate-clasping at base, with narrowly callose and continuously fine-denticulate or spinulose margins. Heads numerous in a terminal, rather open panicle. Involucre c. 5 – 15 mm long (in ours mostly 5 – 10 mm), with lanceolate, somewhat glandular bracts. Florets (in ours) c. 10 – 15; ligules yellow above, pinkish grey-violet at back, drying purplish. Achenes c. 2.5 – 3.5 mm long (in ours mostly 2.5 mm), brown, compressed, oblong, 5 – 7(9)-striate on each face, terminating in a beak c. 2.5 times as long as the achene; pappus of fine white hairs, spreading.

Habitat: A weed of abandoned farm fields and waste ground around agricultural areas. Occasional to locally frequent.

Central Coastal Lowlands: 1916(BM), 1920(BM), al-Jarudiyah, al-Qatif Oasis; 7770, 4 km S al-Qatif town.

South Coastal Lowlands: 1890(BM), al-Mutayrifi, al-Hasa; 3923, near 'Ayn al-Khadud, al-Hasa.

● 42. **Reichardia** Roth

Annual or perennial herbs with ligulate heads. Involucral bracts imbricate, hyaline-margined; receptacle ebracteate. Florets yellow, often red-purplish at backs. Achenes more or less 4-sided, transversely wrinkled-tubercled, more or less dimorphic, with a dense fine pappus of simple white hairs joined at base in a hollow ring and deciduous together.

Reichardia tingitana (L.) Roth
R. orientalis (L.) Hochr.

Decumbent or short-ascending glabrescent annual, branched from the base with stems 5 – 25 cm long, sometimes flowering in dwarf condition 3 – 4 cm high. Leaves mostly rosulate, oblong to oblanceolate in outline, entire, nearly entire, or pinnatifid with dentate lobes, those below tapering to a petiole, those above sessile, auriculate-clasping. Heads mostly solitary, terminal, on somewhat thickened peduncles, yellow-flowered with ligules reddish purple at backs and sometimes partly above. Involucre 1 – 1.5 cm long with outer bracts rather broadly ovate, broadly white-hyaline-margined, with a dark, extended keel-like appendage at apex. Achenes 4-sided with rounded angles, deeply 4-grooved longitudinally as if composed of 4 coalescent ribs, strongly rugose-tubercled transversely, c. 2 mm long, brownish grey, truncate at apex; pappus very caducous and air-buoyant, of very fine white numerous hairs c. 6 – 8 mm long and coalescent at base in a ring open at center. *Plate 234.*

Rather closely resembling a *Launaea* in outward appearance but differing in its reddish-tinged florets, broad heads, and thickish stems.

Habitats: Sands or sandy-silts; perhaps more common around rocky terrain. Frequent.

Northern Plains: 1587(BM), al-Batin/Wadi al-'Awja junction; 880, 9 km S 'Ayn al-'Abd; 888, Jabal al-'Amudah; 1240, Jabal an-Nu'ayriyah; 1270(BM), 15 km ENE Qaryat al-'Ulya; 1542(BM), 36 km NE Hafar al-Batin.

Northern Summan: 716, 33 km SW al-Qar'ah wells; 782, 69 km SW al-Qar'ah wells; 3340, 5 km WSW an-Nazim; 3834, ash-Shayyitat, 27 – 29N, 47 – 22E.

Central Coastal Lowlands: 403, Dhahran; 3253, 8 km W Qatif-Qaisumah Pipeline KP – 144; 7723, 2 km S Ras Tanaqib, 27 – 49N, 48 – 53E.

Vernacular Names: MAKNĀN (Al Murrah), ḤALAWLA' (Shammar), MURĀR (Kuwait, Dickson 1955).

● 43. **Crepis** L.

Annual or perennial herbs with ligulate heads and leaves often rosulate. Involucre with an outer row of short bracts and an inner 1 or 2 rows of mostly equal larger bracts indurating at maturity and enclosing the outer achenes calyx-like. Achenes striate, dimorphic, beaked or tapering above, with a pappus of fine white hairs.

Crepis aspera L.

Annual herb with stems erect and usually rough with rigid bristles, sometimes hardly prickly, 10–50 cm high. Basal leaves rosulate, oblong to oblanceolate, dentate to runcinate-pinnatifid, with acute lobes, tapering to a petiole, to c. 15 cm long; stem leaves oblong-oblanceolate, sessile, half clasping at base, deeply dentate to pinnatifid, with acute lobes. Heads ligulate, with florets yellow, the involucre 6–8 mm long, constricted above the middle in fruit and somewhat reflexed at apex. Outer involucral bracts short, membranous, the inner ones linear-lanceolate, acute, close-tomentose at back and furnished with scattered, yellowish, rigid bristles, the bracts hardening in maturity and clasping the outer achenes. Outer achenes inconspicuous, often epappose, winged ventrally, the inner ones sublinear, somewhat compressed, striate, 1–2 mm long, with a beak exceeding the achene and terminating in a 3–4 mm-long pappus of equal white hairs.

Our desert specimens have strongly prickly stems (var. *aspera*); the one collected as a garden weed was practically devoid of stem bristles (var. *inermis* (Cass.) Boiss.)

Habitat: Occasional in the northern desert or as a weed around gardens or farms. Rare.

Northern Summan: 3892, 14 km ESE Umm 'Ushar; 3867, 15 km SSE Samudah, 27–44N, 45–00E (slightly outside our area).

Central Coastal Lowlands: 7418(BM), Dhahran, garden weed.

Monocotyledoneae

Plants with one seed leaf (cotyledon) when embryonic and with vascular bundles scattered in the stem, parallel-veined leaves, and floral parts usually in threes or multiples of three.

60. HYDROCHARITACEAE

Aquatic herbs with leaves alternate, paired, or whorled. Flowers unisexual or bisexual, regular, 3-merous, enclosed in a spathe or between 2 bracts, with stamens 1 to numerous. Ovary inferior, with (2)3–6(15) carpels and numerous ovules. Fruit usually a capsule.

● Halophila Thou.

Submerged marine perennial herbs with leaves in pairs. Flowers unisexual, enclosed in a sheath, the male ones with 3 perianth lobes and 3 stamens, the female with 3 (or 2–5) minute perianth lobes and an ovary with 3 (or 2–5) filiform stigmas and numerous ovules. Fruit membranous.

Two species of this genus growing submerged in near-coastal waters of the Arabian Gulf, with *Halodule* (representing a different family), are the main constituents of ecologically and economically important 'seagrass' communities (see chapter 6).

Key to the Species of Halophila

1. Leaves elliptical-oblong, 1–2 cm long, on petioles equalling or exceeding
   the blade . . . . 1. *H. ovalis*
1. Leaves linear-oblong, 4–6 cm long, on petioles much shorter than the
   blade . . . . 2. *H. stipulacea*

1. **Halophila ovalis** (R. Br.) Hook. f.

Perennial herb with spreading, elongating rhizomes in the sea bottom. Leaves ascending in pairs from nodes in the buried rhizome; blades oblong to obovate-elliptical, entire, mostly 10 – 20(22) mm long, 6 – 9(10) mm wide, on petioles (in ours) mostly about equalling or somewhat exceeding the blade, with small sheaths at their insertion at the nodes.

Habitat: Near-coastal, soft, sea-bottom at depths usually between 2 and 15 m. Locally frequent. Detached leaves often seen on beaches, particularly after storms. Less frequent than *H. stipulacea* and *Halodule*, and often found deeper than they are, on softer mud bottoms.

Central Coastal Lowlands: 2918(BM), drifted leaves at base of Dammam Port causeway; 2920(BM), 1 km SE Za'l Island, Tarut Bay, in water 2 m deep; Tarut Bay, specimens collected by Enrivonmental Unit, Saudi Arabian Oil Company.

See chapter 6 for ecological relationships and economic notes.

2. **Halophila stipulacea** (Forssk.) Aschers.

Perennial herb with rhizome creeping in the sea bottom and leaves erect or ascending. Leaves paired, often rather approximate, linear-oblong to lanceolate-oblong, obtuse, mostly (3)4 – 6 cm long, 6 – 9 mm wide, with short flat petioles hidden in basal sheaths on the rhizome nodes.

Habitat: Soft sea bottom in near-coastal waters, mostly at depths of 1 to 15 m. Locally frequent or dominant. Detached leaves often seen on beaches after storms.

Central Coastal Lowlands: 1094(BM), Ras Tanura, detached plants drifted onto beach; 2917(BM), Dammam Port causeway, drifted pieces; 2921(BM), Tarut Bay, 1 km SE Za'l Island, growing in water 2 m deep; Tarut Bay, specimens collected by Environmental Unit, Saudi Arabian Oil Company.

See chapter 6 for ecological relationships and economic importance.

61. NAJADACEAE

Submerged aquatic herbs with linear leaves opposite or appearing whorled. Flowers solitary, unisexual, the male with 2 perianth segments and a single stamen, the female consisting of a single, usually naked, ovary with branched stigmas. Fruit indehiscent.

● Najas L.

Submerged, freshwater annuals with opposite or false-whorled, linear leaves. Flowers unisexual, sessile in the leaf axils. Other characters as for the family.

Najas graminea Del.

Submerged, freshwater herb rooting from the nodes. Leaves soft, appearing whorled, narrowly linear and nearly filiform, mostly (in ours) 10 – 18 mm long, c. 0.7 – 0.9 mm wide, acute, often 2-nerved, nearly entire or faintly antrorsely serrulate-toothed with (in ours) c. 15 – 20 teeth on each margin. Flowers sessile in the axils, unisexual, the male with a single sheathed stamen, the female with a naked, oblong-ellipsoid ovary with stigma branched near the apex.

Triest, in his revision of Old World *Najas* (1988, 141), lists our No. 3937(BM) as this species. No. 3934 (upon which the above leaf description is based) had leaves with marginal teeth characteristic of the species in form but resembling *N. minor* All. in number.

Habitat: Bottoms of freshwater springs in the oases. Occasional.

South Coastal Lowlands: 3934, 'Ayn Mar'ah, al-Hasa Oasis, rooted at 1.3 m depth; 3937(BM), 'Ayn Umm al-Lif, al-Hasa Oasis, submerged.

62. POTAMOGETONACEAE

Submerged, emergent, or floating herbs of fresh or brackish water. Leaves alternate or opposite, sheathing at base. Flowers bisexual, regular, usually with reduced perianth. Stamens 4; ovary superior, of 4 carpels. Fruit 1-seeded.

● Potamogeton L.

Immersed aquatic plants of fresh to brackish water with flowers usually emergent at anthesis and leaves sheathed at base. Flowers bisexual, in spikes; perianth lobes 4; stamens 4; carpels 4. Fruits of 1 – 4 somewhat drupaceous nutlets.

Key to the Species of Potamogeton

1. Leaves elliptical-oblong, petiolate, 1.5 – 2.5 cm wide; spike dense,
   continuous .... 1. *P. nodosus*
1. Leaves very narrowly linear or filiform, under 3 mm wide; flowers clustered
   on an interrupted spike .... 2. *P. pectinatus*

1. **Potamogeton nodosus** Poir.

P. natans L.

Robust, long-stemmed, aquatic perennial of fresh water. Leaves mostly floating, elliptic to oblong, sometimes lanceolate or subovate, mostly acute, sometimes subcordate at base, the blade leathery, mostly 4 – 8(15) cm long, 1.5 – 2.5 cm wide, entire, on rather thick petioles about equalling or exceeding the blade. Flowers in a dense, thick-peduncled, cylindrical spike 2 – 4(5) cm long. Perianth lobes broadly rhomboid. Fruit a short-beaked, turgid nutlet c. 3 mm long.

Habitat: Irrigation channels in the oases. Our one specimen was from a slow-flowing unlined irrigation channel. The status of this species in the concrete irrigation and drainage system later installed in al-Hasa requires investigation.

South Coastal Lowlands: 522(BM), irrigation channel near al-Fudul, al-Hasa Oasis, 1966.

2. **Potamogeton pectinatus** L.

Immersed aquatic perennial herb. Leaves submerged or floating, very narrowly linear to filiform, 4 – 7 cm long, with stipules adnate to the blade at base forming a basal sheath. Flowers clustered in interrupted, slender-peduncled, narrow spikes.

Habitat: Irrigation canals of the oases. Apparently infrequent.

Central Coastal Lowlands: 524(BM), irrigation canal near Darin, Tarut Island.

63. RUPPIACEAE

Submerged aquatic herbs of salt or brackish waters. Flowers bisexual, regular, without perianth. Stamens 2; carpels 4 or more. Fruits indehiscent on elongating stipes.

● Ruppia L.

Submerged aquatic herb with subfiliform leaves. Flowers as for the family.

Ruppia maritima L.
Submerged aquatic herb of the sea or brackish inland waters. Leaves very narrowly linear to filiform, c. 5 – 12(15) cm long, 0.5 mm wide, sheathed at base. Flowers emerging from the sheaths on much elongating peduncles. Fruits obliquely ovoid with a short beak, 2 – 3 mm long on stipes 5 – 15 mm long, 2 – 6 together on a common peduncle, thus appearing umbellate.
Habitat: Brackish springs and drainage channels or shallow, almost intertidal, sea water.
Northern Plains: 521(K,BM), 839(BM), channels from 'Ayn al-'Abd (a brackish, sulfurous spring).
Central Coastal Lowlands: 8819, brackish water hole 8 km W Abqaiq; Tarut Bay, specimens collected by the Environmental Unit, Saudi Arabian Oil Company. Also reported in beach shallows in the vicinity of Manifah and other more or less protected Gulf coastal points.

64. ZANNICHELLIACEAE

Submerged aquatic herbs with entire leaves. Flowers unisexual, axillary, solitary or cymose. Perianth absent or of 3 scales; stamens 1 – 3; carpels 1 – 9, superior. Fruit indehiscent.

● Zannichellia L.

Submerged, monoecious, aquatic herbs with filiform, sheathing leaves. Flowers paired, with 1 staminate and 1 pistillate adjacent in the axil. Male flowers without perianth, 1-stamened, the female with a cupular perianth and 4 – 8 carpels. Fruit falcate.

Zannichellia palustris L.
Submerged aquatic herb of fresh or brackish waters. Leaves narrowly linear to filiform, 2.5 – 6 cm long, 0.5 – 2 mm wide. Fruits (2)4(6), subsessile or pedicelled in an axillary umbel, somewhat curved, beaked, often crenate or dentate at back.
Habitat: Irrigation channels in the oases.
Central or South Coastal Lowlands: Not seen by the author but reported by Chaudhary, Parker and Kasasian (1981) as an aquatic weed of eastern Saudi Arabia.

65. CYMODOCEACEAE

Submerged marine perennials with creeping rhizome and leaves linear, flat or terete, sheathing at base. Flowers unisexual (the plants monoecious or dioecious), the staminate with 2 stamens and threadlike pollen; the pistillate of 1 or 2 sessile or stipitate carpels. Fruit a 1-seeded nutlet.

Key to the Genera of Cymodoceaceae
1. Leaf blades flat; anthers unequal; stigma 1 . . . . 1. *Halodule*
1. Leaf blades terete; anthers equal; stigmas 2 . . . . 2. *Syringodium*

● 1. **Halodule** Endl.

Submerged marine plant with narrow linear leaves. Flowers unisexual, the staminate pedicelled with 2 unequal stamens, 1 clearly exserted beyond the other; the pistillate sessile.

Halodule uninervis (Forssk.) Aschers.
Diplanthera uninervis (Forssk.) Williams

Submerged marine perennial herb, often forming more or less dense beds with spreading rhizomes in the sea bottom. Leaves clustered 2 – 4(5) at the nodes, ascending, narrowly linear, 10 – 15(25) cm long, 2 mm wide, more or less truncate and dentate at apex.

Habitat: Soft, near-shore sea bottoms at depths of 1 – 15 m. Probably the most abundant marine angiosperm in the Gulf and often dominant in 'seagrass' beds. Locally common.

Central Coastal Lowlands: 1095(BM), Ras Tanura beach (detached pieces); 2915(BM), Dammam Port causeway, growing at less than 1 m depth (low tide); 2919(BM), Tarut Bay, 1 km SE Za'l Island, growing with *Halophila* at 2.5 m depth; Tarut Bay, specimens collected by Environmental Unit, Saudi Arabian Oil Company.

See chapter 6 for ecological relationships and economic notes.

● Another species of *Halodule*, *H. wrightii* Aschers., has been recorded from Bahrain (Cornes 1989) and may be found in Saudi Arabian waters. Den Hartog (1970) notes that both species are found in the Gulf. According to his descriptions, *H. uninervis* has leaf blades up to 3.5 mm wide with tridentate tip, anthers 2 – 3 mm long, styles 28 – 42 mm long and fruit 2 – 2.5 mm long. *H. wrightii* has leaves 1 mm or less wide with bicuspidate tip, anthers 3.5 – 5 mm long, style 10 – 28 mm long and fruit 1.5 – 2 mm long. Our specimens have not been examined in a flowering state, but their leaf characteristics place them in *H. uninervis*.

● 2. **Syringodium** Kützing

Submerged marine plant with fine, subulate leaves. Staminate flowers stalked, with 2 anthers connate dorsally below. Pistillate flowers sessile with paired stigmas.

Syringodium isoetifolium (Aschers.) Dandy
Cymodocea isoetifolia Aschers.

Submerged marine dioecious perennial with 2 – 3 leaves, sheathed at base, arising from each node of a creeping rhizome. Leaf sheaths 1.5 – 4 cm long, often tinged reddish; blades subulate, 7 – 30 cm long, 1 – 2 mm wide. Leaves of the inflorescence much shortened. Staminate flowers on stalks c. 7 mm long, with paired, equal anthers c. 4 mm long; pistillate flowers sessile, with ovary ellipsoid, 3 – 4 mm long and stigmas 2, 4 – 8 mm long. Fruit ellipsoid, oblique, 3.5 – 4 mm long, with a 2 mm bifid beak. Description after den Hartog (1970, 178).

Habitat: Mainly on mud bottom in shallow water, sometimes mixed with other marine angiosperms.

Central Coastal Lowlands: Apparently not yet recorded from Saudi Arabian parts of the Gulf but reported by den Hartog (1970) from Bahrain, where it was collected on a coral reef at al-Jufayr by Good in 1950. It will almost certainly be found to occur also in Saudi Arabian waters.

● Another Cymodoceaceous species, *Thalassodendron ciliatum* (Forssk.) den Hartog (= *Zostera ciliata* Forssk., *Cymodocea ciliata* (Forssk.) Ehrenb. ex Aschers.), is listed by den Hartog (1970) from Iran at Chah Bahar on the northern coast of the Gulf of Oman. Its range might be found to extend into the Gulf although records are so far lacking. It has broader, flat, falcate leaves, 7 – 12 mm wide and up to c. 13 cm long, basally sheathed and clustered usually c. 6 on a longish common stalk bearing numerous ringlike leaf scars. The leaves are truncate and finely toothed at apex, serrate on the upper lateral margins.

66. GRAMINEAE (Poaceae)

Annual or perennial herbs, sometimes shrubby, with stems (culms) cylindrical, hollow in the internodes. Leaves distichous, clasping the culms below with split sheaths, spreading above, with a ventral outgrowth (the ligule) at the junction of sheath and lamina. Flowers usually bisexual, with perianth greatly reduced or apparently absent; stamens mostly 3; ovary superior, usually with 2 styles and plumose stigmas. Fruit a caryopsis, consisting of a seed adnate to the pericarp. The basic unit of the inflorescence is the spikelet, typically consisting of 1 to several flowers (florets), each compressed between 2 bracts (the lemma and the palea) and all subtended below by 2 empty bracts (the glumes).

The grasses comprise the largest family of flowering plants in eastern Arabia in terms of species number and are represented in our territory by about 91 species in 52 genera. Only about half of the species, however, are desert plants, the remainder being weeds seldom seen outside cultivated areas or other disturbed lands. The family is of considerable local economic and ecological importance, including many species grazed both by domestic livestock and wildlife. Several grasses, such as *Panicum turgidum*, are dominant perennials of important plant communities. Other perennials prominent in the desert vegetation are *Stipagrostis* spp., *Pennisetum divisum*, *Lasiurus scindicus*, *Stipa capensis*, *Hyparrhenia hirta*, and *Cymbopogon commutatus*. *Schismus barbatus*, *Rostraria pumila*, and *Cutandia memphitica* are abundant and widespread desert annuals.

Key to the Genera of Gramineae

The genera of grasses are keyed here in three artificial groups based on generalized inflorescence forms. Successful use requires careful distinction among the technical terms for the spikelet parts and larger inflorescence units to avoid confusing, for example, glumes for lemmas or a multi-flowered spikelet for a spike. Examination should begin with a circumscription of the spikelet, then of the larger units.

Group Key to the Gramineae

1. Inflorescence composed of single or multiple spikes or of compact, spikelike racemes or panicles
    2. Spikelets in a simple terminal spike or in a dense, more or less cylindrical compact raceme or panicle resembling a spike (including inflorescences with a single spikelet or with few sessile spikelets) ... GROUP A (Fig. 8.1)
    2. Spikelets in 2 or more spikelike groups .... GROUP B (Fig. 8.2)
1. Inflorescence not composed of 1 or more spikelike units; spikelets inserted singly or in pairs in an open or contracted branching, paniculate inflorescence .... GROUP C (Fig. 8.3)

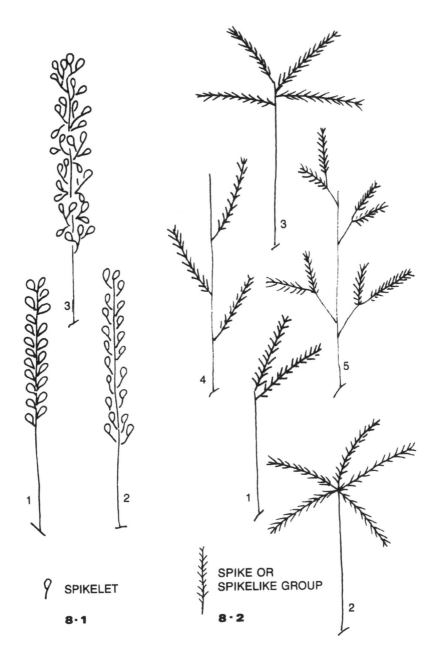

SPIKELET

8·1

**SPIKE OR
SPIKELIKE GROUP**

8·2

Fig. 8.1. Group A grass inflorescences, spikelets in a simple, terminal, spike or dense, spikelike group. Schematics: 1 — simple terminal spike, 2 — spikelike raceme, 3 — spikelike panicle.

Fig. 8.2. Group B grass inflorescences, spikelets in two or more spikes or spikelike groups. Schematics: 1 — spikes in a single terminal pair, 2 — spikes digitate, 3 — spikes subdigitate, 4 — spikes inserted along main axis, 5 — spikes in multiple pairs.

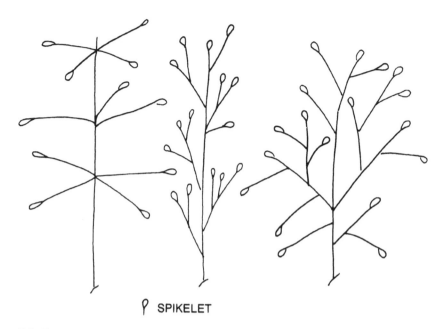

♀ SPIKELET

8.3. Group C grass inflorescences, spikelets single or paired in a panicoid inflorescence.

Keys to the Genera, by Groups
Group A (Fig 8.1)

Spikelets in a simple terminal spike or in a dense, more or less cylindrical compact raceme or panicle resembling a spike.

1. Spike only 2–2.5(3) mm broad, very dense; spikelets 2 mm or less
   long .... 47. *Sporobolus*
1. Spike (or spicoid raceme or panicle) more than 3 mm wide; spikelets longer
   than 2 mm
     2. Inflorescence a peculiar spicate raceme composed of about 10–15 male
        spikelets above an enlarged white bony single false fruit that is c. 9 mm
        long, 6 mm wide .... 9. *Coix*
     2. Inflorescence not as above
          3. Spikelets with a distinct circular pit above the middle of the glossy
             outer glume .... 5. *Dichanthium*
          3. Spikelets without pit on back of glume
               4. Lemmas with 9 plumose awns at apex .... 38. *Enneapogon*
               4. Lemmas not with 9 awns
                    5. Spikelets inserted in notches or steps in the axis
                         6. Spikelets solitary, alternate
                              7. Spikelets immersed in longitudinal cavities of the cylindrical
                                 rachis thus forming a narrow green spike resembling a
                                 jointed culm .... 45. *Parapholis*
                         7. Spikelets standing outside the rachis

    8. Glumes with 3 parallel awns .... 50. *Aegilops*
    8. Glumes awnless or 1-awned
        9. Spikelets (except the terminal one) with lower glume
           absent .... 41. *Lolium*
        9. Spikelets with both glumes present
           10. Spike 3 – 5 cm long, compressed in a single plane; glumes
               acuminate with outer nerves convergent at apex .. 52. *Eremopyrum*
           10. Spike 6 – 18 cm long, squarish in cross-section; glumes ovate
               or oblong with outer nerves separated at apex .... 51. *Triticum*
  6. Spikelets in groups of 2 or 3; spike dense
    11. Spikelets hairy; ligule replaced by a rim of hairs .... 4. *Lasiurus*
    11. Spikelets not hairy; ligule membranous .... 49. *Hordeum*
5. Spikelets not inserted in notches or steps in the axis
  12. Spikelets bearing one or more bristles (not soft hairs) at base of glumes
    13. Bristles at base of glume joined at base into a disc or cup, numerous,
        the inner ones densely ciliate .... 10. *Cenchrus*
    13. Bristles at base of glume free to their bases, one to several, or
        numerous, scabridulous but not ciliate
        14. Bristles single, or in groups of 2 – 3, or in groups of 6 – 8,
           persistent after falling of spikelets; annual weeds of cultivated
           areas or disturbed ground .... 12. *Setaria*
        14. Bristles numerous, falling with the spikelet; shrubby desert
           perennial .... 11. *Pennisetum*
  12. Spikelets not bearing bristles at base of glumes
    15. Glumes bearing long silky hairs on their backs or base
        16. Inflorescence whitish; glumes not awned .... 1. *Imperata*
        16. Inflorescence pale green; glumes awned .... 21. *Polypogon*
    15. Glumes without long silky hairs on their backs
        17. Ligule membranous
           18. Lemmas unawned; glumes with winglike keels .. 46. *Phalaris*
           18. Lemmas awned; glumes not winged
               19. Glumes connate in lower ⅓ to ½; awn inserted near
                  base of the lemma .... 20. *Alopecurus*
              19. Glumes free to base; awns inserted at or near apex of
               the lemma
                 20. Glumes with 3 parallel awns .... 50. *Aegilops*
                 20. Glumes unawned (but sometimes acuminate)
                    21. Awns less than 1.5 cm long
                       22. Spikelets 3 – 6 mm long,
                          numerous .... 27. *Rostraria*
                       22. Spikelets 15 – 30 mm long, only 1 – 3(5)
                          per culm .... 28. *Brachypodium*
                 21. Awns 6 cm or more long .... 48. *Stipa*
        17. Ligule replaced by hairs
           23. Glumes much shorter than the groups of
               lemmas .... 19. *Aeluropus*
          23. Glumes as long as the lemmas .... 32. *Schismus*

Group B (Fig. 8.2)

Spikelets in 2 or more spikes or spikelike racemes (the latter sometimes referred to in the key below as 'spikes').

1. Spikes terminal, digitate or near-digitate
    2. Spikes 2 (if spikelets strongly laterally compressed and 3 – 6-flowered follow 2nd lead of this couplet)
        3. Spikelets awnless .... 15. *Paspalum*
        3. Spikelets awned .... 6. *Ischaemum*
    2. Spikes 3 or more
        4. Spikelets with awns
            5. Awns kneed .... 5. *Dichanthium*
            5. Awns straight .... 30. *Chloris*
        4. Spikelets awnless (sometimes with short aristate point)
            6. Spikes 1 – 2(3) mm wide, flexible
                7. Lower glume reduced to a minute, triangular scale .. 13. *Digitaria*
                7. Both glumes at least ⅓ to ½ as long as the spikelet
                    8. Spikelets 2- (appearing 1-) flowered .... 31. *Cynodon*
                    8. Spikelets 3 – 5-flowered .... 35. *Eleusine*
            6. Spikes wider, stiff or hardened
                9. Spikes surpassed by naked extremity of axis; ligule membranous .... 34. *Dactyloctenium*
                9. Spikes not exceeded by axis; ligule a rim of hairs .... 36. *Ochthochloa*
1. Spikes spaced along principal axis of the inflorescence
    10. Spikelets awnless (or with awn less than 2 mm long)
        11. Spikelets several-flowered
            12. Ligule membranous .... 37. *Leptochloa*
            12. Ligule of hairs .... 19. *Aeluropus*
        11. Spikelets 1-flowered
            13. Apparent pedicel immediately below the spikelet (actually the lowest rachilla node) enlarged as a whitish, short-cylindrical callus somewhat broader than long .... 18. *Eriochloa*
            13. Pedicel of usual aspect, not enlarged as a callus
                14. Glumes acuminate or pointed; ligule totally absent .... 17. *Echinochloa*
                14. Glumes not acuminate; ligule a ring of hairs ... 14. *Brachiaria*
    10. Spikelets awned
        15. Awns kneed; desert plants
            16. Spicate racemes single on the inflorescence branches .... 5. *Dichanthium*
            16. Spicate racemes paired on the inflorescence branches
                17. Spicate racemes of each pair diverging at an obtuse angle (often nearly 180°); plant aromatic .... 2. *Cymbopogon*
                17. Spicate racemes of each pair nearly parallel or diverging at an acute angle; plant not aromatic .... 3. *Hyparrhenia*
        15. Awns straight; weeds or ruderals .... 17. *Echinochloa*

Group C (Fig. 8.3)

Inflorescence not composed of 1 or more spikelike units; spikelets single or in pairs in an open or contracted branching, paniculate inflorescence.

1. Lemma with 3 awns or with 1, 3-branched awn
  2. All 3 awns (or awn branches) naked
    3. Spikelets 1-flowered; awns nearly equal, emerging apically from a single point .... 22. *Aristida*
    3. Spikelets several-flowered; central awn longer; awns parallel, inserted separately below the apex .... 29. *Bromus*
  2. Central branch of the awn plumose, the 2 lateral branches naked, much shorter .... 23. *Stipagrostis*
1. Lemma awnless or with 1 simple awn
  4. Glumes surpassed by the lemmas or groups of lemmas
    5. Lemma awnless
      6. Ligule a ciliate rim or rim of hairs
        7. Lemma or rachilla joint with long silky hairs, 7 – 10 mm long .... 24. *Phragmites*
        7. Lemma and rachilla joint without hairs
          8. Spikelets with 1 floret .... 47. *Sporobolus*
          8. Spikelets of 3 or more florets
            9. Coarse seashore perennial, to 1 m or more high .... 40. *Halopyrum*
            9. Desert or weed annuals less than 50 cm high
              10. Glumes equalled or only slightly exceeded in length by the group of lemmas; dwarf annual less than 10 cm high .... 32. *Schismus*
              10. Glumes several times shorter than the groups of lemmas; plant usually exceeding 15 cm .. 39. *Eragrostis*
      6. Ligule membranous
        11. Glumes less than ¼ as long as the spikelet .... 44. *Sphenopus*
        11. Glumes at least ½ as long as the spikelet
          12. Pedicels capillary, flexible .... 43. *Poa*
          12. Pedicels not capillary, or absent
            13. Pedicels absent, with spikelets subsessile on straight, spikelike branches .... 37. *Leptochloa*
            13. Pedicels thickened, stiff; inflorescence intricately and dichotomously branched .... 42. *Cutandia*
    5. Lemmas awned
      14. Awn terminal; reed over 1 m high .... 24. *Phragmites*
      14. Awn inserted slightly below apex of lemmas; plants under 30 cm high
        15. Awns shorter than lemma .... 27. *Rostraria*
        15. Awns at least as long as lemma .... 29. *Bromus*
  4. Glumes, at least one, as long as the lemmas or exceeding them (disregarding awns on lemmas when present)
    16. Stout reedlike perennial to 3 m or more; culms 1 cm or more thick; leaf blades 2 – 5 cm broad .... 25. *Arundo*

16. Grasses not reedlike, with smaller culms, narrower leaves
   17. Spikelets with only 1 fertile floret
      18. All spikelets awnless
         19. Glumes smooth, membranous, rounded at back, the upper equalling the spikelet, the lower much shorter ... 16. *Panicum*
         19. Glumes scabridulous-keeled, not membranous, equal or subequal and both c. twice as long as the floret . 21. *Polypogon*
      18. Spikelets, at least some, with awns
         20. Spikelets grouped in 3s, 1 sessile, 2 peduncled .. 7. *Chrysopogon*
         20. Spikelets grouped in pairs (1 sessile, 1 peduncled), or single
            21. Spikelets paired: 1 sessile, awned; the other peduncled, awnless; awns less than 3 cm long .... 8. *Sorghum*
            21. Spikelets inserted separately, all alike; awns more than 6 cm long .... 48. *Stipa*
   17. Spikelets with more than one fertile floret
      22. Lemmas unawned .... 32. *Schismus*
      22. Lemmas awned
         23. Ligule a rim of hairs; awn terminal .... 33. *Centropodia*
         23. Ligule membranous; awn inserted on back of the lemma .... 26. *Avena*

● 1. **Imperata** Cyr.

Perennial grasses with long leaf blades and erect culms. Inflorescence a solitary, terminal, erect spicate panicle of short-pedicelled, long silky-hairy spikelets. Spikelets 2-flowered, paired, the upper of each pair fertile. Glumes subequal, as long as the spikelet.

Imperata cylindrica (L.) P. Beauv.

Glabrescent perennial grass with scaly rhizome and erect culms to over 1 m high. Leaf blades convolute below, flat above, narrowly linear, c. 3 mm wide (in ours), to over 60 cm long, antrorsely scabridulous at margins; junction with the sheath indistinct, low in the plant, the sheath often free of the culms above; ligule membranous. Panicle spikelike, cylindrical, silvery-silky, 8 – 15(20) cm long, c. 1.5 – 2 cm wide. Spikelets c. 4 mm long, linear-lanceolate, grouped in pairs in which the upper is longer-pedicelled than the lower, all long silky-hairy, mostly from the base but also from the glumes, the hairs straight, ascending, 2 – 3 times as long as the spikelet. Glumes nearly equal, membranous. Stigmas exserted; anthers c. 2.5 mm long.

  Easily recognized by the very soft, silky, spicate inflorescence.

Habitat: A weedy species found around cultivated areas or waste ground, abandoned gardens, and banks. Locally frequent.

Central Coastal Lowlands: 440(BM), Dhahran; 547(K), S al-Qatif town; C. Parker No. 55(BM), al-Qatif Oasis.

South Coastal Lowlands: 8200, near 'Ayn Umm Sab', al-Hasa Oasis.

Vernacular Names: ḤALFA' (al-Hasa farmer).

● 2. **Cymbopogon** Spreng.

Aromatic perennial grasses with a paniculate inflorescence consisting of paired racemes, one sessile and one peduncled. Spikelets also paired, one sessile and one pedicelled, inserted on the joints of the racemes. Sessile spikelet bisexual, fertile, awned from the lemma; pedicelled spikelet male or neuter, awnless. Glumes subequal, exceeding the lemma.

Cymbopogon commutatus (Steud.) Stapf
C. parkeri Stapf

Perennial grass, strongly aromatic in vegetative parts, densely tufted at base with erect culms up to 1 m high. Leaf blades linear, subfiliform, somewhat curled; ligules scarious. Panicle terminal, erect, more or less spatheate at base, with rather distant pairs of diverging racemes. Lowest joint of the sessile raceme prominently swollen and hardened. Awn kneed, with a spirally twisted column, exserted for about the length of the spikelet. *Plate 235.*

● Always recognizable, even when not flowering, by the distinct, sweet, lemony odor of the crushed foliage.

Bor (in Townsend and Guest 1968) maintains *C. parkeri* distinct from *C. commutatus* primarily on the basis of differences in vegetative parts. Some Arabian records of this plant have apparently been much confused in the literature with *C. schoenanthus* (L.) Spreng., which also occurs in Arabia but which seems to be limited to Sudanian floristic territory. The following differential characters are listed by Dr T. Cope, Royal Botanic Gardens, Kew (personal communication):

> *C. commutatus:* lower glume of sessile spikelet with the margins sharply inflexed above, but rounded below; awn well developed, geniculate, with a distinct, spirally twisted column.
>
> *C. schoenanthus:* lower glume of sessile spikelet with the margins sharply inflexed throughout; awn poorly developed, flexuous but not really geniculate, scarcely differentiated into a column and limb.

Bor *(op. cit.)* separates *C. schoenanthus* from *C. parkeri* by the glabrous basal sheaths and glabrous lower glumes in the sessile spikelets of the former.

To date, apparently only *C. commutatus* has been collected in the Eastern Province. *C. schoenanthus*, if present, would be more likely found in the southern parts of our territory.

Habitat: Rocky ground, especially along beds of rocky channels in the Summan. Occasional to locally frequent.

Northern Plains: 3120(BM), tributary to al-Batin, 28–01N, 45–33E.

Northern Summan: 1090(BM), edge of the Dahna SW ash-Shumlul (leg. Longhitano); 2772(BM), 25–50N, 48–00E; 2838(BM), 32 km ESE ash-Shumlul; 2858(BM), 26–01N, 48–27E; 2874(BM), 20 km W 'Uray'irah.

Southern Summan: 2240(BM), 11 km ENE al-Hunayy; 1874, 32 km S Jaww al-Hawiyah; 8372, al-Ghawar, 38 km S 'Uthmaniyah GOSP–13.

Vernacular Names and Uses: SAKHBAR ('Utaybah, ad-Dawasir), IDHKHIR (ad-Dawasir), KHAṢAB (Qahtan), ḤAMRA' (Rawalah). Various species of *Cymbopogon* are collected or cultivated in some parts of the world, including India and Pakistan, for distillation extraction of the aromatic oil known as citronella, or lemon grass oil. This is used in perfumery and as an insect repellent. *C. commutatus* certainly contains this aromatic principle, but there is no record of its collection for extraction in northeastern Arabia. The author, however, has seen small bunches of *Cymbopogon* sold in an herbalist shop in Dammam, and it certainly has a history of medicinal use, probably as an infusion or tea, in Arabia.

● 3. **Hyparrhenia** Anderss.

Annual or perennial grasses. Inflorescence a panicle, spatheate at base and branch points, composed of few to numerous peduncled pairs of racemes. Spikelets paired in the racemes, with one sessile, bisexual, awned, the other pedicellate, male, unawned. Glumes equal, equalling the spikelet; lemma of the fertile spikelet awned from an apical sinus.

Hyparrhenia hirta (L.) Stapf

Perennial grass, densely tufted at base, with erect culms to c. 75 cm high. Leaf blades linear, long-tapering, to c. 30 cm long, 2 – 3 mm wide, mostly glabrous except at and immediately above the ligule, where ciliate-hairy; ligule membranous, more or less laciniate. Panicle loose to moderately dense, 8 – 20 cm long, spatheate at base and branches, contracted, with raceme pairs peduncled on ascending, often 1-noded branches to c. 7 cm long. Racemes c. 1.5 – 2.5 cm long, c. 2 mm wide (excluding awns), 1 of each pair sessile, 1 shortly peduncled. Spikelets c. 5 mm long, white-hirsute on the glumes and pedicels, the fertile one with a fine, brown twisted awn c. 25 mm long and often kneed twice. *Plate 236.*

Habitat: Usually on rocky ground, in shallow sand in limestone crevices. Locally frequent.

Northern Plains: 3108(BM), al-Batin, 28 – 00N, 45 – 28E.

Northern Summan: 7941, 25 – 39N, 48 – 48E.

Central Coastal Lowlands: 166(K,BM), 176(K,BM), Dhahran.

● 4. **Lasiurus** Boiss.

Perennial grasses, somewhat woody at base, with stiff leaf blades and ligule a rim of hairs. Inflorescence a solitary, terminal, easily disarticulating spikelike raceme. Spikelets in 2s or 3s, sessile and pedicellate, unawned, 2-flowered, ciliate-hirsute. Glumes unequal, acuminate.

Lasiurus scindicus Henr.

L. hirsutus (Vahl) Boiss.

Erect perennial grass to c. 1 m high with culms simple or branched, woody-rhizomatous below. Leaf blades hard, to 12 cm or more long; ligule a rim of hairs. Racemes terminal, erect, spicate, dense, breaking at joints, 5 – 10 cm long, silky-hirsute. Spikelets c. 6 – 10 mm long, the pedicelled ones sometimes shorter; rachis joints densely hirsute. Glumes unequal, cilitate-hirsute from the margins and keels.

This plant has been long and widely known as *L. hirsutus* (Vahl) Boiss. Cope (1980) has shown that this name combination is invalid and that the grass is in fact conspecific with Henrard's *L. scindicus*.

Habitat: Usually on rocky ground or shallow silty-sandy soils. Frequent and widespread but seldom dominant except very locally.

Central Coastal Lowlands: 160(K), 228, 429, 1079(BM), Dhahran.
Vernacular Names and Uses: ḌAʻAH (gen.), ḌUʻAYY (Al Wahibah), HAḌĪD ('Utaybah, Sulabah, Qahtan). Grazed by livestock.

● 5. Dichanthium Willem.

Annual or perennial grasses with an inflorescence of terminal, subdigitate or single-spicate racemes. Spikelets in pairs of one sessile, one pedicelled, the sessile bisexual and awned. Glumes equal or subequal.

Key to the Species of Dichanthium
1. Raceme solitary, terminal, or single-terminal on each of several branches spaced along the main axis; spikelets with a distinct circular pit above middle of the outer glume .... 2. *D. foveolatum*
1. Racemes 3–9 together, inserted subdigitately in a terminal umbel; glumes without pit on back .... 1. *D. annulatum*

1. **Dichanthium annulatum** (Forssk.) Stapf
Tufted perennial grass up to c. 1 m high with hair tufts at nodes of the erect or ascending culms. Ligule scarious. Inflorescence usually of 3–9 spicate, often purplish, racemes inserted terminally and subdigitately on the culm. Spikelets subimbricate, in pairs of 1 sessile, 1 pedicellate, the sessile c. 4 mm long with a fine kneed awn several times as long as the spikelet. Glumes with longish, fine spreading hairs from the angles.
Habitat: A grass primarily of disturbed ground such as field edges or ditch banks. Probably occasional to very locally frequent.
Central or South Coastal Lowlands: Not collected by the author within the strict limits of the Eastern Province but seen in Qatar on the edges of our territory and reported as common there by Batanouny (1981). S. A. Chaudhary has collected it in east-central Arabia, and it probably occurs in the east. It is likely to be seen on somewhat disturbed ground.

2. **Dichanthium foveolatum** (Del.) Roberty
Eremopogon foveolatum (Del.) Stapf
Tufted perennial grass, 30–60 cm high, with fine, ascending-erect culms kneed below, almost capillary above, shortly and finely pubescent at the nodes. Leaf blades fine, linear, with more or less inrolled margins, mostly under 5 cm long; ligule scarious, fimbriate, and appearing hairy toward the apex. Inflorescence of solitary, erect, spicate racemes c. 4–5 cm long, terminal on the upper culm branches, the young racemes enclosed at base in a spathelike sheath. Spikelets c. 4 mm long, subimbricate, sometimes purplish above, inserted parallel to the rachis; rachis and pedicels densely pubescent with fine, whitish, ascending straight hairs. Glumes equal, the lower flattish, glabrous, glossy-yellowish, with a circular glossy pit above the middle (deeper and more distinct on the sessile spikelet). Sessile spikelet bisexual, fertile, with a kneed glabrescent awn exserted c. 15–20 mm above the spikelet. Pedicelled spikelets awnless.
Habitat: Very shallow sand on limestone slopes. Apparently rather rare.
Central Coastal Lowlands: 169(K), Dhahran.
Southern Summan: 8388, Darb Mazalij, 25–02N, 49–00E.

● 6. **Ischaemum** L.

Annual or perennial grasses with an inflorescence of 2 or several terminal, spicate racemes. Spikelets usually in pairs of 1 sessile, 1 pedicelled. Rachis joints thickened. Spikelets awned from the upper lemma; lower glume often transversely furrowed.

Ischaemum molle Hook. f.

Ascending to erect grass with inflorescence a terminal pair of spicate racemes c. 5 – 6 cm long, 2 – 3 mm wide. Spikelets sessile or subsessile, pale greenish yellow, or purplish near apex, c. 4 mm long, lanceolate-oblong, pubescent on the rachis and lower glumes. Glumes nearly equal, the lower with a distinct transverse furrow near the base and pubescent at lower back. Upper lemma membranous, awned from an apical sinus. Awn brown, twisted, kneed, 2 – 2.5 times as long as the spikelet. Anthers purplish, 2 – 3 mm long.

Habitat: Collected only once, as a weed in a hotel lawn. Rare.

Central Coastal Lowlands: 7076, Dhahran (det. Cope).

● 7. **Chrysopogon** Trin.

Perennial grasses with inflorescence an open panicle of awned spikelets in threes, with one sessile and two pedicelled. Spikelets dorsally compressed; glumes subequal, exceeding and including the lemmas, awned from the apex.

Chrysopogon plumulosus Hochst.

C. aucheri var. *quinqueplumis* auct.

Densely tufted, fine-culmed perennial grass, usually 20 – 50 cm high. Sheaths and lamina more or less finely pubescent, the lamina usually less than 8 cm long, sometimes ciliate below with coarser, transparent, bulbous-based hairs; ligule a short rim of fine hairs. Inflorescence a terminal panicle 5 – 9(10) cm long, with branches whorled. Spikelets c. 7 mm long (not including the awns), linear-terete, sometimes purplish, terminal on the branches in groups of 3, 1 central and sessile and 2 lateral, pedicelled. Pedicels of lateral spikelets and base of sessile spikelet densely and conspicuously bearded with antrorse, golden-tawny hairs. Glumes of the lateral spikelets nearly equal, each with a terminal plumose awn longer than the spikelet; awn of the lemma shorter or sometimes nearly obsolete. Sessile spikelet with 1 glume and the lemma awned; awn of the lemma longer than all others in the spikelet triad, kneed near middle, twisted below the knee.

Habitat: Nearly always on rocky ground, on elevated terrain or along rocky wadi bottoms. Infrequent to locally frequent.

Northern Summan: 2773, 25 – 50N, 48 – 00E; 7942, 25 – 39N, 48 – 48E.

Southern Summan: 1051(BM), 24 km N al-'Udayliyah.

South Coastal Lowlands: 8321, 5 km SE al-Hufuf.

Vernacular Names: GHARAZ (Al Murrah, Qahtan).

● 8. **Sorghum** Moench

Stout, erect, annual or perennial grasses with flat leaf blades. Inflorescence a terminal panicle, loose to strongly contracted, with numerous awned or awnless spikelets in pairs of one sessile and fertile, one pedicellate and reduced. Spikelets 2-flowered, the upper bisexual in the sessile spikelet. Glumes equal; lemma of the fertile floret with a kneed awn, or awnless.

Key to the Species of Sorghum
1. Panicle loose, pyramidal, verticillate, with elliptical-lanceolate spikelets; grain not exposed at maturity; perennial . . . . 2. *S. halepense*
1. Panicle tight, contracted, with ovate to subglobose spikelets; grain protruding at maturity; annual . . . . 1. *S. bicolor*

1. **Sorghum bicolor** (L.) Moench
S. vulgare Pers.

Stout annual grass with erect, glabrous culms up to c. 2 – 3 m tall in cultivation but usually much shorter in escapes. Leaf blades up to c. 50 cm long, 5 cm wide; ligule ciliate along upper margin. Panicle erect, usually rather narrow, contracted, with numerous ovoid to subglobose spikelets, the grains protruding from the spikelet at maturity, the fertile spikelets with kneed awns from the lemma. Pedicelled spikelets narrower, grainless, awnless.

No attempt is made here to place our one specimen among the numerous varieties, subspecies or even segregate species into which the grain sorghums have been divided. Bor (in Townsend and Guest 1968) is followed here in taking a very broad view of the species.

Habitat: Rarely as a weed around cultivated areas as an escape, or introduced with seeds of other crops. Rare.

South Coastal Lowlands: 7811, adh-Dhulayqiyah, al-Hasa.

Vernacular Names and Uses: DHURAH (gen.). Grain sorghum. Commonly and widely cultivated as a staple grain in southern and western Arabia but much less common in the northeast.

2. **Sorghum halepense** (L.) Pers.

Stout perennial grass sending up erect culms to c. 1.5 m high from spreading rhizomes. Leaf blades flat, to c. 2.5 cm wide. Panicle open with verticillate branches, 15 – 40 cm long. Sessile spikelet c. 5 mm long, oblong-elliptical, pubescent, with a kneed awn 1 – 3 times as long as the spikelet. Pedicelled spikelet narrower, lanceolate, awnless. Glumes equal or nearly so in both spikelets.

Habitat: A weed around cultivated areas or at roadside in towns. Occasional to rare.

Central Coastal Lowlands: 7765, Sayhat, nursery plots; 8268, al-Khubar.

South Coastal Lowlands: 7804, al-Mubarraz, al-Hasa, in irrigated roadside lawn.

Uses: Johnson grass. Cultivated as forage in many parts of the world.

● 9. **Coix** L.

Annual grasses with broad leaves and a peculiar terminal, spicate inflorescence consisting of several male spikelets above a female spikelet that is enclosed in a bony, beadlike involucre interpreted as a modified leaf sheath. Male spikelets 2-flowered.

Coix lacryma-jobi L.

Stout, glabrous, annual grass with culms to over 1 m tall and leaf blades 2 – 4(5) cm wide, to 30 cm long. Racemes 1 to several from the upper sheaths, each with several to numerous male spikelets c. 8 – 9 mm long, both sessile and short-pedicellate on the upper axis and with a single female spikelet below, enclosed in a glossy, hard-bony, white or greyish ovoid to pyriform, beadlike involucre c. 9 mm long, 6 mm wide.

'Job's tears', cultivated in tropical countries for its grains and its hard female spikelet involucres sometimes used as beads.

Habitat: Known in the Eastern Province only as a rare weed in cultivated lawns or gardens, where it is occasionally introduced with imported seeds.

Central Coastal Lowlands: 7079(BM), Dhahran, hotel lawn.

● 10. Cenchrus L.

Annual or perennial grasses with flat leaf blades and ligule a rim of hairs. Inflorescence a contracted, spikelike raceme or panicle. Spikelets awnless, solitary or several together, surrounded at base by a sessile involucre of bristles or spines, these connate below in a disc or cup and falling with the spikelets as a burrlike dispersal unit.

Cenchrus ciliaris L.

Perennial grass, sometimes shrubby, with culms ascending from a stout, somewhat woody rhizome, often geniculate below, to c. 100 cm high. Inflorescence a terminal, cylindrical, false spike 5 – 10(13) cm long, 0.7 – 1.6 cm broad with involucred spikelets densely crowded or sometimes rather loose. Spikelets solitary or clustered 2 – 3, within an involucre of 2 sorts of bristles: the outer, lower ones fine, terete, shorter, minutely scabridulous; the inner ones terete and scabridulous above but broadened, flattened and fine-ciliate below, often with 1 bristle longer and broader than the others, the inner bristles often about 10 mm long, but length of all bristles rather variable (2 – 16 mm). *Plate 237.*

● The spikelike inflorescence somewhat resembles that of *Pennisetum divisum*, with which it may be found growing. The spikelet involucres of *Cenchrus ciliaris* differ distinctly, however, in having some bristles flattened and ciliate, and concrete at base in a disc.

Habitat: More common on disturbed ground than in natural habitats, but when found in the latter often associated with rocky terrain. Occasional to frequent.

Northern Plains: 3087(BM), al-Batin, 28 – 01N, 45 – 29E; 1234(BM), Jabal an-Nu'ayriyah.

Northern Summan: 2785, 25 – 50N, 48 – 00E; 3136(BM), 6 km SE ad-Dabtiyah.

Central Coastal Lowlands: 25(K), 28(K), 7869, Dhahran; 1066, Dhahran stables farm; 4067, az-Zulayfayn, 26 – 51N, 49 – 51E.

South Coastal Lowlands: 7777, 7 km SE al-Hufuf; 8314, 8 km SE al-Hufuf.

Vernacular Names and Uses: KHAḌIR (Qahtan), THUMŪM (Bani Hajir), GHARAZ (N tribes, Musil 1928b, 357). Certainly a useful fodder grass although not abundant enough in the desert to be of great importance.

● Chaudhary and Zawawi (1983) report several other species of *Cenchrus* as occurring in Saudi Arabia as weeds. These have been collected in western or central parts of the Kingdom but have not been observed by the author in the Eastern Province: *Cenchrus biflorus* Roxb. (= *C. barbatus* Schumach.), which differs from all other local species in having the bristles of the involucre retrorsely rather than antrorsely scabrid; *Cenchrus setigerus* Vahl, which has the inner bristles rigid, flat, toothlike, and connate for half or more of their length, and the outer bristles entirely or nearly absent; and *Cenchrus pennisetiformis* Hochst. et Steud., with inner bristles connate below and forming a cup 1 – 3 mm long, the connate ones very slender and setiform above, the outer ones fine and numerous. The last is known to occur in Qatar and near Riyadh and might be found in eastern Saudi Arabia. Another species, *C. echinatus* L., has the bristles united to form almost a complete, cuplike structure, spiny at the faces and margins. It has been reported from Riyadh (Chaudhary and Cope 1983).

● 11. Pennisetum L. Rich.

Annual or perennial grasses with inflorescence a dense, spikelike, terminal panicle of awnless, sessile or pedicellate spikelets. Spikelets with an involucre of hairy or glabrous bristles at base, the bristles free to the very base (cf. *Cenchrus*, where coalescent below in a disc). Spikelets 2-flowered, with glumes equal or the lower smaller.

Key to the Species of Pennisetum
1. Spicate panicle 3 – 10 cm long; spikelets mostly solitary in the involucre; glumes unequal; desert native .... 1. *P. divisum*
1. Spicate panicle 20 – 40 cm long; spikelets paired or in 3s in the involucre; glumes subequal; cultivated species .... 2. *P. glaucum*

1. Pennisetum divisum (Gmel.) Henr.

P. dichotomum (Forssk.) Del.

Glabrous, glaucous, shrubby perennial grass, somewhat woody below, with culms stiff, many-branched, forming bushes to c. 1.5 m high. Leaf blades mostly 3 – 8 cm long, 1 – 2 mm wide, longitudinally folded, deciduous, leaving yellowish-brown empty sheaths at the nodes; ligule a rim of fine hairs. Spicate panicle terminal, dense to somewhat loose, cylindrical-lanceolate, 3 – 10 cm long, c. 1 – 1.5 cm broad. Spikelets c. 4 mm long, mostly solitary, seated in an involucre of numerous, antrorsely scabridulous bristles, the bristles unequal, from shorter than the spikelet to sometimes c. 16 mm long. Glumes distinctly unequal, the lower about ⅔ as long as the upper.

This plant, when not in flower, might be confused with *Panicum turgidum* (q.v.), which is of somewhat similar habit. *Pennisetum divisum* is taller, with more elongate culms, and with the culm branches diverging at lesser angles.

Habitat: Coastal sands and sands around rocky ground. Widespread and frequent to common, but not extensively dominant.

Central Coastal Lowlands: 2(K), 170(K), 428, 1078, Dhahran; 1987, ad-Dawsariyah.

Vernacular Names and Uses: The species' close resemblance in habit to *Panicum turgidum* has led to similar names being applied to both plants. The more experienced herdsmen of the Eastern Province always distinguish them carefully with different name forms of the same linguistic root. Thus: THAYMŪM (Bani Khalid, Bani Hajir, as-Sulabah), THUMŪM (Qahtan) = *Pennisetum divisum*, while THUMĀM (often pronounced *'thmām'* or *'thamām'*) is regularly applied to the *Panicum*. Some northern tribes, such as Rawalah, appear to use *thumām* regularly for the *Pennisetum*, perhaps because of the general absence of *Panicum turgidum* from much of their territory. *Thumām* for *Pennisetum* has also been heard among the oasis villagers of al-Ajam, al-Qatif area. An informant of Al Rashid reported that southern tribes from the steppes immediately south of the Rub' al-Khali use the name ḤILĀM for *Pennisetum divisum*. This is a useful grazing plant, much browsed by camels.

2. **Pennisetum glaucum** (L.) R. Br.

    P. typhoides (Burm.) Stapf et Hubbard

    P. americanum (L.) K. Schum.

Stout, robust, erect annual grass to c. 2 m high, with finely pubescent sheaths and leaf blades. Leaf blades to c. 1 – 3(5) cm wide; ligule a dense rim of silky hairs. Panicle cylindrical, dense, terminal, very spikelike, mostly 20 – 40 cm long, 2 – 2.5 cm wide, with densely pubescent central axis. Spikelets c. 4 mm long, in pairs or 3s, subsessile above a common involucre of naked or finely plumose bristles equalling or slightly exceeding the group. Spikelets turgid at maturity. Anther cells with a fine, tufted beard at apex.

Habitat: A casual around edges of cultivation as an escape or an introduction with other crop seeds. Rare.

Northern Plains: 7855, al-Batin, farm near plot of alfalfa, 28 – 15N, 45 – 51E.

Vernacular Names and Uses: DUKHN (gen.). Pearl millet. Rather common in cultivation in southern and western Arabia but rare in the Eastern Province and to the north.

● 12. **Setaria** P. Beauv.

Annual or perennial grasses with spicate inflorescences. Spikelets awnless, short-pedicelled, with 1 to several, usually scabrid, bristles at the base, 2-flowered with the upper one bisexual. Glumes unequal.

Key to the Species of Setaria

1. Spikelets 3 mm long; exposed lemma wrinkled transversely ..... 1. *S. pumila*
1. Spikelets c. 2 mm long, not transversely wrinkled
    2. Bristles at base of spikelet retrorsely scabridulous .... 2. *S. verticillata*
    2. Bristles antrorsely scabridulous .... 3. *S. viridis*

1. **Setaria pumila** (Poir.) Roem. et Schult.

    S. glauca (L.) P. Beauv.

Annual grass, glabrous to sparsely pubescent, with thin culms ascending or erect, often kneed near the base, to c. 75 cm tall. Leaf blades flat or longitudinally folded, to c. 20 cm long, 7 mm wide; ligule a fringe of hairs. Inflorescence a dense, narrowly cylindrical, spikelike panicle, usually 4 – 9 cm long, c. 8 mm wide. Spikelets c. 3 mm long, very short-pedicelled on the shortly pubescent central axis, each subtended at base by a group of c. 5 – 10 ascending bristles. Glumes unequal, the upper shorter than the transversely wrinkled upper lemma.

Habitat: A weed of farm or gardens. Occasional.

South Coastal Lowlands: 3930, near 'Ayn al-Khadud, al-Hasa Oasis; 7779, adh-Dhulayqiyah, al-Hasa Oasis area.

2. **Setaria verticillata** (L.) P. Beauv.

Tufted annual grass, glabrous to sparingly pubescent, with ascending-erect culms often kneed at the base, to c. 75 cm high. Leaf blades thin, flaccid, flat, lanceolate to linear-attenuate, mostly 4 – 20 cm long, 4 – 10 mm wide; ligule a short ciliate rim. Inflorescence a dense, terminal, narrowly cylindrical spikelike panicle usually 2 – 10 cm long, 4 – 8 mm wide, sometimes at length lobed or interrupted. Spikelets c. 2 mm long, subsessile on the angular central axis, each subtended by one retrorsely scabridulous bristle somewhat exceeding the spikelet. Upper glume as long as the spikelet, 5 – 7-nerved; lower glume much shorter. Upper lemma finely punctate.

Habitat: A weed of farm or garden. Occasional to locally frequent.

Central Coastal Lowlands: 916, Sayhat.

South Coastal Lowlands: 1882(BM), Abqaiq; 2728, ar-Ruqayyiqah, al-Hasa; 2899(BM), al-'Arfaj farm, al-Hasa.

Vernacular Names: LUZZAYQ ('sticky-grass', referring to the scabrid inflorescence, al-Hasa farmers).

3. Setaria viridis (L.) P. Beauv.

Tufted annual grass, usually sparsely hairy, with ascending culms to c. 40 cm high. Leaf blades flat, flaccid, usually less than 15 cm long, 9 mm wide, tapering to apex and often also somewhat toward the base; ligule a short ciliate rim. Inflorescence a dense, cylindrical, spikelike panicle 2 – 6(10) cm long, c. 7 mm wide, often tapering toward the apex. Spikelets c. 2 mm long, subsessile on short branches from the scabrid-striate main axis, each subtended at base by 1 – 3 antrorsely scabridulous bristles mostly 2 – 3 times as long as the spikelet. Upper glume about as long as the spikelet; lower glume about half as long.

Habitat: A weed of gardens and farms. Occasional.

Central Coastal Lowlands: 1912(BM), 1913(BM), al-Jarudiyah, al-Qatif Oasis.

South Coastal Lowlands: 1889, al-Mutayrifi, al-Hasa; 1904, ash-Shuqayq, al-Hasa; 7786, adh-Dhulayqiyah, al-Hasa.

Vernacular Names: LUZZAYQ ('sticky-grass', referring to the scabrous inflorescence, al-Hasa farmers).

● 13. **Digitaria** Heist. ex Fabr.

Annual or perennial grasses with softish leaves, scarious ligules and ascending culms. Inflorescence mostly of spicate racemes paired, whorled, or (as in ours) more or less digitate. Spikelets 2-flowered, inserted in groups of 2 – 3 against the rachis. Glumes very unequal, the lower minute or absent, the upper much larger.

Key to the Species of Digitaria

1. Upper glume (at back of spikelet) 0.3 – 0.5 times as long as the spikelet; lower lemma with minute glassy scabridities on the nerves, at least near the apex (sometimes obscure) .... 2. *D. sanguinalis*
1. Upper glume 0.8 – 1 times as long as the spikelet; lower lemma entirely devoid of scabridities .... 1. *D. ciliaris*

1. **Digitaria ciliaris** (Retz.) Koel.

D. adscendens (Kunth) Henr.

D. sanguinalis var. *ciliaris* (Retz.) Rendle

Annual grass with culms branching and more or less geniculate at base, often rooting from the somewhat swollen lower nodes, up to c. 50 cm high. Leaf blades tapering to apex, to c. 12 cm long (ours c. 6 cm), up to c. 8 mm wide, somewhat hairy near mouth of the sheath, glabrous above, sometimes wavy at margins; ligule scarious. Inflorescence of 4 – 9 (ours c. 6) spicate, digitate or subdigitate racemes 5 – 15 cm long (ours 8 – 13 cm), 1.5 – 2 mm broad, the rachis shortly and finely pubescent at point of insertion. Spikelets lanceolate, acute and subacuminate at apex, c. 3.5 mm long, 0.8 – 1 mm wide, in pairs of 1 subsessile and 1 on a 2.5 – 3 mm-long pedicel. Lower glume reduced to a minute triangular or truncate scale; upper glume (in ours) nearly as long as the spikelet, acuminate, hairy on the lower back and strongly ciliate above. Lower lemma nerved, flat to longitudinally concave, more or less densely fine-ciliate on the marginal nerves.

Habitat: A weed of garden, farm, or well watered waste ground.

Central Coastal Lowlands: 7823, Dhahran.

South Coastal Lowlands: 8661, 1 km E al-Hufuf, al-Hasa Oasis.

2. **Digitaria sanguinalis** (L.) Scop.

Annual grass with culms decumbent at base, often rooting at the lower nodes, ascending to c. 30 – 40 cm high or occasionally taller. Leaf blades up to c. 20 cm long, 0.8 cm wide, sometimes somewhat undulate or repand at margins and with clear, bulbous-based hairs below; ligule scarious. Inflorescence usually of 4 – 6 flexible, spicate, racemes, 6 – 13 cm long and 1 – 2 mm wide, spreading digitately at the ends of the culms or subdigitately in approximate whorls. Spikelets lanceolate, acute, c. 3 mm long, appressed to the flattened, angled rachis in pairs of 1 subsessile, 1 on a pedicel 2 – 2.5 mm long. Lower glume reduced to a minute triangular scale, the upper c. 0.3 – 0.5 times as long as the spikelet. Lower lemma prominently nerved and glabrescent at back; sometimes shortly ciliate at margins, with minute glassy scabridities on the nerves distally (sometimes obscure).

Habitat: A weed of farm lands or other disturbed ground, usually in shaded, well watered, sites. Frequent.

Central Coastal Lowlands: 1914(BM), al-Jarudiyah, al-Qatif Oasis.

South Coastal Lowlands: 1885(BM), Abqaiq; 1907(BM), ash-Shuqayq, al-Hasa Oasis.

● 14. **Brachiaria** Griseb.

Annual or perennial grasses with ascending or decumbent culms. Inflorescence paniculate or racemose, with spikelets in spikes or racemes. Spikelets somewhat dorsally compressed, with very unequal glumes, the lower much shorter, with back to the rachis; lower lemma resembling the upper glume and about as long.

Brachiaria leersioides (Hochst.) Stapf

Ascending or decumbent grass, sometimes rooting at nodes, nearly glabrous except around mouths of the sheaths. Leaf blades 3 – 8(10) cm long, 4 – 8 mm wide, tapering gradually to the apex; ligule a rather inconspicuous and deep-set rim of hairs. Inflorescence terminal or (less often) axillary, consisting of 2 to c. 8 spicate racemes 2 – 3.5 cm long, spaced along the main axis. Spikelets pale green often flushed with purple, c. 3 – 3.5 mm long and 1.2 mm wide, closely spaced on very short pedicels along a somewhat sinuous, flattened or 3-angled rachis. Lower glume much shorter than the upper, clasping the spikelet at base with back to the rachis.

Habitat: Collected only once, as a weed in a lawn. Rare and probably recently introduced.

Central Coastal Lowlands: 7075, 7821(K), Dhahran.

● 15. **Paspalum** L.

Annual or perennial grasses with inflorescence of solitary, digitate, or racemed spikes. Spikelets subsessile, mostly 2-ranked on a flattened rachis, plano-convex, 2-flowered. Glumes very unequal, the lower absent or reduced to a small scale.

Key to the Species of Paspalum

1. Spikelets shortly fine-pubescent over the upper glume; racemes usually over 5 cm long .... 1. *P. distichum*
1. Spikelets entirely glabrous; racemes mostly 3 – 4 cm long .... 2. *P. vaginatum*

1. **Paspalum distichum** L.

P. paspaloides (Michx.) Scribn.

Perennial grass with spreading rhizomes, glabrous except around mouths of the sheaths, with decumbent to ascending culms to over 60 cm long. Leaf blades to c. 12 cm long, flat, tapering to apex, 3 – 7 mm wide; ligule short, membranous. Inflorescence a pair of terminal, spikelike racemes mostly 5.5 – 7 cm long, 3 mm wide, with numerous, imbricate, 2-ranked, short-pedicelled spikelets on a flattened rachis. Spikelets elliptical, plano-convex, c. 3 – 3.5 mm long, c. twice as long as broad. Lower glume obsolete but often present in some spikelets as a rudimentary scale; upper glume equalling the spikelet, rounded at back, shortly and finely pubescent over most of its surface.

● The Committee for Spermatophyta of the International Bureau for Plant Taxonomy and Nomenclature, by declining a proposal to reject the name *P. distichum* (*Taxon* 32(2): 281), has happily resolved the nomenclatural issue that led temporarily to application of the name *P. paspaloides* (Michx.) Scribn. to this plant and to application of its earlier long and widely used epithet to *P. vaginatum*.

Habitat: A weed of wet ground in the oases. Our single record was from a very wet site near rice cultivation. Little collected but may be frequent locally at some farm sites.

South Central Lowlands: 3941(BM), 2.5 km ENE al-Hufuf.

Vernacular Names: None recorded, but the name MARRĀNĪ would doubtlessly be applied to it by oasis folk by analogy with the very similar and much more common *P. vaginatum*.

2. **Paspalum vaginatum** Sw.

Creeping perennial grass, spreading widely by numerous rhizomes and stolons, rooting at the lower nodes and sending up ascending to erect culms to c. 60 cm high. Leaf blades linear, tapering to apex, flat, to c. 10 cm long, 1 – 3 mm wide, glabrous except around mouths of the sheaths; ligule a membranous rim. Inflorescence a pair of terminal, spicate racemes mostly 3 – 4(5) cm long, c. 2 mm wide, with numerous imbricated, subsessile spikelets inserted in 2 ranks on a flattened rachis. Spikelets lanceolate, subacuminate, plano-convex, c. 3.5 – 4 mm long and 3 times as long as broad. Lower glume entirely lacking; upper glume as long as the spikelet, quite glabrous.

● This species is more common than the preceding, which it closely resembles but from which it is readily separated by hand lens examination of its entirely glabrous spikelets. It is common in urban and agricultural areas but is seldom seen flowering and does so only in particularly favorable circumstances. Collections of flowering specimens by the author indicate that it has both spring and autumn flowering periods.

When not flowering it has been confused with *Cynodon dactylon*, which it somewhat resembles in habit. It has stouter and fleshier rhizomes than *Cynodon*, however, and its membranous ligules (*Cynodon* has a rim of hairs) are a good aid to identification. **Habitat:** Wet ground around canal banks and garden edges in oases and towns. Common.

Central Coastal Lowlands: 924(BM), 7818, 7820, Dhahran; 7825, 1 km S al-Qatif town; C. Parker No. 57(BM), 75(BM), al-Qatif Oasis.

Vernacular Names and Uses: MARRANI ('smooth grass', or 'supple-grass', gen. in the oases). A plant of considerable economic importance in oasis and urban areas. In the oases it can be a troublesome and difficult-to-eradicate weed, but it also serves as a useful soil binder along water channels. There it is also the main source of uncultivated fodder for household sheep and goats, and herdsmen or farmers can often be seen cutting and carrying away great bundles of it for this use. Landscapers of the Arabian American Oil Company (now the Saudi Arabian Oil Company) discovered as early as the 1940s that *Paspalum*, identified simply as 'Qatif grass', made an excellent lawn grass in the newly developing oil field communities. It is widely used for this purpose now and is considered the best all-around grass for lawns wherever sufficient water can be provided. It is established easily in sand by planting pieces of rhizome and keeping them well watered until roots take hold from the nodes. A dense, easy-to-care-for turf results. Lawns of *Paspalum* sometimes suffer damage, visible as spreading brown spots, from larvae of a Scarabaeid beetle, *Pentodon algerinum*, which eat the roots. They can be controlled by periodic spraying with insecticide.

• 16. **Panicum** L.

Annual or perennial grasses with inflorescence a diffuse, or sometimes contracted, panicle. Spikelets pedicelled, unawned, 2-flowered, usually rounded, smooth, grainlike. Glumes usually very unequal with the lower smaller, the upper equalling the spikelet. Lower lemma glumelike.

Key to the Species of Panicum

1. Glumes very nearly equal, both as long as the spikelet; shrubby desert perennial . . . . 4. *P. turgidum*
1. Glumes very unequal, the lower ⅕ to ¾ as long as the spikelet; weeds or ruderals of farm or inhabited areas
    2. Spikelets 4 – 6 mm long; lower glume ½ to ¾ as long as the spikelet, acute-acuminate . . . . 2. *P. miliaceum*
    2. Spikelets 2 – 3 mm long; lower glume ⅕ to ½ as long as the spikelet, mostly obtuse or subobtuse
        3. Lower glume c. ⅕ as long as the spikelet, encircling the spikelet base like a cup . . . . 3. *P. repens*
        3. Lower glume nearly half as long as the spikelet, mostly unilateral . . . . 1. *P. antidotale*

1. **Panicum antidotale** Retz.

Stout perennial grass with thick rootstock; culms erect, smooth-woody and glossy at least below, swollen at the nodes, up to c. 1.7 m high. Leaves (in ours) mostly 15 – 40 cm long, to c. 1.5 cm wide, linear-acuminate, flat, antrorsely scabridulous to the touch;

ligule indistinct, appearing along its upper margin as a rim of very fine hairs. Panicle ovate to pyramidal in outline, terminal, 20 – 30 cm long, with branches sometimes up to 12 cm long, ascending and somewhat contracted when young, effuse with spreading branches when mature. Spikelets numerous, subsessile or pedicellate, glabrous, 2 – 3 mm long. Lower glume c. ½ as long as the spikelet, obtusish to subacute but not acuminate.

Habitat: A weed or ruderal of abandoned farm plots or ditches around towns or oases. Usually on ground somewhat wet with runoff water. Occasional.

Central Coastal Lowlands: 633(K), Dhahran stables farm; 8143, Dhahran.

South Coastal Lowlands: 7782(BM), adh-Dhulayqiyah, al-Hasa; 7807(BM), al-Mutayrifi, al-Hasa; 8675, 1 km N 'Ayn al-Luwaymi, al-Hasa Oasis.

2. **Panicum miliaceum** L.

Stout annual with thickish fascicled culms, ascending to erect, in ours mostly 30 – 50 cm high, sometimes much taller. Leaf blades to c. 25 cm long, 0.5 – 1.5 cm wide, flat, often somewhat ascending, antrorsely scabridulous to the touch at margins; sheaths hairy with fine, spreading, bulbous-based hairs; ligule a rim of fine, silky hairs. Panicle terminal, contracted, ascending, in ours mostly 10 – 20 cm long, to c. 3.5 cm wide. Spikelets numerous, pedicellate, ovate to lanceolate, acute, 4 – 6 mm long. Lower glume ovate, acuminate, distinctly nerved, ½ to ¾ as long as the spikelet.

This is the common millet of Europe, and as might be expected in such a cultivated species, even our two specimens vary distinctly in some characters. No. 7816 has plump, ovoid spikelets little over 4 mm long, while No. 7783 has lanceolate spikelets up to 6 mm long.

Habitat: A rather rare weed around farm fields or road edges. Not known to the author to be cultivated locally.

Central Coastal Lowlands: 7816(K), al-Khubar.

South Coastal Lowlands: 7783(BM), adh-Dhulayqiyah, al-Hasa.

3. **Panicum repens** L.

Perennial grass with culms ascending from spreading rhizomes, to c. 70 cm high. Leaves 2-ranked, rather distant except in frequently cut plants, the blades mostly 10 – 20 cm long, long-tapering, to c. 7 mm wide, glabrous except immediately above the ligule; upper margins of the sheath soft-ciliate; ligule indistinct, short-membranous but more or less fine-ciliate along the upper margin. Panicle 10 – 20 cm long, usually quite loose with branches ascending, or sometimes effuse-spreading. Spikelets pedicellate, elliptical-lanceolate, acute, somewhat membranous, glabrous, 2 – 3 mm long. Lower glume c. ⅕ as long as the spikelet, encircling the base as a cupule. *Plate 238.*

Habitat: A weed of gardens, lawn edges. Locally frequent.

Central Coastal Lowlands: 635(K), 7808a, Dhahran; 7828, 1 km S al-Qatif town.

4. **Panicum turgidum** Forssk.

Glabrous, glaucescent, shrub-like perennial grass with ascending, tangled culms often multiple-branched upwards at swollen, knotty nodes, forming rounded bushes up to c. 1 m high. Roots with feltlike sheaths densely covered with adherent sand. Leaf blades often 6 – 8 cm long, 2 – 4 mm wide, flat; ligule a short rim of hairs. Panicle rather irregular, sparse, often 4 – 7 cm long, with 1 to several racemes of pedicelled spikelets.

Spikelets ovoid, turgid, glabrous, c. 4 mm long. Glumes about equal, the lower one thus nearly equalling the spikelet. Stigmas densely plumose; anthers rust-colored, slightly over 2 mm long. *Plate 239.*

● This plant in sterile condition somewhat resembles *Pennisetum divisum* but has a more tangled habit with denser, knottier branch points in the culms. A very important plant in the coastal lowlands, dominant over hundreds of square kilometers in the widespread Panicetum of the coastal and subcoastal sands. It also occurs as a codominant or associate in communities with *Rhanterium epapposum* and *Calligonum comosum.* Active rapid growth is made after winter and spring rains; the bushes become mostly dormant in summer and autumn, dying back to a dry, brown condition. Each bush is usually on a sand hummock resulting from its wind shadow.

Habitat: Coastal and subcoastal sands, where often dominant; also not infrequent deeper inland. Common.

Central Coastal Lowlands: 171(K), 427, 1077(BM), Dhahran; 1753(BM), Abu Hadriyah; 1988, ad-Dawsariyah.

Northern Summan: 3140(BM), 6 km SE ad-Dabtiyah.

South Coastal Lowlands: 7664, 8 km S Wadi as-Sahba, 23 – 54N, 49 – 08E.

Vernacular Names and Uses: THUMĀM (gen., often pronounced 'thamām', 'thmām'), ḤABBAY (villagers of al-Ajam, al-Qatif area, apparently a diminutive of ḥabbī, 'grainy', referring to the grainlike spikelets). The superficial resemblance, when the plants are not flowering, to *Pennisetum divisum* is reflected in the common name of the latter, THAYMŪM. Herdsmen, however, never confuse the two. See the description of *Pennisetum divisum* for additional notes on vernacular names. This grass is a very important grazing plant and a mainstay of camels, goats and sheep in the coastal lowlands of the Eastern Province. Even in dry summer condition it provides grazing as standing hay. Elderly tribesmen report that Bedouins, in times of famine, used to gather the grains of this grass and roast them for food.

● 17. Echinochloa P. Beauv.

Annual or perennial grasses with soft leaves and (in ours) ligule absent. Inflorescence consisting of simple, short, dense, spicate racemes along a main axis. Spikelets subsessile, awned or awnless, rounded dorsally, flattened ventrally. Glumes unequal, the lower smaller.

Key to the Species of Echinochloa
1. Spikelets 2 – 2.8(3) mm long; inflorescence loose, open, with racemes not
   or only moderately imbricate .... 1. *E. colona*
1. Spikelets 3 – 3.5 mm long; inflorescence dense, closed, with racemes
   strongly imbricate (but sometimes loose at very base) ...... 2. *E. frumentacea*

A third, widely-known species, *E. crusgalli* (L.) P. Beauv., has been collected in central Arabia and may well occur on disturbed ground in the Eastern Province although it is certainly not common. It has spikelets 3 – 4 mm long, hispid on the nerves, in spaced racemes usually over 2 cm long (in the 2 species above, mostly less than 2 cm). It usually is conspicuously awned from the apex of the lower lemma.

1. Echinochloa colona (L.) Link
Annual, nearly glabrous grass with culms procumbent to ascending, 20 – 40 cm long.

Leaf blades soft, glabrous, to c. 25 cm long, 3–6(8) mm wide; ligule absent. Inflorescence of simple, dense, spicate racemes spaced along the main axis, appressed or ascending, not or partially overlapping, mostly 1–2 cm long. Spikelets 2–3 mm long, subsessile, closely packed, sometimes purplish, pubescent to hispidulous at back, unawned. Glumes much shorter than the spikelet.

Habitat: A weed of farm, gardens, lawn edges and other moist disturbed grounds. Frequent.

Central Coastal Lowlands: 1064, Dhahran; 7813, al-Khubar; 7829, 1 km S al-Qatif town.

South Coastal Lowlands: 1893(BM), al-Mutayrifi, al-Hasa; 2911(BM), al-'Arfaj farm, al-Hasa; 7787, adh-Dhulayqiyah, al-Hasa.

Vernacular Names: WAGHL (al-Hasa farmer).

2. Echinochloa frumentacea Link

E. crusgalli var. *frumentacea* (Roxb.) W. F. Wight
E. colona var. *frumentacea* Ridley

Rather stout, glabrous, annual grass with culms to c. 0.5 cm wide, to over 50 cm long. Leaves soft, mostly 20–25 cm long, to c. 12 mm wide; ligule absent. Inflorescence dense, paniculate, terminal, 10–15 cm long, consisting of numerous, c. 1.5 cm-long, overlapping racemes. Spikelets 3–3.5 mm long, grey-green or purplish, densely packed, subsessile, pubescent to weakly hispidulous at back on the nerves. One glume nearly half as long as the spikelet; apices of the lemmas laterally compressed forming short mucros.

● This species is rather weakly differentiated from *E. colona*, of which it has been treated as a variety. It is considered by some to be a variety of *E. crusgalli*. It is overall stouter and lusher than *E. colona*, and the larger spikelets and denser inflorescence in our specimens are distinctive. It is cultivated in some tropical countries as a fodder grass.

Habitat: Lawns, farms, or gardens. Found only once in our area, in an irrigated street median in al-Khubar, where it was probably introduced with imported seed. Rare.

Central Coastal Lowlands: 7814(K), al-Khubar.

● 18. Eriochloa Kunth

Annuals or perennials with flat leaf blades. Inflorescence a panicle of racemes, or racemes spaced along a central axis. Spikelets with one fertile flower, distinctive in having the lowest rachilla node (appearing as the pedicel immediately below the spikelet) swollen into a depressed-cylindrical callus. Lower glume obsolete; the upper as long as the spikelet. Lower lemma resembling the upper glume.

Eriochloa fatmensis (Hochst. et Steud.) W. D. Clayton

E. nubica (Steud.) Hack. et Stapf ex Thell.

Annual with culms spreading and ascending, to 80–100 cm long. Leaf blades (in our specimen) 25–30 cm long, 10–15 mm wide. Inflorescence c. 15–20 cm long, consisting of several c. 5–8 cm-long racemes spaced along a central axis. Spikelets grouped by 2s along the longitudinally ridged raceme axis, 1 in each pair longer-pedicellate than the other. Spikelets lanceolate, c. 3–3.5 mm long (excluding the basal callus), appressed hairy, terminating in an awnlet c. 1–1.3 mm long; pedicels shorter than the spikelet, hairy, apparently terminating at the spikelet base in a whitish, smooth, cylindrical callus, the latter somewhat broader than long and sometimes tinged reddish-purple above.

Habitat: A weed of cultivated, relatively well watered ground. Apparently infrequent.
Central Coastal Lowlands: 8643(K), Dhahran, irrigated lawn.

● 19. Aeluropus Trin.

Perennial grasses with more or less convolute or folded leaf blades. Inflorescence a dense, 1-sided, headlike panicle or an elongated compound spike. Spikelets laterally compressed, of 4 – 18 florets. Glumes somewhat unequal, keeled; lemmas 7 – 11-nerved, mucronate.

Key to the Species of Aeluropus
1. Inflorescence capitate, with a single dense terminal spike; spikelets hairy .... 1. *A. lagopoides*
1. Inflorescence elongate, with several spikes on a main axis; spikelets glabrous .... 2. *A. littoralis*

1. Aeluropus lagopoides (L.) Trin. ex Thwaites
Pubescent perennial grass with widely spreading, wiry stolons or rhizomes and shoots ascending from the rooting nodes, sometimes densely tufted. Culms to c. 15 cm high. Leaves distichous, usually channeled or infolded longitudinally, the blades and sheaths with fine hairs; ligule a rim of hairs. Inflorescence a very dense, 5 – 15 mm-long, subglobose or oblong, terminal head of hairy spikelets.
Habitat: A halophyte of *sabkhah* edges or saline ground around cultivated areas. A definite indicator of salty soil conditions. Most frequent on waste ground but also found around desert salt flats.
Northern Plains: 836, 30 km SW al-Mish'ab.
Central Coastal Lowlands: 3(K), C. Parker 61(BM), Dammam; C. Parker 53(BM), al-Qatif.
South Coastal Lowlands: 3803, 10 km NW Salwah; 8291, 6 km N al-Mutayrifi.
Vernacular Names and Uses: ʿIKRISH (gen.). Grazed by livestock on saline ground where other plants are not available.

2. Aeluropus littoralis (Gouan) Parl.
Glabrescent perennial grass with spreading rhizomes and creeping, somewhat woody stolons; stems decumbent, terminal or ascending from the rooting nodes, to c. 25 cm high. Leaf blades inrolled, channeled, more or less distichous. Inflorescence elongate, terminal, spicate, mostly 3 – 6(8) cm long, consisting of c. 4 – 8 sessile secondary spikes appressed to the main axis, each 5 – 10(15) mm long. Spikelets 2-rowed, closely packed, glabrous.
Habitat: Saline ground around cultivated areas with irrigation run-off. Less common than the above species.
South Coasal Lowlands: 515(K), 1 km N 'Ayn Umm Sab', al-Hasa Oasis.
Vernacular Names and Uses: Local names not recorded, but almost certainly known by the same name as the above species, ʿIKRISH. Probably of some value as a grazing plant where other species are unavailable.

● 20. Alopecurus L.

Annual or perennial grasses. Inflorescence a terminal, terete, dense ovoid or cylindrical spike (actually a very dense, spicate, panicle). Spikelets compressed, 1-flowered. Glumes nearly equal, often connate at the lower margins. Lemma awned from the back or from near the base.

Alopecurus myosuroides Huds.

Glabrous annual grass with erect or ascending culms up to c. 45 cm high. Leaf blades long-tapering to apex, mostly 8 – 15 cm long, 3 – 6 mm wide; ligule membranous. Inflorescence a narrow, dense, cylindrical, terminal spike (actually a strongly contracted panicle) 5 – 12 cm long, c. 5 mm wide. Spikelets laterally compressed, c. 5 mm long, pubescent at the dorsal angle of the glumes. Glumes connate marginally in their lower ⅓ to ½; lemma awned from near the base. Styles long, sometimes prominent and exserted and exceeding the awns.

Habitat: A weed found among field crops or associated with spilled grains. Rare.

Northern Summan: 8089, N Batn al-Faruq, 25 – 43N, 48 – 53E, at Bedouin camp site; seeds introduced with spilled grain.

● 21. **Polypogon** Desf.

Annual or perennial grasses with inflorescence a dense panicle of very numerous awned spikelets. Spikelets with one bisexual floret; glumes equal or subequal, rounded at back, awned (as in our common species) or awnless. Lemma shorter than the glumes, awnless or shortly awned.

Polypogon monspeliensis (L.) Desf.

Glabrous annual grass with culms ascending or erect, sometimes kneed near the base, to c. 40 cm high. Leaf blades flat, soft, to c. 15 cm long, 8 mm wide, very finely and antrorsely scabridulous to the touch; ligule membranous. Panicles solitary, terminal, very dense, spikelike, sometimes weakly lobed, oblong to cylindrical-lanceolate, 2 – 12 cm long, 0.8 – 2 cm wide. Spikelets subsessile or pedicellate, disarticulating below the glumes, c. 2 mm long (excluding the awns). Glumes nearly equal, rounded at back, finely pubescent, longer than the floret, with a whitish awn at apex 2 – 3 times as long as the glume. Lemma usually shortly awned.

Habitat: A weedy species, usually on disturbed ground around farms or gardens. Occasional to frequent.

Central Coastal Lowlands: 179(K), 432, Dhahran; 1067, Dhahran stables farm; 8122, al-Wannan, roadside.

South Coastal Lowlands: C. Parker 6(BM), 25(BM), al-Hasa; 8199, near 'Ayn Umm Sab', al-Hasa Oasis.

Vernacular Names: ZARRI' (Qahtan).

● 22. **Aristida** L.

Perennial or annual grasses with narrow, convolute leaves and spikelets 1-flowered, awned, pedicellate in a panicle. Glumes persistent, equal or unequal. Lemma teretish, convolute, with bearded callus, furnished at apex with 3 awns.

Aristida adscensionis L.

Glabrous grass, usually annual but sometimes perennating, with culms erect or ascending, 10–50 cm high, sometimes dwarfed. Leaves narrow, the lamina mostly 5–12 cm long, less than 2 mm broad, convolute at least distally; ligule a rim of very fine, short hairs. Panicle elongate, generally more or less contracted. Glumes unequal. Lemmas exceeding the glumes, bearded at base, mostly 9–12 mm long, acuminate, convolute-teretish, tipped with a 3-branched, 10–25 mm-long, antrorsely scabridulous awn, with 1 branch sligthly longer. *Plate 240.*

● Our specimen from Dhahran (1076,BM), a robust plant to c. 40 cm high and with awns up to 25 mm long, can be assigned to var. *caerulescens* (Desf.) Hack. This is treated by some authors as a separate species, *A. caerulescens* Desf. The specimens from al-Ghawar area are much smaller annuals of the typical variety.

Habitat: Silts or silty-sands over limestone. Apparently rather rare in the Eastern Province but perhaps sometimes overlooked.

Central Coastal Lowlands: 1076(BM), Dhahran, var. *caerulescens* (Desf.) Hack.

Southern Summan: 8371, al-Ghawar, 54 km S 'Uthmaniyah GOSP-13; 8345, al-Ghawar, 18 km N Harad (dwarf forms 7–15 mm high).

● 23. **Stipagrostis** Nees

Mostly perennial grasses with convolute leaf blades and ligule a rim of hairs. Inflorescence a contracted or open panicle of short-pedicellate spikelets. Spikelet 1-flowered, bisexual, disarticulating above the glumes. Glumes equal or unequal, persistent. Lemmas terete, convolute, terminating in a column with a 3-branched awn, at least the central branch of which is plumose and often longer. An important genus of desert grasses in Arabia, including two species (*S. plumosa* and *S. drarii*) widespread and abundant and of high grazing value. *Stipagrostis*, with about 50 species, is widely distributed from Africa through the Middle East to Central Asia and Pakistan. One of our species, *S. ciliata*, represents the Section *Schistachne*, with the three-branched awn articulated near the middle of the lemma. The others belong to Sect. *Stipagrostis*, with the awn jointed at the lemma apex. All of our species have the central awn branch plumose and longer, and the lateral branches naked, shorter, quite fine and often inconspicuous.

Key to the Species of Stipagrostis
1. Culm nodes densely and conspicuously bearded with tufts of straight hairs; column of the awn deciduous with upper part of the lemma, thus hollow, semi-cylindrical at base .... 1. *S. ciliata*
1. Culm nodes glabrous (but sometimes woolly immediately below); column deciduous above the lemma, solid
   2. Panicles mostly 20–30(35) cm long, 10–15(20) cm wide, open with spreading branches; central branch of the awn 12–17(22) mm long and plumose entirely or nearly entirely to the trifurcation .... 2. *S. drarii*
   2. Panicles mostly 8–20 cm long, under 3 cm wide, with branches erect-contracted; central branch of the awn 22–55 mm long, naked for a distance of at least 7 mm above the trifurcation
      3. Column of awn plumose near the trifurcation .... 4. *S. hirtigluma*
      3. Column entirely naked
         4. Apex of plumose awn branch acutish, with axis prolonged as a short, fine mucro; lower culm internodes closely white-grey woolly .... 6. *A. plumosa*

4. Apex of plume obtuse, without exserted mucro; culms glabrous throughout
    5. Leaves glabrous (except at mouths of sheaths); central awn branch 20 – 25 mm long; column 3 – 6 mm long .... 5. *S. obtusa*
    5. Leaves furnished throughout with scattered long hairs; central awn branch 30 – 40 mm long; column 7 – 12 mm long .... 3. *S. foexiana*

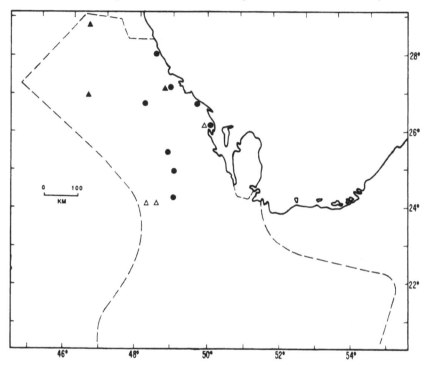

Map 8.15-A. Distribution records for three of the less common species of *Stipagrostis* in the coverage area of this flora (dashed line): *Stipagrostis ciliata* (●), *Stipagrostis obtusa* (▲) and *Stipagrostis foexiana* (△).

1. **Stipagrostis ciliata** (Desf.) de Winter
    Aristida ciliata Desf.

Perennial grass, densely tufted at base with culms erect, often kneed at base, extending well above most of the leaves, 30 – 80 cm high. Culms with conspicuous spreading hair tufts at the nodes; internodes glabrous. Leaf blades narrower than the sheaths, tightly involute, thus appearing only c. 1 mm wide, linear, circinate, to c. 17 cm long; ligule a short rim of hairs with longer hairs around mouth of the sheath; sheaths otherwise glabrous. Panicle erect, terminal, often rather contracted, 10 – 15(20) cm long. Spikelets pedicellate, 10 – 14 mm long, glabrous, pale but often darkened purplish near the base, the upper glume slightly longer than the lower. Awn disarticulating with the upper part of the lemma, leaving the column hollow and half-cylindrical at base; central, plumose branch of the awn usually 45 – 50 mm long above the trifurcation, the plume with a fine exserted mucro at apex. *Plate 241, Map 8.15-A.*

Habitat: Nearly always in very shallow sand over limestone, or in the sand-filled crevices of rocks. Infrequent.

Northern Plains: 891, 1736(BM), Jabal al-'Amudah.

Northern Summan: 3341, 5 km WSW an-Nazim; 7923, 25 – 35N, 48 – 50E.

Central Coastal Lowlands: 161(K), 8275, Dhahran; 444, 5 km SSW Dhahran; 1752(BM), Abu Hadriyah; 4070, az-Zulayfayn.

Southern Summan: 8390, Darb Mazalij, 25 – 02N, 49 – 00E; sight record, 32 km N Harad.

Vernacular Names: KHAṢĀB (ad-Dawasir), SAḤAM (Qahtan), ṢULLAYYĀN (N Arabia, Musil 1928b, 363).

2. Stipagrostis drarii (Täckh.) de Winter
S. arabiaefelicis Bor
Aristida drarii Täckh.

Perennial grass with few to several stiff, erect culms, sometimes branched once or twice, arising from a somewhat woody rootstock, to 120(150) cm high; internodes densely and closely woolly; nodes glabrous. Leaf blades narrower than the sheaths, tightly involute with margins connivent and thus cylindrical-linear, less than 1 mm wide, 10 – 20(25) cm long; ligule a short rim of very fine hairs. Leaves unusual in that the sheaths and lamina often appear to outgrow some of the culm branches terminally, giving the appearance of 2 – 3 fine, terminally exserted blades. Panicles terminal, lanceolate-pyramidal, contracted with ascending branches (when young) to spreading and open, 10 – 30(35) cm long, up to 20 cm wide, sheathed at base when young. Spikelets pedicelled, narrowly lanceolate, 8 – 10(11) mm long (excluding the awn), glabrous or very finely scabridulous. Glumes equal or the upper slightly longer; upper glume acuminate; one or both glumes sometimes with a very short mucro at apex. Central branch of the awn (12)14 – 16(22) mm long from the trifurcation, plumose from apex entirely down to the trifurcation or to within less than 5 mm of it. *Plate 242.*

A very distinctive Arabian grass, subendemic to the Peninsula and ranging from the southeastern Rub' al-Khali as far northwest as Sinai, whence it was described by Täckholm. The Arabian plants were found to be conspecific with *S. drarii* after Bor's (1967) description of them as *S. arabiaefelicis* from specimens collected by V. Dickson and the present author.

Habitat: Deep sands; often widespread and common on semi-stabilized sands and even on somewhat mobile dunes. Usually rather widely spaced.

Northern Dahna: 606(BM), 5 km SE ath-Thumami; 342(BM), 38 km W Khurays.

Central Coastal Lowlands: 7761, 10 km WSW Dhahran; 2708, al-'Alah; 8334, 26 km SW Ghunan.

Southern Dahna: 475(K), eastern Dahna, 24 – 10N; 2144, eastern Dahna, 23 – 07N; 3819, 24 – 26N, 48 – 05E; 8357, 8 km ENE Rawdat at-Tawdihiyah; 8379, 7 km SE Qirdi GOSP.

South Coastal Lowlands: 197(K), 549, 5 km E Abu Shidad; 1863, Abqaiq.

Rub' al-Khali: 1021, 22 – 45N, 50 – 20E; 7629, ash-Shuwaykilah, 19 – 47N, 48 – 44E; 7674, as-Sanam, 22 – 44N, 50 – 16E; 7641, Abu Shidad, 18 – 22N, 46 – 30E.

Vernacular Names and Uses: SABAṬ (gen.). Well known among the Bedouins as a useful grazing plant and in some hyperarid parts, such as the Rub' al-Khali, virtually the only Graminaceous fodder available.

3. **Stipagrostis foexiana** (Maire et Wilczek) de Winter
Aristida foexiana Maire et Wilczek

Densely tufted perennial grass with several to numerous erect culms mostly 20 – 35(40) cm high. Culm nodes with a narrow purplish band; internodes glabrous. Leaves densely crowded at base, subfiliform, tightly involute, the blades 5 – 13 mm long, 0.5 mm or less wide, furnished throughout with scattered, long, fine white hairs. Panicles erect, to c. 15 cm long, 2(3) cm wide; spikelets mostly 10 – 13 mm long, with glumes subequal. Central awn branch (30)35(40) mm long from the trifurcation, with a purplish axis; column 7 – 12 mm long. Plume of the central branch rounded-obtuse at apex, without exserted axis. *Map 8.15-A.*

● A species quite close to *S. obtusa*. It has been treated as a variety (*pubescens*) of *S. obtusa* by some authors, but is now generally taken to be distinct. The plant is overall more robust than *S. obtusa*. Records to date suggest that in our territory it replaces *S. obtusa* from Dhahran to the south.

Habitat: Rocky shelves around limestone hillsides. Infrequent.

Central Coastal Lowlands: 168(K), 8339, Dhahran.

Southern Summan: 8354, western Manakhir ad-Dakakin; 8364, hills near Khashm az-Zaynah; 8387(K), Darb Mazalij, 25 – 02N, 49 – 00E.

4. **Stipagrostis hirtigluma** (Steud. ex Trin. et Rupr.) de Winter
Aristida hirtigluma Steud. ex Trin. et Rupr.

Perennial grass, densely tufted at base with culms erect or ascending, to c. 40 cm high. Culms with glabrous or glabrescent, non-woolly internodes. Leaf blades mostly 4 – 10 cm long, involute, subfiliform and hardly 1 mm wide; sheaths glabrous or finely scabridulous on the nerves, hairy at the mouth; ligule a short ciliate rim. Panicle to c. 20 cm long with branches contracted to ascending. Spikelets 8 – 12 mm long with glumes pubescent, somewhat unequal. Lemma c. 4 mm long, villous at base of the callus. Column of the awn pubescent-plumose in upper part near the trifurcation; central awn branch up to c. 50 mm long, plumose in the upper half or more, with axis shortly prolonged, naked, beyond the plume.

Likely to be confused only with *S. plumosa*, from which it can be separated by its awn column being pubescent above (entirely glabrous in *S. plumosa*) and by its non-woolly culm internodes (distinctly woolly in *S. plumosa*).

Habitat: Vesey-Fitzgerald's specimen (see below) was noted as collected in 'soft silt'. Rare in our area and probably to be found only in the extreme southeast near the United Arab Emirates border.

South Coastal Lowlands: The author knows of no records of this species definitely from within the strict confines of the Eastern Province. It is known to the east in the United Arab Emirates, however, and Zohary (1957) reported a specimen collected by Vesey-Fitzgerald near the present Saudi Arabia – UAE border. It may, therefore, be found in these extreme southeastern reaches of our territory. It is not clear whether Vesey-Fitzgerald's mention of the species (1957) as occurring in the gravel plain west of Ba'ja was based on a specimen or a sight record. His specimen (No. 13454, HJ) collected during the same field trip was noted 'Sila, Sabakha Mutti' (Zohary 1957). As-Sil' lies on the coast at the western edge of Sabkhat Matti, almost due south of Ba'ja.

Vernacular Names: None recorded, but it would most probably be referred to as NAŞĪ, by analogy with *S. plumosa*.

5. Stipagrostis obtusa (Del.) Nees
Aristida obtusa Del.

Perennial grass, densely tufted at base, with several erect, 1-noded culms 5 – 25 (30) cm high; nodes glabrous, with a narrow, dark-purplish ring; internodes glabrous. Leaf blades very short and narrow, crowded at base of plant, virtually filiform and only 0.5 mm wide, up to c. 8 cm long, tightly involute, mostly circinate, glabrous except at mouths of the sheaths; ligule a rim of hairs. Panicle terminal, erect, with ascending branches, 4 – 10(15) cm long, 0.5 – 2 cm wide. Spikelets pedicelled, narrowly lanceolate, pale, 10 – 12 mm long. Glumes subequal, with the upper slightly longer. Central awn branch 20 – 25 mm long from the trifurcation, plumose in the upper half to two-thirds; axis of the awn filiform, purplish. Plume rounded-obtuse at apex, without exserted mucro. *Map 8.15-A.*

Habitat: Usually in very shallow sand over limestone, or among rocks. Infrequent.

Northern Plains: ravine to al-Batin, 28 – 50N, 46 – 24E.

Northern Summan: 759, 69 km SW al-Qar'ah wells.

Central Coastal Lowlands: 1781(BM), Abu Hadriyah.

Vernacular Names: ṢULAYLAH (N Arabia, Musil 1928b, 363; 1927, 624).

6. Stipagrostis plumosa (L.) Munro ex T. Anders.
Aristida plumosa L.

Perennial grass, densely tufted at base with culms rather thin, flexible, mostly 2 – 3-noded, erect or geniculately ascending, (8)15 – 45 cm high. Internodes closely woolly at least below; basal and middle sheaths woolly or glabrous, tight. Leaf blades tightly convolute, subfiliform-cylindrical, linear, circinate or flaccid, mostly 4 – 10 cm long, less than 1 mm wide; ligule a ciliate rim. Panicles solitary-terminal on numerous culms, contracted with ascending branches, mostly lanceolate or oblanceolate in outline, (6)10 – 15(20) cm long, 1 – 4 cm wide. Spikelets linear-lanceolate, pedicellate, 10 – 15 mm long, with glumes subequal or unequal, the upper aristate and often somewhat longer, the lower acuminate. Awn varying rather widely in length, the central branch from c. 25 to 55 mm long in our specimens with the lower, naked portion (6)9 – 15(17) mm long; plume 15 – 41 mm long. Awn kneed at the trifurcation, or (particularly in younger spikelets) straight.

● Our collections include both specimens with woolly basal sheaths (var. *plumosa*) and those glabrous (var. *brachypoda* (Tausch) Bor, = *Aristida brachypoda* Tausch). Apparently contrary to Bor's experience with Iraq plants (Townsend and Guest 1968) our glabrous-sheathed specimens have some of the longest awns. As noted by Bor, however (op. cit., 392), the sheath wool is evanescent. In our experience, blasting by sand-laden winds sometimes removes this feature.

Overall this is a rather variable species, both in habit and anatomical features. On shallow rockier soils or under dry conditions it is sometimes few-culmed and dwarfed, only c. 10 cm high. Many of our specimens have some sparse, short pubescence on the glumes.

Habitat: Shallow, stable sand, or sandy-silts. Common.

Northern Plains: 12(K), as-Saffaniyah; 835, 30 km SW Ras al-Mish'ab; 892, Jabal al-'Amudah; 1012, 12 km N Jabal al-'Amudah.

Northern Summan: 812, 37 km NE Jarrarah; 2837, 32 km ESE ash-Shumlul.

Central Coastal Lowlands: 168a(BM), 922, 8274, 8340(K), Dhahran; 1858, 5 km W 'Ain Dar GOSP–2; 1976, ad-Dawsariyah; 1813(BM), 49 km SE Abu Hadriyah.

South Coastal Lowlands: 1946, 28 km S al-'Uqayr; 1967, near Abqaiq salt pits; 175(K), Jawb Abu 'Arzilah; 8315, 5 km SE al-Hufuf; 8326, 25 km NW al-Hufuf; 8332, 28 km SW Abqaiq.

Southern Summan: 7708, Wadi as-Sahba, 24–00N, 49–05E; 8347, al-Ghawar, 12 km N Harad; 8353, western Manakhir ad-Dakakin; 8389(K), Darb Mazalij, 25–02N, 49–00E.

Rub' al-Khali: 7699, al-Mulayhah al-Gharbiyah, 22–53N, 49–29E.

Vernacular Names and Uses: NAṢĪ (often pronounced 'NUṢĪ', gen.), RĀHIM (Al Rashid, S tribes). A very important grazing plant over much of Arabia. Like other such species, it has variant names denoting specific growth stages or condition. Thus: TUBAYNĪ ('little straw-grass', Al Murrah, referring to the dwarfed form found on hard ground), ḌA'WĪT (when very large, S tribes, Al Rashid), THAGHĀM (after it has dried and gone pale in color, Bani Hajir). The author has seen the grass being cut in the desert in the northern Hijaz and brought into market at al-Wajh woven into long braids for sale. Tribesmen in the central north also, particularly in the early summer, go out on *naṣī* cutting expeditions, bringing in the grass bundles to livestock on summer wells where local grazing is soon depleted.

● 24. **Phragmites** Adans.

Stout, tall perennial grasses with creeping rhizome, flat leaf blades and hollow, reedlike culms. Inflorescence a large panicle. Spikelets several-flowered, the rachilla long silky-hairy, disarticulating between the florets above the glumes. Glumes equal or (as in ours) unequal.

Phragmites australis (Cav.) Trin. ex Steud.
P. communis Trin.
Stout, glabrous, perennial reed with culms arising from spreading rhizomes, erect, 1–3 m high. Leaf blades flat, long-tapering, to c. 50 cm long, 3 cm wide; ligule a rim of hairs. Panicle plume-like, compound with ascending branches, purplish to silvery-whitish, erect or nodding, 15–40 cm long. Spikelets somewhat laterally compressed, c. 10–14 mm long, gaping, with exserted straight white hairs from the rachilla. Glumes unequal, the lower shorter, both exceeding the lemmas, some of which are narrowly acuminate and thus appearing awned. *Plate 243.*

Easily recognized by its stout, canelike habit, this grass forms many square kilometers of 'reed forest' around the edges of al-Hasa Oasis where run-off water accumulates.

Habitat: Swamps, ditches, wherever fresh to brackish water may stand, particularly around irrigation or drainage run-off from the oases and springs. Common.

Central Coastal Lowlands: 117(K), Ras Tanura; C. Parker 62(BM), al-Qatif.

South Coastal Lowlands: 1899(BM), 1.5 km SSE ash-Shuqayq, al-Hasa; C. Parker 17(BM), al-Hasa.

Vernacular Names: 'AQRABAN (gen.), QAṢBA' (Shammar, Rawalah).

● 25. **Arundo** L.

Tall, stout, perennial reed with hollow culms and broad leaf blades. Spikelets laterally compressed with rachilla disarticulating above the glumes. Glumes equal or subequal; lemma bifid at apex with a short awn in the sinus.

Arundo donax L.

Very stout perennial reed with woody rhizome and erect culms often 4 – 6 m high. Leaf blades broad, flat, 2 – 5 cm wide; ligule a very short, hyaline rim, sometimes short-ciliate. Inflorescence a dense, terminal, plumelike panicle sometimes exceeding 50 cm in length and 10 – 15 cm in width. Spikelets c. 12 mm long; glumes lanceolate-acuminate, about equalling the lemmas; lemmas long-hairy at back, 3-nerved above, with a short awn from the sinus of the bifid apex.

Readily recognizable by its great size and hardly confusable with any other species except perhaps *Phragmites*. The latter, however, has glumes much shorter than the lemmas.

Habitat: Sometimes a cultivated plant but seen growing quite spontaneously along canals in al-Hasa Oasis. Occasional.

Central Coastal Lowlands: 1925, Dhahran Stables farm.

South Coastal Lowlands: 8666, 6 km NE al-Hufuf.

● 26. **Avena** L.

Annual grasses with spikelets on thin, often capillary, pedicels in an open panicle. Spikelets rather large, 1 – 4-flowered. Glumes (in ours) equal or subequal. Lemmas coriaceous or indurated, bifid or entire at apex, with a conspicuous kneed awn inserted near middle of the back.

Avena is represented in the Eastern Province by 6 or 7 species found as weeds around cultivated areas and also occasionally in the desert, where several species appear to be spontaneous. The following key follows Bor in Townsend and Guest (1968).

Key to the Species of Avena

1. Lemmas having, in addition to the dorsal awn, 2 much shorter, fine bristles or awns from the apex
    2. Callous blunt, with a cushionlike articulation, or not articulated; glumes subequal; spikelets 18 – 30 mm long .... 1. *A. barbata*
    2. Callous long-obconical-acuminate, 5 – 7 mm long, with a sharp articulation; glumes distinctly unequal; spikelets 30 – 40 mm long .... 5. *A. ventricosa*
1. Lemmas without apical bristles or awns in addition to the dorsal awn (but sometimes bifid-acuminate)
    3. Lemmas densely bearded at base and often hairy above on back; lowest lemma readily disarticulating, with a cushionlike callus
        4. All lemmas readily disarticulating at maturity with a cushionlike callus .... 2. *A. fatua*
        4. Only lowest lemma readily disarticulating, callused; those above persistent .... 4. *A. sterilis*
    3. Lemmas glabrous or with a few sparse hairs at base; lowest lemma not disarticulating, firmly attached .... 3. *A. sativa*

1. Avena barbata Pott ex Link
  incl. *A. hirtula* Lag.

Culms glabrous, solitary or several, erect, to c. 1 m tall. Spikelets 15 – 30 mm long, in a rich, narrow or broad panicle, often pendulous, 2 – 3-flowered. Glumes subequal; lemmas with short apical bristles in addition to the much longer dorsal awns, densely hairy at least in lower half, all articulated with, and eventually deciduous from, the rachilla.

The majority of our specimens have spikelets mostly 15 – 20 mm long with the apical bristles of the lemmas slightly exceeding the glumes, a form that has been segregated as a separate species, *A. hirtula* Lag.

Habitat: A weed around cultivation or occasionally in desert, where it is usually associated with old camp sites.

Northern Plains: 8756, ravine to al-Batin, 28 – 50N, 46 – 24E.

Northern Summan: 2779(BM), 25 – 50N, 48 – 00E; 1336(BM), 42 km W Qaryat al-'Ulya; 3873, 15 km WSW Umm 'Ushar.

Central Coastal Lowlands: 7427(BM), Dhahran; 7515, Ras Tanaqib; 7714, Ras az-Zawr, 27 – 29N, 49 – 16E.

2. Avena fatua L.

Culms usually several, glabrous, erect, to c. 1 m high. Sheaths and lower leaf blades more or less pubescent. Spikelets 15 – 30 mm long, 2 – 3-flowered with subequal glumes, in a spreading panicle. Lemmas hairy at base and more or less so throughout the lower half, bifid at apex, each disarticulating and falling separately from above.

Habitat: A weed around cultivation and also occasionally in desert, particularly on disturbed ground. Occasional.

Northern Plains: 1725(BM), 40 km E Abraq al-Kabrit; 1728(BM), Jabal al-'Amudah; 3750(BM), 24 km W as-Saffaniyah.

Northern Summan: 216(K), 25 km S Qaryat al-'Ulya.

Central Coastal Lowlands: 8649, Ras Tanura.

3. Avena sativa L.

Culms erect, glabrous. Leaf blades glabrous. Spikelets 2 – 3-flowered, c. 30 mm long (in ours), with nearly equal glumes, pedicellate in a spreading or nodding panicle. Lemmas usually entirely glabrous, even at base, all toughly persistent; only 1 lemma of the spikelet awned, the awn straight and rather short, or all lemmas awnless.

Habitat: This is the cultivated oat, sometimes found growing spontaneously around farms. Rare.

Central Coastal Lowlands: 434, 1070(BM), Dhahran stables farm.

Vernacular Names: ḤAṢĀD (Qahtan), ZARRIʿ (Shammar).

4. Avena sterilis L.

Coarse grass with erect culms to 1 m or more high. Leaf blades up to c. 30 cm long, 1 cm wide, somewhat scabrid at the margins or ciliate near the base. Spikelets 2 – 3-flowered, 18 – 50 mm long, pedicelled in a spreading panicle. Lemmas 15 – 40 mm long, glabrous or hairy at back, those above persistent, not articulated, falling as a unit with the lowest.

Our two specimens are assignable to subsp. *ludoviciana* (Dur.) Gillet et Magne

363

(sometimes treated as a separate species), characterized by spikelets 18 – 25 (rather than 30 – 50) mm long and the lower lemma 15 – 25 (rather than 25 – 40) mm long.

Habitat: A weed or ruderal around cultivated areas; rarely in the desert at abandoned camp sites. Both of our specimens were from disturbed sites with unusual numbers of introduced Mediterranean weeds.

Central Coastal Lowlands: 8255, Dhahran (spikelets 25 mm long).

Northern Summan: 8098, Batn al-Faruq, 25 – 43N, 48 – 53E, old Bedouin camp site. This specimen, apparently associated with spilled grain, had spikelets c. 20 mm long and lemmas densely bearded at base, sparsely hairy on backs, the lowest articulated but somewhat persistent.

5. **Avena ventricosa** Bal. ex Coss. et Dur.

A. bruhnsiana Gruner

Culms grouped, glabrous, to c. 40 cm tall. Leaf blades very narrow. Spikelets 30 – 40 mm long, 2-flowered, on pedicels shorter than the spikelet, in a racemose inflorescence. Glumes distinctly unequal, the lower c. 30 mm long, the upper c. 37 mm. Lower lemma 25 – 30 mm long, at base with a long, sharp, bearded callus 6 – 7 mm long and at apex with 2 bristles in addition to a stout dorsal awn twisted and pubescent below the knee, minutely and antrorsely scabridulous above the knee.

Habitat: Collected only once, in shallow sand over limestone at a coastal desert site. Rare.

Northern Plains: 1737(BM), Jabal al-'Amudah.

● 27. **Rostraria** Trin.

Annual grasses with inflorescence paniculate, often dense, contracted, spikelike. Spikelets compressed, 2- to several-flowered, disarticulating above the glumes. Glumes equal, or unequal with the upper larger. Lemmas bifid, awned from below the apex or from the sinus. An often revised group, with our species previously last assigned to the genus *Lophochloa* and still found there in many published works.

Key to the Species of Rostraria

1. Lemmas unequal; rachilla glabrous to very short-pubescent ..... 1. *R. cristata*
1. Lemmas subequal; rachilla long-hairy between the florets .... 2. *R. pumila*

1. **Rostraria cristata** (L.) Tzvelev

Lophochloa phleoides (Vill.) Reichenb.

Koeleria phleoides (Vill.) Pers.

Annual grass with culms erect or ascending from knees below, to c. 30 cm high. Leaf blades and sheaths more or less pubescent; ligule membranous. Panicle dense, spikelike, 1 – 8 cm long, obtusish, sometimes lobed. Spikelets nearly sessile, with glumes unequal, glabrescent to somewhat pubescent. Lemmas finely papillose, awned from slightly below the dorsal apex; awn usually shorter than the lemma.

Habitat: Apparently more common as a weed around cultivation but also present in the desert. Occasional.

Northern Plains: 3756(BM), 10 km W as-Saffaniyah.

Northern Summan: 732, 33 km SW al-Qar'ah wells, 27 – 14N, 46 – 38E.

Central Coastal Lowlands: 99(K), Dhahran.

South Coastal Lowlands: C. Parker No. 11(BM), farm near al-Hufuf.

2. **Rostraria pumila** (Desf.) Tzvelev
   Lophochloa pumila (Desf.) Bor
   Koeleria pumila (Desf.) Domin
   Trisetum pumilum (Desf.) Kunth

Annual grass with culms single to densely tufted, ascending to erect, 5 – 40 cm high, sometimes dwarfed. Leaf blades 2 – 6(9) cm long, 2 – 4 mm wide; ligule scarious. Panicle dense, spikelike, sometimes somewhat lobed, usually lanceolate in outline, 1.5 – 4(8) cm long. Spikelets 3 – 4 mm long; rachilla long-hairy between the florets; glumes subequal in length, but the lower usually broader, more or less densely short-pubescent, sometimes nearly glabrous. Lemma with a straight awn at back inserted slightly below the apex; awn nearly equalling or somewhat shorter than the lemma. *Plate 244.*

Habitat: Widely occurring in many different desert habitats on shallow to deep sands or sandy-silts. Also sometimes found on disturbed ground. Frequent to common.

Northern Plains: 1160(BM), 1161(BM), Jabal al-Ba'al; 1568(BM), al-Batin/Wadi al-'Awja junction; 1185(BM), Jabal an-Nu'ayriyah; 1253(BM), 15 km ENE Qaryat al-'Ulya; 1321(BM), 8 km W Jabal Dab'; 1480(BM), 50 km WNW an-Nu'ayriyah; 1647(BM), 2 km S Kuwait border in 47 – 14E; 1724(BM), 40 km E Abraq al-Kabrit; 8720, 20 km SW ar-Ruq'i wells, 28 – 51N, 46 – 25E.

Northern Summan: 1385(BM), 1359(BM), 42 km W Qaryat al-'Ulya; 2763, 25 – 49N, 48 – 01E; 3914, 27 – 41N, 46 – 24E; 7478, Batn Sumlul, 26 – 26N, 48 – 15E.

Northern Dahna: 4029, 27 – 35N, 44 – 51E.

Central Coastal Lowlands: 1059, 7894, Dhahran; 1772(BM), Abu Hadriyah; 1982, ad-Dawsariyah, 26 – 55N, 49 – 45E; 3277(BM), 3287, 8 km W Qatif-Qaisumah Pipeline KP – 144; 3301, 2 km W Sabkhat Umm al-Jimal; 4135, BM – 426, S Abu Hadriyah.

Southern Summan: 2059, Jaww ad-Dukhan, 24 – 48N, 49 – 05E; 2084, al-Ghawar, 24 – 27N, 49 – 05E; 2178, 24 – 25N, 48 – 41E; 2181, 24 – 34N, 48 – 45E; 2261, 29 km ENE al-Hunayy; 2148, 23 – 15N, 48 – 33E.

Vernacular Names and Uses: SHU'AYYIRAH ('little barley', az-Zafir), SAJIL (N Arabia, Musil 1927, 624). A useful and often plentiful spring grazing plant.

● 28. **Brachypodium** P. Beauv.

Annual or perennial glabrescent or sparingly pubescent grasses with flat leaf blades. Spikelets compressed, terminal and subterminal, many-flowered, with unequal glumes shorter than the spikelet. Lemmas awned.

Brachypodium distachyon (L.) P. Beauv.
Trachynia distachya (L.) Link

Annual grass, somewhat glaucous and sparingly pubescent, with culms spreading or ascending from the base, often kneed below, mostly 5 – 25 cm long. Leaf blades flat, rather broad for their length, mostly 1.5 – 5(8) cm long, c. 2 mm broad, glabrous or shortly and finely pubescent; ligule membranous but finely ciliate along upper margin. Spikelets terminal and subterminal, solitary or 2 – 3 together, laterally compressed, lanceolate, mostly 12 – 18-flowered, 2 – 3 cm long (excluding the awns), c. 5 mm wide, with awns 7 – 11 mm long. Glumes unequal, the upper c. ¼ as long as the spikelet, the lower shorter. Lemmas (in ours) lanceolate, glabrescent to thinly short-pubescent, ciliate along the upper margins, tapering above into an awn 7 – 11 mm long.

Habitat: Shallow sand among rocks. Infrequent. Also occurs on disturbed ground or at roadsides.

Northern Plains: 899, 1742(BM), Jabal al-'Amudah; 1000(BM), 12 km N Jabal al-'Amudah.

● 29. **Bromus** L.

Annual or perennial grasses with closed leaf sheaths and ligules membranous. Inflorescence paniculate. Spikelets cuneate or lanceolate in outline, several-flowered, with unequal glumes shorter than the spikelet. Lemmas bifid at apex, 1 – 2-awned from below the apex, or awnless.

Key to the Species of Bromus
1. Spikelets elliptical or lanceolate, tapering at apex
   2. Lemmas, at least the upper ones, with 3 awns .... 1. *B. danthoniae*
   2. Lemmas all 1-awned .... 3. *B. scoparius*
1. Spikelets narrowly cuneate, somewhat spreading at apex
   3. Panicle 1-sided, with spikelets secund, more or less parallel-horizontal
      4. Lemmas 10 – 16 mm long; lower glume 1-nerved .... 4. *B. tectorum*
      4. Lemmas 20 – 25 mm long; lower glume 3-nerved .... 5. *B. sericeus*
   3. Panicle and spikelets erect-ascending
      5. Panicle densely contracted, more or less cuneate, of 1 – 10 spikelets, the rachis internodes (except the lowest) 2 mm or less long .... 2. *B. fasciculatus*
      5. Panicle more or less elliptical-oblong, often somewhat interrupted, usually of more than 10 spikelets, with at least 2 lower rachis internodes 5 mm or more long .... 6. *B. madritensis*

1. **Bromus danthoniae** Trin.
Annual grass with 1 to several erect or ascending culms, sometimes kneed below, 20 – 30(40) cm tall. Leaf blades narrowly linear, mostly 1.5 – 3 mm wide, to 8(10) cm long. Panicle erect, contracted, narrowly ovate to oblanceolate in outline, mostly of 4 – 10 spikelets. Spikelets (10)20 – 25(40) mm long (not including awns), lanceolate in outline, glabrous or hairy. Lemmas (at least the upper ones) each with 3 often purplish awns from the back below the apex, the central awn longer; lower lemmas often with only a single medial awn.
● A grass known to be very variable in size of the spikelets and in other features. Varieties with glabrous spikelets (var. *danthoniae*) and hairy spikelets (var. *lanuginosus* Rozhev. in Fedtsch.) are recognized. Further collections are required to determine which form is predominant in our territory.
Habitat: So far only found on silty soil in the north. Rare to occasional.
Northern Plains: 1624(BM), 3 km ESE al-Batin/Wadi al-'Awja junction (var. *danthoniae*).

2. **Bromus fasciculatus** Presl
Pubescent annual with solitary or few-grouped erect culms, often dwarf and sometimes reddish-tinged, 5 – 10(20) cm high. Leaf blades mostly 3 – 6 cm long, 1 – 2 mm wide. Panicle erect, dense, tapering acutely at base, 3 – 6 cm long, 1.5 – 3.5 cm wide, with

ascending spikelets. Spikelets cuneate with acute base, c. 2.5 – 4 cm long (including awns), glabrous to pubescent. Lower glume 1-nerved; upper glume 3-nerved; both hyaline-margined. Lemma margins narrowly hyaline, often involute at least below. **Habitat:** Probably, if present at all, only in silty soils of the north.

● A few of our *Bromus* specimens with small, erect panicles have been dubiously attributed to this species. A recent determination of a similar example of ours by Prof. H. Scholz indicates, however, that such plants are more likely to be depauperate specimens of *B. madritensis* subsp. *haussknechtii* (see p.368). *B. fasciculatus* is described here for reference.

3. **Bromus scoparius** L.
Annual with culms solitary or grouped, sometimes weakly geniculate at base, 15 – 30 cm tall. Leaf blades more or less ciliate-hairy, at least toward the base. Panicles very dense and contracted, oblong in outline, mostly 3 – 6 cm long. Spikelets very short-pedicelled, mostly 15 – 18(20) mm long (not including awns), lanceolate in outline, glabrous or hairy. Lemmas with an awn inserted below the apex, the awn sometimes weakly recurved.
Habitat: A weed in cultivated areas or rarely in desert, where it probably is associated with disturbed ground. Rare.
Northern Plains: 615(K), al-Batin, 30 km NE Umm 'Ushar.
Central Coastal Lowlands: 7408(BM), Dhahran.

4. **Bromus tectorum** L.
Annual with tufted or solitary culms mostly 10 – 30 cm tall in our area, sometimes flowering as single-culmed dwarf plants only 5 – 8 cm high. Leaf blades and sheaths more or less pubescent. Panicle mostly 4 – 10 cm long, often somewhat silvery-shining from the exserted membranous lemma tips, with crowded spikelets on capillary pedicels usually bent to one side. Spikelets narrowly cuneate in outline when mature, broadening toward the apex, mostly 25 – 30 mm long (including the awns). Lemma acuminate, hairy or glabrous, with a minutely, antrorsely scabridulous awn inserted below the apex; the exserted portion of the awn often 10 mm or more long. *Plate 245.*

● Our specimens listed below were separated from *B. sericeus* by the traditional features: shorter spikelets and lemmas, and fewer glume nerves. Some of them, however, clearly have a tendency to bear supernumerary nerves on the glumes. Their indumentum ranges from subglabrous to rather densely hairy. Based on information received at time of going to press from Prof. H. Scholz of Berlin, it is apparent that some, if not the majority, of these specimens fit the characterization of the *B. sericeus* subspecies *fallax*, described by him in 1989 (see additional notes, under *B. sericeus*, p.368). No. 8786, however, was confirmed by Scholz as *tectorum*. The *B. tectorum/sericeus* complex certainly requires further study on a broader specimen base for our area and adjoining regions.
Habitat: Silty soils of basins and wadi bottoms in the northern plains and Summan. Occasional to locally frequent.
Northern Plains: 594(K), al-Batin, 16 km ENE Qulban Ibn Busayyis; 1199(BM), Jabal an-Nu'ayriyah; 8709, 20 km SW ar-Ruq'i wells, 28 – 51N, 46 – 25E; 8776 – 77, 18 km WSW ar-Ruq'i Post, 28 – 59N, 46 – 31E; 8786 – 87, 30 km SW ar-Ruq'i Post, 28 – 54N, 46 – 27E.

Northern Summan: 3896, 27 – 41N, 45 – 19E; 3343(BM), 5 km WSW an-Nazim; 719, 33 km SW al-Qar'ah wells; 769, 69 km SW al-Qar'ah wells; 7480, Batn Sumlul, 26 – 26N, 48 – 15E; 7550, 65 km ESE by E Umm 'Ushar; 8555, Humr Mathluth, 26 – 57N, 47 – 53E; 8572, 12 km WNW Ma'qala; 8591, 13 km NE Umm al-Hawshat; 8632, 2 km NE Dahl Abu Harmalah, 26 – 30N, 47 – 37E.

5. **Bromus sericeus** Drobov

Pubescent annual with fascicled, erect culms, 15 – 30 cm high. Leaf blades to c. 10 cm long; sheaths softly pubescent; ligule lacerate. Panicle with capillary, subpubescent branches, 1-sided with spikelets more or less horizontal in parallel groups. Spikelets c. 40 mm long (including awns), cuneate, more or less pubescent. Lower glume c. 14 mm long; upper glume c. 18 mm long; lemma 20 – 25 mm long (Bor in Townsend and Guest 1968).

● The two specimens listed below approach the typical variety of *B. sericeus*; No. 8777-A has lemmas up to 20 mm long. Further study, however, may prove that the subsp. *fallax* H. Scholz, described in 1989 (*Willdenowia* 19:133 – 36) is more common and may in fact encompass the majority of the specimens we attribute above to *B. tectorum*. Subsp. *fallax* differs from subsp. *sericeus* by its shorter lemmas and spikelets, which fall in the length range for those of *B. tectorum*. According to Scholz it differs from *B. tectorum* sensu stricto by having the mature florets coherent and falling as a unit from the glumes rather than singly.

Habitat: Silty soils of the northern plains. Apparently rare.

Northern Plains: 1625(BM), 3 km ESE al-Batin/Wadi al-'Awja junction; 8777 – A, 18 km WSW ar-Ruq'i border post, 28 – 59N, 46 – 31E.

6. **Bromus madritensis** L. subsp. **haussknechtii** (Boiss.) H. Scholz
    B. haussknechtii Boiss.

Annual with culms several, ascending to erect, occasionally 1-geniculate near the base, 12 – 25(40) cm high. Leaf blades long-tapering to a fine tip, flat or channeled, 4 – 12(18) cm long, 1 – 3(5) mm wide; sheaths with spreading short, and some longer, hairs. Panicle mostly elliptical to oblong in outline, somewhat cuneate at base, fairly contracted but sometimes interrupted, 5 – 10(12) cm long, 2 – 4.5 cm wide. Spikelets in our specimens 30 – 40 mm long (including awns); glumes keeled at back, hyaline-margined; lemmas mostly rounded at back, hyaline-margined except near the base, with a finely scabrid awn c. 15 mm long.

● Our nomenclature for this taxon is based on determinations of two specimens (Nos 8603 and 8724) by Prof. H. Scholz of Berlin. The spikelets on our plants vary from virtually glabrous (some individuals under No. 8746) to rather densely hairy (No. 8603). Feinbrun-Dothan's characters separating this species from *B. fasciculatus* (*Flora Palaestina* 4) are followed in part in our key.

Habitat: Shallow sand-silt along bases of rocky ravines. Occasional to fairly frequent locally in the far north.

Northern Plains: 8711, 20 km SW ar-Ruq'i wells, 28 – 51N, 46 – 25E; 8724, 30 km SW ar-Ruq'i wells, 28 – 47N, 46 – 21E; 8746(K), al-Batin tributary, 28 – 50N, 46 – 24E; 8773, 49 km NE Hafar al-Batin.

Northern Summan: 8603, 17 km ESE Jabal al-Burmah, 26 – 41N, 47 – 45E.

● 30. Chloris Sw.

Annual or perennial grasses with inflorescence of 3 or more digitate spikes. Spikelets 2 – 4-flowered with usually only the lowest fertile. Glumes persistent, unequal; lemmas, at least the fertile one, awned from near the apex. A genus with two (or possibly three) species in eastern Saudi Arabia, both introduced weeds unlikely to be seen at any distance from cultivation.

Key to the Species of Chloris

1. Awn of the lowest lemma c. as long as the lemma; 2 – 3 sterile florets in each spikelet .... 1. *C. gayana*
1. Awn of the lowest lemma more than twice as long as the lemma; sterile florets 1 .... 2. *C. virgata*

C. gayana normally has distinctively longer spikes in the inflorescence, but the variation encountered (5 – 11 cm) leads to an overlap of this character in the two species.

A third species, *C. barbata* Sw., having purplish spikelets with 3 rather than 2 awns, is well known in western Saudi Arabia and has been seen by the author around the city of Jiddah. It might also eventually be found as a weed in the east.

1. Chloris gayana Kunth

Perennial or annual grass with erect culms sometimes over 1 m high. Ligules membranous but very short and inconspicuous, long hairy above. Spikes 6 – 15, 5 – 11 cm long, inserted digitately in a terminal, umbellate inflorescence. Spikelets 3.5 – 4 mm long, laterally compressed, 2-awned; glumes very unequal, membranous, persistent, the longer 1-nerved and somewhat keeled at back, mucronate, somewhat shorter than the spikelet. Lowest lemma more or less hairy along the infolded margins but otherwise glabrous, awned from near the apex with the awn about equalling the lemma.
Habitat: An introduced weed around irrigated crops. Occasional to locally frequent.
South Coastal Lowlands: 7491, 7707, Harad Agricultural Project, Wadi as-Sahba; 7780, 7794(BM), adh-Dhulayqiyah, al-Hasa; 8202, 8203, al-Mutayrifi, al-Hasa Oasis; 8306, al-Hufuf, King Faisal University.

2. Chloris virgata Sw.

Annual grass with culms decumbent at base or erect, to c. 50 cm high; ligules short-membranous, ciliate above. Spikes 4 – 12(15), (3)3.5 – 5(6) cm long, inserted digitately in a terminal umbellate inflorescence, usually quite pale and tending to remain erect or even connivent, not spreading widely as in *C. gayana*. Spikelets c. 3 mm long, 2-awned, in 2 dense rows along the rachis. Glumes membranous, straw-colored, with a green medial nerve. Lemmas long-hairy along the upper margins and thus appearing to have a hair-tuft nearly as long as the spikelet, awned from near the apex with a straight awn more than twice as long as the lemma.
Habitat: A weed of irrigated croplands or gardens, usually on moist ground. Occasional.
South Coastal Lowlands: 8300, al-Hufuf, King Faisal University grounds; 8316, 5 km SE al-Hufuf, waste ground.

• 31. **Cynodon** L. Rich.

Perennial grasses of spreading growth with creeping rhizomes and stolons. Inflorescence of digitate spikes inserted terminally on the culms. Spikelets 1-flowered, awnless, in 2 ranks on the rachis with glumes more or less equal, shorter than the lemmas.

Cynodon dactylon (L.) Pers.
Perennial grass spreading widely by creeping rhizomes and stolons, rooting at the nodes and sending up shoots to c. 30 cm high. Leaves more or less distichous, the lamina to c. 7 cm long, sometimes white-hairy below around the mouth of the sheath; ligule a rim of short, fine hairs. Spikes (3)4 – 5, flexible, bearing crowded, imbricated, compressed, sessile or subsessile spikelets c. 2 mm long inserted somewhat obliquely or almost parallel to the rachis; glumes weakly keeled, acute to mucronate.

Desert and garden weed forms of this plant often differ greatly in size and somewhat in habit, the former having shorter spikes, more approximate leaves and being generally smaller overall.

Habitat: A very common, widespread and troublesome weed in gardens and farms, and also found occasionally in remote desert, especially around camp sites and other disturbed places.

Northern Plains: 3085(BM), al-Batin, 28 – 01N, 45 – 29E.

Northern Summan: 14(K), Rawdat Ma'qala; 811, 37 km NE Jarrarah; 3132(BM), 6 km SE ad-Dabtiyah; 3178(BM), 13 km SSW Hanidh.

Central Coastal Lowlands: 3061(BM), 8145, Dhahran.

Vernacular Names and Uses: THAYYIL (Qahtan), NAJM ('star-grass', perhaps referring to the star-like digitate inflorescence, al-Hasa farmer), NAJIL (Rawalah). *Cynodon* often gradually invades lawns in Eastern Province communities such as Dhahran. In such situations it provides some useful greenery but is inferior to *Paspalum vaginatum*, a valuable local lawn species, in lushness of growth. Around villages it provides forage for household goats but is not frequent enough in the desert to be of much importance for grazing.

• 32. **Schismus** P. Beauv.

Annual grasses with fine leaf blades. Inflorescence a more or less contracted panicle with spikelets laterally compressed, awnless, several-flowered, disarticulating above the glumes. Glumes subequal; lemmas bilobed at apex.

Key to the Species of Schismus
1. Lowest lemma with apical lobes acuminate and the fissure ⅓ to ½ the
   length of the lemma; palea reaching base of the fissure or a little
   beyond .... 1. *S. arabicus*
1. Lowest lemma with lobes acute or obtuse, but not acuminate, and the
   fissure ⅙ to ¼ the length of the lemma; palea exceeding the base of the
   fissure, often as long as the lemma .... 2. *S. barbatus*

1. **Schismus arabicus** Nees
Description, apart from the key characters above, generally as for the following species, which is much more common. Reliable identification of the two species (assuming they

202 *Pulicaria incisa* x 1/5

203 *Rhanterium epapposum* x 1/20

205 *Anvillea garcinii* x 1/4

204 *Rhanterium epapposum* x 3/4

206 *Asteriscus pygmaeus*
x 7/10

207 *Eclipta alba* x 1 1/5

209 *Anthemis melampodina* x 1 3/5

208 *Anthemis melampodina* x 1/5

211 *Aaronsohnia factorovskyi* x 1/2

210 *Aaronsohnia factorovskyi* x 3/10

212 *Senecio glaucus*   x 1/4

213 *Senecio glaucus*   x 1/2

214 *Calendula tripterocarpa*   x 2

215 *Echinops* sp. A x 1/6

216 *Echinops* sp. A x 5/6

217 *Echinops* sp. B x 1/4

218 *Echinops* sp. B x 1/2

219 *Acantholepis orientalis* (heads)   x 1/2

221 *Silybum marianum*   x 5/6

220 *Carduus pycnocephalus*   x 1/5

222 *Onopordum ambiguum*
x 1/20

223 *Onopordum ambiguum*
x 1

225 *Centaurea pseudosinaica*
x 1 1/3

224 *Centaurea pseudosinaica*
x 1/6

226 *Picris babylonica*   x 2

227 *Scorzonera tortuosissima*   x 1/4

228 *Scorzonera papposa*   x 3/10

229 *Scorzonera papposa*   x 2

230 *Launaea angustifolia* x 1/2

231 *Launaea angustifolia* x 1 2/5

233 *Launaea nudicaulis* x 1/4

232 *Launaea mucronata* x 1/8

234 *Reichardia tingitana* x 3/5

235 *Cymbopogon commutatus* x 1/5

236 *Hyparrhenia hirta* x 1/9

237 *Cenchrus ciliaris* x 1/10

238 *Panicum repens* x 1/2

239 *Panicum turgidum* x 1 1/4

240 *Aristida adscensionis* x 1/2

241 *Stipagrostis ciliata* x 1/12

242 *Stipagrostis drarii* x 1/20

243 *Phragmites australis* x 1/20

244 *Rostraria pumila* x 1/3

245 *Bromus tectorum* x 1/3

246 *Schismus barbatus* x 1/3

247 *Centropodia fragilis* x 1/15

248 *Dactyloctenium aegyptium* x 2/3

249 *Ochthochloa compressa* x 1/4

250 *Eragrostis barrelieri* x 1/3

251 *Cutandia memphitica* x 3/8

252 *Stipa capensis* x 1/5

254 *Aegilops kotschyi* x 3/5

253 *Hordeum murinum* x 3/5

255 *Eremopyrum confusum* x 1

256 *Cyperus arenarius* x 1/6

257 *Cyperus conglomeratus* agg. Form A   x 1/10

258 *Cyperus conglomeratus* agg. Form A   x 2/3

259 *Cyperus laevigatus* (c. 30 cm high)

260 *Asphodelus tenuifolius*
x 1/6

261 *Gagea reticulata* x 1 1/3

262 *Allium sindjarense* x 2/3

264 *Allium sphaerocephalum*
(with umbel in spathe) x 1/15

263 *Allium sindjarense* x 1

are specifically distinct) requires dissection of a spikelet although, as pointed out by Bor in Townsend and Guest (1968), *S. arabicus* usually has the glumes of the terminal spikelet over 5 mm long, while those of *S. barbatus* are usually shorter.

● This taxon has been immersed in *S. barbatus* by some workers, such as Täckholm (1974). Bor kept them separate, as does T. Cope (personal communication), who feels that not enough intermediates are known to justify their union. Detailed field studies of the distribution of the two in relation to microhabitat might well be helpful to understanding their relationship.

Habitat: Apparently essentially the same as the following species although not nearly as common.

Northern Plains: 1320(BM), 8 km W Jabal Dab'; 1708(BM), ash-Shaqq, 28–30N, 47–42E; 8721, SW ar-Ruq'i, 28–51N, 46–25E.

Northern Summan: 2801(BM), 15 km ESE ash-Shumlul.

Vernacular Names and Uses: as for the following.

2. **Schismus barbatus** (L.) Thell.

Dwarf, tufted annual grass, glabrous except around the ligules, with several to numerous culms prostrate to ascending, usually 5–15 cm long. Leaf blades involute, or sometimes flat, 2–5 cm long, 1–1.5 mm wide; ligule a line of somewhat exserted hairs. Panicles terminal, ovate-oblong in outline, mostly 1–3(4) cm long, 0.7–2 cm wide, contracted or somewhat loose, sometimes lobed. Spikelets green or purplish, pedicellate, laterally compressed, 4–5(6) mm long. Glumes subequal, lanceolate-acuminate, longitudinally nerved, white-margined, c. ⅔ as ⅞ as long as the spikelet, or equalling it. Lemmas pubescent, bifid at apex. Seeds ovoid, c. 0.75 mm long, smooth and somewhat transparent, resembling clear drops of amber. *Plate 246.*

Habitat: Deep or shallow sands or sandy silts. Very widespread and probably the most common of all the annual grasses of the Eastern Province.

Northern Plains: 1196(BM), 1200, Jabal an-Nu'ayriyah; 664, Khabari Wadha; 890, Jabal al-'Amudah; 1002, 12 km N Jabal al-'Amudah; 1167, Jabal al-Ba'al; 1255, 15 km ENE Qaryat al-'Ulya; 1603(BM), 3 km ESE al-Batin/Wadi al-'Awja junction; 1648(BM), 2 km S Kuwait border in 47–14E; 1684(BM), al-Hamatiyat, 28–50N, 47–30E; 1699(BM), ash-Shaqq, 28–30N, 47–42E; 3068(BM), ad-Dibdibah, 28–45N, 47–01E; 3078, al-Musannah, 28–43N, 46–47E; 3102(BM), al-Batin, 28–00N, 45–28E.

Northern Summan: 691, 20 km WSW Khubayra; 778, 69 km SW al-Qar'ah wells; 3243, 5 km N Umm al-Hawshat; 3228, 74 km S Qaryat al-'Ulya on Riyadh track; 1360(BM), 42 km W Qaryat al-'Ulya; 7476, Batn Sumlul, 26–26N, 48–15E; 2764, 25–49N, 48–01E; 3160(BM), 11 km SSW Hanidh.

Northern Dahna: 798, 15 km SE Hawmat an-Niqyan.

Central Coastal Lowlands: 91(K), 96(K), 99a(K), 417, Dhahran; 928, 10 km S al-Khafji/as-Saffaniyah junction; 3314, 5 km S Nita'; 1408(BM), 7 km NW Abu Hadriyah; 1461(BM), 2 km E al-Ajam; 1979, ad-Dawsariyah; 3280, 8 km W Qatif-Qaisumah Pipeline KP–144; 4063, az-Zulayfayn, 26–51N, 49–51E; 7728, 2 km S Ras Tanaqib, 27–49N, 48–53E; 7897, 15 km W Jabal Haydaruk.

Southern Summan: 2058, Jaww ad-Dukhan, 24–48N, 49–05E; 2136, 20 km WNW Jabal Dab'; 2241, 26 km ENE al-Hunayy; 2256, 29 km ENE al-Hunayy.

South Coastal Lowlands: 1441(BM), 2 km E Jabal Ghuraymil; 2116, 3 km SW al-Jawamir.

Vernacular Names and Uses: ṢUMAYMA' (Al Murrah, Qahtan, Bani Khalid, Shammar, Bani Hajir), KHAFŪR (N Arabia, Musil 1928b, 344; 1927, 601). An important and often very abundant spring grazing plant.

● 33. Centropodia (R. Br.) Reichb.

Perennial desert grasses with more or less pubescent leaves and ligule consisting of a rim of hairs. Inflorescence a dense, contracted, spiciform panicle. Spikelets 2 – 3-flowered with subequal, nerved glumes including the florets; lemmas pubescent, bifid at apex, with a straight awn inserted at base of the sinus.

Key to the Species of Centropodia

1. Anthers oblong, 0.7 – 1 mm long; awn of the lower floret 3 – 4 mm long; panicle 5 – 10 cm long .... 1. *C. forsskalii*
1. Anthers linear, 1.5 – 2 mm long; awn of the lower floret c. 2 mm long; panicle usually 15 – 20(25) cm long .... 2. *C. fragilis*

1. Centropodia forsskalii (Vahl) Cope
Asthenatherum forsskalii (Vahl) Nevski
Danthonia forsskalii (Vahl) R. Br.

Ascending or decumbent perennial grass, 10 – 30(40) cm high, with roots felt-covered and often with adherent sand. Leaf blades short-pubescent; sheaths with silky hairs; ligule a rim of hairs. Panicle terminal, densely contracted, partly sheathed at base by sheath of the highest leaf, which has a reduced blade. Spikelets c. 8 mm long, with acuminate, subequal glumes nerved and rounded at backs. Lemma pubescent, bifid at apex, with a (3)3.5 – 4 mm-long straight awn inserted slightly above the middle at base of the apical sinus and distinctly exceeding the lemma tip. Anthers oblong in outline, 0.7 – 1 mm long.

For further characters useful in distinguishing this from the following, rather similar, species, see notes with *C. fragilis*, below.

Habitat: Stable or mobile sands, sometimes even in active dune areas. Occasional to frequent.

Northern Plains: 834, 30 km SW al-Mish'ab.

Central Coastal Lowlands: 167(K), 178(K,BM), 234(K,BM), 439, Dhahran; 192(BM), al-Khursaniyah road junction; 3183(BM), 9 km N Shadqam junction.

Northern Summan: 7959, 25 km WSW 'Ain Dar GOSP–1.

South Coastal Lowlands: 388(BM), 7 km W Ghunan; 8024, Qalamat Dab'.

Southern Dahna: 2143, E Dahna edge, 23 – 07N.

Rub' al-Khali: 1015(BM), 23 – 03N, 49 – 55E; 7698, 8028, al-Mulayhah al-Gharbiyah, 22 – 53N, 49 – 29E.

Vernacular Names and Uses: QAṢBA' (Al Murrah). This name, closely related to the general Arabic name for reeds, *qaṣab*, is used quite specifically for *Centropodia* by Al Murrah and other southern tribes. The grass is never very abundant but is grazed by camels. The author's field experience confirms reports that its occurrence marks soft spots in the sand where camels find hard going and automobiles are liable to get stuck. This suggests that it is more frequent in former dune slip faces where sand stratification is steeply inclined.

2. **Centropodia fragilis** (Guinet et Sauvage) Cope
   Asthenatherum fragilis (Guinet et Sauvage) Monod
   Danthonia fragilis Guinet et Sauvage

Rather shrubby, ascending, perennial grass with several to numerous culms mostly 50 – 70(100) cm high. Leaf blades short-pubescent, 8 – 20 cm long; sheaths with silky hairs; ligule a rim of silky hairs. Panicle terminal, narrow-contracted, spiciform, 12 – 20(30) cm long, often weakly sheathed at base by the uppermost leaf sheath. Spikelets c. 8 mm long, with glumes equal, parallel-nerved and rounded at back. Lemmas distinctly shorter than the glumes, furnished with white silky hairs, bifid at apex, with a straight yellowish awn c. 2 mm long inserted c. ⅓ below the apex at base of the sinus, the awn hardly exceeding the lemma tip. *Plate 247.*

● Rather similar to the preceding, from which it usually but not always differs by its stouter, more erect habit and longer panicle. A clear feature in dissection are the anthers of *C. fragilis*, fully twice as long and distinctly narrower than those of *C. forsskalii*. The awn of *C. fragilis*, besides being shorter, only slightly exceeds the lemma apex while in *C. forsskalii* it distinctly does so. The florets of *C. fragilis* are shorter, with the lemma and awns well below the apices of the glumes; in *C. forsskalii* the lemmas or awns often reach the apex of the glumes or may even be slightly exserted.

Habitat: Deep sands, sometimes even on mobile dunes. Occasional.

Northern Dahna: 7541, 8 km NE Umm al-Jamajim.

Central Coastal Lowlands: 8333, 26 km SW Ghunan.

South Coastal Lowlands: 7913, 3 km S Jabal Ghuraymil, 25 – 46N, 49 – 32E.

Rub' al-Khali: 8035, al-Jawb, 23 – 07N, 49 – 53E; 7662, ash-Shuwaykilah, 19 – 55N, 48 – 58E.

Vernacular Names and Uses: QAṢBA' (see notes for the preceding species, which apply equally here).

● 34. Dactyloctenium Willd.

Annual or perennial grasses with an inflorescence of digitate, terminal spikes. Spikelets laterally compressed, 3 – 5-flowered, crowded in 2 rows along the rachis. Glumes subequal, the upper mucronate or awned.

Dactyloctenium aegyptium (L.) P. Beauv.

Annual grass, ascending or with creeping stems rooting at the nodes and sending up erect culms to c. 40 cm high. Leaf blades flat, thin, 3 – 8 mm wide, to c. 30 cm long, somewhat undulate or wavy at margin, often ciliate below; ligule a very short rim of fine hairs. Spikes 2 – 6, digitately spreading, 2 – 4 cm long, 4 – 7 mm wide, glossy yellow-grey to greenish, sometimes purplish below, the rachis produced in a short terminal mucro. Spikelets compressed, imbricate, 3 – 4 mm long, several-flowered, inserted in 2 lateral rows and set nearly at right angles to the axis. Upper glume strongly compressed, the keel produced in a curved mucro and short awn; lemmas acuminate to submucronate. *Plate 248.*

Habitat: A weed of garden and farm, and of roadside and walk edges in towns. Usually in shaded locations. Occasional to frequent.

South Coastal Lowlands: 1110(BM), 1881, Abqaiq.

Vernacular Names: 'IJLAH ('heifer-grass', Qahtan), RUSHAYDI (al-Hasa farmer), ZAHHAF ('creeper-grass', al-Hasa gardener).

● 35. Eleusine Gaertn.

Annual or perennial grasses with an inflorescence of 2 to several terminal, digitate or subdigitate spikes. Spikelets biseriate, few to several-flowered, disarticulating above the glumes. Glumes unequal, shorter than the first lemma.

Eleusine indica (L.) Gaertn.
Smooth erect or ascending annual grass, up to c. 50 cm high, with compressed culms and sheaths, nearly glabrous except around the fringed or nearly obsolete ligules. Leaf blades rather narrowly linear. Spikes 2 – 4(6), or even up to 13 in lush specimens, or occasionally single, digitately spreading, compressed, 4 – 10(15) cm long, c. 3 mm wide, often with 1 or 2 separate spikes inserted below the digitate ones. Spikelets 4 – 6 mm long, acute, imbricate, sometimes purplish, biseriate on one side of the rachis.
Habitat: A weed of irrigated gardens, farms. Apparently rather rare.
Central Coastal Lowlands: 7815(K), al-Khubar.
South Coastal Lowlands: 8297, al-Hufuf, King Faisal University.

● 36. Ochthochloa Edgew.

Perennial grass with an inflorescence of several digitate spikelike racemes. Spikelets several-flowered, imbricate, compressed; glumes unequal, shorter than the spikelet. A genus allied to *Eleusine*, in which our species has long been placed, but differing in several characters, chiefly that unlike *Eleusine*, the racemes fall entire.

Ochthochloa compressa (Forssk.) Hilu
Eleusine compressa (Forssk.) Aschers. et Schweinf. ex C. Christ.
Tough perennial grass, spreading by wiry, creeping, branched stolons rooting at the nodes, with erect culms up to c. 50(70) cm high. Basal sheaths woolly below. Leaf blades up to c. 4 cm long, 3 mm wide, much shorter above, rigid and somewhat recurved. Racemes spikelike, mostly 4 – 6 together, digitately spreading, compressed, 2 – 4(5) cm long, 4 – 8 mm wide, with a slender, pubescent rachis. Spikelets ovate, compressed, 3 – 8 mm long, imbricate, subsessile to short-pedicellate, 4 – 6-flowered, sometimes purplish; glumes somewhat unequal, distinctly shorter than the spikelet. *Plate 249.*
Habitat: Sandy wadi beds, often in rocky country. Occasional.
Northern Summan: 3137(BM), 6 km SE ad-Dabtiyah; 221(K), 10 km SSW al-'Uwaynah.
Southern Summan: 2263, 29 km ENE al-Hunayy; 8343, al-Ghawar, 69 km N Harad.
Central Coastal Lowlands: 3769, 30 km NE Abu Hadriyah.
South Coastal Lowlands: 8322, 5 km SE al-Hufuf.
Vernacular Names: 'IJLAH (Qahtan).

● 37. Leptochloa P. Beauv.

Annual or perennial grasses with inflorescence of several to numerous, simple, elongated spikes or racemes inserted along the main axis, with sessile or short-pedicelled spikelets. Spikelets 2-several-flowered, with unequal or subequal glumes. Lemmas (in ours) 2-toothed at tip with mid-nerve prolonged in a short mucro.

Leptochloa fusca (L.) Kunth
Diplachne fusca (L.) P. Beauv.

Tufted, glabrous, perennial grass with culms erect or ascending, often branched, to c. 1.5 m high. Leaf blades to c. 25 cm long, rather narrow and with convolute margins; ligule scarious, more or less coarsely laciniate. Panicle mostly 20 – 30 cm long, c. 6 cm wide, with numerous ascending, simple branches usually 4 – 7 cm long. Spikelets greyish-green, 5 – 10 mm long, acutish, short-pedicellate, weakly compressed, appressed parallel to the rachis. Glumes unequal, the upper c. half or less as long as the spikelet. Lemmas bifid at apex with a greenish, scabridulous midnerve prolonged in a short mucro.

Habitat: A weed of wet, irrigated lands. So far known in the Eastern Province only as a weed of gardens or lawns. Rare to infrequent.

Central Coastal Lowlands: 7462, Dhahran.

South Coastal Lowlands: 7810, adh-Dhulayqiyah, al-Hasa; 8205, al-Mutayrifi, al-Hasa Oasis; 8292, al-Hufuf, King Faisal University.

● 38. **Enneapogon** Desv. ex P. Beauv.

Annual or perennial grasses with narrow leaves; ligule a rim of hairs. Inflorescence a compact, spicate, terminal panicle of lanceolate to ovate, short-pedicelled spikelets. Spikelets 3 – 6-flowered, with only the lower florets bisexual and fertile, disarticulating above the glumes. Glumes membranous, unequal, exceeding the lemmas (except for the awns); lemmas 9-nerved, the nerves produced above as 9 scabrous to plumose awns often longer than the lemma.

Enneapogon desvauxii P. Beauv.
E. brachystachyus (Jaub. et Spach) Stapf

Dwarfish, finely silky-pubescent grass, our specimen apparently annual although the species is often described as perennial, with culms kneed below, 5 – 15 cm high. Leaf blades narrowly linear, up to c. 10 cm long, c. 2 mm wide. Panicles lanceolate in outline, 2.5 – 3.5 cm long, c. 1 cm wide, dense, spikelike, glossy pale greenish, often mottled with grey. Spikelets c. 7 – 8 mm long (including the awns), laterally compressed, finely pubescent, the membranous glumes unequal, exceeding all the lemmas (except for the awns). Awns of the fertile lemma 9, short-plumose in the lower half, in ours c. 5 – 6 mm long and 3 times as long as the lemma, well exserted from the spikelet.

Habitat: Our single record was in shallow sand in a rocky ravine. Apparently quite rare in the Eastern Province but inconspicuous and possibly sometimes overlooked among other small grasses.

Northern Plains: 3107(BM), al-Batin, 28 – 00N, 45 – 28E. The species appears to be generally restricted to Sudanian territory. Its presence in al-Batin is not wholly surprising, as that channel is known to be a northern route for the penetration of some Sudanian species into the northeast.

● 39. **Eragrostis** P. Beauv.

Annual or perennial grasses with narrow leaves and sheaths usually hairy at mouth; ligule a rim of hairs. Inflorescence paniculate, mostly open, rarely contracted, of laterally compressed, pedicellate spikelets. Spikelets several- to numerous-flowered, unawned, the glumes more or less unequal and distinctly shorter than the spikelet.

Key to the Species of Eragrostis
1. Lower panicle branches verticillate, the lower whorls with spreading whitish hairs at the axis branch point; panicles all terminal .... 2. *E. pilosa*
1. Panicle branches mostly all alternate, none with hairs at base; some panicles emerging laterally from the leaf sheaths .... 1. *E. barrelieri*

1. **Eragrostis barrelieri** Dav.
Annual grass, 20 – 40 cm high, glabrous except around mouths of the sheaths, which are hairy; culms ascending, kneed below or nearly straight, sometimes branching. Leaf blades to c. 8(10) cm long, c. 3 mm wide; ligule a rim of hairs. Panicles effuse, lanceolate to oblong in outline, mostly 4 – 11 cm long, 2 – 6 cm wide, terminal, usually with some smaller ones partially exserted from the upper or middle sheaths. Spikelets laterally compressed, linear-lanceolate, numerous, short-pedicellate on the subfiliform panicle branches, 1 – 1.5 mm wide, variable in length (3 – 17 mm, but generally fairly constant in individual plants), with 3 to 20 flowers. Glumes unequal, keeled, hardly reaching the middle of the first lemma. *Plate 250.*
Habitat: Found on ruderal sites and as a weed, also sometimes locally in the desert but then generally indicating somewhat disturbed conditions. Occasional.
Northern Summan: 3162(BM), 11 km SSW Hanidh.
Central Coastal Lowlands: 3768, 5 km N Abu Hadriyah.
South Coastal Lowlands: 7793, adh-Dhulayqiyah, al-Hasa; 7844, 5 km SE al-Hufuf; 8191, near ash-Shuqayq, al-Hasa Oasis; 8303, al-Hufuf, King Faisal University.

2. **Eragrostis pilosa** (L.) P. Beauv.
Annual with few to numerous ascending culms, to c. 55 cm high. Leaf blades flat, to c. 19 cm long, 4 mm wide, but often smaller; sheaths bearded at mouth with white hairs; ligule a dense rim of shorter hairs. Panicle usually very open, 15 – 25 cm long, 5 – 12 cm wide, with branches capillary, lax, verticillate below but alternate above; the lower whorls with some spreading white hairs 3 – 5 mm long at the branch point. Spikelets linear-elliptic, 3 – 5 mm long, c. 1 mm wide, 5 – 9-flowered, often purple-tinged, on pedicels often equalling or exceeding the spikelet. Lemmas 1 – 1.4 mm long.

No. 8805 is a short (25 cm), densely tufted form collected with No. 8804 but in a drier, exposed situation. Its panicles are smaller (6 – 11 cm long) and only c. 1 – 1.5 cm wide, contracted, with the branches appressed to the main axis even when mature. The other characters are generally typical.
Habitat: Only in nurseries and other cultivated areas. Infrequent.
Central Coastal Lowlands: 8804(K), 8805(K), nursery 3 km E Dhahran.

● 40. **Halopyrum** Stapf

Coarse, stout perennial seashore grasses with narrow, convolute leaves. Inflorescence a narrow, terminal panicle of laterally compressed, numerous-flowered spikelets. Glumes nearly equal, slightly shorter than the first lemmas; rachilla densely bearded between the florets at the base of each lemma.

Halopyrum mucronatum (L.) Stapf

Coarse perennial grass with erect culms and woody rhizome, to c. 1.5 m tall. Leaf blades narrowly linear, convolute, channeled, tapering to a fine subulate apex, to 30 cm or more long; ligule a dense rim of hairs. Inflorescence a narrow, contracted, subspicate panicle, linear to linear-lanceolate in outline, 10 – 30 cm long, with spikelets sessile to short-pedicellate on ascending to appressed lateral branches. Spikelets lanceolate, laterally compressed, mostly c. 1.5 cm long, 4 – 5 mm wide, c. 10-flowered. Glumes mostly equal or subequal, not quite reaching the apex of the first lemmas, acute, rounded at back or weakly keeled, 5-nerved. Lemmas 3-nerved, lanceolate, acute; rachilla densely bearded at base of each lemma with straight, white, ascending hairs ⅓ to ½ as long as the lemma. *Map 8.16.*

Habitat: Seashore sands, often on hummocks not far above the highest tide line. Widely but spottily distributed along the Gulf coast.

Northern Plains: 7975, Ras al-Mish'ab.

Central Coastal Lowlands: 118(K,BM), Ras Tanura; 3772, Ras az-Zawr.

South Coastal Lowlands: 7068, Ras as-Sufayra, 25 – 35N, 50 – 15E; 1938, 3 km S al-'Uqayr.

This monotypic genus is widely distributed around the shores of the Indian Ocean and extends up the Gulf coast as far as Kuwait.

Vernacular Names and Uses: 'IJLAH (Bani Khalid). The plant is much grazed by livestock, including horses, along the coast.

• 41. Lolium L.

Annual or perennial grasses with flat leaf blades and ligule membranous. Inflorescence a terminal spike with spikelets inserted alternately in 2 ranks in excavations of the axis. Spikelets laterally compressed, awned or awnless, several-flowered, with edge to the axis. Inner glume much reduced except in the terminal spikelet.

Key to the Species of Lolium

1. Glume from ¾ as long to equalling or exceeding the spikelet (ignoring awns when present)
    2. Glume nearly equalling the spikelet; awns conspicuous, longer than their lemmas; spikelets not turgid at maturity .... 2. *L. persicum*
    2. Glumes mostly well exceeding the spikelets; awns weak, shorter than their lemmas, or absent; spikelets swollen, turgid at maturity .. 4. *L. temulentum*
1. Glume ½ to ¾ as long as the spikelet; spikelets entirely awnless or with weak awns mostly shorter than their lemmas
    3. Lemmas entirely awnless; spikelets c. 10 – 15 mm long ...... 3. *L. rigidum*
    3. Lemmas, at least some of the upper ones, weakly awned; spikelets 15 – 25(30) mm long .... 1. *L. multiflorum*

1. Lolium multiflorum Lam.

Annual, nearly glabrous grass with culms to c. 1 m high. Leaf blades flat, to c. 20 – 25 cm long, 4 – 8 mm wide, auriculate at mouth of the sheath. Spike 12 – 25(30) cm long. Spikelets (12)15 – 25(30) mm long, 8 – 15-flowered. Glume about half as long as the spikelet. Lemmas c. 6 – 7 mm long, at least some of the upper ones with fine, straight, sometimes inconspicuous awns 2 – 10 mm long.

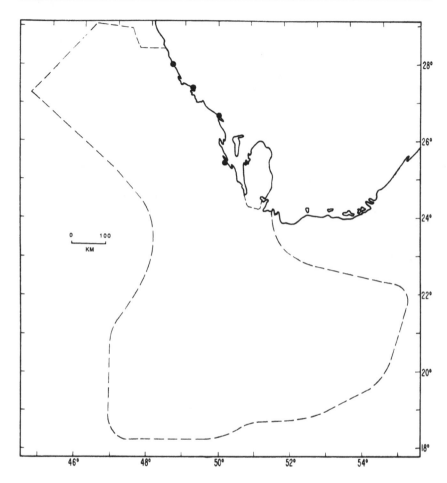

Map 8.16. Distribution as indicated by collection sites of *Halopyrum mucronatum* (●) in the coverage area of this flora (dashed line), eastern Saudi Arabia. Each symbol represents one collection record.

Habitat: A weed of roadside or gardens. Occasional.
Central Coastal Lowlands: 7074(BM), Dhahran; 7891(K), al-Khubar.

2. **Lolium persicum** Boiss. et Hohen.

Glabrous annual grass with culms simple or branching below, to c. 75 cm high. Leaf blades flat, 8 – 20(30) cm long, to c. 8 mm wide. Spikes mostly 15 – 30 cm long, sometimes somewhat interrupted near the middle. Spikelets 10 – 15(20) mm long (excluding the awns), closely appressed. Glume nearly or fully equalling, or exceeding, the spikelet. Lemmas 6 – 7 mm long, weakly bifid at apex, with a straight awn 6 – 13 mm long.

Habitat: A weed of disturbed ground along roadsides or around cultivated areas; also occasionally in the desert around old camp sites. Occasional.

Northern Plains: 1491(BM), 2 km NW an-Nu'ayriyah; 1716(BM), Abraq al-Kabrit; 3748(BM), 24 km W as-Saffaniyah.
Central Coastal Lowlands: 8254(K), Dhahran; 1089(BM), as-Sarrar (leg. Longhitano); 7890(K), al-Khubar.
South Coastal Lowlands: 8215(K), near ash-Shuqayq, al-Hasa Oasis.

3. Lolium rigidum Gaud.

Glabrous annual with culms ascending, somewhat kneed at base, usually 30 – 60 cm high. Leaf blades long-tapering to apex, mostly 5 – 20 cm long, to c. 8 mm wide. Spike 10 – 30 cm long, erect, more or less stiff, with spikelets appressed-ascending. Spikelets awnless, usually 10 – 15 mm long. Glume ½ to ¾ as long as the spikelet, rarely equalling it in a few spikelets.
Habitat: A weed of agricultural land, particularly around grain fields; also occasionally found at roadsides or on other disturbed ground.
Central Coastal Lowlands: 3060(BM, K), 8253, Dhahran.
South Coastal Lowlands: 8219(K), near ash-Shuqayq, al-Hasa Oasis; 8310, al-Hufuf.
Vernacular Names: ZUWĀN (N Arabia, Musil 1927, 631).

4. Lolium temulentum L.

Annual grass with several culms, up to c. 60 cm high. Leaf blades auricled above the ligule, to c. 30 cm long. Spikes erect, rigid, to c. 30 cm long. Outer glume well exceeding the spikelet, 7 – 9-nerved. Lemmas 6 – 8 mm long, usually weakly awned, becoming swollen, hard and turgid when mature. (Description largely after Bor in Townsend and Guest 1968).
Habitat: To be expected as a weed of grain fields.

Not yet recorded with certainty from the Eastern Province but has been collected on the edge of our territory at al-Kharj (S. A. Chaudhary, personal communication) and to be expected as a weed in cereal fields of the east.

● 42. Cutandia Willk.

Glabrous annual grasses with numerous, dichotomously branched panicles. Spikelets 3 – 8-flowered, bisexual, laterally compressed; glumes unequal or subequal, shorter than the spikelet; lemmas 3-nerved.

Cutandia memphitica (Spreng.) Benth.

Low, ascending or decumbent annual grass, 10 – 30 cm high, sometimes purplish-flushed. Leaf sheaths usually wider than the lamina; ligule scarious. Inflorescence of numerous, richly and dichotomously branched, elongated, zig-zagged panicles; panicle branches and pedicels flattened or angled, finely ciliate in the angles. Spikelets (6)7 – 9(10) mm long, several-flowered, with rachilla visible between the flowers; glumes somewhat unequal, about half as long as the spikelet, narrowly lanceolate-acuminate, keeled and puberulent at back; lemmas lanceolate-acuminate, weakly keeled, terminating in a mucro. *Plate 251.*
Habitat: Sand or sandy silts. A widely distributed and common spring annual.
Northern Plains: 1254(BM), 15 km ENE Qaryat al-'Ulya; 893, Jabal al-'Amudah; 996, 12 km N Jabal al-'Amudah; 1223, Jabal an-Nu'ayriyah; 8775, 18 km WSW ar-Ruq'i Post, 28 – 59N, 46 – 31E.

Northern Dahna: 797, 15 km SE Hawmat an-Niqyan; 2317, 22 km ENE Rumah; 4028, 27 – 35N, 44 – 51E; 7535, 8 km NE Umm al-Jamajim.

Northern Summan: 762, 69 km SW al-Qar'ah wells; 2830, 26 – 21N, 47 – 25E; 2850, 26 – 18N, 47 – 56E; 7479, Batn Sumlul, 26 – 26N, 48 – 15E; 1342(BM), 42 km W Qaryat al-'Ulya; 8442, 'Irq al-Jathum, 25 – 54N, 48 – 36E.

Central Coastal Lowlands: 100(K), 416, Dhahran; 927, 10 km S al-Khafji/as-Saffaniyah junction; 3282, 8 km W Qatif-Qaisumah Pipeline KP – 144.

South Coastal Lowlands: 2273, 13 km S 'Ain Dar GOSP 4.

Vernacular Names: ṢUMAYMA' ('hard grass', Al Murrah), ZARRI' ('seed-grass', Shammar, Rawalah).

● 43. Poa L.

Annual or perennial grasses with a paniculate inflorescence of laterally compressed spikelets. Spikelets awnless, with 2-several florets, the uppermost rudimentary, the lower bisexual. Glumes more or less unequal.

Key to the Species of Poa
1. Desert perennial with culms swollen, bulb-like at base . . . . 1. *Poa sinaica*
1. Annual weed; culms not bulb-like at base . . . . 2. *Poa annua*

1. Poa sinaica Steud.

Densely tufted perennial with several to numerous rather fine, smooth culms with basal bulb-like swellings clothed in pale sheaths. Culms erect, sometimes weakly 1 – 2-geniculate near the base, 15 – 30(50) cm high. Leaf blades fine, folded, almost capillary, 4 – 5 cm long. Panicle terminal, rather dense, sometimes somewhat interrupted, oblong to linear-lanceolate in outline, mostly 2 – 4 cm long. Spikelets 4 – 6 mm long, elliptical, compressed; glumes subequal, more or less keeled, acute, membranous above, exceeded by the lemmas; lemmas acute, membranous above, densely ciliate in the lower half. Anthers 1.5 – 1.7 mm long.

None of the spikelets in our specimens are viviparous or proliferating.

Habitat: Silty ledges along rocky ravines. Occasional in the far north.

Northern Plains: 8751, tributary to al-Batin, 28 – 50N, 46 – 24E; 8706, near ar-Ruq'i wells, 29 – 01N, 46 – 32E; 8783, 30 km SW ar-Ruq'i post, 28 – 54N, 46 – 27E.

Vernacular Names and Uses: NAZA' (northern Arabia, Musil 1927, 615); NIZZA' (Gillett and Rawi in *Flora of Iraq* 9:120). Known as a useful grazing plant in northern Arabia and the Syrian Desert but not plentiful enough in our territory to be of much importance.

2. Poa annua L.

Rather delicate, glabrous, annual grass, tufted, with fine culms 10 – 25 cm high. Leaf blades linear, flaccid, flat, to c. 10 cm long, 1 – 2.5 mm wide; ligule scarious, extending above the mouth of the sheath. Panicle terminal, more or less ovate in outline, open, 4 – 8 cm long, 2 – 5 cm wide, with capillary branches. Spikelets laterally compressed, c. 4 mm long, with the glumes unequal, exceeded by the lemmas. Glumes and lemmas white-margined; lemmas more or less fine-pubescent along the nerves. Anthers 0.5 – 0.8 mm long, about 4 times as long as broad.

Habitat: So far known only as a weed of lawns in urban areas; possibly introduced with lawn seed. Occasional to locally frequent.

Central Coastal Lowlands: 4075, 8767, Dhahran.

● 44. **Sphenopus** Trin.

Slender annual grasses with branched, open, paniculate inflorescence. Spikelets small, laterally compressed, several-flowered, awnless, on club-shaped pedicels, disarticulating above the glumes. Glumes unequal, short, well exceeded by the florets.

Sphenopus divaricatus (Gouan) Reichb.

Glabrous, rather delicate, often purplish annual grass with culms ascending, often kneed near the base, 10 – 30 cm long. Leaf blades 2 – 5(7) cm long, 1 – 2 mm wide, mostly involute; ligule membranous, well exserted above mouth of the sheath. Panicle terminal, ovate-oblong in outline, up to 12 cm long, 8 cm wide but usually smaller, open, with very fine capillary branches, the main branches in opposite pairs on the main axis. Spikelets pedicelled on the final branches, mostly 1.5 – 2.5 mm long, 4 – 5(7)-flowered, laterally compressed, on 2 – 5 mm-long clublike pedicels thickening upward. Glumes membranous, inconspicuous, minute, unequal, shorter than the first floret. Lemmas c. 0.7 – 1 mm long.

Habitat: Usually on damp, saline ground around the margins of *sabkhahs*; often associated with *Frankenia pulverulenta*. Very locally frequent but not widespread.

Northern Plains: 7978, *sabkhah* along Khawr al-Khafji; 7981, 24 km W al-Khafji; 7995, Sabkhat an-Nu'ayriyah.

Central Coastal Lowlands: 7961, 12 km S as-Saffaniyah.

● 45. **Parapholis** C. E. Hubbard

Annual grasses with flat leaf blades and erect or curved culms. Inflorescence a narrow, terminal spike with alternate, awnless spikelets inserted in excavations of the rachis. Spikelets 1-flowered, disarticulating above the glumes; glumes equal, inserted side-by-side in front of the floret.

Parapholis incurva (L.) C. E. Hubbard

Glabrescent, tufted annual grass, usually with characteristically incurved, decumbent to ascending numerous culms, to c. 20 cm high. Leaf blades mostly 3 – 5 cm long, sometimes up to 10 cm; ligule scarious. Spike linear, narrow, of jointed aspect, to c. 10 cm long, 2 – 3 mm wide. Spikelets alternate, appressed, immersed in longitudinal cavities of the rachis, 6 – 11 mm long, covered dorsally by the 2 lanceolate-linear glumes inserted side-by-side.

Habitat: Known only as a weed of farm or garden. Infrequent to rare.

South Coastal Lowlands: Chaudhary/Aramco H34, al-Hasa Oasis area.

● 46. **Phalaris** L.

Annual or perennial grasses with flat leaf blades. Inflorescence a spicate or capitate panicle with all spikelets bisexual and fertile, or with some spikelets reduced, sterile. Spikelets laterally compressed, 3-flowered with only the uppermost fertile. Glumes equal or subequal, often winged on the keels.

Key to the Species of Phalaris

1. Spikelets in clusters of 1 fertile with several sterile, empty ones below; fertile spikelets with glumes acuminate at apex and with acute appendages on wings .... 2. *P. paradoxa*
1. Spikelets all fertile; glumes shortly acute or mucronate but not acuminate, and with wings unappendaged, erose .... 1. *P. minor*

1. Phalaris minor Retz.

Glabrous, tufted, annual grass with culms erect or kneed near base, 30 – 70(100) cm high. Leaf blades flat, flaccid, long-tapering, to c. 25 cm long, 8 mm wide; ligule membranous, embracing the culm above the mouth of the sheath. Inflorescence a dense, terminal, spikelike, ovoid to cylindrical-lanceolate panicle, 2 – 7 cm long, 1 – 1.5 cm wide. Spikelets clustered in groups, short-pedicellate, strongly compressed, c. 4 – 5 mm long. Glumes equal, exceeding the lemmas, whitish with green nerves, the keel with a longitudinal, narrow, erose-dentate wing. Lemma finely appressed-pubescent, with a linear scale at base.

Habitat: A weed around gardens, farms; also seen in desert around old camp sites and apparently sometimes on quite undisturbed desert ground. Occasional.

Northern Plains: 3088(BM), al-Batin, 28 – 01N, 45 – 29E; 3751, 24 km W as-Saffaniyah.

Northern Summan: 1343(BM), 42 km W Qaryat al-'Ulya; 3342, 5 km WSW an-Nazim; 7940, 25 – 39N, 48 – 48E.

Central Coastal Lowlands: 438, 442, 553(K), 7422(BM), Dhahran; 1061, Dhahran stables farm.

South Coastal Lowlands: 8189, near ash-Shuqayq, al-Hasa Oasis.

2. Phalaris paradoxa L.

Glabrous annual grass with culms ascending-erect, or kneed at base, to c. 50 cm tall. Leaf blades linear, to c. 15 cm long, 4 mm wide; ligule membranous, extending above mouth of the sheath. Panicle dense, spikelike, terminal, oblong, mostly 4 – 6 cm long, c. 1.5 cm wide, often embraced at base by an expanded sheath. Spikelets strongly laterally compressed, pedicellate, in groups of 5 – 7, of which only one is fertile, the others reduced, empty. Fertile spikelet c. 7 mm long, with glumes equal, acuminate-subaristate at apex and with acute appendage on the dorsal wing. Lemma glabrescent.

Habitat: Our single specimen was growing as a weed in a newly-established lawn. Apparently rare.

Central Coastal Lowlands: 7421(BM), 8250(K), Dhahran.

● 47. Sporobolus R. Br.

Annual or perennial grasses with small spikelets in open or contracted panicles. Spikelets awnless, 1-flowered, disarticulating above the glumes. Glumes unequal, the upper often as long as the spikelet. Caryopsis free from the lemma and palea, falling free at maturity, the pericarp free from the seed.

Key to the Species of Sporobolus

1. Panicle open, verticillate-pyramidal, 2 – 10 cm wide .... 1. *S. ioclados*
1. Panicle linear-cylindrical, spikelike, only 2 – 3 mm wide .... 2. *S. spicatus*

1. **Sporobolus ioclados** (Nees ex Trin.) Nees

S. arabicus Boiss.

S. marginatus auct.

Perennial grass, densely tufted at base and often spreading by stolons, with erect culms to c. 75 cm high. Leaf blades stiff, narrowly linear, up to c. 20 cm long, tightly involute and thus appearing only 1 – 2 mm wide, glabrous at back but pubescent and somewhat mealy within the channel; ligule a rim of very fine hairs. Panicle terminal, contracted when young but later open with spreading verticillate, simple or shortly branched capillary branches, the panicle up to c. 16(22) cm long, 6(10) cm wide. Spikelets subsessile to shortly pedicellate, approximate to somewhat imbricate in the upper ⅔s of the panicle branches, 2.5 – 4 mm long, lanceolate, membranous. Glumes very unequal, the upper slightly shorter than the spikelet, the lower ¼ to ⅓ as long as the spikelet. Anthers pinkish, 1.2 – 2 mm long.

Our name for this species follows Cope (personal communication 1988); his opinion is that *S. arabicus* Boiss. and *S. ioclados* (Nees ex Trin.) Nees are conspecific.

Habitat: Found in a wide variety of terrain types but most commonly seen in coastal areas on low-lying, poorly drained, somewhat saline ground, or around *sabkhah* margins. It often spreads there with running stolons, establishing at intervals tight tussocks filled around the sheath bases with blown silt. Also found as a weed. Locally frequent.

Northern Plains: 837, 30 km SW Ras al-Mish'ab; 7973, Ras al-Mish'ab.

Central Coastal Lowlands: 18(K), Dammam; 165(K), Dhahran; 3730(BM), Ras az-Zawr.

Vernacular Names and Uses: RASHĀD ('straight-grass', gen.), HAWR (Bani Khalid). Apparently sometimes grazed in coastal zones but tough and probably saline, and considered to be poor forage.

2. **Sporobolus spicatus** (Vahl) Kunth

Stoloniferous perennial grass, creeping, rooting at the nodes and sending up ascending to erect culms to c. 35 cm high. Leaf blades rather stiff, mostly involute, to c. 8 cm long, pubescent in the channel above the ligule; ligule a short rim of hairs. Inflorescence a very narrow, tight, cylindrical-linear false spike 5 – 11 cm long, only 2 – 3 mm wide. Spikelets c. 1.5 mm long, lanceolate, membranous, subsessile or pedicellate on the closely appressed panicle branches. Glumes unequal, the upper c. ¾ as long as the spikelet, the lower c. ¼ to ⅓ as long as the spikelet. Anthers purplish, c. 0.5 mm long.

Habitat: A weed of lawn edges or waste ground around inhabited areas. Occasional to very locally frequent. Apparently a rather recent introduction, first appearing at Dhahran about 1974.

Central Coastal Lowlands: 4087, 7077, Dhahran.

● 48. **Stipa** L.

Perennial or annual grasses with leaf blades usually convolute. Inflorescence a panicle of pedicellate, narrow, one-flowered spikelets. Glumes membranous, long; lemmas convolute, narrow, acuminate, with an awn, often long, from the apex.

Stipa capensis Thunb.

S. tortilis Desf.

Annual, tufted, very prominently awned grass with numerous culms kneed at base, then erect, darkened at the nodes, 15 – 30(45) cm high. Leaf blades mostly 4 – 10 cm long, very narrow, convolute; ligule a very small ciliate rim; sheaths often expanded around the base of emerging panicles. Panicles erect, terminal, 8 – 30 cm long, emerging from sheaths at base, with awns erect and sometimes twisted together above. Spikelets crowded, narrowly lanceolate, acuminate, 15 – 20 mm long (not including the awn). Glumes membranous, glabrous, somewhat widening above the base then acuminate, subequal to somewhat unequal. Floret with sharp callus at base; lemma very narrow, 6 – 7 mm long, pubescent, jointed at apex with a long-exserted awn 6 – 11(13) cm long; awn 2-kneed, twisted and short-pubescent in lower half. *Plate 252.*

Habitat: Widespread, at least in the north, on firmer, silty soils but particularly characteristic and seasonally dominant in the shrubless *qar'ah* flatlands of the central northern plains.

Northern Plains: 1664(BM), 2 km S Kuwait border in 47 – 14E; 1514(BM), 30 km ESE al-Qaysumah; 1537(BM), 36 km NE Hafar al-Batin; 844, 9 km S 'Ayn al-'Abd; 1269(BM), 15 km ENE Qaryat al-'Ulya; 1330(BM), 8 km W Jabal Dab'.

Northern Summan: 3917, 27 – 41N, 46 – 24E; 2765, 25 – 49N, 48 – 01E; 1399(BM), 42 km W Qaryat al-'Ulya.

Central Coastal Lowlands: 101(K), Dhahran; 3743, 7715, Ras az-Zawr; 3269, 8 km W Qatif-Qaisumah Pipeline KP – 144.

Southern Summan: 1876(BM), 33 km N Harad.

Vernacular Names and Uses: ṢAMʻAʼ ('sharp-grass', Mutayr, Al Murrah, Rawalah, gen.). Grazed by all livestock when young but, as pointed out by Musil (1928b) and Dickson (1955), avoided and somewhat dangerous when maturing because of the sharply callused awned florets which may penetrate the mouth parts of animals. Dickson notes that the dry plants provide useful forage as standing hay after the fruits have fallen. In Iraq the presence of *Stipa* has been taken as an indicator of severe overgrazing. This may be at least partially true also in northeastern Arabia, where it seems to multiply at the expense of *Stipagrostis plumosa* or other preferred grasses on disturbed ground. Vesey-Fitzgerald (1957) provides a useful descriptive summary of *Stipa* habitats in northern and northeastern Arabia.

● 49. Hordeum L.

Annual or perennial grasses with flat leaf blades more or less auricled at base. Inflorescence a dense terminal spike with the spikelets grouped in 3s, the central sessile, bisexual, the laterals pedicelled or sessile, male or neuter (or sometimes fertile in cultivated species). Spikelets 1-flowered. Glumes subulate-linear; lemmas awned or awnless.

Key to the Species of Hordeum
1. Wild species with spikelets ultimately deciduous, on a fragile axis; lateral spikelets of the triad pedicellate
   2. Some glumes in each triad ciliate . . . . 2. *H. murinum*
   2. Glumes all non-ciliate (although somewhat scabrous) . . . . 1. *H. marinum*
1. Cultivated species with spikelets persistent on a tough axis; lateral spikelets of the triad sessile . . . . 3. *H. vulgare*

1. **Hordeum marinum** Huds. subsp. **gussoneanum** (Parl.) Thell.

H. geniculatum All.

Annual glabrous grass with numerous culms, kneed or decumbent at base, up to c. 30 cm high. Leaf blades flat, to c. 8 cm long. Spike to c. 3 cm long, 5 – 6 mm wide (excluding the awns). Central spikelet of the triad sessile, fertile, c. 5 mm long (not including the awn), with an awn c. 12 mm long from apex of the lemma. Lateral spikelets pedicelled, much reduced, not or hardly awned. Glumes of both the central and lateral spikelets awnlike throughout, not or hardly broadened below.

Habitat: The single record was from an open gritty plain with somewhat silty soil. Rare.

Northern Plains: 3747(BM), 24 km W as-Saffaniyah.

Vernacular Names: The name ZARRI' (Shammar) would probably be used in the north by analogy with *H. murinum*.

2. **Hordeum murinum** L. subsp. **glaucum** (Steud.) Tzvelev

H. glaucum Steud.

Annual grass with culms ascending to erect, mostly 10 – 35 cm high. Leaf blades flat, linear-acuminate, mostly 3 – 8 cm long, 2 – 3 mm wide, glabrous to thinly pubescent with scattered fine hairs; ligule scarious. Spike linear-oblong, 3 – 7 cm long, c. 8 mm wide (excluding the awns). Central spikelet of each triad c. 6 – 7 mm long, sessile, but with the floret short-stipitate above the awnlike glumes; lateral spikelets pedicelled, c. 9 mm long. Awn of the central spikelet c. 22 – 25 mm long, slightly exceeded by those of the acuminate lateral spikelets. At least some of the glumes in each triad distinctly ciliate at the margins. *Plate 253*.

At least some of our specimens have dimensionally dimorphic anthers, those of the central spikelet being c. 0.4 mm long while those of the laterals approach 1 mm. All are enclosed at anthesis.

Habitat: A weedy species usually indicative of habitat disturbance and found around desert camp sites. Also occasionally in apparently undisturbed habitats in the north. Occasional to locally frequent.

Northern Plains: 595(K), al-Batin, 16 km ENE Qulban Ibn Busayyis; 582, 15 km SSW Hafar al-Batin; 1606(BM), 3 km ESE al-Batin/Wadi al-'Awja junction; 3089(BM), al-Batin, 28 – 01N, 45 – 29E; 3101(BM), 3103(BM), al-Batin, 28 – 00N, 45 – 28E; 3911, 27 – 41N, 46 – 02E; 8789, 30 km SW ar-Ruq'i Post, 28 – 54N, 46 – 27E.

Northern Summan: 8592, 13 km NE Umm al-Hawshat.

Central Coastal Lowlands: 7420(BM), Dhahran; 8125, 10 km SSE al-Wannan.

Vernacular Names and Uses: ZARRI' (Shammar). Grazed by livestock when young but rough and less palatable when maturing because of its hard, sharp awns.

3. **Hordeum vulgare** L.

Stout, glabrous, annual grass with erect culms to c. 90 cm high. Leaf blades flat, tapering to apex, with small auricles at base; ligule a membranous rim. Spike oblong-lanceolate, 6 – 10 cm long, up to c. 2 cm wide (excluding the awns), dense, with 2 or 6 rows of fertile spikelets. In cultivated form *vulgare* ('six-rowed barley'), all 3 spikelets of the triad are fertile, sessile, 8 – 12 mm long, with the lemma produced into a straight, erect awn 12 – 15 cm long, and the awn of the central spikelet slightly longer than those of the laterals. In cultivated form *distichon* ('two-rowed barley', sometimes treated as a separate species, *H. distichon* L.), only the central spikelet of each triad is fertile,

c. 8 – 10 mm long, antrorsely scabridulous, with an awn up to c. 15 cm long, and the lateral spikelets are reduced, inconspicuous, and awnless.

Habitat: Occasional in springtime along roadsides, desert tracks, or around desert camp sites, arising from spilled grain.

Northern Plains: 1727(BM), Jabal al-'Amudah; 3749, 24 km W as-Saffaniyah.

Central Coastal Lowlands: 435, Dhahran Stables farm (two-rowed form); 441, 8251, 8252, Dhahran (two-rowed form).

Vernacular Names and Uses: SHA'IR (gen.). Cultivated barley, often used by Bedouins for livestock fodder in summer or when grazing is otherwise unavailable.

● 50. Aegilops L.

Annual grasses with culms often kneed below. Inflorescence a spike of terete, solitary, 2 – 8-flowered spikelets with the upper florets often sterile. Glumes awned or awnless; lemmas toothed or awned at apex.

Aegilops kotschyi Boiss.

Annual grass with numerous ascending culms, often kneed below, mostly 20 – 35 cm long. Leaf blades mostly 2 – 3 mm wide, ciliate on the lower margins and particularly around the short-membranous ligule and apex of the sheath; sheaths often with fine spreading hairs below. Spikes solitary, terminal, densely awned, lanceolate to oblong in outline and 2 – 3 cm long (excluding the awns), mostly with c. 4 spikelets. Glumes c. 5 – 6 mm long, parallel-nerved, truncate at apex, with 3(4) flat, straight, antrorsely scabrous awns c. 2.5 – 5 cm long. Lemmas with 1 – 3 awns. *Plate 254.*

Habitat: Rocky, somewhat elevated ground. Rather rare and sometimes found on disturbed ground.

Northern Plains: 7967, Jabal al-'Amudah.

Northern Summan: 3874, 15 km WSW Umm 'Ushar.

● 51. Triticum L.

Annual grasses with flat leaf blades and inflorescence a 2-rowed spike. Spikelets compressed, 3 – 9-flowered. Glumes nearly equal, usually shorter than the spikelet and asymmetrically keeled, with a tooth or awn at apex. Lemmas awned or awnless.

Triticum aestivum L.
T. sativum Lam.
T. vulgare Vill.

Annual grass with culms to c. 1 m high and leaf blades to 2 cm broad in cultivation, but spontaneous plants and escapes usually smaller, often around 40 cm high. Leaf blades flat, pubescent when young; ligule a membranous rim. Spike 5 – 12(18) cm long, c. 1 – 1.5 cm wide (excluding awns, if present). Spikelets inserted alternately in steps on a flattened, ciliate-margined rachis, approximate to subimbricate, dense, oblong, c. 1 – 1.2 cm long. Glumes slightly shorter than the lemmas, keeled off center and toothed or awned toward one side of the apex. Lemmas awnless (in beardless varieties) or with an awn to c. 7 cm long.

Habitat: The common cultivated wheat. Sometimes found as an escape or as spontaneous individuals along roadsides or in the desert, arising from spilled grain. Infrequent.

Central Coastal Lowlands: 7892, al-Khubar, roadside (a bearded variety).

Vernacular Names and Uses: ḤINṬAH, QAMḤ (gen.). Wheat cultivation and production in Saudi Arabia increased enormously in the early 1980s with Government support through guaranteed purchases at several times the world market price. Circular plots with center-pivot, overhead sprinkler irrigation became a widespread technique.

● 52. Eremopyrum (Ledeb.) Jaub. et Spach

Annuals with flat leaf blades and membranous ligules. Inflorescence a compact, compressed spike with spikelets 2 – several-flowered. Glumes equal or subequal, keeled, tapering to a point or into an awn; lemmas keeled, muticous or awned.

Eremopyrum confusum Meld.

Annual, 8 – 20(30) cm high, with culms ascending, 1 – 2-geniculate near the base. Leaf blades linear-lanceolate, to 8(10) cm long, 3 – 8(12) mm wide; ligule membranous, erose. Spike oblong-elliptic, or sometimes ovate, compressed, 2.5 – 5 cm long, 1.5 – 3 cm wide. Spikelets 2 – 3-flowered; glumes c. 15 mm long (including awn), keeled at back, glabrous or hairy, finely and antrorsely scabrid along the upper margins, long-tapering into a short awn; lemmas c. 20 mm long (including awn), rounded below at back, strongly keeled in the upper ¾, long-tapering into a finely scabrid awn c. 5 mm long. *Plate 255.*

The plants in our single collection of this species belong to var. *glabrum* Meld. ex Bor, with spikelets virtually glabrous except for the fine scabridities along the upper margins of the glumes and lemmas.

Habitat: Shallow sand-silt along edges of rocky ravines; rare.

Northern Plains: 8784(K), 30 km SW ar-Ruq'i border post, 28 – 54N, 46 – 27E.

67. JUNCACEAE

Herbs, mostly perennial, with round stems and cylindrical or grasslike, often basal, leaves. Inflorescence mostly cymose, of small bisexual flowers with 6 brownish, scarious, glumelike perianth lobes. Anthers 6 or 3. Ovary superior, with 1 style and 3 stigmas. Fruit a capsule, completely or partially 3-loculed.

Juncus L.

Perennial or annual herbs with simple stems and leaves terete or channeled, sheathing at base, or absent. Inflorescence of terminal or false-lateral cymes. Flowers with a perianth of 6 equal or unequal tepals, each with a stamen inserted at base. Capsule 3-valved with several to numerous seeds.

Juncus rigidus Desf.

J. arabicus (Aschers. et Buch.) Adams
J. maritimus Lam. var. *arabicus* Aschers. et Buch. ex Boiss.
Stout, tufted, glabrous perennial rush with creeping rhizome and numerous rigid erect stems up to c. 1.5 m high. Leaves (or sterile stems) arising from the base, terete, sharp-pointed, erect and mostly parallel, 2 – 3 mm in diameter, mostly somewhat shorter than the flowering stems. Inflorescence a false-lateral, contracted or somewhat loose, often lobed panicle of numerous flowers, up to c. 25 cm long, 8 cm wide, but usually smaller. Lower bract erect, appearing to continue the stem, pungent, usually not as long

as the inflorescence; upper bract much shorter, acuminate. Flowers in numerous peduncled clusters of 1–5. Perianth c. 4–5 mm long, the outer tepals lanceolate-acuminate, the inner ones oblong, acutish or obtusish, hyaline-margined at apex. Capsule lanceolate-ovoid, somewhat triquetrous, tapering to an acute apex, c. 3–4 mm long, 2 mm broad, smooth, brown, 1–1.7 times as long as the perianth, dehiscing by 3 valves above, persistent after dehiscence.

● Reports of *Juncus acutus* L. sensu stricto from eastern Saudi Arabia are probably erroneous. That species differs in having the inner tepals very broadly obtuse, truncate or retuse, with the hyaline margins prolonged into auricles. The capsule is 4–6 mm long, subspherical, and fully twice as long as the perianth. Täckholm (1974) has also noted a characteristic difference in habit: *J. acutus* has pungent stems radiating at many angles from the base and thus forming globose tufts, while *J. rigidus* has stems erect and parallel. The *arabicus* form of *J. rigidus* prevalent in our area has capsules sometimes well exceeding the perianth, and it is probably this character that has led to some attributions to *J. acutus*.

Habitat: Poorly drained and somewhat saline ground around the oases near irrigation run-off; also sometimes in coastal salt marsh. Locally very common.

Central Coastal Lowlands: 123(K), Ras Tanura; 1886(BM), Sayhat; 7403, al-Qatif; 8641, Sabkhat Khuwaysirah, 26–49N, 48–43E (at water hole); also seen near well on Jana Island, Arabian Gulf.

South Coastal Lowlands: 1911(BM), 1.5 km SSE ash-Shuqayq, al-Hasa Oasis; 8211, Aramco farm near al-Qarn, al-Hasa Oasis; 3800(BM), 10 km NW Salwah at water well.

Vernacular Names and Uses: NAMAṢ (gen.), WASAL (Bani Khalid), ḤALFA' (sic, al-Hasa farmer). Sometimes used in the oases as a material for weaving mats.

68. CYPERACEAE

Herbs or somewhat shrubby plants, mainly perennial and of wet habitats, with stems often trigonous and leaves grasslike but with sheaths closed. Flowers in spikelets with one to many florets and arising in the axil of a glumelike scale. Spikelets grouped in racemes, panicles or umbels, or sometimes heads, often bracteate. Perianth absent or reduced to bristles or scales; stamens 1–3; style branches 2–3. Fruit a nutlike achene.

The term 'glume' as applied here in this family refers to any of the undifferentiated scales of the spikelet.

Key to the Genera of Cyperaceae

1. Leaves flat, linear, rigid, harshly serrulate at margins and lower midrib,
     5–10 mm wide and over 1 m long; a 'sawgrass' 1–2(3) m tall ... 6. *Cladium*
1. Leaves not as above; plants usually (except *Schoenoplectus*) under 1 m high
     2. Spikelets more or less compressed; glumes distichous
          3. Spikelets in a dense, capitate inflorescence 1 cm or less broad;
             spikelets 2–7-flowered, of which only 1–3 are fertile, with glumes
             weakly and untidily distichous .... 5. *Schoenus*
          3. Spikelets in an umbellate inflorescence, or if capitate the head over
             1 cm broad; spikelets 8–numerous-flowered, fertile throughout,
             usually with glumes neatly distichous .... 1. *Cyperus*
     2. Spikelets terete, with glumes spirally disposed
          4. Style flattened and fimbriate below .... 4. *Fimbristylis*

4. Style not flattened or fimbriate
    5. Inflorescence terminal; spikelets mostly over 1 cm long in a simple
       umbel .... 3. *Bolboschoenus*
    5. Inflorescence false-lateral; spikelets under 1 cm long in a compound
       umbel .... 2. *Schoenoplectus*

● 1. Cyperus L.

Annual or perennial herbs with stem usually trigonous and leaves sheathing at base. Inflorescence mostly an umbel or head of spikelets, often subtended by 1 or more leaflike bracts. Spikelets compressed with glumes 2-ranked, imbricate. Fertile flowers bisexual, with stamens 3(1 – 2); style usually 3-cleft; perianth absent. Achenes mostly trigonous.

Key to the Species of Cyperus
1. Style bifid; plant of wet ground with spikelets in a dense head 1. *C. laevigatus*
1. Style trifid; farm or garden weed with spikelets in umbelled clusters, or
   plants of deserts or coast with spikelets variously disposed
    2. Leaves and bracts flat; weed of damp soil in gardens, lawns; rhizomes
       tuberous .... 2. *C. rotundus*
    2. Leaves and bracts channeled or teretish; rhizome not tuberous
        3. Plants under 30 cm high; roots glabrous to sparsely pubescent,
           not sheathed in dense, woolly tomentum holding sand
            4. Plant tufts linked by wire-like rhizome .... 3. *C. arenarius*
            4. Tufts not rhizome-connected .... 4. *C. jeminicus*
        3. Plants 30 – 70 cm high; roots sheathed in dense, woolly tomentum
           holding sand
            5. Achene obovate, unwinged; spikelets mostly 10 – 25 mm long with
               glumes persistent .... 5. *C. conglomeratus*, Form A
            5. Achene oblong, winged; spikelets mostly 25 – 55 mm long with
               glumes deciduous from base .... 6. *C. conglomeratus*, Form B

1. Cyperus laevigatus L.
Juncellus laevigatus (L.) C. B. Clarke
Perennial sedge with numerous terete, erect, dark green rushlike culms arising from a creeping rhizome, 30 – 100 cm high. Leaves inconspicuous, mostly reduced to thin basal sheaths. Inflorescence a dense, false-lateral, subglobular head 1 – 4 cm in diameter with few-many sessile spikelets; bracts 2, 1 erect and appearing to continue the culm, about equalling or slightly exceeding the head, 1 much shorter, sometimes covered by the head. Spikelets lanceolate-linear, 5 – 15(20) mm long, c. 2.5 mm broad, with numerous, close-packed, oblong glumes; glumes often chestnut and green-nerved medially, weakly carinate above. Achene obovate, compressed, whitish, c. 1.2 – 1.5 mm long. *Plate 259.*

Our specimens appear to belong to var. *laevigatus*. Var. *distachyos* (All.) Coss. et Dur. (*C. distachyos* All.) has also been observed and may be more common.
Habitat: Usually in or near fresh to brackish water along irrigation channels, wet spots or pools in the oases. Locally frequent.
Central Coastal Lowlands: 338, 1887(BM), Sayhat; 3799(BM), 3801(BM), 10 km NW Salwah, at well; 8208, near Umm al-Hamam, al-Qatif Oasis; 8206, Dhahran; C. Parker 60(BM), Dammam; V. Dickson, Sayhat (as *C. distachyos*, Dickson 1955, 99).
Vernacular Names: ḤAṢAL (Bani Hajir).

2. Cyperus rotundus L.

Perennial sedge with rhizome bearing oblong-ellipsoid blackish tubers; culm striate, trigonous, erect, leafy at base, up to c. 60 cm high. Leaves flexible, flat, to c. 40 cm long, 4 – 7 mm wide. Inflorescence umbelliform-paniculate, with unequal rays, often branched, subtended by several unequal, leaflike bracts well exceeding the inflorescence. Spikelets numerous, loosely clustered on the inflorescence branches, linear, mostly (in ours) 4 – 6 mm long, c. 1.5 mm wide, reddish-brown, green-nerved on the keel.

A variable species; varieties are known in the Middle East with spikelets up to 5 cm long.

Habitat: A weed of wet ground around gardens, fields, lawns. Frequent and often troublesome.

Central Coastal Lowlands: 1071(BM), Dhahran stables farm; 8256, Dhahran; 8207, 8264, Sayhat; C. Parker 59(BM), al-Qatif.

South Coastal Lowlands: 8302, al-Hufuf.

Vernacular Names: SI'D (al-Hasa farmer), SI'ID (al-Qatif farmer).

● S. A. Chaudhary (personal communication) reports collecting one or more plants in al-Qatif Oasis which he would assign to *C. esculentus* L., a widespread tropical weed. This species is differentiated from *rotundus* in the literature by its subspherical tubers, which are solitary and terminal (while *rotundus* has them elongated or irregular, and chained). We have seen no plants definitely attributable to *C. esculentus*, although it might not be unexpected as an introduced weed.

3. Cyperus arenarius Retz.

Glabrous, pale green perennial with culms often solitary but sometimes grouped 2 to several, often growing in straight lines as small, distant surface tufts arising from a wirelike horizontal rhizome running 10 – 15 cm below ground; fibrous roots very sparse at plant base but nodes of the rhizome with filiform rootlets; culms terete when fresh, 8 – 25(30) cm high. Leaves very narrow-linear to subfiliform, mostly 5 – 15 cm long, c. 1 mm wide, channeled above and rounded below, the channel closing near the subulate apex, mostly shorter than the culm except in very young plants. Inflorescence a dense subglobose head of radiating sessile spikelets, 1 – 2.5(3) cm in diameter, subtended by an erect, leaflike bract longer than the head and by a shorter bract below. Spikelets ovate to broadly lanceolate, 6 – 12(15) mm long, 3 – 5 mm wide, straw-colored to light brown, 8 – 25-flowered; glumes tightly imbricate, ovate-oblong, rounded at back and boat-shaped, short-mucronate. Achene obovate, c. 1.5 mm long, somewhat concave ventrally and obtusely keeled at back, grey to light brown (also reported black). *Plate 256.*

Habitat: Coastal sands, usually within c. 1 km of the shore. Locally frequent.

Central Coastal Lowlands: 17(K), Dammam; 8258, 2 km SE Dammam Port; 8271, northern al-Khubar.

4. Cyperus jeminicus Rottb.

Tufted perennial, (ours) only 10 – 20 cm high with glabrous to thinly pubescent roots and tufts not interconnected by rhizome. Leaves linear, channeled, c. 1 mm broad, mostly shorter than the culms. Spikelets 5 – 9 mm long, 1 – 2.5 mm wide, brown to reddish brown, scarious margined, more or less keeled at back, mucronate, in congested heads or clustered on short rays.

Our few specimens of this locally distinctive sedge are provisionally assigned this name at the recommendation of Dr Ilkka Kukkonen, Helsinki. Several coastal specimens of this plant were seen growing in the form of a peculiar hollow 'crown' or ring of culms.

Habitat: Sand or sand-filled interstices in limestone. So far seen at coastal sites or elevated spots not far from the coast.

Central Coastal Lowlands: 3793(BM), 5 km W Dhahran; 7750, 13 km NE Abu Hadriyah.

Cyperus conglomeratus Rottb. agg. ———— General Notes
Our plants of this long confused complex appear to be separable into at least two taxa. Their levels are still indeterminate but appear to approach species; the two primary segregates are described briefly below as Forms A and B. We are indebted to Dr Ilkka Kukkonen of the Botanical Museum, University of Helsinki, for comments on duplicates of some of our specimens during his studies, not complete at press time, of Arabian *Cyperus* of this group.

5. **Cyperus conglomeratus** Rottb. agg., Form A
C. aucheri auct.
Perennial 30 – 60(70) cm high with stout rhizome and roots woolly-tomentose, holding sand. Culms erect or ascending, more or less striate-sulcate or angular. Leaves narrowly linear, involute-channeled, teretish, to c. 50 cm long, mostly not reaching the inflorescence; bracts leaflike, 10 – 30(40) cm long. Spikelets clustered on rays in an umbellate inflorescence or sometimes in a single dense head, pale straw-colored or somewhat greenish when fresh, mostly 10 – 30 mm long, 2.5 – 4 mm broad, smoothish, with glumes densely imbricated; glumes generally persistent. Achene obovate, 2.5 – 3 mm long, 1.5 – 2 mm wide, concave ventrally, convex-rounded at back, the margins rather sharply angled but unwinged. *Plates 257, 258.*

● This is the most common form of the complex over most of our territory, being found from coastal sands to deep inland and the Dahna. It is the *C. conglomeratus* or *C. aucheri* referred to by the majority of authors dealing with our region. According to Kukkonen (personal communication), *C. conglomeratus* Rottb. sensu stricto, interpreted by him as a plant having unwinged achenes only 1.5 mm long, is not represented by any of our specimens. He feels rather that our form has strong affinities with *C. macrorrhizus* Nees and may well prove referable to that species. The question of the presence of *C. aucheri* Jaub. et Spach among our plants requires further study.

The specimens we place here are quite variable with respect to spikelet size and disposition. There is a continuum of variation in the achene: from plump, full fruits shallowly concave ventrally, filled with endosperm and ripening darkish brown, to very hollow, thin forms deeply concave, without endosperm and remaining yellowish. The latter appear to be abortive and infertile; variations of both forms may be present in the same spikelet, or one or the other may predominate. Specimens apparently of this species from the northwestern Rub' al-Khali have very small spikelets bearing the 'empty' fruit form exclusively. Such achene abortion could of course be a seasonal or drought effect. Such a habit would conserve moisture and energy and might be of adaptive significance in hyper-arid habitats. Form B of this complex (below), found in the most arid regions, regularly bears very thin, almost papery achenes with little

391

endosperm even when fertile. Such forms are also highly drought-resistant; achenes from Rub' al-Khali specimens were found to still contain considerable moisture 19 years after collection.

Habitat: Sands of all types. Extremely common over much of the Eastern Province except in parts of the Rub' al-Khali and perhaps adjoining regions, where it is replaced by Form B (below). Tolerates disturbed ground well and sometimes colonizes clean-graded sites in one or two seasons, providing useful, if not very dense, cover against wind erosion.

Northern Summan: 8562, al-Malsuniyah, 25 – 31N, 48 – 13E.

Central Coastal Lowlands: 1055, Dhahran; 1860, Abqaiq; 8637, 33 km W al-Jubayl, 27 – 00N, 49 – 20E; 8265, 8266, 5 km NE Dhahran.

Vernacular Names and Uses: 'ANDAB (Al Murrah, Bani Hajir), QAṢIṢ (Al Rashid, other tribes of the southeastern Rub' al-Khali), THUNDA' (N Arabia, cf. THUDDA' used by Al Murrah for the species when very small), MUSSAY' (Shammar), DAMDĪM (when dry, Al Murrah). Livestock graze the plant, which may be very abundant and virtually the only available forage at some times. Generally, however, it is considered to be coarse and inferior to most grasses and many other plants.

6. Cyperus conglomeratus Rottb. agg., Form B

Perennial to c. 60 cm high with culms erect or ascending. Leaves narrowly linear, involute-channeled, to c. 50 cm long, mostly not reaching the inflorescence; bracts leaflike, 10 – 30(40) cm long. Inflorescence rather diffuse, with spikelets sometimes only (1)2 – 4 together, spreading, pale, smooth with tightly imbricated glumes, (15)25 – 55(60) mm long, 2 – 3 mm wide; glumes rounded at back, finely mucronate, deciduous with the achenes from below often leaving the lower rachis naked. Achene oblong, 3 – 4(4.7) mm long, 1.5 – 2 mm wide, longitudinally hollow ventrally, rounded at back, the lateral margins with pale wings 0.5 – 0.7 mm wide and sometimes somewhat expanded near the middle (See notes with Form A, above).

Habitat: Deep sands. Characteristic of the eastern Rub' al-Khali but also found in the southern Dahna and may be more widely distributed in deep sands.

Southern Dahna: 8358, 24 – 15N, 48 – 10E.

Rub' al-Khali: 1935, Zumul Camp, 22 – 20N, 54 – 57E; 2890(BM), 7860, Ramlah Camp, 22 – 10N, 54 – 21E; 7880, Camp S – 3, 21 – 37N, 54 – 46E; 8406, Camp G – 3, 20 – 26N, 55 – 07E; 8422, G – 1 Airstrip, 20 – 44N, 55 – 09E.

Vernacular Names and Uses: 'ANDAB (Al Murrah, Bani Hajir), QAṢIṢ (Al Rashid, other tribes of the southeastern Rub' al-Khali), DAMDĪM (when dry, Al Murrah). A useful grazing plant in remote sand regions where better forage is not available.

● 2. Schoenoplectus (Reichb. ex Benth.) Palla

Annuals or perennials with terete or trigonous culms and leaves basal, often reduced. Inflorescence false-lateral with an erect lower bract; spikelets several, sessile or pedunculate; glumes numerous, all similar. Stamens 3; style base not enlarged, persistent as a fruit beak. Achene smooth or transversely ridged.

Schoenoplectus litoralis (Schrad.) Palla
Scirpus litoralis Schrad.

Rhizomatous perennial with culms erect, smooth, dark green, trigonous above, teretish and spongy at base, mostly 1 – 2 m high. Leaves present only at base, flat, thin and flaccid, often reduced to sheaths on the older culms but sometimes persisting when submerged, up to c. 55 cm long, 12 mm wide. Inflorescence a compound, rarely simple, umbel with unequal rays up to c. 3 cm long, each ray with 2 – 6 spikelets; bract stiff, erect, trigonous, sharp-pointed, appearing to continue the culm, exceeding the inflorescence. Spikelets narrowly ovate or lanceolate in outline, terete, mostly 6 – 9 mm long, c. 3 mm wide. Glumes oblong, membranous, glabrous, reddish-brown medially, keeled, bifid at apex, mucronate from apex of the keel. Florets with a perianth of bristles; filaments compressed; style bifid.

Generally much more robust than the following species.

Habitat: Wet ground or standing fresh to brackish water. Rare.

Central Coastal Lowlands: 8210, Dhahran, golf course pool.

● 3. **Bolboschoenus** Aschers. ex Palla

Perennials with culms leafy and with tuberous swollen bases. Flowers bisexual with or without hypogynous bristles, in pedunculate spikelets. Bracts several, leaflike. Style base not enlarged, persistent. Achene smooth.

Bolboschoenus maritimus (L.) Palla
Scirpus maritimus L.
Scirpus tuberosus Desf.
Perennial with rhizome swollen in tubers and culms erect, smooth, distinctly trigonous, 20 – 80 cm high. Leaves keeled below and channeled above at least in lower part, less than 5 mm wide, usually shorter than the culm. Inflorescence a simple terminal umbel with unequal rays up to c. 2.5 cm long, the umbel subtended by 1 to several linear bracts shorter or longer than the inflorescence. Spikelets terete, lanceolate in outline, tapering to an acute apex, mostly c. 15 mm long, 3 – 5 mm wide, pale brown. Glumes oblong to lanceolate, membranous except at the weakly keeled medial nerve, more or less bifid at apex, mucronate from the nerve. Florets with a perianth of bristles; filaments compressed. Achenes obovate, dark brown to black, smooth and shining, flat ventrally and rounded at back, c. 2.5 mm long.

Habitat: Swampy ground around fresh water in the oases. Infrequent to rare.

South Coastal Lowlands: 516(BM), 1 km N 'Ayn Umm Sab', al-Hasa Oasis.

● 4. **Fimbristylis** Vahl

Annual or perennial herbs with stems leafy below. Inflorescence generally umbellate, sometimes capitate or reduced to a solitary spikelet. Spikelets terete with glumes imbricate. Perianth absent; stamens 3(1 – 2). Style with a flattened and dilated base, usually ciliate or fimbriate, bifid or trifid at apex. Achene lenticular or trigonous.

Fimbristylis sieberiana Kunth
F. ferruginea (L.) Vahl var. *sieberiana* (Kunth) Boeck.
Tufted perennial, 20 – 40(60) cm high, with culms erect, finely striate and somewhat angular in cross-section. Leaves ascending from the base from broadened, brownish sheaths, narrowly linear, flexible, 10 – 25 cm long, only 1 – 1.5 mm wide, sometimes reduced and inconspicuous. Inflorescence a terminal (or rarely false-lateral) umbel of 2 – 6 spikelets with pedicels unequal, mostly 2 – 8 mm long; or spikelets rarely 1 – 2,

sessile (and then false-lateral). Spikelets ascending, ovoid, mostly acute, terete, brownish, finely and often obscurely tomentellous, 5 – 8(10) mm long, 3 – 4 mm wide. Glumes ovate, weakly keeled and short-mucronate above, tomentellous in upper part; style compressed, somewhat dilated at base, bifid at apex and ciliate at margins. Achene ovate, c. 1.5 mm long, flat ventrally and rounded at back, whitish to pale tan, very finely reticulate.

Habitat: Wet ground around fresh water in the oases; sometimes a weed in lawns or gardens. Infrequent.

Central Coastal Lowlands: 1923(BM), near al-Jarudiyah, al-Qatif Oasis; 1928, 3062(BM), Dhahran.

South Coastal Lowlands: 1894(BM), near 'Ayn Umm Sab', al-Hasa Oasis.

● 5. Schoenus L.

Perennial herbs with leaves reduced to basal sheaths. Spikelets capitate (as in ours) or spiciform or panicled, the inflorescence often bracteate. Spikelets more or less compressed, few-flowered, with glumes imbricate, 2-ranked (sometimes indistinctly so). Perianth of several bristles, or absent; stamens usually 3; style trifid at apex. Fruit a nutlike achene.

Schoenus nigricans L.

Tufted perennial with creeping rhizome and culms teretish, more or less striate, erect, 20 – 60(80) cm high. Leaves subfiliform, grooved, about half as long as the culms, arising basally from broadened, brownish sheaths. Inflorescence a very dense, contracted, terminal or somewhat false-lateral head 10 – 15 mm long, up to c. 10 mm broad, subtended by 1 or 2 erect, triquetrous bracts exceeding the head and partly sheathing it below with their broadened bases. Spikelets several, congested, erect-ascending, c. 10 mm long, few-flowered, more or less compressed but appearing rather irregular. Glumes lanceolate, keeled, acute-acuminate, light to dark reddish-brown, glabrous or sparsely tomentellous. Achene ovoid, obscurely trigonous, c. 1.5 mm long, smooth and shining, conspicuously milk-white.

Habitat: Swampy ground around fresh water in the oases. Infrequent to rare.

South Coastal Lowlands: 514(BM), 3 km N 'Ayn Umm Sab', al-Hasa Oasis.

● 6. Cladium P. Browne

Rhizomatous herbs with leaves flat or cylindrical, usually rough-serrulate at margins and keel. Inflorescence a panicle of 1 – 3-flowered, bisexual, more or less terete spikelets with imbricate glumes. Perianth absent; stamens 2 – 3; stigmas 2 – 3; achene ovoid or trigonous.

Cladium mariscus (L.) Pohl
C. jamaicense Crantz

Stout 'sawgrass' with heavy rhizome and culms erect, terete, hollow, leafy from the base to the inflorescence, up to c. 2 m high. Leaves linear, keeled, rather rigid, with tough sheaths below, up to c. 80 cm or more long, 3 – 10 mm wide, antrorsely rough scabrous-serrulate at margins and keel, with a long triquetrous point. Inflorescence a compound panicle up to 50 cm long, the branches ending in clusters of 3 – 20 spikelets. Spikelets terete, acute, 3 – 4 mm long, reddish brown, with 5 – 7 glumes.

Habitat: Wet, swampy ground in the oases. Apparently rare.

South Coastal Lowlands: So far known to have been collected only once: in 1953 by V. Dickson in the al-Hasa Oasis, on swampy ground near 'Ayn Umm Sab' (reported as *C. jamaicense* Crantz, Dickson 1955, 98). Her specimen (No. 734) at Kew (seen by the author) has the following collection data: 'near Um Saba Spring, Hofuf, 16 March 1953, swampy ground near stream. Tall rush 7′ high with clusters of brown flowers up the top of stem.' The specimen, which does not include the lowest parts, has culms 5–6 mm in diameter and leaves 4–6 mm wide and about 70 cm long. The panicle subgroups are c. 5 cm long and 5 cm broad, with spikelets 3–4 mm long, 1 mm wide, clustered at the panicle branch ends. The swampy area of this collection location was drained in the 1970s, and the habitat has grossly changed. The author visited the site in 1984 and found it dry, without sign of *Cladium*. A tour of adjoining lands also failed to disclose any specimens.

69. TYPHACEAE

Herbs of marsh or aquatic habitat with linear leaves sheathing at base. Flowers unisexual, in a spadix or globular heads, the staminate ones above. Perianth of bristles; floral envelopes absent. Staminate flowers with 2–5 stamens, the pistillate with a single free ovary; style and stigma 1. Fruit a nutlet.

● Typha L.

Stout, tall aquatic herbs with rhizomatous or stoloniferous rootstock. Leaves linear, leathery. Flowers in 2(3) very dense cylindrical, spadixlike spikes on the same axis, the staminate above. Staminate flowers without envelope, subtended by hairs, with 2–5 stamens; pistillate flowers subtended by numerous hairs. Ovary short-stipitate with hairs at base; fruit a nutlet.

Typha domingensis Pers.
T. australis Schum. et Thonn.
T. angustata Bory et Chaub.

Stout, erect, glabrous herb with teretish stems usually 1.5–3 m high. Leaves narrowly linear, leathery, sheathing at base, as long as or longer than the stem, narrower immediately above the sheath, 4–10 mm wide, flat above, rounded-convex below. Spikes cylindrical, extremely dense, terminal or subterminal, the pistillate one pale brown or pale greyish-brown, 15–40 cm long, c. 17–20 mm wide, the staminate one above on the same axis, usually separated by a few cm from the pistillate one below. Pistillate flowers c. 6 mm long, subtended by numerous white-silky hairs.

● Our single specimen has a very long (36 cm) pistillate spike and the staminate spike above apparently completely absent or fallen, represented above a 5 cm thicker interval by a completely naked axis c. 35 cm long.

Habitat: Relatively fresh standing water, as in run-off areas near springs. Rare.

Central Coastal Lowlands: 636(K), 'Ayn as-Subayghawi. Also seen near a leaking water well on Jana Island in the Gulf.

Vernacular Names: BARDĪ (gen.).

70. PALMAE

Woody trees, shrubs or vines with leaves ('fronds') in a terminal cluster or alternate, petiolate, with simple or pinnately compound blade. Inflorescence usually paniculate, subtended by 1 or more bracts, often enclosed when young in a spathe. Flowers small, actinomorphic, mostly unisexual, with a perianth of 6 segments in 2 series. Stamens usually 6. Ovary superior, 1 – 3-loculate; fruit a berry or drupe.

● Phoenix L.

Dioecious unbranched trees with rough trunk covered with persistent dead leaves or leaf bases. Leaves spreading from a terminal crown, pinnately compound with distinct woody midrib and leaflets numerous, pointed, induplicate, those below reduced to spines. Inflorescence paniculate, enclosed in spathes when young. Flowers with cupuliform calyx and corolla of 3 petals, the staminate with 6 or 3 anthers, the pistillate with an apocarpous ovary of 3 carpels, only one developing into the fruit. Seed grooved longitudinally on the ventral side.

A genus of 10 – 12 species (Corner 1966), often hybridizing and intergrading at the edges of their ranges, represented in eastern Saudi Arabia by a single, economically important species, the date palm.

Phoenix dactylifera L.

Dioecious tree with single trunk or with offshoots arising from the base, up to c. 15(25) m high, covered with reflexed persistent dead leaves or (when pruned) leaf bases. Leaves glaucous, spreading from a fibrous terminal crown on strong petioles broadened at base, mostly 3 – 5 m long, with a woody midrib, the leaflets numerous, induplicate, grading below into strong spines. Inflorescences several, richly paniculate, emerging from bases of the upper leaves, enclosed when young in reddish-brown to greenish compressed spathes c. 50 – 75 cm long which split longitudinally as the panicles emerge. Flowers sessile, inserted alternately on the numerous strands of the panicle (or so-called spadix), yellowish, glabrous and waxy-smooth; staminate flowers c. 8 mm long with calyx reduced to a very short, 3-toothed cup and petals 3, oblong-lanceolate, more or less acute, valvate, exceeding the 6 stamens; pistillate flowers c. 5 mm long, the calyx an obscurely lobed cup, the petals obtuse, tightly imbricate, with the 3 carpels slightly exserted. Fruit a drupelike 'date' developing from one carpel of each flower, sessile, generally ovoid to cylindrical-ellipsoid, highly variable in shape, color and size according to variety, ripening yellow to red, mostly 30 – 50 mm long, 20 – 28 mm wide, with the calyx persistent at base. Seed cylindrical-oblong, ventrally furrowed longitudinally, with micropyle on dorsal side.

Habitat: Intensively cultivated; also frequent and apparently quite spontaneous as scattered clumps in coastal desert tracts, in sand where fresh or brackish water is within reach of its roots (see below).

Central Coastal Lowlands: 'Wild' trees are most common within c. 30 km of the Gulf coast, particularly between Dhahran and al-Jubayl.

Vernacular Names and Uses: NAKHL, NAKHĪL (collective and plural), NAKHLAH (sing. unitat.). An extensive anatomical, agricultural and varietal vernacular vocabulary is associated with this plant, as would be expected in a species so important economically. The date palm is intensively cultivated in the larger oasis regions, particularly al-Hasa (where the date palm population has been estimated at over two million trees), and in the greater al-Qatif area. Vidal (1954), in a useful account of local date culture, notes that over 40 varieties are recognized in the Gulf region and lists 35 of eastern Saudi Arabia. Among the more important are the *khulāṣ* (primarily of al-Hasa and considered to be highest in quality), the *ruzayz* (of al-Hasa), and the *khunayzī* (of al-Qatif). Different varieties ripen at different times, and the market is thus provided with a cycle of different fruit types through the summer harvesting period.

Date production began declining in the late 1950s as the market weakened and farmers moved into more profitable truck gardening. The situation was complicated in the 1970s by a developing labor shortage and the frequent abandonment of this labor-intensive, low-profit activity for more rewarding enterprises. The Saudi Arabian Government began taking supportive measures in the early 1980s through a guaranteed purchasing program at subsidized prices, but it is questionable whether the earlier intensive monoculture of dates will (or should) be restored.

● Desert Populations of *Phoenix*

Some discussion is required about the widespread populations of uncultivated date palms so evident to any traveler in the coastal lowlands of eastern Saudi Arabia. These are perhaps most frequent in the sands, overlying *sabkhahs*, in the tracts between Dhahran and al-Jubayl within c. 30 km of the coast. Such trees, or clumps of trees slowly enlarging by basal offshoots from a 'mother' plant, are generally called *ḥīsh* (sing. *ḥīshah*). They are quite fertile, and 'wild-type' dates, generally inferior to those of the cultivated clones, are sometimes gathered from them by Bedouins, who refer to such untended fruits as *wahalān*. The populations appear to consist of staminate and pistillate trees in roughly equal numbers, indicating propagation from seed. They are pollinated naturally by the wind or, rarely by hand by Bedouins who may sometimes camp for long periods in their vicinity. The author annually collected staminate flowers from such plants near Dhahran for successful pollination of cultivated trees in the town.

Climatic and, in some areas, edaphic and hydrological conditions in coastal east Arabia are obviously suitable for the spontaneous establishment and growth of *Phoenix*. The controlling factor is the availability of subsurface water, and uncultivated palms today are generally found only in a rather narrow elevation range within a few meters of sea level, where sweet to brackish water lies within the root zone. This distribution is clearly evident on an aerial photo-mosaic of the Dhahran area, where the Dammam Dome can be seen to be surrounded just above *sabkhah* level by a relatively dense circular band of spontaneous scrub date palms.

Such palms have sometimes been described as relics of former cultivation, from an earlier period when the date palm oases were larger, more numerous, and better watered. Known areas of Medieval or earlier cultivation, however, do not appear to have appreciably higher numbers of such palms than do other parts quite devoid of archeological evidence. Other explanations would include chance establishment through date seeds discarded by man and the possible existence of a relict, truly wild population. The 'seed-drop' theory is tenable and generally consistent with today's botanical evidence. Germinating date seeds are often seen around inhabited areas and cultivation, although not very often in the desert.

The author, however, believes a 'relict wild population' hypothesis is worthy of consideration and further study. Various authorities point out that truly wild populations of the date palm are unknown. One wonders, however, how a spontaneous date palm can be with certainty said to be 'not wild'. One would in fact expect such a wild plant to resemble the cultivated date, but without some of the fruit virtues for which it has been selected. The situation is complicated by wind-assisted cross-pollination, whereby almost any remaining wild population would be inextricably linked with the gene pools of nearby cultivated stock.

The origin of the date palm has never been satisfactorily explained, and Corner (1966) concluded that the question was 'insoluble as ever' after giving arguments against the theory that it originated in India from *Ph. sylvestris*. Popenoe (1973) lists various suppositions about its geographical area of origin including the belief by Beccari, the late-19th Century monographer of *Phoenix*, that it was in our Gulf region.

Fossil palm wood of Miocene age is not infrequently collected in the central Arabian coastal plain. This cannot be attributed with any certainty to genus, but samples seen by the author appear to be structurally consistent with the trunks of *Phoenix*. Given the fact that coastal environmental conditions since the late Tertiary have probably remained within the tolerance range of *Phoenix*, one must consider the possibility of a date palm homeland here in one of the most important date-producing areas of the world.

71. LILIACEAE

Mostly perennial herbs, rarely shrubs or trees, with bulbous or rhizomatous rootstock and leaves alternate or whorled. Inflorescence racemose or (as in *Allium*) cymose-umbelliform; flowers usually bisexual and actinomorphic, 6-merous. Stamens mostly 6, in 2 series, inserted opposite the perianth segments. Ovary superior, sometimes adherent below, usually with axile placentation. Fruit a capsule or berry. As treated here, the family includes the Alliaceae, sometimes treated as a separate family largely on the basis of its inflorescence type.

Care is required in the field study or collecting of this group not to miss the bulb when present. This may be rather deep and difficult to extricate without breaking the tender stems above.

Key to the Genera of Liliaceae
1. Root without bulb .... 1. *Asphodelus*
1. Root with bulb
    2. Flowers yellow or greenish-yellow, solitary or 2−6 in a cymose or
       umbellike inflorescence .... 2. *Gagea*
    2. Flowers white, pink, or brownish, numerous or at least somewhat
       more than 6, umbelled or racemed
       3. Flowers white, purple-reddish, or both, numerous in a cymose
          umbel enclosed when young in a spathe .... 3. *Allium*
       3. Flowers brown, brownish-green, or brownish-coral, up to c. 25 in
          a raceme, bracteate but not spatheate .... 4. *Dipcadi*

● 1. **Asphodelus** L.

Herbs with non-bulbous root, sometimes with tubers, and a leafless scape. Leaves arising from the base. Flowers racemose or paniculate. Perianth regular with 1-nerved, petaloid segments. Ovary with 3 biloculate cells; capsule 3-valved, loculicidal.

Key to the Species of Asphodelus

1. Leaves not viscid; fruit pedicels jointed ⅓ to ½-way from the
   base .... 1. *A. tenuifolius*
1. Leaves viscid, usually with adherent sand at least below; fruit pedicels
   jointed ¹⁄₂₀ to ¼-way from the base
   - 2. Fruit pedicels ascending, jointed c. ⅙ to ⅕-way from their
     bases .... 3. *A. viscidulus*
   - 2. Fruit pedicels reflexed at base, jointed ¹⁄₂₀ to ¹⁄₁₀-way from
     base .... 2. *A. refractus*

1. **Asphodelus tenuifolius** Cav.

A. fistulosus L. var. *tenuifolius* (Cav.) Baker

Erect, glabrous annual herb with scapes solitary or several, often branched above with branches ascending-erect. Leaves numerous, ascending from base, narrowly linear-teretish, subulate above, mostly ⅓ to ½ as long as the scapes, only 1–2 mm wide, entirely smooth or sometimes very finely glandular or scabridulous in lower to middle parts. Flowers loosely racemed on the upper scape or branches; flowering pedicels more or less equal to the flower; bracts triangular, acuminate, whitish-scarious with a keeled medial nerve. Perianth c. 3 mm long, campanulate; lobes white with a purple medial nerve. Capsule nearly globose, c. 3 mm in diameter, on ascending pedicels mostly somewhat exceeding the capsule and jointed slightly below the middle. Valves of capsule sculptured with a series of parallel horizontal furrows and ridges. Seeds shaped like segments of an orange, sharply angular, c. 2–3 mm long, pitted on the flat inner faces and pitted and furrowed on the rounded back, brown to grey, very finely and shortly papillose. *Plate 260. Map 8.17.*

Habitat: Usually on somewhat silty soil but also found on sands. Common.

Northern Plains: 1299(BM), 15 km ENE Qaryat al-'Ulya; 666, Khabari Wadha; 682, 20 km WSW Khubayra; 1316(BM), 8 km W Jabal Dab'; 1507(BM), 30 km ESE al-Qaysumah; 856, 9 km S 'Ayn al-'Abd; 900, Jabal al-'Amudah; 1675(BM), 2 km S Kuwait border in 47–14E; 1162(BM), Jabal al-Ba'al; 1214(BM), Jabal an-Nu'ayriyah.

Northern Summan: 706, 33 km SW al-Qar'ah wells; 765, 69 km SW al-Qar'ah wells; 2771, 25–49N, 48–01E; 3330, 12 km WSW an-Nazim; 1394(BM), 42 km W Qaryat al-'Ulya.

Central Coastal Lowlands: 1041, 1832(BM), Dhahran; 7735, 2 km S Ras Tanaqib.

Southern Summan: 2186, 24–34N, 48–45E; 2055, Jaww ad-Dukhan, 24–48N, 49–05E; 2214, 26 km SSE al-Hunayy; 2230, 15 km WSW al-Hunayy.

Vernacular Names: BARWAQ (gen.), BAYRAQ (Al Murrah). Bedouins report that livestock avoid grazing on the plant. Dickson (1955) and others note that it is used by the Bedouins in making *iqt* (dried cakes of boiled curdled milk).

2. **Asphodelus refractus** Boiss.

Annual herb with one to several branching scapes ascending outside the center of the leaf rosette, 20–35(45) cm high. Leaves numerous, ascending from base, viscid and usually with adherent sand, narrowly linear and teretish with subulate apex, to c. ½ as long as the scapes, c. 1–2 mm wide. Flowers distant in very loose racemes on the upper scape branches; pedicels about equalling the flower; bracts scarious, triangular, acute, with medial nerve. Perianth 2–3 mm long, with lobes narrowly oblong to sublinear, white with greenish to reddish medial nerve. Fruiting pedicels reflexed at the base, jointed within c. 0.5 mm of the base, mostly 3–4 times as long as the capsule. Capsule subglobose or somewhat 3–lobed, to c. 4 mm broad, obscurely ridged and furrowed horizontally. Seeds c. 2.5–3 mm long, dark grey, sharply angular with the inner faces concave and 1 distinctly broader, the back furrowed. *Map 8.17.*

Habitat: Relatively deep sands. Apparently rare but perhaps often overlooked.

Northern Summan: 8103, 33 km NNW 'Uray'irah; 8449, 8 km W Nita'.

Northern Dahna: 8566, near 'Urayq al-Huwaymil, 25–25N, 47–37E.

3. Asphodelus viscidulus Boiss.

Herb with 1 to several erect scapes, often branched above and 10–25 cm high. Leaves rosetted, ascending, narrowly linear and terete, viscid and usually with adherent sand, ⅓ to ½ as long as the scapes and c. 1–1.5 mm wide. Scapes bearing loosely racemed, distant flowers, often for half or more of their lengths; bracts scarious, triangular, with a faint medial nerve and keel. Flowers 2–2.5 mm long on pedicels about equalling the flower; tepals oblong, white with reddish medial nerve. Capsule globose or depressed globose, 2.5–3 mm broad, faintly furrowed horizontally. Seeds c. 2 mm long, grey, sharply angled, the inner faces concave with 1 broader, furrowed at back. *Map 8.17.*

Habitat: Silty soils in basins; sometimes on rocky terrain. Occasional.

Northern Plains: 1484(BM), 50 km WNW an-Nu'ayriyah; 1739(BM), Jabal al-'Amudah.

Northern Summan: 2840, 32 km ESE ash-Shumlul; 8576, Dahl al-Furayy, 26–39N, 47–05E.

Central Coastal Lowlands: 1410(BM), 8 km NW Abu Hadriyah; 1802(BM), 5 km NE al-Khursaniyah; sight records at Dhahran.

Southern Summan: 24–00N, 49–00E.

● 2. Gagea Salisb.

Bulbous herbs, often dwarf, with narrowly linear leaves. Flowers solitary or in an umbellike inflorescence, usually yellow or greenish-yellow. Perianth funnel-shaped, divided into 6 flat segments. Ovary 3-angled and 3-lobed; capsule membranous, stipitate or not.

Gagea reticulata (Pall.) Schult. et Schult. f.

Sparsely pubescent dwarf herb, 4–11 cm high, with small oblong bulb c. 1–1.5 cm long; bulb with reticulate fibrous coats usually somewhat ascending the scape and with a dense matted network of fibrous rootlets at base. Leaves 1–2, very narrowly linear to filiform, longer than the inflorescence, channeled above. Flowers erect, solitary or 2–4(6) in an umbellike inflorescence, subtended by very narrow, leaflike bracts exceeding the flowers. Perianth 8–10 mm long at anthesis, enlarging with age to c. 20 mm long, with 6 flat, lanceolate, acute-acuminate tepals in 2 sets of 3, the outer slightly longer. Tepals spreading, bright yellow or greenish-yellow within, parallel-nerved, brownish-green at back. Capsule oblong, obtuse, 10–12 mm long, 5–7 mm wide, brownish. Seeds more or less half-orbicular, flat with a fine, raised, marginal rim, brown, c. 2 mm across. *Plate 261.*

A showy and attractive lily despite its small size. The author has seen one flower with supernumerary tepals (total of 9) and stamens (total of 8). The condition is apparently rare.

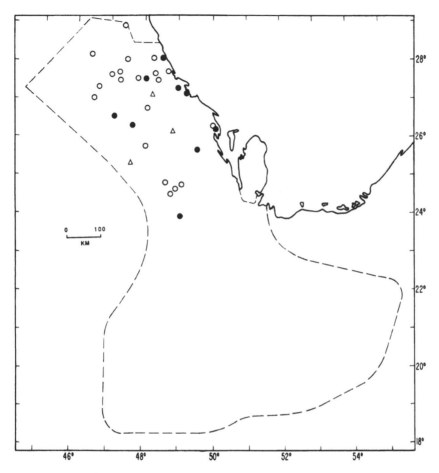

Map 8.17. Distribution as indicated by collection sites of *Asphodelus tenuifolius* (○), *Asphodelus viscidulus* (●) and *Asphodelus refractus* (△) in the coverage area of this flora (dashed line), eastern Saudi Arabia. Each symbol represents one collection record.

Habitat: Silts or silty sands, usually on rocky terrain. Occasional.
Northern Plains: 1176(BM), Jabal al-Ba'al; 1217, Jabal an-Nu'ayriyah; 8749, ravine to al-Batin, 28–50N, 46–24E.
Northern Summan: 736, 33 km SW al-Qar'ah wells; 1333(BM), 42 km W Qaryat al-'Ulya; 2803, 15 km ESE ash-Shumlul; 3233, 74 km S Qaryat al-'Ulya; 3335, 12 km WSW an-Nazim.
Vernacular Names and Uses: SHAḤḤŪM ('fat-wort', Musil 1927, 621; 1928a, 706). Musil notes that the plant was eaten raw by Bedouins.

● 3. Allium L.

Perennial herbs with tunicated bulb and scape sheathed below by leaves arising from the base, of characteristic onion odor. Inflorescence a terminal umbel of few to many flowers, enclosed when young in a membranous spathe. Flowers with 6 petaloid perianth segments in 2 series. Ovary free, 3-celled, or 1-celled by abortion; fruit a membranous capsule; seeds angular.

Key to the Species of Allium

1. Umbel densely globular; plant 30 – 110 cm high .... 2. *A. sphaerocephalum*
1. Umbel loose; plant 8 – 20(25) cm high .... 1. *A. sindjarense*

1. Allium sindjarense Boiss. et Hausskn. ex Regel

Herb (6)8 – 20(25) cm high with scape arising from an oblong to ovoid bulb 2 – 3(4) cm long, 1 – 2(3) cm wide with outer tunics brown and strongly fibrous-reticulate, the inner ones pale, shining, hardly fibrous; sometimes smaller bulblets with smoothish, hardly fibrous tunics also present. Leaves usually 3(1 – 4), with whitish membranous sheaths at very base, very narrowly linear to subfiliform, involute-channeled above, mostly 5 – 15(25) cm long, only 1 – 2 mm wide, well exceeding the scape or sometimes shorter. Umbel rather loose, with 5 – 35(52) rays spreading to ascending, somewhat unequal, (0.5)1 – 3(4) cm long, each terminating in a single flower. Spathe with nerves green when young, later becoming reddish or colorless. Flowers c. 5 – 6(7) mm long with tepals broadly to narrowly lanceolate, acute, white or weakly suffused with pink, with a prominent, sometimes keeled, medial nerve that is green when young, becoming reddish. Filaments simple, somewhat broadening at base; stamens often somewhat exserted. Fruit a membranous capsule c. 3 mm long with 3 narrow, almost winglike lobes. *Plates 262, 263. Map 8.18.*

The umbel on our No. 8790 has 52 rays, the longest measuring 4 cm. This plant has sometimes been confused in the literature with the more western species, *A. desertorum* Forssk., which differs in having non-reticulate bulb tunics and somewhat larger flowers.

Habitat: Almost always on silty soil. Occasional to locally frequent, mainly in our far north.

Northern Plains: 1272, 15 km ENE Qaryat al-'Ulya; 1190, Jabal an-Nu'ayriyah; 1421, 2 km E Qaryat al-'Ulya; 1611, 2 km ESE al-Batin/Wadi al-'Awja junction; 1634, 1663, 1678, 2 km S Kuwait border in 47 – 14E; 1715, 11 km NW Abraq al-Kabrit; 1721, 40 km E Abraq al-Kabrit; 1729, Jabal al-'Amudah; 7575, Jabal Dab'; 8755, ravine to al-Batin, 28 – 50N, 46 – 24E; 8772, 49 km NE Hafar al-Batin; 8766, ravine to al-Batin, 28 – 49N, 46 – 23E; 8790, 30 km SW ar-Ruq'i Post, 28 – 54N, 46 – 27E.

Northern Summan: 1354, 42 km W Qaryat al-'Ulya; 3880, 15 km WSW Umm 'Ushar; 8594, 38 km NE Umm al-Hawshat, 27 – 20N, 47 – 26E.

Central Coastal Lowlands: 1748, 31 km NNW Abu Hadriyah; 1784, Abu Hadriyah; 3723, 8 km N Abu Hadriyah; 3779, 3 km E Ras al-Ghar; 7742, 5 km NW Ras Tanaqib.

Vernacular Names and Uses: ṬīṬ (N tribes, Dickson 1955; cf. classical ṬīṬ, ṬīṬĀN used for wild onions). The inner bulb is most probably edible, but the plant is not apparently known to be used for food or flavoring.

2. Allium sphaerocephalum L.

Herb arising from bulb with brown, membranous to somewhat fibrous, tunics. Scape erect, 30 – 110 cm high, leafy somewhat above ground level with pale green leaf sheaths. Leaves pale green, smooth, linear, nearly flat below and involute-channeled or nearly

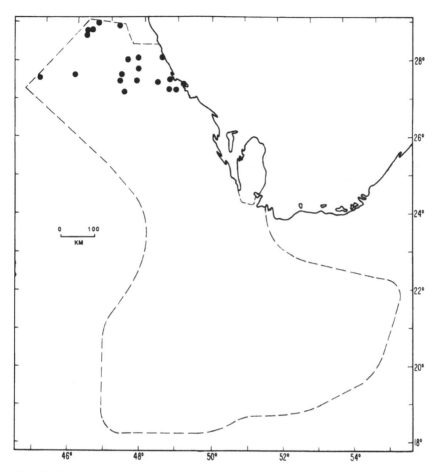

Map 8.18. Distribution as indicated by collection sites of *Allium sindjarense* (●) in the coverage area of this flora (dashed line), eastern Saudi Arabia. Each symbol represents one collection record.

terete above, c. 2 – 4 mm wide, rather flaccid and often somewhat pendulous and lying on the ground. Spathe tightly enveloping the young umbel, erect, terete and acuminate above, as if extending the scape. Umbel dense, globose, numerous-flowered, 3 – 5 cm in diameter. Flowers c. 5 mm long, purple or pinkish; tepals with a darker medial nerve. Inner 3 filaments 3-pointed, the central point bearing the anther; anthers often somewhat exserted, pinkish at least when young. *Plates 264, 265.*

Habitat: Shallow to deep sand; often in thin sand cover over limestone with the bulb in rock fissures. Occasional to locally common, with numerous waist-high scapes conspicuous in some areas in late spring.

Central Coastal Lowlands: 172, 1058(BM), Dhahran; 1788(BM), 13 km E Abu Hadriyah; 8634, 40 km NW Abqaiq.

Northern Dahna: 3251, S edge of 'Irq Jaham, 25 – 30N, 47 – 50E.

403

Vernacular Names: KURRÂTH, KURRAYTH (Mutayr, Shammar, 'Utaybah, Rawalah), BAŞAL ('onion', Al Murrah). Not known to be used for food although the bulb is edible.

● **4. Dipcadi** Medik.

Herbs with bulbs and linear, fleshy leaves. Flowers in a 1-sided raceme, brownish or greenish. Perianth tubular to funnel-shaped, deeply cleft into 6 acute lobes. Ovary 3-celled; capsule 3-lobed, opening by 3 valves.

Dipcadi erythraeum Webb et Berth.
D. unicolor (Stocks) Baker
Glabrous herb arising from an ovoid, 2 – 4 cm-long bulb with whitish scarious tunics. Scape single, usually ascending obliquely, 10 – 20 cm long. Leaves 3 – 4, narrowly linear, smooth, shining green, often exceeding the scape and accumbent on the ground, 3 – 5 mm wide, channeled above and rounded below, arising from the base with sheaths mostly below ground level. Flowers 5 – 15 in a somewhat 1-sided raceme along nearly the full length of the scape; pedicels shorter than the flowers and exceeded by the scarious, acute bracts. Perianth 12 – 15 mm long, campanulate, divided c. ½ way to base into elliptic-oblong, somewhat recurved lobes, greenish-brown to brownish-coral. Capsule ascending, broadly oblong in outline, obtuse, 12 – 15 mm long, with 3 rounded lobes, smooth-shining, greenish or brownish-grey. Seeds orbicular, flat-faced, black. *Plate 266.*
Habitat: Stable sands. Frequent.
Northern Plains: 1267(BM), 15 km ENE Qaryat al-'Ulya; 1203(BM), Jabal an-Nu'ayriyah; 1420, 2 km E Qaryat al-'Ulya; 1598, 3 km ESE al-Batin/Wadi al-'Awja junction; 932, 10 km S Khafji/as-Saffaniyah junction.
Northern Dahna: 2038, 9 km E Khurays.
Central Coastal Lowlands: 50(BM), 90, 1815(BM), Dhahran; 3317, 5 km S Nita'; 3721, 8 km N Abu Hadriyah.
Southern Summan: 2206, 26 km SSE al-Hunayy.
South Coastal Lowlands: 2130, Barqa as-Samur.
Vernacular Names and Uses: 'ANŞALÂN (gen.), 'AŞANŞAL (Rawalah, Shammar). The bulb is sometimes said to be edible; the author has found it bitter.

72. IRIDACEAE

Perennial herbs with corm, rhizome or bulb; leaves radical or radical and cauline, equitant. Flowers mostly regular, bisexual, rather showy, subtended by 2 spathelike bracts. Perianth of 6 petaloid segments in 2 series, often contracted below into a tube. Stamens 3, separate or monadelphous. Ovary inferior; style 1; stigmas 3. Fruit a loculicidal capsule with numerous seeds.

● **Gynandriris** Parl.

Perennial herbs with a corm with reticulate-fibrous tunic and channeled leaves. Flowers usually showy, of 6 petaloid, clawed segments, the outer 3 (the 'falls') spreading or reflexed, the inner 3 (the 'standards') more or less erect. Anthers 3, situated beneath the 3 petaloid style branches. Ovary 3-celled; fruit a capsule.

Gynandriris sisyrinchium Parl.

Iris sisyrinchium L.

Perennial herb, solitary or several-clustered, 10 – 30(50) cm high with stems erect, arising from an ovoid corm enveloped in brown, fibrous tunics. Leaves usually 2, ascending from the base, falcate, linear, channeled above with margins folded inward, rounded at back, tapering to an acuminate apex, parallel-nerved and leathery, somewhat glossy. Flowers 1 to several, racemed, short-pedicelled, 2 – 4 cm long, arising from papery, parallel-nerved, weakly inflated subcylindrical spathes with acuminate apex. Flowers in our plants usually with a ground color of purple to purplish-pink, faintly and radially veined darker. Falls spathulate with ascending-spreading claws, the limb reflexed, usually obtuse, with a white to yellowish central spot flecked below with purple; standards erect, oblanceolate, obtuse or acutish. V. Dickson (1955) notes that a form with very pale to white perianth ground color has also been seen; in the author's experience the flowers lighten somewhat with age. Stigmas 2-lobed, with the stamens adherent. Capsule to c. 3.5 cm long. *Plates 267, 268.*

One of the showiest desert flowers, sometimes growing in striking masses.

Habitat: Silty basins of the northern Summan and plains. Not common but sometimes very abundant locally, forming dense colonies. It is more frequent inland and northwest of our area in silt bottoms of the Hajarah and in Najd.

Northern Plains: 851, 9 km S 'Ayn al-'Abd. Vesey-Fitzgerald (1957) reported wide colonies in the floor of al-Batin near Hafar al-Batin, but this population has probably been destroyed or much reduced by recent urban expansion.

Northern Summan: 1422(BM), 2 km E Qaryat al-'Ulya.

Vernacular Names and Uses: ʿUNṢAYL (N tribes), SUʿʿAYD (N tribes, Musil 1928b, 360; 1927, 621). V. Dickson (1955) notes that the plant is bitter and seldom grazed by livestock.

73. ORCHIDACEAE

Terrestrial or epiphytic, sometimes saprophytic, herbs with leaves mostly alternate. Flowers racemed, in spikes, or solitary, bisexual, zygomorphic, with 6(5) petaloid perianth segments of which the outer 3(2) are referred to as sepals. Medial petal often forming a lip or 'labellum'. Functional stamens only 1(2), adnate to the style to form a column or 'gynandrium'. Ovary inferior, usually twisted in flower, 1-loculed. Fruit a capsule with numerous, dust-fine seeds. One of the largest families of flowering plants, widely distributed in the world but particularly characteristic of the tropics. They are (or once were) represented in eastern Saudi Arabia by one Mediterranean species, recorded only once.

● Orchis L.

Perennial herbs, usually with tuberous roots, with basal leaves rosulate and stem leaves sheathing. Flowers bracteate, in racemes or spikes, with subequal, free sepals. Labellum usually more or less trilobate, spurred.

Orchis laxiflora Lam. subsp. **palustris** (Jacq.) Aschers. et Graeb.

O. palustris Jacq.

Perennial herb with erect, stiff stem 30 – 60(80) cm high. Leaves linear, stiff, 10 – 20 cm long, 3 – 8(12) mm wide, tapering gradually from the base. Inflorescence a rather loose, erect spike 5 – 10 cm long, the flowers with acute, 3-nerved bracts. Flowers c. 15 – 20 mm long, pink to magenta, the lip often whitish at center with purplish markings, prominently spurred, with the central lobe often retuse, slightly exceeding the lateral ones. A Mediterranean and European species also known from Turkey, Iran and Afghanistan.

Habitat: Swamps or wet fields. Very rare and possibly now absent from our area. If still present, to be found in oasis swamps near fresh water.

South Coastal Lowlands: Collected and recorded once (as *O. palustris* Jacq.) by R. E. Cheesman, who found it blooming 28 January 1924 in a reed swamp c. 1 – 2 km north of 'Ayn Umm Sab', al-Hasa Oasis (Cheesman 1926). Cheesman reported seeing several specimens 'among the rushes along the marshy edge nearest the inlet of fresh water'. He described it as 'flower spike 1½ feet' (certainly referring to the total height of the plant), 'petals pale puce, leaves green and unspotted, several specimens growing in marshy land among thick rushes. In flower January 28. Faint, sweet smell.' Cheesman's specimen from this site, No. 5151(BM), seen by the author, has spikes 8 – 10 cm long and c. 4 cm broad. A small color illustration of a flower, probably sketched from life by the collector, is mounted with it.

The author has searched the find location of this plant several times without success. A general search of other areas within a few kilometers of 'Ayn Umm Sab' also failed to turn up any specimens. The reed and rush swamp described by Cheesman was drained in the early 1970s with construction of the al-Hasa irrigation and drainage system, and the local habitat (and possibly the entire population) of the orchid has been destroyed. Other swamps in the al-Hasa and al-Qatif areas ought to be checked carefully before writing it off as an Eastern Province species.

GLOSSARY OF BOTANICAL TERMS

Definitions are provided here only for botanical terms occurring in this book, in the senses used there. Readers seeking broader vocabulary guidance may consult the standard works of Featherly (1954) and Jackson (1928), both of which were used freely in the preparation of this list. Terms with the commonly used prefix 'sub-', which adds the meaning 'somewhat', 'partly' or 'nearly', are given only in their unmodified form.

ABORTION. The imperfect or arrested development of an organ.

ACAULESCENT. Stemless.

ACCUMBENT. Lying against another body.

ACHENE. A small, dry, indehiscent one-seeded fruit.

ACTINOMORPHIC. Radially symmetrical; regular.

ACUMINATE. Tapering to a prolonged point.

ACUTE. Distinctly pointed, the margins meeting at less than a 90-degree angle.

ADHERENT. Attached or joined though normally separate.

ADNATE. Joined or grown together, in reference to unlike parts.

AESTIVAL. Occurring in summer; in *Tamarix*, with the inflorescence on branches of the current year's growth.

AGGREGATE. Growing, or collected, in groups.

ALTERNATE. Placed singly at different heights on an axis or stem.

ALVEOLATE. With honeycomb-like pits or pattern.

AMPLEXICAUL. Clasping the stem.

ANDROECIUM. The stamens; the male parts of the flower.

ANGIOSPERM. A plant with seeds enclosed in an ovary.

ANNUAL. Completing a life cycle in one year.

ANNULATE. Ring-shaped.

ANTERIOR. On the front side.

ANTHER. The pollen-bearing portion of a stamen.

ANTHESIS. The time of flower opening, blooming.

ANTHOCYANIN. The blue or reddish pigment of some flowers or other plant organs.

ANTRORSE. Directed upward, toward the apex.

APETALOUS. Without petals.

APEX. The summit, tip of a part.

APICAL. At or pertaining to the tip or summit.

APICULATE. Having a minute pointed tip.

APOCARPOUS. Having carpels separate.

APPENDICULATE. With a small appendage.

APPRESSED. Lying flat against a part or organ.

APPROXIMATE. Close together but not united.

AQUATIC. Living in water.

ARACHNOID. Cobwebby, with fine entangled hairs.

ARISTATE. Awned or with a fine, awn-like apex.

ARMED. With defensive structures such as spines or prickles.

ARMAMENT. Collectively, the spines, prickles or other defensive structures.

AROMATIC. With a spicy, sweet or pungent smell or taste.

ARTHROPHYTA. An ancient division of non-flowering plants, mainly of the Paleozoic era, characterized by jointed stems.

ARTICULATED. Jointed, or appearing so.

ASCENDING. Rising somewhat obliquely.

ASSOCIATION. As used in this book, nearly synonymous with COMMUNITY; also, any regular occurrence together of two or more species.

ASYMMETRICAL. Irregular in shape, without radial symmetry.

AUGER-SHAPED. Screw-shaped.

AURICLE. A rounded or pointed earlike lobe.

AURICULATE. With auricles.

AUTECOLOGY. The study of the life history and environmental relations of an individual plant species; also, those aspects themselves.

AWN. A bristlelike appendage, especially in a grass spikelet.

AWNED. With an awn or awns.

AXIL. The upper angle formed by an attached part and an axis, especially a leaf on a stem.

AXILE. In or on the axis.

AXILLARY. In the axils.

AXIS. The main or central line of a plant or plant part.

BASAL. At the base.

BEAKED. With a prolonged, pointed projection.

BEARDED. With longish hairs or a hair tuft.

BERRY. A pulpy fruit with immersed seeds.

BIENNIAL. Completing a life cycle in two years.

BIFID. Cleft or divided into two terminal parts.

BIGIBBOUS. With two swellings.

BILABIATE. Two-lipped.

BILOBED. Two-lobed.

BILOCULAR. With two cells or compartments.

BIOMASS. The weight of living matter.

BIOTOPE. A region of more or less uniform habitat, with its living organisms.

BIPINNATE. With both primary and secondary divisions pinnate, as some leaves.

BIPINNATISECT. Deeply cut or divided pinnately into primary divisions and again into secondary ones, as some leaves.

BISACCATE. With two distended, baglike protuberances.

BISERIATE. In two rows.

BISEXUAL. Having reproductive parts of both sexes; perfect; hermaphrodite.

BIVALVED. With two valves.

BLADE. The broadened portion of a leaf.

BOSS. A rounded protuberance.

BRACT. A leaflike organ subtending a flower or other inflorescence unit.

BRACTEATE. With bracts.

BRACTEOLE. A small bract; the bract of a secondary inflorescence unit.

BRACTEOLATE. With bracteoles.

BRACTLET. A small bract.

BRISTLE. A stiff hair.

BULB. A modified, usually underground, bud.

BULBIFEROUS. Bearing bulbs.

BULBLET. A small bulb.

BULBOUS. Bulb-shaped, enlarged and rounded.

BULLATE. With blistered or puckered surface.

CADUCOUS. Falling early or easily.

CALLOSE. Thickened and hard-textured.

CALLOSITY. A hardened thickening.

CALLUS. A thickened or hardened part; in grasses, the extension, often pointed, at the base and point of insertion of a lemma.

CALYCULATE. Calyxlike; surrounding another part like a calyx.

CALYX. The outermost floral envelope; the sepals collectively.

CAMPANULATE. Bell-shaped.

CANESCENT. Finely and closely grey- or whitish-pubescent.

CAPILLARY. Hairlike, very fine and slender.

CAPITATE. Head-shaped; grouped in a dense, headlike cluster.

CAPITULUM. A dense head of sessile flowers.

CAPSULE. A dry fruit splitting open along lines or releasing seeds through pores.

CAPSULAR. With capsules; capsulelike.

CARPEL. A simple pistil or one unit of a compound pistil.

CARPELLATE. Having carpels, the term often with numerical prefix indicating their number.

CARPOPHORE. In a flower, the part of a receptacle prolonged between the carpels.

CARTILAGINOUS. Hardened, tough.

CARUNCLE. A protuberance near the hilum of some seeds.

CARUNCULATE. With a caruncle.

CARYOPSIS. The one-seeded fruit, or grain, of grasses, with the seed coat adnate to the pericarp.

CATKIN. A scaly, often pendulous spike bearing apetalous, unisexual flowers.

CAUDAL. Pertaining to, or like, a tail.

CAUDATE. Tailed; with a tail-like appendage.

CAULESCENT. Having a stem or stems.

CAULINE. On or pertaining to a stem.

CELL. A cavity or compartment, especially in an ovary; a locule.

CHAFFY. With small membranous scales.

CHAMAEPHYTE. A perennial plant with resting buds borne above but near ground level.

CHARTACEOUS. Papery or parchmentlike.

CHLOROPHYLL. The green coloring matter of plants.

CILIATE. Fringed with hairs.

CIRCINATE. Coiled into a ring, or partially so.

CIRCUMSCISSILE. Opening along a horizontal line around the circumference, especially a capsular fruit.

CLAVATE. Club-shaped.

CLAWED. With a claw, especially a petal.

CLAW. The narrowed, elongate base of some petals.

CLEFT. Divided about half-way or more to base into lobes.

CLEISTOGAMOUS. With fertilization taking place within the unopened flower.

CODOMINANT. Joining one or more other species in dominance of a plant community.

COALESCENT. Grown together, usually in reference to similar parts.

COHERENT. Joined, especially in reference to similar parts.

COLUMN. A central body formed from joined stamens or styles; in grasses, the lower, sometimes twisted portion of an awn.

COLUMNAR. Shaped like a column.

COMMISSURAL. At or pertaining to a place of joining.

COMMUNITY. As used in this book: a unit of vegetation, not formally ranked, characterized by one or more particular dominant, usually perennial, species and usually identifiable over a relatively broad geographical area or in different areas.

COMOSE. Bearing a tuft of hairs.

COMPOUND. Having similar parts aggregated in a common whole; in reference to leaves, a single leaf composed of two or more blades (leaflets).

COMPRESSED. Flattened.

CONCOLOROUS. Of the same color.

CONE. A dense, often ovoid, inflorescence composed of overlapping scales, characteristic of gymnosperms.

CONNATE. United.

CONNIVENT. Converging or coming together, but not connected.

CONOID. Cone-shaped.

CONTORTED. Twisted or bent.

CONVOLUTE. Rolled up longitudinally from the sides, as some leaves; with edges overlapping in the bud, as floral envelopes.

CORDATE. Heart-shaped and, in leaves, with attachment at the broader, notched end.

CORIACEOUS. With a leathery texture.

CORM. A bulblike but solid stem, usually underground.

CORNIGEROUS. With radiating spines, in reference to the globular head of *Echinops*.

COROLLA. The inner floral envelope consisting of petals; the petals collectively.

CORONA. A crownlike growth form or structure; in flowers a crownlike appendage standing between the stamens and corolla, or on the corolla.

CORTEX. The bark or rind of a stem or root.

CORYMB. A more-or-less flat-topped, broadened group of flowers with the outer flowers opening first.

CORYMBOSE. With, or in the form of, corymbs.

CREEPING. Running along the ground and rooting.

CRENATE. Edged with rounded teeth or lobules; scalloped.

CRENULATE. Finely crenate.

CRISPED. More or less curled.

CROWN. A corona; the juncture of roots with the stems; the spreading, leafy part of a tree.

CRUSTACEOUS. Hardened and brittle.

CRYSTALLINE. Shining crystal-like.

CULM. The stem of grasses and sedges.

CUNEATE. Wedge-shaped and, in leaves, with the point of attachment at the narrow end.

CUPULE. A cuplike structure.

CUPULIFORM. Cup-shaped.

CUSPIDATE. With a sharp, rigid point.

CYATHIUM. The specialized, ultimate inflorescence of *Euphorbia*, consisting of a trilocular ovary surrounded by single stamens equivalent to male flowers.

CYMBIFORM. Boat-shaped.

CYME. A group of flowers, often broad and flat-topped, with the inner flowers opening first.

CYMOSE. Cymelike; with cymes.

CYTOTAXONOMY. The classification of organisms based on cellular structure, especially the morphology and number of the chromosomes.

DECIDUOUS. Falling off naturally; not persistent.

DECUMBENT. Reclining but with ascending tips.

DECURRENT. With part of the leaf extending down the stem from the point of insertion.

DECURVED. Bent downward.

DECUSSATE. With pairs alternating at right angles.

DEFINITE. Of certain, usually invariable, number; not exceeding 10 or, as sometimes used, not exceeding 20.

DEFLEXED. Bent abruptly outward or downward.

DEHISCING. Opening when mature.

DELTOID. Triangular.

DENTATE. Edged with outward-pointing teeth.

DENTICULATE. Finely dentate.

DEPRESSED. More or less flattened from above.

DIADELPHOUS. With stamens in two groups.

DICHASIUM. A cyme branching in opposite directions from beneath an earlier-opening flower.

DICHOTOMOUS. Branching repeatedly in forks by pairs.

DICOTYLEDONOUS. With two seed 'leaves'; belonging to the Dicotyledoneae.

DIDYNAMOUS. Four-stamened with the stamens in two pairs of unequal length.

DIFFUSE. Loosely spreading or branching.

DIGITATE. Fingerlike, with similar parts spreading from a common point of insertion.

DIMORPHIC. Occurring in two different forms.

DIOECIOUS. Unisexual with the male and female flowers on different plants.

DISARTICULATING. Separating at a joint or joints.

DISC. A disk-shaped flower part; the central part of the capitulum in some Compositae.

DISCIFORM. Circular and depressed; disk-shaped.

DISCOID. Disk-shaped; in Compositae, a flower head with tubular disc florets but without radiate marginal florets.

DISSECTED. Cut or divided into many segments.

DISTAL. Toward the apex; at or toward the end opposite the point of attachment.

DISTANT. More or less widely spaced.

DISTICHOUS. Disposed in two ranks or rows.

DIVARICATE. Diverging widely.

DIVERGING. Inclining away from each other; separating gradually.

DIVIDED. Separated or segmented to the base or midrib.

DOMINANT. In plant communities, a species, often perennial and relatively large, that is predominant and which by itself or with other species defines the vegetation unit; also, having such characteristics.

DORSAL. On or at the back.

DORSO-VENTRALLY. From back to front.

DRUPE. A fleshy, one-seeded fruit with the seed enclosed in a stony endocarp, or pit.

DRUPACEOUS. Resembling, or bearing, drupes.

DWARF. Of much less than normal stature.

ECARUNCULATE. Without a caruncle.

ECHINATE. Furnished with prickles.

EDAPHIC. Of or related to the soil or soil conditions.

EFFUSE. Loosely spreading or branching; diffuse.

EGLANDULAR. Without glands.

ELLIPSOID. A solid body of elliptical outline.

ELLIPTIC, ELLIPTICAL. Ellipse-shaped, tapering to evenly rounded ends with the sides not parallel.

EMARGINATE. Notched at the apex.

EMBRYO. The rudimentary plant formed in the seed.

EMBRYONIC. In embryo form; at a very early developmental stage.

EMERGENT. Growing out; also, rising out of the surface of water.

ENDEMIC. Found only in a given region.

ENDEMISM. The state of being endemic; the degree of this state among the plants of a region.

ENDOCARP. The inner part, sometimes stony, of the wall of a mature ovary or a fruit.

ENSIFORM. Sword-shaped.

411

ENTIRE. Without teeth or lobes and evenly margined.

ENVELOPE. A surrounding part or group of parts, such as the corolla or calyx of a flower.

EPAPPOSE. Without pappus.

EPHEMERAL. Short-lived.

EPICALYX. A series of bracts near and resembling the calyx.

EPIDERMIS. The true cellular skin or covering of a plant below the cuticle.

EPIGYNOUS. Borne on the ovary; having flower parts inserted on the ovary.

EPIPHYTIC. Growing upon a plant but not parasitically.

EQUAL. Alike in length, size or number.

EQUITANT. Folded together and overlapping lengthwise.

ERECT. Upright, standing straight without inclination.

EROSE. Irregularly eroded, especially at the margins.

EVANESCENT. Short-lasting, soon disappearing.

EXSERTED. Protruding, sticking out.

EXSTIPULATE. Without stipules.

EXUDATE. A secretion from a surface or gland.

FALCATE. Sickle-shaped.

FALLS. The spreading and reflexed outer three, petaloid perianth segments in *Iris*.

FASCICLED. In bundles or clusters.

FERTILE. Capable of producing fruit; in reference to a vegetative part, bearing flowers or other reproductive organs.

FETID. With an unpleasant odor.

FIBROUS. With numerous woody fibers.

FIDELITY. The degree to which a plant is specific to a given vegetation unit or community.

FIELD. The background color of a part.

FILAMENT. The stalk of an anther.

FILIFORM. Threadlike; long and slender.

FIMBRIATE. Fringed; bordered with slender processes.

FISSURE. A line of splitting or a deep, narrow indentation.

FLACCID. Limp.

FLEXUOUS. Bent alternately in opposite directions; zig-zag.

FLORAL. Of or pertaining to the flower.

FLORET. A small flower in a head or cluster, as in the Compositae; the flower, including the lemma and palea, of a grass.

FOETID. With an unpleasant odor.

FOLLICLE. A dry fruit formed from a single carpel and dehiscing along a suture.

FOSSE. An excavation or groove.

FREE. Not joined.

FROND. The foliage of ferns; the leaf of a palm.

FRUTESCENT. Becoming shrubby, shrublike.

FUGACEOUS. Soon falling or disappearing.

GAMETOPHYTE. The plant generation bearing sexual organs producing gametes.

GAMOPETALOUS. Having the petals united.

GAMOSEPALOUS. Having the sepals united.

GAPING. Open-mouthed.

GENICULATE. Bent kneelike.

GIBBOUS. Swollen, often in one side or direction.

GLABRESCENT. Nearly glabrous; becoming glabrous.

GLABROUS. Devoid of any pubescence.

GLAND. A secreting or non-secreting protuberance, often minute, on a surface or terminating a hair.

GLANDULAR. With glands.

GLAUCESCENT. Somewhat glaucous.

GLAUCOUS. Covered with a very fine, powdery or waxy bloom.

GLOBOSE. Spherical, globular.

GLOMERULE. A cluster of capitula or heads in a common involucre, as in the inflorescence of some Compositae.

GLUME. In a grass inflorescence, the scale subtending the spikelet, not enclosing a floret; in sedges, any of the scales of the spikelet.

GRANULATE. As if covered with numerous fine granules; fine-grainy.

GYNANDRIUM. A structure composed of one or more stamens joined with the pistil.

GYNOECIUM. The pistil or pistils; the female parts of the flower.

GYNOPHORE. The stalk of an ovary, often prolonged through the perianth.

GYNOSTEGIUM. A structure formed by the joined stamens and pistil, as in the Asclepiadaceae.

HABIT. The general growth form or appearance of a plant.

HABITAT. The kind of locality in which a plant grows.

HALF-INFERIOR. Partly inferior, in reference to an ovary.

HALOPHILOUS. Preferring or tolerating saline soils.

HASTATE. Arrow head-shaped with the basal lobes turned outward.

HAUSTORIUM. The rootlike sucker attaching a parasitic plant to its host plant.

HEAD. A dense cluster of sessile flowers on a common receptacle.

HELICOID. Spirally curved, coiled or twisted.

HEMICRYPTOPHYTE. A perennial plant having resting buds at ground level.

HERBACEOUS. Not woody or perennial; also, green and opaque as opposed to scarious, non-green.

HETEROCARPOUS. Having fruits of more than one form.

HETEROSTYLOUS. Having flowers differing in the styles and, sometimes, in reproductive capability.

HILUM. The scar on a seed marking the point of attachment.

HIRSUTE. With rather long and sometimes somewhat stiff hairs.

HIRTULOUS. Finely and slightly hairy.

HISPID. With rough, stiff hairs or bristles.

HISPIDULOUS. Finely or slightly hispid.

HOMOCARPOUS. Having fruits all of the same form.

HYALINE. Thin and translucent.

HYGROSCOPIC. Changing form or position with changes in humidity.

HYPANTHIUM. An enlarged, often cuplike structure in some flowers, surrounding the ovary and supporting the stamens and floral envelopes.

HYPODISCAL. Inserted below the edge of the disc, as stamens in the flowers of some *Tamarix*.

HYPOGYNOUS. Inserted beneath and free from the pistil or pistils; having the parts so constructed (flower).

IMBRICATE. Overlapping.

IMMERSED. Imbedded; also, growing completely under water.

IMPARIPINNATE. Having an odd number of leaflets, with a single terminal leaflet.

IMPRESSED. Furrowed or depressed below a surface, as if by pressure.

INCISED. Cut sharply and somewhat deeply at the margins.

INCLUDED. Not protruding; wholly surrounded.

INDEHISCENT. Not opening at maturity.

INDUMENTUM. Any hairy or downy covering.

INDUPLICATE. Bent or folded upward lengthwise, forming a channel above.

INDURATE. Hardened.

INFERIOR. In reference to an ovary, situated below and often supporting the other flower parts.

INFLATED. Hollow, swollen, and often thin.

INFLEXED. Turned or bent inward.

INFLORESCENCE. The mode of flowering or its arrangement; also, the flowers of a plant collectively.

INSERTED. Attached.

INTRICATE. More or less densely and irregularly branched and entangled.

INVOLUCEL. The bracts at the base of a secondary structure, such as the secondary rays of an umbel.

INVOLUCRE. A group or ring of bracts surrounding several flowers or an inflorescence branch or unit.

INVOLUCRAL. Of or pertaining to the involucre.

INVOLUTE. With edges rolled upward and inward.

IRREGULAR. Not regular in shape; in reference to flowers, zygomorphic, not radially symmetrical.

KEELED. Ridged more or less sharply, usually longitudinally at back.

KEEL. A ridge, more or less sharp, usually longitudinal and at back.

KNEE. An abrupt bend.

LABELLUM. In orchids, the third, usually enlarged petal, becoming anterior through twisting of the ovary.

LACERATE. Irregularly cut or torn at the margins.

LACINIATE. Cut into lobes by deep, narrow incisions.

LACTIFEROUS. Producing white, milky sap.

LAMINA. The blade, or broadened portion, of a leaf or petal.

LANATE. With woolly, interwoven hairs.

LANCEOLATE. Rather narrow and tapering to each end, with the broadest part below the middle.

LATERAL. On or at the side.

LATEX. White milky sap.

LAX. Loose, distant, not crowded.

LEGUME. A one-celled fruit developing from a single carpel, usually dehiscing into two valves, with the seeds attached along the ventral suture; also, a member of the family Leguminosae.

LEMMA. The lower of the two bracts or scales immediately enclosing the flower or floret in grasses.

LENTICULAR. Convexly lens-shaped.

LIGULE. A strap-shaped body such as the limb of ray flowers in the Compositae; an appendage at the mouth of the leaf sheath in grasses.

LIGULATE. Having ligules.

LIMB. In gamopetalous flowers, the expanded terminal portion of the corolla; also, the lamina or expanded part of a leaf or individual petal.

LINEAR. Elongated and narrow with margins parallel or nearly so.

LOBED. Having segments, parts or divisions, the term sometimes with a numerical prefix indicating their number.

LOBULE. A small lobe.

LOCULE. A cell or compartment of an ovary.

LOCULAR. Having one or more compartments or cells, the term often with a numerical prefix indicating their number.

LOCULICIDAL. Opening at back along the dorsal suture into the cavity.

LOMENT. A legume or other fruit breaking at maturity into one-seeded segments.

LOOSE. With parts more or less distant, not crowded.

LYRATE. Pinnatifid with the terminal lobe enlarged and rounded.

MARGIN. The edge, border.

MARGINAL. On or pertaining to the edge.

MEDIAL. At the middle.

MEDIFIXED. Attached by the middle or somewhat off center, as some hairs.

MEMBRANOUS. Thin and translucent.

MERICARP. A fruit part that splits away, such as one of the two carpels in the Umbelliferae.

MICROPYLE. The aperture in the skin of a seed.

MIDNERVE. The main, central vein in a leaf or other part.

MONADELPHOUS. With stamens in a single group, all united by their filaments.

MONOCHASIAL. Of or pertaining to a monochasium, or single-branched cyme.

MONOECIOUS. Having unisexual flowers, with both male and female flowers on the same plant.

MUCRO. A short fine point.

MUCRONATE. With a short fine point.

MUCRONULATE. With a very fine, short point.

MULTICARPELLATE. Having, or derived from, more than one carpel.

MURICATE. Rough with short, hard protuberances or rounded points.

NERVED. Veined.

NERVATION. The mode of veination; the nerves or veins collectively.

NET-VEINED. With veins branching in a netlike pattern.

NEUTER. Sexless, having neither male nor female parts; reproductively non-functional.

NODE. A point or region on a stem that normally bears a leaf or leaves.

NUMEROUS. Not readily counted and often somewhat variable; in number more than 10, or as sometimes used, more than 20.

NUT. A dry, indehiscent, usually one-celled and one-seeded fruit, often with a hardened wall or husk.

NUTLET. A small nut.

OBCONICAL. Conical and attached at the narrower end.

OBDELTOID. Triangular and attached at the narrow end.

OBLANCEOLATE. Reverse-lanceolate, lanceolate but having the broadest part nearer the apex than the base.

OBLIQUE. Slanting; also, unequal-sided, such as the base of a leaflet.

OBLONG. Longer than broad with the margins nearly parallel.

OBOVATE. Reverse-ovate; egg-shaped in outline and attached at the narrower end

OBPYRIFORM. Reverse-pyriform; pear-shaped and attached at the narrower end.

OBSOLETE. Not or hardly apparent, or entirely absent.

OBTRIANGULAR. Obdeltoid; triangular and attached at the narrow end.

OBTUSE. Blunt and rounded at end, with the margins meeting at an angle of more than 90 degrees.

OCHREA, OCREA. A short stem sheath formed at the nodes from stipules, as frequent in the Polygonaceae.

OCHREOLE. A small or secondary ochrea.

OPERCULATE. Furnished with a lid.

OPPOSITE. On both sides at the same level, as leaves on a stem; standing opposite another part as opposed to being disposed alternately.

ORBICULAR. Flat with circular outline.

OVARY. The part of the pistil containing the ovules; an immature fruit.

OVATE. Egg-shaped in outline and attached at the broader end.

OVOID. Egg shaped as a solid body.

OVULE. The immature seed within the ovary.

PALE. A chaffy, scarious scale.

PALEA. The upper of the two bracts or scales immediately enclosing the floret in grasses.

PALMATE. With lobes radiating from a common central area or point, as fingers on a hand.

PALYNOLOGY. The study of pollen and other plant spores, especially their fossilized remains.

PANICLE. An open, or sometimes congested, branching group of flowers; a compound or branched raceme.

PANICLED. Disposed in a panicle.

PANICULATE. Panicle-like.

PAPILIONACEOUS. Like or having the pea-like flowers characteristic of many legumes, with the corolla consisting of a standard, wings and keel.

PAPILLA. A minute, nipple-like gland or protuberance.

PAPILLOSE. Having papillae.

PAPPOSE. Having a pappus.

PAPPUS. The tuft of hairs, bristles or chaffy appendages on some fruits or seeds, particularly the achenes of Compositae.

PARASITIC. Deriving nourishment from another living organisim.

PARIETAL. Borne on or belonging to the wall of an organ, such as the wall of an ovary.

PARIPINNATE. Pinnate with an equal number of leaflets, the leaf terminating in a pair of leaflets.

PARTED. Divided nearly to the base or midrib.

PARTITE. Parted, often with a numerical prefix indicating the number of divisions.

PECTINATE. Comblike, with narrow approximate segments or long, narrow, even teeth.

PEDICEL. The stalk of a single flower or fruit, or, in grasses, of a single spikelet.

PEDICELLATE. Borne on a pedicel.

PEDUNCLE. The stalk of a group of flowers or fruits, or sometimes of a single one.

PEDUNCLED, PEDUNCULATE. Borne on a peduncle.

PELLUCID. Transparent or nearly so.

PELTATE. In reference to leaves, attached within the margin on the surface rather than at the margin; shield-shaped.

PENDULOUS. Hanging downward.

PENICIL. The outer partial involucre of the individual headlets in the inflorescence of *Echinops*.

PENTAMEROUS. With parts in fives.

PEPO. The fruit of the gourd family, Cucurbitaceae, berry-like, from an inferior ovary and usually with an outer rind.

PERENNATING. Living for more than one year or season, especially in reference to plants normally annual.

PERENNIAL. Continuing to live for more than two years.

PERFECT. In reference to flowers, having functioning parts of both sexes; bisexual.

PERFOLIATE. With the stem appearing to pass through the leaves.

PERIANTH. The floral envelope of whatever form, often used in reference to flowers without clearly defined calyx and corolla, or with one envelope absent.

PERICARP. The wall of a mature ovary.

416

PERIDISCAL. Inserted around the edge of the disc, as the stamens in some species of *Tamarix*.

PERIGYNOUS. Arising around the ovary and not beneath it, as when the stamens and floral envelopes arise from a hypanthium surrounding the ovary; also, having such construction (flower).

PERSISTENT. Remaining attached, not falling off.

PETAL. One unit of the inner floral envelope, or corolla, of a polypetalous flower, often colored and showy.

PETALOID. Petal-like in color or texture.

PETIOLE. The stalk of a leaf.

PETIOLAR. Borne on or pertaining to a petiole.

PETIOLATE. Borne on a petiole.

PETIOLULE. The stalk of a leaflet or secondary unit of a compound leaf.

PHYLLODE. An expanded petiole without blade.

PHYSIOGNOMY. The general habit and growth form, and other gross features, of a plant or vegetation unit.

PHYTOCHEMICAL. Pertaining to, or derived from, a study of the chemistry of plants.

PILOSE. With soft, rather long hairs.

PINNA. A primary division of a pinnately compound leaf.

PINNATE. With divisions, segments, or other units side-by-side on both sides of an axis, featherlike, as the leaflets of a pinnately compound leaf.

PINNATIPARTITE. Pinnately divided more than half way to the base or midrib.

PINNATISECT. Pinnately cut down to the base or midrib.

PINNULE. A secondary pinna; the secondary division of a pinnately compound leaf; in a bipinnate leaf, the final leaflet.

PISTILLATE. Having pistils and not stamens; female.

PLACENTATION. The place or mode of disposition of the placenta in the ovary.

PLANO-CONVEX. Flat on one side and rounded convexly on the other.

PLICATE. Folded once or more.

PLUMOSE. With hairs or bristles spreading laterally from the axis; featherlike but the hairs often spreading in all directions, in three dimensions.

PNEUMATHODIUM. An emergent, erect, stemlike aerating structure arising from the roots, as in mangroves.

POD. A dry, many-seeded, dehiscent fruit; the fruit of a legume, or sometimes a crucifer.

POLLEN. The fertilizing, dustlike powder produced in the anthers of flowering plants.

POLYGAMODIOECIOUS. Polygamous but mainly dioecious.

POLYGAMOUS. Bearing bisexual and unisexual flowers on the same or different plants.

POLYMORPHIC. Taking different forms; variable in form.

POME. A fleshy fruit produced from a compound pistil, such as an apple.

PRICKLE. A small spinelike body arising from the epidermis or bark.

POSTERIOR. At or toward the back or the axis.

PRISMATIC. Prism-shaped, with edge-parallel, more or less flat faces meeting at angles.

PROCUMBENT. Prostrate, lying more or less flat on the ground.

PROLIFERATING. In grass spikelets, producing small, bulb-like dispersal units.

PROSTRATE. Lying more or less flat on the ground; procumbent.

PSAMMOPHIL. A plant found mainly in sand.

PSAMMOPHILOUS. Preferring sandy habits.

PUBERULENT. Very finely or minutely pubescent.

PUBESCENT. Hairy; bearing hairs of any kind.

PULVERULENT. With a somewhat powdery or dustlike surface.

PUNCTATE. Marked with fine, often depressed, dots.

PUNGENT. Terminating in a sharp, rigid point.

PYRAMIDAL. Pyramid-shaped.

PYRENE. A nutlet; a small stone of a drupe.

PYRIFORM. Pear-shaped.

QUADRANGULAR. Four-angled, the angles more or less right angles.

RACEME. An elongating inflorescence consisting of a central rhachis bearing a number of flowers on pedicels of approximately equal length.

RACEMED. Grouped in racemes.

RACEMOSE. Racemelike; arranged in racemes.

RACHIS, RHACHIS. An axis bearing flowers or, in a compound leaf, leaflets; the stalk of a fern frond.

RACHILLA, RHACHILLA. A secondary or small rachis; in grasses and sedges, the axis bearing the florets.

RADIATE. Spreading from a common center; in some inflorescences of Compositae and other families, having the marginal flowers or florets elongated and spreading.

RADICLE. The lower portion of an embryo representing a rudimentary root.

RAY. One of the radiating branches of an umbel; in the inflorescences of some Compositae, a marginal, often elongated floret.

RECEPTACLE. The expanded terminal portion of the axis bearing the organs of one flower or providing a common support for the florets of a compound flower, as in the Compositae.

RECEPTACULAR. On or pertaining to the receptacle.

RECURVED. Bent or curved downward or backward.

REDUCED. Smaller than normal in size and often simpler in form.

REFLEXED. Abruptly bent downward or backward.

REMOTE. Distant; not close together.

RENIFORM. Kidney-shaped.

REPAND. Undulate or wavy, especially at the margins.

REPENT. Creeping and prostrate, often rooting at the nodes.

RETICULATE. With a netlike pattern.

RETRORSE. Directed downward or backward.

RETUSE. With a shallow notch at the apex.

REVOLUTE. With edges rolled downward, toward the back side.

RHIZOME. A rootlike, usually underground, stem.

RHIZOMATOUS. Having rhizomes.

RHOMBOID. Rhombic in outline, four-sided with some angles obtuse.

ROSETTE. A group of leaves spreading or radiating from the base of a plant.

ROSULATE. In the form of a rosette (leaves).

ROTATE. Circular, wheel-shaped (a flower), applied especially to a gamopetalous corolla with a spreading limb and short tube.

RUDERAL. A plant growing in waste places or disturbed ground.

RUDIMENTARY. Imperfectly or incompletely developed, usually of smaller than normal size.

RUGOSE. Wrinkled with sunken veins, such as some leaves.

RUGULOSE. Finely rugose.

RUGULOSITY. Fine wrinkles of a surface.

RUNCINATE. Sharply incised with backward-pointing teeth.

SACCATE. Bag-shaped.

SAGITTATE. With two pointed lateral lobes at base, like the base of an arrow head.

SALTBUSH. A shrublet usually growing on saline ground, especially one of the family Chenopodiaceae.

SALVERFORM. With a slender tube and abruptly expanding limb (a corolla).

SAMARA. A winged, achene-like fruit.

SAPROPHYTE. A plant deriving its nourishment from the decaying bodies of other organisms.

SCABRID, SCABROUS. Rough.

SCABRIDITY. A rough protrusion.

SCABRIDULOUS. Finely or minutely rough.

SCALE. Any thin, scarious body.

SCANDENT. Climbing.

SCAPE. A leafless peduncle, often long, arising from the base.

SCARIOUS. Thin, dry and membranous; not green.

SCHIZOCARP. A pericarp, or fruit wall, which splits into one-seeded portions or mericarps.

SCURFY. Covered with small, often irregular scales.

SECUND. Turned to one side.

SEGETAL. A weed of farm fields, particularly of cereal crops.

SEGMENT. One of several generally similar parts or sections into which an organ is naturally divided, such as a lobe or division of a leaf.

SEPAL. One of the individual segments of the calyx, usually green and herbaceous.

SEPALOID. Sepal-like.

SEPTICIDAL. Opening along the line or lines of junction, especially a capsular fruit.

SEPTUM. A partition.

SERRATE. With forward pointing teeth.

SERRULATE. Finely serrate.

SESSILE. Attached without any stalk.

SETA. A bristle, or a fine, bristle-like body.

SETIFORM. Bristle-shaped.

SETOSE. Bristly; covered with bristles.

SHRUB. A low, usually several-stemmed woody plant; a bush.

SHRUBLET. A small shrub.

SILICLE, SILICULA. The shorter fruit form of the Cruciferae, less than three times as long as broad but similar in structure to a SILIQUE.

SILIQUE, SILIQUA. The longer characteristic fruit of the Cruciferae, three times or more as long as broad, consisting of two valves falling away from a frame upon which the seeds grow and across which a false partition is formed.

SIMPLE. Of a single unit, not compound.

SINUS. A more or less deep, often rounded indentation of the margin; the space between two lobes of a leaf or other organ.

SINUATE. With a more or less deeply wavy margin.

SOLITARY. Single, with only one in the same place.

SORUS. A cluster of sporangia, or spore cases, in ferns.

SPADIX. A thickened or fleshy flower spike.

SPATHE. A large bract subtending, surrounding or enveloping a flower group.

SPATHEATE. Having a spathe.

SPATHULATE. Generally oblong but broader toward the apex.

SPICATE. Spikelike; arranged in a spike.

SPICIFORM. Spike-shaped.

SPICOID. Similar to, but technically not, a spike.

SPIKE. An inflorescence, usually elongated, consisting of sessile flowers on a common axis.

SPIKELET. The basic inflorescence unit of grasses, consisting of one or more florets subtended by two glumes.

SPINE. A sharp woody outgrowth from the stem, usually a modified branch, petiole or stipule.

SPINELET. A small spine.

SPINESCENT. Ending in a spine or spikelike structure; bearing a spine or spines.

SPINOUS. Spiny.

SPINULE. A little spine.

SPINULOSE. With small spines.

SPORE. A cell which becomes free and capable of developing directly into a new individual.

SPORANGIUM. A sac or case producing spores.

SPORANGIFEROUS. Bearing sporangia.

SPOROPHYTE. A plant of the generation producing spores.

SPUR. A hollow, slender or sacklike projection from a petal or sepal.

SPURRED. Bearing a spur.

SQUAMULOSE. With small scales.

STAMEN. The pollen-bearing organ in the flower; the male organ.

STAMINATE. Having stamens and not pistils; male.

STAMINODE. A sterile, antherless stamen, sometimes strongly modified in form.

STANDARD. In a papilionaceous flower, the upper, broadened, more or less erect petal; in *Iris*, one of the three upper, erect petaloid perianth segments.

STELE. The axial cylinder of stem tissue in which vascular tissue is developed.

STELLATE. Star-shaped; radiating like the points of a star.

STERILE. Barren; not producing fruit.

STIGMA. That part of the pistil or style which receives the pollen.

STIGMATIC. Pertaining to the stigma; bearing a stigma.

STIPE. A stalk or support, specifically the 'leafstalk' of a fern or the stalk of an ovary or carpel.

STIPITATE. Having a stipe; mounted on a stipe.

STIPULE. One of the paired appendages borne at the base of a leaf in some plants.

STIPULAR. Pertaining to or derived from stipules.

STIPULATE. Having stipules.

STOLON. A runner or any basal branch that tends to root.

STOLONIFEROUS. Having stolons.

STRIATE. Finely grooved or ridged, usually longitudinally.

STRICT. Stiffly upright, very straight.

STRIGOSE. With sharp, stiff, appressed straight hairs or bristles.

STYLE. That part of the pistil between the ovary and the stigma.

STYLOPODIUM. The enlargement at the base of the styles in Umbelliferae.

SUB-. A prefix adding to a term the meaning 'slightly', 'nearly', or 'somewhat'.

SUBEROUS. Corky in texture.

SUBTENDING. Standing immediately below.

SUBULATE. Awl-shaped.

SUCCULENT. Fleshy and juicy.

SUFFRUTESCENT. Somewhat or slightly shrubby or woody.

SULCATE. Grooved or furrowed lengthwise.

SUPERIOR. In reference to an ovary, having the floral envelopes inserted beneath it.

SUPERNUMERARY. Of more than normal number.

SUTURE. A seam, or line of union; a line of opening.

SYMPATRIC. Living in the same area.

SYMPETALOUS. With petals wholly or partially fused.

SYNECOLOGY. The study of the mutual and environmental relationships of groups of organisms; also, those relationships themselves.

TAXON. A unit of classification, of any rank.

TENDRIL. A threadlike extension of a plant, grasping and clinging for support.

TEPAL. A segment or division of the perianth; term used where petals and sepals are not readily differentiated.

TERETE. Circular in cross-section.

TERMINAL. At or from the apex or end.

TERNATE. In threes, or divided into three parts or segments.

TERRESTRIAL. Growing on land, not in water.

TETRADYNAMOUS. Having four long stamens and two short ones.

TETRAGONAL, TETRAGONOUS. Four-angled.

THALLOID. Flattish and leaflike.

THEROPHYTE. A plant completing its life cycle in one season; an annual.

TOMENTELLOUS. Slightly or thinly tomentose.

TOMENTOSE. Densely woolly or pubescent with soft, short matted hairs.

TORUS. The receptacle of a flower; the area of the axis, sometimes extended, upon which the floral parts or the fruit are inserted.

TORULOSE. Cylindric with swellings or contractions at intervals.

TRIPINNATE. Pinnately compound with three levels of division.

TRIAD. A group of three similar parts or organs.

TRIFID. Divided into three parts.

TRIFOLIOLATE. Having three leaflets.

TRIFURCATION. A point of division into three branches.

TRIGONOUS. Three-angled.

TRILOBATE. Three-lobed.

TRILOCULAR. With three locules or cells.

TRIPARTITE. Divided into three parts.

TRIQUETROUS. Three-edged or -angled.

TRISECT. Cut or divided into three levels of division, or into three parts.

TRUNCATE. Ending abruptly, usually as if cut more or less straight across.

TUBE. The lower, more or less tubular portion of a gamopetalous corolla as opposed to the expanded terminal portion, or limb.

TUBERCLED, TUBERCULATE. Having tubercles, or small warty or knobby protusions.

TUBER. A thickened or expanded portion of a subterranean stem or root.

TUBEROUS. Bearing tubers.

TUBIFLORATE. Having, or pertaining to, a tubular flower or flowers, especially the tubular as opposed to the ligulate florets in Compositae.

TUBIFORM. Tube-shaped.

TUBULAR. Tubular in form; in Compositae, tubular as opposed to ligulate (a flower or floret).

TUFTED. Growing or occurring in clumps or compact groups.

TUNICATED. With concentric layers or coats.

TURGID. Swollen (but not with air) and usually firm.

TWINING. Winding spirally around a supporting structure.

UMBEL. A group of several flowers with pedicels attached to, and often spreading from, a common point.

UMBELLATE, UMBELLED. In or having umbels.

UMBELLET. An umbellule.

UMBELLIFORM. Umbel-shaped.

UMBELLULE. A secondary or ultimate umbel arising from a primary umbel branch or ray.

UNARMED. Without prickles or spines.

UNDULATE. Wavy.

UNEQUAL. Differing in length, size or number.

UNIFOLIOLATE. With a single leaflet, the other or others absent by abortion.

UNILATERAL. One-sided.

UNILOCULAR. With one locule or cell (an ovary).

UNISERIATE. In a single row.

UNISEXUAL. Of one sex; with either stamens or pistils but not both.

UTRICLE. A small, bladder-shaped fruit or fruit covering.

VAGINATE. Sheathed.

VALVE. A section or piece into which a capsule, legume, or some other fruits naturally separate at maturity.

VALVATE. Opening by valves.

VASCULAR. Having or pertaining to vessels or ducts.

VEGETATIVE. Concerned with bodily growth as opposed to reproduction.

VEIN. A strand of vascular tissue in a flat organ such as a leaf.

VELUTINOUS. Velvety with fine soft hairs.

VERNAL. Occurring in or pertaining to the spring season; in *Tamarix*, developing on branches of the previous year's growth (the inflorescence).

VERRUCOSE. With wartlike protuberances.

VERTICILLASTER. A false whorl, composed of a pair of opposed cymes, as in Labiatae.

VERTICILLATE. Whorled, inserted around a common node at the same level on a stem or other axis, as some leaves or flowers.

VILLOUS. With rather long hairs.

VIRGATE. Rather straight, slender and wandlike.

VISCID. Sticky with a tenacious secretion.

VIVIPAROUS. Having seeds or bulbils germinating or sprouting while still attached to the mother plant.

VIVIPARY. The state of being viviparous.

WHORL. An arrangement of organs in a circle at the same level on an axis.

WHORLED. Disposed in a whorl or whorls.

WING. Any flattened, often membranous expansion attached to an organ; in papilionaceous flowers, one of the two, usually long-clawed, lateral petals.

ZYGOMORPHIC. Capable of division into equal parts by only one plane of symmetry.

GAZETTEER

This list of geographic names includes only places referred to in the text. They are alphabetized by the first (or only) element of each name ignoring the Arabic definite article, which is appended when present. The definite article may be recognized in names of the text as a two-letter prefix beginning with an 'a' and followed by a hyphen, e.g. 'al-', 'as-'. Each name is followed in parentheses by its transliteration, with full diacritical marks, in the unified system of the U.S. Board on Geographic Names (BGN) and the (British) Permanent Committee on Geographical Names (PCGN). This is the system followed on the best available maps of the area (see Bibliographic Notes), and it allows recovery of the Arabic spelling if required.

The area abbreviations included with each entry refer to the eight topographical subunits described in Chapter 2, as follows: NP – Northern Plains, NS – Northern Summan, SS – Southern Summan, ND – Northern Dahna, SD – Southern Dahna, CCL – Central Coastal Lowlands, SCL – South Coastal Lowlands, RK – the Rub' al-Khali. Geographical coordinates are provided to the nearest whole minute. For area names they refer to the approximate geographic center of the feature.

Definitions of Arabic topographic terms, as used in eastern Saudi Arabia, are included in the place name sequence. They are italicized and identified as 'topgr. term'.

'Aba, al- (Al 'Abā), village and farms northwest of al-Qatif Oasis; CCL, 26° 44'N, 49° 45'E.

Abqaiq (Buqayq), town and oil production center; SCL, 25° 56'N, 49° 40'E.

Abqaiq Salt Pits, open pit salt mine near Abqaiq; SCL, 25° 49'N, 49° 43'E.

Abraq (abraq), topgr. term: rocky hill with sand on slopes.

Abraq al-Kabrit (Abraq al Kabrīt), hill; NP, 28° 07'N, 47° 56'E.

Abraq al-Khaliqah (Abraq al Khalīqah), = Abraq al-Kabrit, q.v.

Abu 'Ali Island (Abū 'Alī), island in the Arabian Gulf; CCL, 27° 20'N, 49° 30'E.

Abu Bahr (Abū Baḥr), gravel plain in the western Rub' al-Khali; RK, 21° 20'N, 48° 00'E.

Abu Hadriyah (Abū Ḥadrīyah), oil production center, CCL, 27° 19'N, 48° 58'E.

Abu Ma'n (Abū Ma'n), village and farms northwest of al-Qatif Oasis; CCL, 26° 40'N, 49° 49'E.

Abu Shidad (Abū Shidād), hill west of Abqaiq; SCL, 25° 52'N, 49° 22'E.

Abu Zahmul (Abū Zahmūl), well and watch tower near al-'Uqayr; SCL, 25° 38'N, 50° 12'E.

Aden Protectorates, former British-protected states in southern Arabia, now the People's Democratic Republic of South Yemen.

Aflaj, al- (Al Aflāj), district of farms and villages in southern Najd; 22° 15'N, 46° 40'E.

'Ain Dar GOSP 1, gas-oil separation plant in the 'Ain Dar oil field area; CCL, 25° 52'N, 49° 15'E.

'Ain Dar GOSP 2, gas oil separation plant in the 'Ain Dar oil field area; CCL, 25° 58'N, 49° 14'E.

'Ain Dar GOSP 3, gas-oil separation plant in the 'Ain Dar oil field area; SCL, 25° 46'N, 49° 15'E.

'Ain Dar GOSP 4, gas-oil separation plant in the 'Ain Dar oil field area; SCL, 25° 38'N, 49° 16'E.

Ajam, al- (Al Ājām), village in al-Qatif Oasis; CCL, 26° 34'N, 49° 56'E.

'Alah, al- (Al 'Alāh), hill northwest of Dhahran, CCL, 26° 28'N, 49° 50'E.

'Amad 2 Camp (Al 'Amad), oil drilling

site in the northern Rub' al-Khali; RK, 22° 38'N, 52° 02'E.

'Amad 2 well (Al 'Amad), = 'Amad 2 Camp, q.v.

'Anik ('Anik), village in al-Qatif Oasis; CCL, 26° 31'N, 50° 01'E.

'Ar'ar ('Ar'ar), town in northern Saudi Arabia; 30° 59'N, 41° 02'E.

'Arabiyah Island, al- (Al 'Arabīyah), Saudi Arabian island in mid-Arabian Gulf; CCL, 27° 46'N, 50°11'E.

'Asir ('Asīr), district in southwestern Saudi Arabia; 18° 30'N, 42° 30'E.

Awarik, al- (Al Awārik), dune area in the southwestern Rub' al-Khali; RK, 20° 30'N, 48° 42'E.

'Awja, al- (Al 'Awjā'), = Wadi al-'Awja, q.v.

'Ayn (*'ayn*), topgr. term: spring.

'Ayn, al- (al 'Ayn), wells and Bedouin camping grounds on the southern edge of the Rub' al-Khali; RK, 19° 35'N, 54° 54'E.

'Ayn al-'Abd ('Ayn al 'Abd), sulfurous spring on the southern edge of the Saudi Arabia-Kuwait Partitioned Neutral Zone; NP, 28° 14'N, 48° 16'E.

'Ayn Dar ('Ayn Dār), village west-northwest of Abqaiq; CCL, 26° 00'N, 49° 23'E.

'Ayn Jadidah ('Ayn al Jadīdah), former spring and farm near al-Hasa Oasis; SCL, 25° 39'N, 49° 32'E.

'Ayn al-Khadud ('Ayn al Khadūd), spring in al-Hasa Oasis; SCL, 25° 22'N, 49° 37'E.

'Ayn Mar'ah ('Ayn Mar'ah), spring in al-Hasa Oasis; SCL, 25° 24'N, 49° 38'E.

'Ayn Najm ('Ayn Najm), sulfurous spring near al-Hasa Oasis; SCL, 25° 24'N, 49° 33'E.

'Ayn as-Subayghawi ('Ayn aş Şubayghāwī), spring; CCL 26° 40'N, 49° 45'E.

'Ayn Umm al-Lif ('Ayn Umm al Līf), spring in al-Hasa Oasis; SCL, 25° 22'N, 49° 38'E.

'Ayn Umm Sab' ('Ayn Umm Sab'ah), spring in al-Hasa Oasis; SCL, 25° 28'N, 49° 35'E.

'Aziziyah, al- (Al 'Azīzīyah), industrial site south of al-Khubar; CCL, 26° 11'N, 50° 13'E.

Badanah (Badanah), town in northern Saudi Arabia; 31° 00'N, 40° 59'E.

Badrani Junction, al- (Al Badrānī), road junction near al-Qatif Oasis; CCL, 26° 31'N, 49° 58'E.

Bahath, al- (Al Baḥath), wells in the northern Rub' al-Khali; RK, 22° 58'N, 50° 00'E.

Bahrain (Al Baḥrayn), island state in the Arabian Gulf; 26° 05'N, 50° 33'E.

Ba'ja (Al Ba'jā'), wells in the western United Arab Emirates; 24° 06'N, 51° 45'E.

Bani Ma'n (Banī Ma'n), village in al-Hasa Oasis; SCL, 25° 23'N, 49° 38'E.

Bani Tukhman (Banī Ṭukhmān), dune and plains area in the northern Rub' al-Khali; RK, 22° 32'N, 49° 35'E.

Barqa (*barqā'*), topgr. term: rocky hill with sand on slopes, often broader than an *abraq*.

Barqa as-Samur (Barqā' as Samur), hills near Yabrin; SCL, 23° 36'N, 49° 03'E.

Batin, al- (Al Bāṭin), prominent, elongated wadi-depression; NP, 28° 30'N, 46° 00'E.

Batn (*baṭn*), topgr. term: depression or hollow.

Batn al-Faruq (Baṭn al Farūq), elongated depression; SS, 25° 45'N, 49° 02'E.

Batn Sabsab (Baṭn as Sabsab), depression; NS, 26° 20'N, 48° 24'E.

Batn Sumlul (Baṭn aş Şumlūl), depression; NS, 26° 30'N, 48° 15'E.

Bay of Salwah (Dawḥat Salwah), bay

south of Bahrain; SCL, 25° 00'N, 50° 45'E.

Bidah, al- (Al Bayḍah), area of rocky desert; SS, 24° 40'N, 48° 40'E.

Bir (bi'r), topgr. term: hand-dug well.

Bir Fadil (Bi'r Fāḍil), well in the northern Rub' al-Khali; RK, 22° 04'N, 49° 46'E.

Bir Harad (Bi'r Ḥaraḍ), well in southern Ghawar; SS, 24° 14'N, 49° 11'E.

Birkat al-Haytham (Birkat al Haytam), ruined cistern on Darb Zubaydah, northern Saudi Arabia; 29° 45'N, 43° 38'E.

BM T-42, survey bench mark in the northern Rub' al-Khali; RK, 23° 07'N, 49° 56'E.

Bu'ayj (Bu'ayj), wells near Salwah; SCL, 24° 40'N, 50° 38'E.

Bunayyan (Bunayyān), well in the northern Rub' al-Khali; RK, 23° 10'N, 50° 56'E.

Buraydah (Buraydah), city in al-Qasim district, north-central Saudi Arabia; 26° 20'N, 43° 58'E.

Burayman (Buraymān), village in western Saudi Arabia near Jiddah, 21° 40'N, 39° 14'E.

Burq Aba ad-Dalasis (Burq Abā ad Dalāsīs), hills north of al-Hasa; SCL, 25° 44'N, 49° 30'E.

Burq (burq), topgr. term: pl. of *barqā',* q.v.

Burq ash-Sharif (Barqā' Sharīf), hills in the south bank of al-Batin; NP, 28° 05'N, 45° 40'E.

Bushire (Bushehr), Gulf coastal town in Iran, 28° 58'N, 50° 50'E.

Camp Ramlah 1, oil drilling site in the northeastern Rub' al-Khali; RK, 22° 10'N, 54° 21'E.

Camp S-3, oil exploration camp in the eastern Rub' al-Khali, moved periodically but for long periods in the vicinity of 21° 22'N, 54° 21'E.

Camp Shaybah 9, oil drilling site in the northeastern Rub' al-Khali; RK, 22° 32'N, 54° 03'E.

Camp Tumaysha 1 (Aṭ Ṭumayshā'), oil drilling site in the eastern Rub' al-Khali; RK, 21° 09'N, 54° 52'E.

Dabtiyah, ad- (Aḍ Ḍabṭīyah), well and *sabkhah;* NS, 26° 28'N, 48° 33'E.

Dahl (dahl), topgr. term: sink hole; solution cavity.

Dahl al-Furayy (Daḥl al Furayy), sinkhole; NS, 26° 47'N, 47° 05'E.

Dahna, ad- (Ad Dahnā'), elongated, arcuate dune belt; ND-SD, 25° 00'N, 48° 00'E.

Dalwah, ad- (Ad Dalwah), village in al-Hasa Oasis; SCL, 25° 24'N, 49° 41'E.

Dammam (Ad Dammām), principal city of the Eastern Province, Saudi Arabia; CCL, 26° 26'N, 50° 06'E.

Dammam Port (Ad Dammām), port on the Arabian Gulf; CCL, 26° 28'N, 50° 11'E.

Darb (darb), topgr. term: route, trail.

Darb al-Kunhuri (Darb al Kunhurī), trail from al-Jubayl inland toward Riyadh; CCL-NS, 26° 40'N, 47° 58'E.

Darb Mazalij (Darb Mazālīj), trail from al-Hasa to Riyadh; SCL-SS—SD, 25° 00'N, 48° 50'E.

Darb Zubaydah (Darb Zubaydah), historic pilgrims' route from Iraq to Makkah; 29° 00'N, 43° 25'E.

Darin (Dārīn), village on Tarut Island; CCL, 26° 33'N, 50° 04'E.

Dawhah [-t] (dawhah [-t]), topgr. term: bay, inlet.

Dawhat Duwayhin (Dawḥat Duwayhin), bay on the Gulf coast east of Qatar; SCL, 24° 25'N, 51° 25'E.

Dawmat al-'Awdah (Dawmat al 'Awdah), hill and escarpment; NS, 27° 11'N, 48° 10'E.

Dawsariyah, ad (Ad Dawsarīyah), Gulf coastal site; CCL, 26° 55'N, 49° 45'E.

Dhabhah (Dhabḥah), village in al-Batin; NP, 27° 56'N, 45° 23'E.

Dhahran (Aẓ Ẓahrān), hills and oil company town; CCL, 26° 19'N, 50° 08'E.

Dhufar (Ẓufār), province of the Sultanate of Oman, southern Arabia, 17° 20'N, 54° 30'E.

Dhulayqiyah, adh- (Adh Dhulayqīyah), spring and farms near al-Hasa Oasis; SCL, 25° 16'N, 49° 41'E.

Dibdibah, ad- (Ad Dibdibah), extensive gravel plain; NP, 28° 00'N, 46° 30'E.

Eastern Province (Al Minṭaqah ash Sharqīyah), the Eastern Province of Saudi Arabia, bounded approximately: on the north by al-Batin, on the west by the western edge of the Dahna, on the east by the Arabian Gulf, and extending south and east through the Rub' al-Khali to the Saudi frontiers with the United Arab Emirates and the Sultanate of Oman.

Fadili, al- (Al Fāḍilī), oil field area west of al-Jubayl; CCL, 26° 58'N, 49° 10'E.

Farajah (Farajah), well in the north-central Rub' al-Khali; RK, 21° 36'N, 50° 36'E.

Farshah, al- (Al Farshah, or Farshat al Muzayri'), silt flat in the Dahna north of Riyadh near Rumah; ND, 25° 33'N, 47° 16'E.

Faruq, al- (Al Farūq), = Batn al-Faruq, q.v.

Faw al-Janubi, al- (Al Fāw al Janūbī), wadi tributary to al-Batin; NP, 28° 05'N, 45° 50'E.

Faylaka Island (Faylakā), Gulf island off Kuwait, 29° 25'N, 48° 23'E.

Fudul, al- (Al Fuḍūl), village in al-Hasa Oasis; SCL, 25° 22'N, 49° 41'E.

Fulayj al-Janubi, al- (Al Fulayj al Janūbī), wadi tributary to al-Batin; NP, 28° 16'N, 46° 03'E.

Ghawar, al- (Al Ghawār), elevated tract and oil field west of al-Hasa Oasis; SCL-SS, 25° 00'N, 49° 12'E.

Ghunan (Ghūnān), former police post between Dhahran and Abqaiq; CCL, 26° 08'N, 49° 54'E.

Great Nafud, the (An Nafūd), large sand area in northern Arabia, 28° 30'N, 41° 00'E.

Habl, al- (Al Ḥabl), area north of 'Ayn Dar; CCL, 26° 30'N, 49° 15'E.

Hadh Bani Zaynan (Ḥādh Banī Zaynān), area in the central Rub' al-Khali; RK, 20° 15'N, 50° 00'E.

Hadidah, al- (Al Ḥadīdah), meteor craters in the central Rub' al-Khali; RK, 21° 30'N, 50° 28'E.

Hafar al-Batin (Ḥafar al Bāṭin), town in al-Batin; NP, 28° 26'N, 45° 59'E.

Hajarah, al- (Al Ḥajarah), rocky district of northern Saudi Arabia and Iraq; 29° 30'N, 43° 45'E.

Hamatiyat, al- (Widyān al Ḥamāṭīyāt), series of small wadis; NP, 28° 56'N, 47° 26'E.

Hamra Judah (Ḥamrā' Jūdah), elevated rocky area; NS, 25° 56'N, 48° 47'E.

Hanidh (Ḥanīdh, sometimes also pronounced Ḥanīn), village; NS, 26° 35'N, 48° 38'E.

Harad (Ḥaraḍ), railroad station and village; SS, 24° 09'N, 49° 04'E.

Harad Agricultural Project, farms project in Wadi as-Sahba near Harad; SS, 24° 04'N, 49° 00'E.

Harad Station, railroad station at Harad, q.v.

Haradh, = Harad, q.v.

Hasa, al- (Al Ḥasā), large oasis area, the name formerly more widely applied to most of the northern Saudi Arabian Gulf coastal tract now known as the Eastern Province; SCL, 25° 24'N, 49° 10'E.

Hasa-'Ayn Dar junction, al-, road junction; SCL, 25° 53'N, 49° 27'E.

Hasa Oasis, al-,= al-Hasa, q.v.

Hawk, al- (Al Hawk), depression on the western edge of the Rub' al-Khali; SS, 22° 06'N, 48° 38'E.

Hawmat an-Niqyan (Ḥawmat an Niqyān), area of high dunes in the northern Dahna; ND, 26° 40'N, 45° 50'E.

Hibakah, al- (Al Ḥibākah), area in the central Rub' al-Khali; RK, 21° 15'N, 50° 10'E.

Hijaz, al (Al Ḥijāz), mountainous geographical area including most of western Arabia.

Hinnah, al- (Al Ḥinnāh), wells, ruins and village; CCL, 26° 56'N, 48° 46'E.

Hufuf, al- (Al Hufūf), city in al-Hasa Oasis; SCL, 25° 22'N, 49° 36'E.

Hufuf-Salwah junction, al-, road junction near al-Hufuf; SCL 25° 20'N, 49° 36'E.

Hunayy, al- (Al Ḥunayy), Government post and village; SS, 24° 58'N, 48° 45'E.

'Irj ('Irj), wells; CCL, 26° 22'N, 48° 51'E.

'Irq ('irq), topgr. term: linear sand ridge or elongate dune.

'Irq al-Ghanam ('Irq al Ghanam), dune area in the northern Rub' al-Khali; RK, 21° 57'N, 49° 42'E.

'Irq Jaham ('Irq Jahām), dune area in the Dahna; ND, 26° 00'N, 47° 15'E.

Jabal (jabal), topgr. term: hill, usually rocky.

Jabal Abu Shidad,= Abū Shidād, q.v.

Jabal Barakah (Jabal Barākah), hill on the Gulf coast of the United Arab Emirates; 24° 00'N, 52° 19'E.

Jabal Dab' (Jabal Ḍab‘), hill near al-Wari'ah; NP, 27° 46'N, 47° 20'E; also, a hill south of Yabrin; SCL, 23° 03'N, 48° 59'E.

Jabal Fazran (Fazrān), hill west of Abqaiq; CCL, 26° 17'N, 49° 16'E.

Jabal Ghuraymil (Jabal Ghuraymīl), prominent hill southwest of Abaqiq; SCL, 25° 47'N, 49° 32'E.

Jabal Haydaruk (Jabal al Haydarūk), hill near al-Fadili; CCL, 27° 06'N, 49° 11'E.

Jabal Madba'ah (Maḍba‘ah), hill near Yabrin; SCL, 23° 26'N, 49° 00'E.

Jabal Mutayrihah (Jabal Muṭayrīḥah), hill; NS, 25° 58'N, 48° 32'E.

Jabal Sha'ban (Jabal Sha‘bān), hill in the al-Hasa Oasis; SCL, 25° 23'N, 49° 38'E.

Jabal Umm ar-Rus (Umm ar Ru'ūs), hill at Dhahran; CCL, 26° 19'N, 50° 08'E.

Jabal al-'Amudah (Al ‘Amūdah), hill on the Gulf coast; NP, 28° 10'N, 48° 36'E.

Jabal al-Ahass (Jabal al Aḥaṣṣ), hill; CCL, 26° 54'N, 48° 39'E.

Jabal al-Ba'al (Jabal al Ba‘āl), hill north of an-Nu'ayriyah; CCL, 27° 43'N, 48° 19'E.

Jabal al-Barri (Al Jubayl al Barrī), hill near al-Jubayl; CCL, 26° 54'N, 49° 38'E.

Jabal al-Qarah (Jabal al Qārah), hill in the al-Hasa Oasis; SCL, 25° 24'N, 49° 41'E.

Jabal an-Nu'ayriyah (Jibāl an Nu‘ayrīyah), hill group near an-Nu'ayriyah village; CCL, 27° 32'N, 48° 24'E.

Jabal ar-Ruhayyah (Jabal Ruḥayyah), hill; SS, 25° 34'N, 49° 05'E.

Jafurah, al- (Al Jāfūrah), large dune district; SCL, 24° 00'N, 50° 00'E.

Jal (jāl), topgr. term: rocky escarpment, low cliff or valley side.

Jal al-Wutayd (Jāl al Wutayd), rocky ridge; SCL, 24° 01'N, 50° 56'E.

Jalmudah (Jalmūdah), ruins on the Gulf coast near al-Jubayl; CCL, 27° 02'N, 49° 38'E.

Jana Island (Janā), island in the Arabian Gulf; CCL, 27° 22'N, 49° 54'E.

Jarrarah (Jarrārah), village; NS, 27° 00'N, 47° 04'E.

Jarudiyah, al- (Al Jārūdīyah), village in al-Qatif Oasis; CCL, 26° 32'N, 49° 59'E.

Jawamir, al- (Al Jawāmīr), hill group near Yabrin; SCL, 23° 21'N, 48° 56'E.

Jawb (jawb), topgr. term: valley, low-lying ground.

Jawb Abu 'Arzilah (Jawb Abū 'Arzilah), depression; SCL, 24° 30'N, 50° 47'E.

Jawb al-'Abd (Jawb al 'Abd), depression; SCL, 24° 10'N, 50° 50'E.

Jawb al-'Asal (Jawb al 'Aṣal), *sabkhah* depression in the northern Rub' al-Khali; RK, 22° 55'N, 49° 59'E.

Jawb, al- (Al Jawb), low-lying district east of Yabrin; SCL, 23° 07'N, 50° 00'E.

Jaww (jaww), topgr. term: depression or basin, usually sandy-bottomed.

Jaww Butayhin (Jaww Buṭayhīn), *sabkhah* depression near base of the Qatar Peninsula; SCL, 24° 07'N, 51° 31'E.

Jaww ad-Dukhan (Jaww ad Dukhān), elongated basin near al-Ghawar; SS, 24° 40'N, 49° 06'E.

Jaww Ghanim (Jaww Ghānim), depression; NS, 26° 21'N, 48° 39'E.

Jaww al-Hawiyah (Jaww al Ḥawīyah), basin in al-Ghawar; SCL, 24° 42'N, 49° 11'E.

Jaww ash-Shanayin (Jaww ash Shanā'in), low-lying area near 'Ayn Dar; SCL, 25° 49'N, 49° 21'E.

Jawwiyah, al- (Al Jawwīyah), wells; NS, 26° 07'N, 48° 44'E.

Jazirah [-t] (jazīrah [-t]), topgr. term: island.

Jazirat al-Farisiyah (Al Fārisīyah), Iranian island in the Gulf; CCL, 27° 59'N, 50° 11'E.

Jibal al-Harmaliyat (Jibāl al Ḥarmalīyāt), hill group; NS, 26° 11'N, 48° 05'E.

Jiban (jībān), topgr. term: pl. of *jawb*, q.v.

Jiban, al- (Al Jībān), region of elongated depressions; SCL, 24° 15'N, 50° 45'E.

Jiddah (Jiddah), Saudi Arabian city on the Red Sea; 21° 29'N, 39° 11'E.

Jirwan (Jirwān), drilled water well in the northern Rub' al-Khali; RK, 23° 27'N, 50° 53'E.

Jishsh, al- (Al Jishsh), village in al-Qatif Oasis; CCL, 26° 30'N, 49° 59'E.

Ju'aymah, al- (Al Ju'aymah), oil terminal on the Arabian Gulf; CCL, 26° 49'N, 49° 59'E.

Ju'uf, al- (Al Ju'ūf), oil drilling site, also written al-Jauf; NP, 28° 07'N, 47° 56'E.

Jubayl, al- (Al Jubayl), town on the Arabian Gulf; CCL, 27° 01'N, 49° 40'E.

Jubail, = al-Jubayl, q.v.

Juhaymi, al- (Al Juhaymī), wells; CCL, 26° 44'N, 49° 20'E.

Jurayd Island (Al Jurayd), island in the Arabian Gulf; CCL, 27° 12'N, 49° 57'E.

Karan Island (Karān), island in the Arabian Gulf; CCL, 27° 43'N, 49° 49'E.

Khabari Wadha (Khabārī Waḍḥā'), silt basins area; NP, 28° 05'N, 47° 30'E.

Khafji, al- (Al Khafjī), oil camp and village on the Arabian Gulf; NP, 28° 25'N, 48° 32'E.

Khafji-as-Saffaniyah rd junct, road junction near as-Saffaniyah; NP, 27° 59'N, 48° 42'E.

Kharj, al- (Al Kharj), agricultural district southeast of Riyadh; 24° 10'N, 47° 20'E.

Khashm (khashm), topgr. term: headland, promontory of an escarpment or hill front.

Khashm al-'Abd (Khashm al 'Abd), plateau promontory south of Salwah; SCL, 24° 15'N, 50° 52'E.

Khashm az-Zaynah (Khashm az Zaynah), hill promontory west of Harad; SS, 24° 12'N, 48° 35'E.

Khawr (khawr), topgr. term: bay or inlet; also, a brackish well.

Khawr al-Khafji (Khawr al Khafjī), inlet on the Arabian Gulf coast; NP, 28° 25'N, 48° 32'E.

Khobar, al-, = al-Khubar, q.v.

Khubar, al- (Al Khubar), town on the Arabian Gulf near Dhahran; CCL, 26° 17'N, 50° 12'E.

Khubayra (Khubayrā'), wells near al-Wari'ah; NP, 27° 46'N, 47° 10'E.

Khurais-al-'Udayliyah junction, road junction; SCL, 25° 23'N, 49° 12'E.

Khurays (Khurayṣ), oil field area and village, in the Dahna; ND, 25° 05'N, 48° 02'E.

Khursaniyah, al- (Al Khursānīyah), oil field center; CCL, 27° 13'N, 49° 13'E.

Khursaniyah road junction, al- (Al Khursānīyah), road junction; CCL, 27° 09'N, 49° 09'E.

Khuwayran, al- (Al Khuwayrān), sandy ridge in the northern Rub' al-Khali; RK, 22° 48'N, 50° 11'E.

Kurayn Island (Kurayn), island in the Arabian Gulf; CCL, 27° 39'N, 49° 49'E.

Kuwait (Al Kuwayt), state and city in northeastern Arabia; 29° 20'N, 48° 00'E.

Lahy ar-Rayyan, al- (Al Laḥy ar Rayyān), hill group; NS, 26° 40'N, 48° 09'E.

Lihabah, al- (Al Lihābah), wells and village; NS, 27° 15'N, 46° 58'E.

Lisafah, al- (Al Liṣāfah), wells and village; NS, 27° 37'N, 46° 52'E.

Mahakik, al- (Al Maḥākīk), sand area; RK, 22° 50'N, 51° 30'E.

Majann, al- (Al Majann), plateau-plain

near base of the Qatar Peninsula; SCL, 24° 05'N, 51° 35'E.

Majma'ah, al- (Al Majma'ah), town in central Arabia; 25° 54'N, 45° 21'E.

Manakhir ad-Dakakin (Manākhir ad Dakākīn), hill group; SS, 24° 13'N, 48° 23'E.

Manakhir, al-, = Manakhir ad-Dakakin, q.v.

Manifah (Manīfah), Gulf coastal oil field center; CCL, 27° 37'N, 48° 59'E.

Medina (Al Madīnah), city in western Arabia; 24° 28'N, 39° 35'E.

Midra ash-Shimali, al- (Al Midrá ash Shimālī), prominent hill near Dhahran; CCL, 26° 21'N, 50° 05'E.

Mish'ab, al- (Al Mish'āb), port on the Arabian Gulf; NP, 28° 07'N, 48° 36'E.

Mishash (mishāsh), topgr. term: shallow hand-dug well.

Mishash al-'Ashawi (Mishāsh al 'Ashāwī), shallow wells on the western edge of al-Ghawar; SS, 24° 16'N, 48° 53'E.

Mishash Ibn Jum'ah (Mishāsh Ibn Jum'ah), wells in the northeastern Summan; NS, 26° 30'N, 48° 11'E.

Mishash Jaww Dukhan (Mishāsh Jaww Dukhān), shallow seasonal wells in al-Ghawar; SS, 24° 47'N, 49° 06'E.

Mubarraz, al- (Al Mubarraz), town in al-Hasa Oasis; SCL, 25° 25'N, 49° 35'E.

Muhassan, al- (Al Muḥassan), road junction near al-Mubarraz, al-Hasa; SCL, 25° 25'N, 49° 34'E.

Muhtaraqah, al- (Al Muḥtaraqah), village in the northern part of al-Hasa Oasis; SCL, 25° 36'N, 49° 34'E.

Mulayhah al-Gharbiyah, al- (Al Mulayḥah al Gharbīyah), wells in the northern Rub' al-Khali; RK, 22° 53'N, 49° 29'E. Note: placed incorrectly c. 7 km too far east on most maps.

Mulayjah (Mulayjah), village south of
an-Nu'ayriyah; CCL, 27° 16'N, 48°
26'E.

Munayzilah, al- (Al Munayzilah),
village in al-Hasa Oasis; SCL, 25°
23'N, 49° 41'E.

Musannah, al- (Al Musannāh), gravel
ridge in the Dibdibah; NP, 28°
40'N, 46° 40'E.

Mustannah, al- (Al Mustannah), gravel
plain extending south from al-Hasa;
SCL, 24° 40'N, 49° 45'E.

Mutayrifi, al- (Al Muṭayrifī), village in
al-Hasa Oasis; SCL, 25° 28'N, 49°
33'E.

Na'ayim, an- (An Na'ā'im), tributary
to al-Batin; NP, 27° 57'N, 45° 28'E.

Nadqan (Nadqān), drilled well in the
northern Rub' al-Khali; RK, 23°
09'N, 50° 08'E.

Nafud (nafūd), topgr. term: area of
dunes.

Nafud al-Jur'a (Nafūd al Jur'ā'), sand
area adjoining the northern Dahna;
ND, 27° 09'N, 45° 02'E.

Najd (Najd), large geographical region
in central Arabia.

Na'lat Shadqam (Na'lat Shadqam),
rocky plateau west of al-Hasa; SCL,
25° 39'N, 49° 26'E.

Naqirah, an- (An Naqīrah), wells; NP,
27° 53'N, 48° 13'E.

Nariya, = an-Nu'ayriyah, q.v.

Nazim, an- (Ẓahr an Naẓīm), hilly
plateau in the northeastern Summan;
NS, 26° 55'N, 48° 12'E.

Neutral Zone, see Saudi Arabia-Kuwait
Partitioned Neutral Zone.

Nibak (Nibāk), village south of Salwah;
SCL, 24° 25'N, 50° 50'E.

Niqa Fardan (Niqā Fardān), large dune
in the western Rub' al-Khali; RK,
21° 59'N, 48° 40'E.

Nita' (Niṭā'), village; CCL, 27° 13'N,
48° 25'E.

Nu'ayriyah, an- (An Nu'ayrīyah),
village and oil pumping station (the

latter often written 'Nariya'); CCL,
27° 28'N, 48° 29'E.

Nu'ayriyah-as-Saffaniyah jnct, an-,
road junction; NP, 27° 55'N, 48°
42'E.

Oman ('Umān), region in eastern
Arabia, now the Sultanate of Oman.

Qa'amiyat, al- (Al Qa'āmīyāt), area in
the southwestern Rub' al-Khali; RK,
18° 10'N, 48° 25'E.

Qaisumah Pump Station (Al
Qayṣūmah), oil pumping station of
the Trans-Arabian Pipe Line
Company (Tapline) near al-
Qaysumah; NP, 28° 19'N, 46° 07'E.

Qalamah [-t] (qalamah [-t]), topgr.
term: machine-drilled well, borehole.

Qalamat Dab' (Qalamat Ḍab'), drilled
water well south of Yabrin; SCL,
23° 03'N, 49° 00'E.

Qalib (qalīb), topgr. term: hand-dug
well, usually deep.

Qannur (Qannūr), wells and *sabkhah;*
CCL, 27° 04'N, 48° 55'E.

Qar'ah, al- (Al Qar'ah), wells in the
northern Summan; NS, 27° 26'N,
46° 52'E; also, an extensive
shrubless plain, NP, 28° 15'N, 47°
30'E.

Qarn, al- (Al Qarn), village in al-Hasa
Oasis; SCL, 25° 31'N, 49° 36'E.

Qarn Abu Wail (Qarn Abū Wā'il),
prominent hillock south of Salwah;
SCL, 24° 40'N, 50° 51'E.

Qaryat as-Sufla (Qaryat as Suflá),
village and farms; NS, 27° 29'N,
47° 53'E.

Qaryat al-'Ulya (Qaryat al 'Ulyā),
village in the eastern Summan; NS,
27° 33'N, 47° 42'E.

Qatar (Qaṭar), peninsular state on the
Arabian Gulf coast.

Qatif Junction, pipeline junction near
al-Qatif; CCL, 26° 31'N, 49° 57'E.

Qatif, al- (Al Qaṭīf), town and oasis on
Tarut Bay; CCL, 26° 33'N, 50°
00'E.

Qatif-Qaisumah Pipeline KP-144,
pipeline kilometer post; CCL, 27°
20'N, 48° 51'E.

Qatif-Safwa road junction, road
junction; CCL, 26° 37'N, 49° 58'E.

Qaysumah, al- (Al Qayṣūmah), village;
NP, 28° 19'N, 46° 07'E.

Qirdi GOSP (Al Qirḏī), gas-oil
separation plant in the Qirdi oil field
area; SD, 24° 45'N, 48° 15'E.

Qulban (qulbān), topgr. term: pl. of
qalīb, q.v.

Qulban Ibn Busayyis (Qulbān Ibn
Buṣayyiṣ), wells in al-Batin; NS, 27°
50'N, 45° 12'E.

Qurayn, al- (Al Qurayn), village in al-
Hasa Oasis; SCL, 25° 29'N, 49°
36'E.

Rahimah (Raḥīmah), town near Ras
Tanura; CCL, 26° 42'N, 50° 04'E.

Railroad KP-350, kilometer post on the
Damman-Riyadh railway (old route
via Harad); SS, 24° 11'N, 48° 37'E.

Rakah, ar- (Ar Rākah), wells near Thaj;
CCL, 26° 49'N, 48° 41'E; also, a
hamlet between al-Khubar and
Dammam, CCL, 26° 21'N, 50°
11'E.

Ras (ra's), topgr. term: coastal
headland, cape.

Ras al-Ghar (Ra's al Ghār), headland
on the Arabian Gulf coast; CCL, 27°
32'N, 49° 13'E.

Ras as-Saffaniyah (Ra's as Saffānīyah),
headland on the Arabian Gulf coast;
NP, 28° 00'N, 48° 47'E.

Ras as-Sufayra (Ra's aṣ Ṣufayrā'),
headland on the Arabian Gulf coast;
SCL, 25° 35'N, 50° 15'E.

Ras Tanaqib (Ra's Tanāqīb), headland
on the Arabian Gulf; CCL, 27°
50'N, 48° 53'E.

Ras Tanura (Ra's Tannūrah), headland
and oil terminal on the Arabian Gulf
coast; CCL, 26° 38'N, 50° 09'E.

Ras az-Zawr (Ra's az Zawr), coastal
headland; CCL, 27° 28'N, 49° 18'E.

Rawakib 1 camp, = Rawakib well site,
q.v.

Rawakib well site (Ar Rawākib), oil
drilling site in the eastern Rub' al-
Khali; RK, 21° 49'N, 55° 21'E.

Rawdah [-t] (rawḍah [-t]), topgr. term:
closed basin, usually silt-floored.

Rawdat Khuraym (Rawḍat Khuraym),
silt basin on the western edge of the
Dahna; ND, 25° 23'N, 47° 16'E.

Rawdat Ma'qala (Rawḍat Ma'qalā'),
basin near ash-Shumlul; NS, 26°
27'N, 47° 19'E.

Rawdat Musay'id (Rawḍat Musay'īd),
basin; NS, 26° 02'N, 48° 42'E.

Rawdat at-Tawdihiyah (Rawḍat at
Tawḍiḥīyah), silt basin on the
western edge of the Dahna; SD, 24°
11'N, 48° 04'E.

Rawdat az-Zu'ayyini (Rawḍat aẓ
Ẓu'ayyinī), basin and wells; NS, 25°
38'N, 48° 04'E.

Rish (rīsh), topgr. term: elongated
rocky ridge.

Rish Ibn Hijlan (Rīsh Ibn Ḥijlān),
rocky ridge west of Abqaiq; SCL,
25° 54'N, 49° 29'E.

Riyadh (Ar Riyāḍ), capital city of Saudi
Arabia, central Arabia; 24° 38'N,
46° 43'E.

Rumah (Rumāḥ), wells and village in
the Dahna north of Riyadh; ND,
25° 35'N, 47° 09'E.

Rumayhiyah (Rumayḥīyah), wells west
of Rumah; 25° 35'N, 46° 58'E.

Rumaylah, ar- ('Urūq ar Rumaylah),
sand tract at the southern end of the
Dahna; RK, 21° 00'N, 47° 30'E.

Ruq'i, ar- (Ar Ruq'ī), wells in al-Batin;
NP, 29° 00'N, 46° 33'E.

Ruqayyiqah, ar- (Ar Ruqayyiqah),
section of al-Hufuf city, al-Hasa
Oasis; SCL, 25° 21'N, 49° 34'E.

Sabkhah [-t] (sabkhah [-t]), topgr.
term: saline flat, often salt-encrusted
(see chapter 2, 'Central Coastal
Lowlands', for detailed description).

Sabkhat al-Mashakhil (Sabkhat al Mashākhīl), *sabkhah* south of Salwah; SCL, 24° 30'N, 50° 57'E.

Sabkhat Matti (Sabkhat Maṭṭī), extensive *sabkhah* on the Gulf coast extending inland; 23° 30'N, 52° 55'E.

Sabkhat an-Nu'ayriyah (Sabkhat an Nu'ayrīyah), *sabkhah* near an-Nu'ayriyah village; NP, 27° 30'N, 48° 26'E.

Sabkhat Umm al-Jimal (Sabkhat Umm al Jimāl), *sabkhah* southeast of an-Nu'ayriyah; CCL, 27° 16'N, 48° 36'E.

Saffaniyah, as- (As Saffānīyah), oil production center on the Gulf coast; NP, 27° 59'N, 48° 46'E.

Safwa (Ṣafwá), village in al-Qatif Oasis; CCL, 26° 39'N, 49° 58'E.

Salwah (Salwah), Government post on the Saudi Arabia-Qatar border; SCL, 24° 44'N, 50° 48'E.

Samamik Island (Jazīrat as Samāmīk), island in the Gulf of Bahrain; SCL, 25° 29'N, 50° 25'E.

Samudah (Sāmūdah), well and village north of al-Batin; NP, 28° 52'N, 44° 54'E.

Sanam, as- (As Sanām), sand area in the northern Rub' al-Khali; RK, 22° 30'N, 51° 00'E.

Sand Mountains, informal name for the terrain in the northeastern Rub' al-Khali with very large sand massifs separated by *sabkhah* floors.

Sarrar, as- (Aṣ Ṣarrār), village on the eastern edge of the Summan; CCL, 26° 58'N, 48° 23'E.

Saudi Arabia-Kuwait Partitioned Neutral Zone, former special administrative zone, now partitioned, betwen Saudi Arabia and Kuwait.

Sayhat (Sayhāt), village near al-Qatif; CCL, 26° 29'N, 50° 02'E.

Sayhat-Ras Tanura road junction, road junction; CCL, 26° 28'N, 50° 01'E.

Shadqam (Shadqam), plateau and oil field area west of al-Hasa; SCL, 25° 40'N, 49° 26'E.

Shadqam Junction, = Shadqam-'Ayn Dar junction, q.v.

Shadqam-'Ayn Dar junction, road junction; SCL, 25° 53'N, 49° 28'E.

Shalfa (Ash Shalfá), interdune floor and drilled well in the northern Rub' al-Khali; RK, 21° 52'N, 49° 43'E.

Shaqq, ash- (Ash Shaqq), elongated depression near the Saudi Arabia-Kuwait Partitioned Neutral Zone; NP, 28° 20'N, 47° 45'E.

Sharjah (Ash Shāriqah), city and state of the United Arab Emirates, east Arabian Gulf; 25° 22'N, 55° 23'E.

Shaybah Camp (Ash Shaybah), oil drilling camp in the northeastern Rub' al-Khali; RK, 22° 32'N, 54° 01'E.

Shayyit al-'Atshan, ash-(Ash Shayyiṭ al 'Aṭshān), elongate, arcuate depression in the northeastern Summan; NS, 27° 35'N, 47° 15'E.

Shayyit, ash- (Ash Shayyiṭ), = ash-Shayyit al-'Atshan, q.v.

Shayyitat, ash- (Ash Shayyiṭāt), collective name for the parallel depressions, ash-Shayyit al-'Atshan and ash-Shayyit ar-Rayyan; NS, 27° 35'N, 47° 15'E.

Shi'ab (Shi'āb), farm hamlet near Ras Tanura; CCL, 26° 49'N, 49° 57'E.

Shu'bah, ash- (Ash Shu'bah), village and oil pump station site in northern Arabia; 28° 53'N, 44° 42'E.

Shumlul, ash- (Ash Shumlūl), village, also known as Ma'qala (Ma'qalá'), in the west-central Summan; NS, 26° 31'N, 47° 20'E.

Shuqayq, ash- (Ash Shuqayq), village in al-Hasa Oasis; SCL, 25° 29'N, 49° 35'E.

Shuqqan, ash- (Ash Shuqqān), area of linear dune ridges in the western Rub' al-Khali; RK, 21° 30'N, 48° 45'E.

Shuwaykilah, ash- (Ash Shuwaykilah), sand area in the central Rub' al-Khali; RK, 20° 05'N, 49° 30'E.

Sikak, as- (As Sikak), village near Salwah; SCL, 24° 39'N, 50° 51'E.

Sil', as- (As Sil'), wells on the Gulf coast east of Qatar; 24° 01'N, 51° 46'E.

Sila, = As-Sil', q.v.

ST-38, oil exploration camp in the eastern Rub' al-Khali; RK, 22° 17'N, 53° 22'E.

Su'ayyirah, as- (As Su'ayyirah), village and wells; NP, 27° 50'N, 47° 29'E.

Sudah, as- (As Sūdah), district in the northeastern part of the Eastern Province; CCL-NP, 27° 40'N, 48° 30'E.

Sulayyil, as- (As Sulayyil), town in southern Najd; 20° 27'N, 45° 35'E.

Summan, as- (Aṣ Ṣummān), extensive rocky plateau east of the Dahna; NS-SS, 26° 00'N, 48° 30'E.

Summan Yabrin (Ṣummān Yabrīn), rocky area west of Yabrin; SS, 23° 00'N, 48° 45'E.

Tapline, the Trans-Arabian Pipe Line, originally extending from Qaisumah Pump Station to the eastern Mediterranean in Lebanon, since 1984 abandoned west of Jordan; also the Trans-Arabian Pipe Line Company.

Tara'iz, at- (Aṭ Ṭarā'īz), sand area in the central Rub' al-Khali; RK, 21° 30'N, 50° 15'E.

Tarut Bay (Dawḥat Tārūt), bay on the Arabian Gulf north of Dammam; CCL, 26° 35'N, 50° 05'E.

Thaj (Thāj), ruins and village; CCL, 26° 52'N, 48° 43'E.

Thumami, ath- (Ath Thumāmī), wells; ND, 27° 35'N, 44° 56'E.

Tuwayq (Ṭuwayq), extensive mountainous plateau in central Arabia; 24° 00'N, 46° 30'E.

'Udayd, al- (Al 'Udayd), ruined village and promontory at the eastern base of the Qatar Peninsula; SCL, 24° 35'N, 51° 26'E.

'Udayliyah, al- (Al 'Uḍaylīyah), oil company town in al-Ghawar; SCL, 25° 08'N, 49° 18'E.

Umm al-Hamam (Umm al Ḥamām), village in al-Qatif Oasis; CCL, 26° 31'N, 50° 00'E.

Umm al-Hawshat (Umm al Hawshāt), village in the northern Summan; NS, 27° 02'N, 47° 17'E.

Umm al-Jamajim (Umm al Jamājim), wells on the western edge of the Dahna; ND, 26° 52'N, 45° 19'E.

Umm Judhay' (Umm Judhay'), wells near the Gulf coast; CCL, 27° 51'N, 48° 49'E.

Umm al-Khis (Umm al Khīs), spring in al-Hasa Oasis; SCL, 25° 23'N, 49° 38'E.

Umm al-Khursan (Umm al Khurṣān), spring near al-Hufuf; SCL, 25° 23'N, 49° 35'E.

Umm Sab', = 'Ayn Umm Sab', q.v.

Umm as-Sahik (Umm as Sāhik), village near al-Qatif Oasis; CCL, 26° 39'N, 49° 55'E.

Umm as-Samim (Umm as Samīm), large *sabkhah* in the northeastern Rub' al-Khali; RK, 21° 30'N, 55° 45'E.

Umm 'Ushar (Rawḍat Umm 'Ushar), basin and wells in al-Batin; NS, 27° 43'N, 45° 03'E.

'Unayzah ('Unayzah), town in north-central Arabia; 26° 06'N, 43° 58'E.

United Arab Emirates, federation of Arab states on the eastern Gulf coast.

'Uqayr, al- (Al 'Uqayr), small port on the Gulf of Bahrain east of al-Hasa; SCL, 25° 38'N, 50° 13'E.

'Uray'irah ('Uray'irah), village on the eastern edge of the Summan west of 'Ayn Dar; NS, 25° 58'N, 48° 53'E.

'Uruq al-Awarik ('Urūq al Awārik), dune area in the western Rub' al-Khali; RK, 19° 00'N, 47° 30'E.

'Uthmaniyah GOSP 13 (Al 'Uthmānīyah), gas-oil separation plant in the 'Uthmaniyah oil field area; SCL, 25° 00'N, 49° 10'E.

'Uwaynah, al- (Al 'Uwaynah), wells; NS 26° 48'N, 48° 20'E.

'Uyun, al- (Al 'Uyūn), northern district of al-Hasa Oasis; SCL, 25° 35'N, 49° 34'E.

Wabrah (Wabrah), wells in the Summan; NS, 27° 26'N, 47° 22'E.

Wadi (wādī), topgr. term: dry water course, valley.

Wadi al-'Awja (Wādī al 'Awjā'), wadi; NP, 29° 07'N, 46° 33'E.

Wadi ad-Dawasir (Wādī ad Dawāsir), wadi system and agricultural area in southern Najd; 20° 25'N, 45° 00'E.

Wadi Hanifah (Wādī Ḥanīfah), wadi near Riyadh; 24° 31'N, 46° 45'E.

Wadi Muqshin (Wādī Muqshin), wadi at the southeastern edge of the Rub' al-Khali; 19° 25'N, 54° 30'E.

Wadi an-Najabiyah (Wādī an Najabīyah), wadi west of 'Uray'irah; NS, 25° 59'N, 48° 33'E.

Wadi ar-Rumah (Wādī ar Rumah, or Wādī ar Rimah), large wadi in north-central Arabia; 25° 40'N, 43° 20'E.

Wadi as-Sahba (Wādī as Sahbā'), large wadi bed anciently flowing from central Arabia eastward to the Gulf; SS-SCL, 24° 00'N, 49° 05'E.

Wajh, al- (Al Wajh), coastal town and port on the Red Sea in northwestern Arabia; 26° 16'N, 36° 28'E.

Wannan, al- (Al Wannān), village; CCL, 26° 54'N, 48° 25'E.

Wari'ah, al- (Al Warī'ah), gravel ridge; NP, 27° 42'N, 47° 32'E; also, a pump station site on the Qatif-Qaisumah Pipeline, 27° 51'N, 47° 26'E.

Widyan (widyān), topgr. term: pl. of wādī, q.v.

Widyan, al- (Al Widyān), geographical district in northern Arabia near the Iraqi border; 31° 00'N, 41° 00'E.

Yabrin (Yabrīn), village and farms on the northern edge of the Rub' al-Khali; SCL, 23° 15'N, 48° 58'E.

Yemen (Al Yaman), geographical region and state, the Yemen Arab Republic, in southwestern Arabia, 15°N, 44°E.

Za'l Island (Za'l), small island in Tarut Bay; CCL, 26° 40'N, 50° 06'E.

Zakhnuniyah Island, az- (Az Zakhnūnīyah), near-coastal island in the Arabian Gulf; SCL, 25° 33'N, 50° 20'E.

Zawr, az- (Az Zawr), village on Tarut Island; CCL, 26° 36'N, 50° 04'E.

Zubalah (Birkat Zubālah), ruined caravan station on Darb Zubaydah; 29° 24'N, 43° 34'E.

Zulayfayn, az- (Aẓ Ẓulayfayn), hillocks near the Gulf coast; CCL, 26° 51'N, 49° 51'E.

Zumul Camp (Az Zumūl), oil drilling camp in the northeastern Rub' al-Khali; RK, 22° 20'N, 54° 57'E.

Zumul, az- (Az Zumūl), general area of the wells, Umm az-Zumul (Umm az Zumūl), in the northeastern Rub' al-Khali; RK, 22° 42'N, 55° 12'E.

BIBLIOGRAPHIC NOTES

A definitive geography of Arabia is yet to be written. The 1968 edition of the *Aramco Handbook* (Arabian American Oil Company 1968) provides a brief but accurate description of regional geology and geography. *Iraq and the Persian Gulf*, one of the volumes in the World War II British Admiralty Geographical Handbook series (Great Britain 1944), is still a useful source of topographical information for parts of our area. Powers, Ramirez, Redmond, and Elberg (1966) provide a detailed technical account of Arabian sedimentary geology. An Aramco manual, *Meteorologic and Oceanographic Data Book* (Arabian American Oil Company n.d.), is perhaps the best single source of weather and climate information for the Eastern Province; it includes both oil company and Saudi Arabian Government data. Eastern Province soils are described and mapped at reconnaissance scale (1:250,000) in the national soil atlas published by the Saudi Arabian Ministry of Agriculture and Water (1986). This, seen by the present author only in Arabic, follows the broad soil-type classification system used by the U.S. Department of Agriculture. Designed mainly to classify lands in terms of potential arability, it is worth reviewing prior to fieldwork even though it does not provide the larger-scale detail preferred by the botanist or ecologist.

For regional plant geography, Zohary's *Geobotanical Foundations of the Middle East* is essential reading. An article by the present author (1984) proposes a revised boundary between the two major floristic regions in the Arabian Peninsula. Guest, in vol. 1 of the *Flora of Iraq*, also provides useful background information on this subject.

With respect to floristics, two definitive, multi-volume floras for neighboring regions are especially valuable tools in studies of our plants: *Flora of Iraq* (Townsend and Guest 1966-; 6 vols published by 1988), and *Flora Palaestina* (Zohary and Feinbrun-Dothan 1966-1986). Many east-Saudi Arabian species are described in these works, and they are recommended for general reference and for their line drawings of some plants not illustrated in the present volume. They are less useful when dealing with plants of our southern subregions where Sudanian elements are more numerous. For these Täckholm's still-handy *Students' Flora of Egypt* is helpful, although with severely abbreviated keys and descriptions. Batanouny's illustrated *Ecology and Flora of Qatar* is also very useful here although its overlap with our species range is far from complete. Another illustrated flora of Qatar, *Wild Plants of Qatar*, compiled under the auspices of the Arab Organization for Agricultural Development, added a few species also found in neighboring Saudi Arabia. Selected Bahrain plants are described by Diana Charles Phillips in her 1988 book, *Wild Flowers of Bahrain*. M. D. and C. D. Cornes give fuller coverage of the same territory in their well-designed *Wild Flowering Plants of Bahrain* (1989). Both of these works feature non-technical descriptions and photographic illustrations; neither is keyed, but the Cornes' book provides non-technical identification tables designed for the amateur naturalist. A. R. Western's valuable book on the United Arab Emirates' flora (1989) completes the circle for our Gulf neighbors.

Rechinger's single-volume *Flora of Lowland Iraq* is still a good supplementary reference for plants of our north not yet covered by the *Flora of Iraq*. In the same category, but more up-to-date and with illustrations, is Al-Rawi and Daoud's *Flora of Kuwait* (1985, 1987). Léonard's series of studies of the Iran desert flora, begun in 1981, should not be overlooked as a source of information about some of our plants that have Irano-Turanian affinities. The ninth and last fascicle of this well-designed and illustrated work was published in 1989.

Among the few modern longer works dealing specifically with Saudi Arabia, special mention must be made of Sheila Collenette's *Illustrated Guide to the Flowers of Saudi Arabia* (1985), an excellent collection of some 1700 color photographs with brief descriptive notes. Its coverage extends to most parts of the Kingdom but is particularly strong for the western mountainous regions. It provides plates for some Eastern Province species not illustrated in the present volume. Blatter's *Flora Arabica* (1919-1936) is discussed in chap. 1. Migahid and Hammouda's *Flora of Saudi Arabia* (2nd ed. 1978) includes many photographs; its nomenclature and distribution data should be checked with other sources.

S. A. Chaudhary has written numerous monographic studies, mostly at the generic level with Saudi Arabia-wide coverage, that have done much to correct and expand earlier identifications and nomenclature of Saudi Arabian plants. His *Grasses of Saudi Arabia* (1989) is a complete, Kingdom-wide treatment of this very important family.

Vesey-Fitzgerald's (1957) paper on the vegetation is the starting point for studies on east Arabian plant communities. Many findings of the more recent, valuable plant-ecological studies in Kuwait by R. Halwagy and colleagues (1974-1982, 1986) are directly applicable to our territory. Guest's summary in vol. 1 of *Flora of Iraq* is also useful, and Thalen's (1981) work on Iraqi expressions of several of our important shrub communities should not be overlooked. Batanouny's book on Qatar is introduced with some plant community data for an area comparable with our southern subregions. Zohary's *Geobotanical Foundations* attempts a broader regional classification of vegetation units and is valuable for that wider context, although his detailed treatment of Arabia was limited by the sparse data then available.

Special note must be made of the important series of papers, 'Studies in the Flora of Arabia', coordinated by Ian Hedge and colleagues at the Royal Botanic Garden, Edinburgh. The first of these, in 1982, was a definitive botanical bibliography of the Arabian Peninsula by Miller, Hedge and King. The third, also appearing in 1982, was a very useful index to plant collectors in Arabia by Wickens. A number of articles dealing with specific floristic groups followed, and these and others in this continuing series must not be overlooked by any investigator.

Workers with a practical interest in agricultural weed identification will find assistance in Chaudhary and Zawawi's *Manual of Weeds of Central and Eastern Saudi Arabia* (1983) and Chaudhary and Akram's *Weeds of Saudi Arabia and the Arabian Peninsula* (1987). The latter includes descriptions and keys in both English and Arabic and is also a very useful reference for the general botanist.

The most useful maps of eastern Saudi Arabia available to the public are the 1:500,000-scale series of quadrangles published in separate geographical and geological versions by the Ministry of Petroleum and Mineral Resources, Kingdom of Saudi Arabia, and the U.S. Geological Survey. The sheets covering the area of this flora (geographical versions) are: *Wadi Al Batin* (GM-203 B), repr. 1982; *Northern Tuwayq* (GM-207 B), repr. 1978; *Western Arabian Gulf* (GM-208 B), 1977; *Central Arabian Gulf* (GM-209 B), 1977; *Northwestern Rub' Al Khali* (GM-213 B), repr. 1976; *Northeastern Rub' Al Khali* (GM-214 B), repr. 1982; *Eastern Rub' Al Khali* (GM-215 B), repr. 1982; *South Central Rub' Al Khali* (GM-219 B), repr. 1981; and *Southeastern Rub' Al Khali* (GM-220 B), repr. 1981. The series, with revisions, is distributed by the Directorate General of Mineral Resources, Jiddah.

BIBLIOGRAPHY

Al-Rawi, Ali, and Hazim S. Daoud. 1985, 1987. *Flora of Kuwait.* vol. 1 (1985) Dicotyledoneae excluding Compositae, by H. S. Daoud, rev. A. Al-Rawi, London: KPI; vol. 2 (1987) Compositae and Monocotyledoneae, by A. Al-Rawi, Kuwait: Kuwait University.

Arab Organization for Agricultural Development. 1983. *Wild plants of Qatar.* N.p.

Arabian American Oil Company. 1968. *Aramco handbook.* Dhahran.

———— N.d. Aramco meteorologic and oceanographic data book for the Eastern Province region of Saudi Arabia. Dhahran: Arabian American Oil Company. Mimeo.

Basson, Philip W., John E. Burchard, Jr., John T. Hardy, and Andrew R. G. Price. 1981. *Biotopes of the western Arabian Gulf.* 2nd ed. Dhahran: Arabian American Oil Company.

Batanouny, K. H. 1981. *Ecology and flora of Qatar.* Doha: University of Qatar.

Baum, Bernard R. 1978. *The genus Tamarix.* Jerusalem: Israel Academy of Sciences and Humanities.

Beccari, O. 1890. Revista monografica delle specie del genere Phoenix Linn. *Malesia* 3:345-416.

Bhandari, M. M. 1978. *Flora of the Indian Desert.* Jodhpur: Scientific Publishers.

Blatter, Ethelbert. 1919-1936. Flora Arabica. I-VI. *Records of the Botanical Survey of India* 8 (no. 1, 1919): 1-123; (no. 2, 1921): 123-282; (no. 3, 1921): 283-365; (no. 4, 1923): 365-450; (no. 5, 1933): 451-501; (no. 6, 1936): 451-519 (sic).

Bor, N. L. 1967. A new *Stipagrostis* from Saudi Arabia. *Österreichische botanische Zeitschrift* 114:100.

Botschantzev, V. 1969. A new species of *Cornulaca* Del. *(Chenopodiaceae)* from Saudi Arabia. *Kew Bulletin* 23:439-40.

———— 1975a. New species of Salsola L. *Botanicheskii Zhurnal* 60:498-505 (in Russian).

———— 1975b. Species subsectionis vermiculatae Botsch. sectionis Caroxylon (Thunb.) Fenzl generis Salsola L. *Novitates Systematicae Plantarum Vascularium* 12:160-94.

———— 1977. The genus *Agathophora* (Fenzl) Bunge (Chenopodiaceae). *Botanicheskii Zhurnal* 62:1447-1452 (in Russian).

———— 1981. Revisio generis Halothamnus Jaub. et Spach (Chenopodiaceae). *Novitates Systematicae Plantarum Vascularium* 18:146-76 (in Russian with key, new species descriptions and some notes in Latin).

Boulos, Loutfy. 1978. Materials for a flora of Qatar. *Webbia* 32(2): 369-96.

Brummitt, R. K. 1982. *Ogastemma*, a new name for *Megastoma* (Boraginaceae). *Kew Bulletin* 36(4): 679-80.

Bunker, D. G. 1953. The south-west borderlands of the Rub' al Khali. *The Geographical Journal* 119(4): 420-30.

Burtt, B. L., and Patricia Lewis. 1949-1954. On the flora of Kuwait. I-III. *Kew Bulletin* 4:273-308 (1949); 7:333-52 (1952); 9:377-410 (1954).

Carter, Humphrey G. 1917. Some plants of the Zor Hills, Koweit, Arabia. *Records of the Botanical Survey of India* 6:171-206.

Chaudhary, Shaukat Ali. 1983. Preliminary studies on genus Convolvulus in Saudi Arabia. *Journal of the Saudi Arabian Natural History Society* 2(3):11-31.

————— 1985. Studies on *Heliotropium* in Saudi Arabia. *Arab Gulf Journal of Scientific Research* 3 (1):33-53.

————— 1986a. Bromeae (Gramineae) of Saudi Arabia. *Proceedings of the Saudi Biological Society* 9:91-113.

————— 1986b. The genus *Eremopyrum* (Gramineae) in Saudi Arabia. *Proceedings of the Saudi Biological Society* 9:115-120.

————— 1986c. The genus *Salsola* (Chenopodiaceae) in Saudi Arabia. *Proceedings of the Saudi Biological Society* 9:57-89.

————— 1987a. The genus *Halothamnus* Jaub. & Spach (Chenopodiaceae) in Saudi Arabia. *Proceedings of the Saudi Biological Society* 10:319-26.

————— 1987b. The genus *Stipagrostis* Nees (Poaceae) in Saudi Arabia. *Proceedings of the Saudi Biological Society* 10:327-41.

————— 1989. *Grasses of Saudi Arabia*. Riyadh: National Agriculture and Water Research Center.

Chaudhary, S. A., C. Parker, and L. Kasasian. 1981. Weeds of central, southern and eastern Arabian Peninsula. *Tropical Pest Management* 27(2): 181-90.

Chaudhary, S. A., and T. A. Cope. 1983. A checklist of grasses of Saudi Arabia (Studies in Flora of Arabia. VI). *Arab Gulf Journal of Scientific Research* 1(2):313-354.

Chaudhary, S. A., and M. A. Zawawi. 1983. *A manual of weeds of central and eastern Saudi Arabia*. Riyadh: Ministry of Agriculture and Water, Regional Agriculture and Water Research Center.

Chaudhary, S. A., and Muhammad Akram. 1987. *Weeds of Saudi Arabia and the Arabian Peninsula*. Riyadh:

Ministry of Agriculture and Water, Regional Agriculture and Water Research Center.

Cheesman, R. E. 1926. *In unknown Arabia*. London: Macmillan.

Cloudsley-Thompson, J. L., and M. J. Chadwick. 1964. *Life in deserts*. London: G. T. Foulis.

Collenette, Sheila. 1985. *An illustrated guide to the flowers of Saudi Arabia* (Flora publication No. 1, Meteorology and Environmental Protection Administration, Kingdom of Saudi Arabia). London: Scorpion.

Cope, T. A. 1980. Nomenclatural and taxonomic notes on *Lasiurus (Gramineae)*. *Kew Bulletin* 35(3): 451-452.

————— 1985. A key to the grasses of the Arabian Peninsula (Studies in the flora of Arabia XV). *Arab Gulf Journal of Scientific Research*, Special Publication No. 1.

Corner, E. J. H. 1966. *The natural history of palms*. London: Weidenfeld and Nicolson.

Cornes, M. D. and C. D. 1989. *The wild flowering plants of Bahrain: an illustrated guide*. London: Immel.

Daoud, Hazim S. 1985. *Flora of Kuwait*. Revised by Ali Al-Rawi. Vol. 1: Dicotyledoneae. London: KPI.

Davis, P. H., and J. Cullen. 1979. *The identification of flowering plant families*. 2nd ed. Cambridge: Cambridge University Press.

De Marco, Giovanni, and Angela Dinelli. 1974. First contribution to the floristic knowledge of Saudi Arabia. *Annali di Botanica* 33:209-36.

Dickson, H. R. P. 1951. *The Arab of the desert*. 2nd ed. London: George Allen & Unwin.

Dickson, Violet. 1938. Plants of Kuwait, north east Arabia. *Journal of the Bombay Natural History Society* 40:528-38.

———— 1955. *The wild flowers of Kuwait and Bahrain.* London: George Allen & Unwin.

Eig, A. 1938. Taxonomic studies on the oriental species of the genus *Anthemis. Palestine Journal of Botany, Jerusalem Series* 1:161-224.

El-Gazzar, A. 1988. Agriophyllum montasiri, a new species of Chenopodiaceae from eastern Arabia. *Mitt. Bot. Staatssamml. München* 27:15-19.

El Hadidi, M. N. 1977. Two new Zygophyllum species from Arabia. *Publications from Cairo University Herbarium,* No. 7 and 8, 327-31.

———— 1978. An introduction to the classification of *Tribulus* L. *Taeckholmia* 9:59-66.

El-Khayal, A. A., W. G. Chaloner, and C. R. Hill. 1980. Palaeozoic plants from Saudi Arabia. *Nature* 285:33-34.

Emboden, William. 1972. *Narcotic Plants.* London: Studio Vista.

Featherly, H. I. 1954. *Taxonomic terminology of the higher plants.* Ames, Iowa: Iowa State College Press.

Giacomini, Valerio, Nunzio Longhitano, and Linneo Corti. 1979a. *Cartography of the vegetation of the Eastern (Hasa) Province of Saudi Arabia.* Pubblicazioni dell'Istituto di Botanica dell'Universitá di Catania. Catania: Coop. Universitaria Libraria Catanese.

———— 1979b. *The vegetation and grazing resources of Wady Al Miyah area, El Hasa Province of Saudi Arabia.* Pubblicazioni dell'Istituto di Botanica dell'Universitá di Catania. Catania: Coop. Universitaria Libraria Catanese.

Good, Ronald. 1955. The flora of Bahrain. In *The Wild Flowers of Kuwait and Bahrain,* by Violet Dickson. London: George Allen & Unwin.

Gray, Jane, Dominique Massa, and A. J. Boucot. 1982. Caradocian land plant microfossils from Libya. *Geology* 10(4):197-201.

Great Britain, Admiralty, Naval Intelligence Division. 1944. *Iraq and the Persian Gulf.* N.p.

Guest, Evan. 1966. *Flora of Iraq.* Vol. 1. Baghdad: Ministry of Agriculture, Republic of Iraq.

Haines, R. Wheeler. 1951. Potential annuals of the Egyptian desert. *Bulletin de L'institut Fouad I du Désert* 1(2):103-18

Halwagy, R. 1986. On the ecology and vegetation of Kuwait. In *Contributions to the vegetation of southwest Asia,* ed. Harald Kurschener, 81-109. Wiesbaden: Dr Ludwig Reichert Verlag.

Halwagy, R., and M. Halwagy. 1974a. Ecological studies on the desert of Kuwait. I. The physical environment. *Journal of the University of Kuwait (Science)* 1:75-86.

———— 1974b. Ecological studies on the desert of Kuwait. II. The vegetation. *Journal of the University of Kuwait (Science)* 1:87-95.

———— 1977. Ecological studies on the desert of Kuwait. III. The vegetation of the coastal salt marshes. *Journal of the University of Kuwait (Science)* 4:33-74.

Halwagy, R., A. F. Moustafa and Susan M. Kamel. 1982. On the ecology of the desert vegetation in Kuwait. *Journal of arid environments* 5:95-107.

Hamilton, W. R., P. J. Whybrow, and H. A. McClure. 1978. Fauna of fossil mammals from the Miocene of Saudi Arabia. *Nature* 274:248-49.

Hartog, C. den. 1970. *The sea-grasses of the world.* Amsterdam: North-Holland Publishing Company.

Hedge, I. C., A. Kjaer, and O. Malver. 1980. Dipterygium — Cruciferae or Capparaceae? *Notes from the Royal Botanic Garden Edinburgh* 38(2): 247-50.

Hedge, I. C., and R. A. King. 1983. The Cruciferae of the Arabian Peninsula: A check-list of species and a key to genera (Studies in the flora of Arabia IV). *Arab Gulf Journal of Scientific Research* 1(1):41-66.

Heyn, Chaia C., Ora Dagan, and Bilha Nachman. 1974. The annual *Calendula* species: taxonomy and relationships. *Israel Journal of Botany* 23:169-201.

Heywood, V. H., ed. 1978. *Flowering plants of the world.* Oxford: Oxford University Press.

Hosni, Hasnaa A. 1978. A new Tribulus species with winged carpels. *Botaniska Notiser* 130:261-62.

Hötzl, H., F. Krämer, and V. Maurin. 1978. 3.1. Quaternary sediments. In *Quaternary period in Saudi Arabia,* ed. S. S. Al-Sayari and J. G. Zötl, 264-301. Vienna: Springer-Verlag.

Hötzl, H., H. J. Lippolt, V. Maurin, H. Moser, and W. Rauert. 1978. Quaternary studies on the recharge area situated in crystalline rock regions. In *Quaternary period in Saudi Arabia,* ed. S. S. Al-Sayari and J. G. Zötl, 230-239. Vienna: Springer-Verlag.

Huschke, R. E., R. R. Rapp, and C. Schutz. 1970. *Meteorological aspects of Middle East water supply.* N.p.: The Rand Corporation.

Ismail, A. M. A. 1983. Some factors controlling the water economy of *Zygophyllum quatarense* (Hadidi) growing in Qatar. *Journal of Arid Environments* 6:239-246.

Jackson, B. D. 1928. *A glossary of botanic terms.* 4th ed. London: Gerald Duckworth & Co. Ltd.

Kassas, M. 1966. Plant life in deserts. In *Arid lands, a geographical appraisal,* ed. E. S. Hills, 145-180. London: Methuen.

King, R. A., and K. J. Kay. 1984. The Caryophyllaceae of the Arabian Peninsula: a checklist and key to taxa (Studies in the flora of Arabia XII). *Arab Gulf Journal of Scientific Research* 2(2):391-414.

Kingdom of Saudi Arabia. Ministry of Agriculture and Water. 1406/1986. *Al kharīṭah al 'āmmah lit turbah.* In Arabic. N.p.

Kutzbach, John E. 1981. Monsoon climate of the early Holocene: climate experiment with the Earth's orbital parameters for 9000 years ago. *Science* 214:59-61.

Lack, Hans-Walter. 1975. *Die Gattung Picris L. sensu lato im ostmediterran-westasiatischen Raum* (vol. 116, *Dissertationen der Universität Wien).* Vienna.

Léonard, J. 1981-1989. *Contribution a l'étude de la flore et de la vegetation des déserts d'Iran.* Fasc. 1-9. N.p.: Jardin Botanique National de Belgique.

——— 1985. Note sur *Prosopis cineraria* (L.) Druce et *P. koelziana* Burkart (Mimosacées asiatiques). *Bulletin du jardin botanique national de Belgique* 55:491-492.

——— 1986a. *Neotorularia* Hedge & J. Léonard, nom générique nouveau de Cruciferae. *Bulletin du jardin botanique national de Belgique* 56:389-395.

——— 1986b. Une variété nouvelle de *Prosopis koelziana* Burkart (Mimosacée de la péninsule arabique). *Bulletin du jardin botanique national de Belgique* 56:483-485.

Mandaville, James P. 1965. Notes on the vegetation of Wadi as-Sahbā', eastern Arabia. *Journal of the Bombay Natural History Society* 62(2): 330-32.

———— 1972. Some experiments with solar ground stills in eastern Arabia. *The Geographical Journal* 138(1):64-66.

———— 1984. Studies in the flora of Arabia XI: Some historical and geographical aspects of a principal floristic frontier. *Notes from the Royal Botanic Garden Edinburgh* 42(1):1-15.

———— 1986. Plant life in the Rub' al-Khali (the Empty Quarter), south-central Arabia. In *Plant life of south-west Asia (Proceedings of the Royal Society of Edinburgh,* Section B, vol. 89), ed. I. C. Hedge, 147-157.

McClure, H. A. 1976. Radiocarbon chronology of late Quaternary lakes in the Arabian desert. *Nature* 263:755-56.

———— 1978. Ar Rub' Al Khali. In *Quaternary period in Saudi Arabia,* ed. Saad S. Al-Sayari and Josef G. Zötl, 252-63. Vienna: Springer-Verlag.

Migahid, A. M., and M. A. Hammouda. 1974. *Flora of Saudi Arabia.* 2nd ed. 1978, rev. A. M. Migahid, 2 vols. Riyadh: Riyadh University.

Miller, A. G. 1984. A revision of Ochradenus. *Notes from the Royal Botanic Garden Edinburgh* 41(3):491-504.

Miller, A. G., I. C. Hedge and R. A. King. 1982. Studies in the flora of Arabia: I, A botanical bibliography of the Arabian Peninsula. *Notes from the Royal Botanic Garden Edinburgh* 40(1):43-61.

Muschler, Reno. 1912. *A manual flora of Egypt.* 2 vols. Berlin: R. Friedlaender. Repr. 1970, J. Cramer.

Musil, Alois. 1926. *The Northern Ḥeğâz.* New York: American Geographical Society.

———— 1927. *Arabia Deserta.* New York: American Geographical Society.

———— 1928a. *The manners and customs of the Rwala Bedouins.* New York: American Geographical Society.

———— 1928b. *Northern Neğd.* New York: American Geographical Society.

Novikova, N. M. 1970. Drawing up a preliminary vegetation map of Arabia. *Geobotanicheskoe Karto-Gratifovanie,* 61-71. RTS 12072, English tr. by Russian Translation Service, the British Library.

Paris, R., and G. Dillemann. 1960. Part two, with particular reference to the pharmacological aspects, of *Medicinal plants of the arid zones* (UNESCO arid zone research - XIII), 55-96. Paris: UNESCO.

Parker, C. 1973. Weeds in Arabia. *PANS* 19:345-52.

Pelly, Lewis. 1866. *Report on a journey to the Wahabee capital of Riyadh in central Arabia.* Bombay.

Philby, H. St.J. B. 1922. *The heart of Arabia.* 2 vols. London: Constable.

———— 1933. *The Empty Quarter.* London: Constable.

———— Papers (unpublished). Private Papers Collection, St. Antony's College, Oxford.

Phillips, Diana Charles. 1988. *Wild flowers of Bahrain: a field guide to herbs, shrubs and trees.* Manama, Bahrain: Privately published.

Popenoe, Paul. 1973. *The date palm.* Coconut Grove, Miami: Field Research Projects.

Popov, George N.d. The vegetation of Arabia south of the Tropic of Cancer. N.p. Unpublished typescript.

Popov, George, and W. Zeller. 1963. Ecological survey report on the 1962 survey in the Arabian Peninsula. F.A.O. Progress Report No. UNSF/DL/ES/6. Rome: Food and Agriculture Organization of the United Nations. Mimeo.

Powers, R. W., L. F. Ramirez, C. D. Redmond, and E. L. Elberg, Jr. 1966. *Geology of the Arabian Peninsula: Sedimentary geology of Saudi Arabia* (Geological Survey Professional Paper 560-D). Washington: U.S. Geological Survey.

Raunkiaer, Barclay. 1913. *Gennem Wahhabiternes Land paa Kamelryg.* Copenhagen. English tr. printed for official use by the Arab Bureau, Cairo, 1916 as *Through Wahhabiland on Camel-back.* Retranslation edited and slightly abridged by Gerald de Gaury as *Through Wahhabiland on Camelback,* London, Routledge and Kegan Paul, 1969.

Rechinger, K. H. 1962. Revision einiger Typen von Velenovsky's Plantae Arabicae Musilianae. *Botaniska Notiser* 115(1): 35-48.

———— 1964. *Flora of lowland Iraq.* Weinheim: J. Cramer.

Sa'ad, F. 1967. *The Convolvulus species of the Canary Isles, the Mediterranean region and the Near and Middle East.* Rotterdam: Bronder-offset.

Schweinfurth, G. 1899. Sammlung arabisch-aethiopischer Pflanzen. *Bulletin de l'Herbier Boissier,* année 7, Appendix 2, 267-340.

Scott, A. J. 1977. Reinstatement and revision of Salicorniaceae J. Agardh (Caryophyllales). *Botanical Journal of the Linnean Society* 75:357-74.

Shimwell, David W. 1971. *The description and classification of vegetation.* London: Sidgwick & Jackson.

Soskov, G. 1973. Generis Calligonum L. species nova ex Aravia. *Novitates Systematicae Plantarum Vascularium* 10:134-35.

———— 1975. Sectio Calligonum generis Calligonum L. (Polygonaceae). *Novitates Systematicae Plantarum Vascularium* 12:147-59.

Stebbins, G. L. 1974. *Flowering plants: evolution above the species level.* Cambridge, Mass.: Harvard University Press.

Täckholm, Vivi. 1974. *Students' flora of Egypt.* 2d. ed. Cairo: Cairo University.

Thalen, D. C. P. 1979. *Ecology and utilization of desert shrub rangelands in Iraq.* The Hague: Dr. W. Junk B.V.

Thesiger, Wilfred. 1946. A new journey in southern Arabia. *The Geographical Journal* 108:129-45.

———— 1948. Across the Empty Quarter. *The Geographical Journal* 111:1-21.

———— 1949. A further journey across the Empty Quarter. *The Geographical Journal* 113:21-46.

———— 1950. Desert borderlands of Oman. *The Geographical Journal* 116:137-71. Map in *Geogr. J.* 117 (1951), Jan.-Mar.

Thomas, Bertram. 1932. *Arabia Felix.* London: Jonathan Cape.

Townsend, C. C., and E. Guest, eds. 1966 –. *Flora of Iraq.* vol. 1 (1966), Introduction to the flora; vol. 2 (1966), Pteridophyta, Gymnospermae, Angiospermae - Rosaceae; vol. 9 (1968), Gramineae; vol. 3 (1974), Leguminales; vol. 4 (1980), part 1 Cornaceae to Rubiaceae, part 2 Bignoniaceae to Resedaceae; vol. 8 (1985), Monocotyledones excluding Gramineae. Baghdad: Ministry of Agriculture, Republic of Iraq.

Trease, G. E. 1961. *A textbook of pharmacognosy.* 8th ed. London: Ballière, Tindale and Cox.

Triest, Ludwig. 1988. *A revision of the genus Najas L. in the Old World* (fasc. 1, tome 22, Mémoires in-8°, nouvelle série, Académie Royale des Sciences d'Outre-mer, Classe des Sciences Naturelles et Médicales). Brussels.

UNESCO. 1960. *Medicinal Plants of the Arid Zones (Arid Zone Research - XIII)*. Paris: UNESCO.

Uvarov, B. P. 1951. *Locust research and control 1929-1950*. Colonial Research Publication No. 10. London: Colonial Office.

Velenovský, J. 1912. Plantae arabicae Musilianae. In *Sitzungsberichte der Königl. Böhmischen Gesellschaft der Wissenschaften, mathematisch-naturwissenschaftliche Classe, Jahrgang 1911*, 11, 1-17.

———— 1923. Arabske rostliny z posledni cesty Musilovy r. 1915: Plantae arabicae ex ultimo itinere A. Musili a. 1915. In *Mémoires de la Société royale des sciences de Bohême, Classe des Sciences, année 1921-1922*, 6, 1-9.

Vesey-Fitzgerald, Desmond. 1951. From Hasa to Oman by car. *The Geographcial Review* 41:544-60.

———— 1957. The vegetation of central and eastern Arabia. *The Journal of Ecology* 45:779-98.

Verdcourt, Bernard, and E. C. Trump. 1969. *Common poisonous plants of East Africa*. London: Collins.

Vidal, F. S. 1954. Date culture in the oasis of al-Hasa. *The Middle East Journal* 8(4):416-28.

Vita-Finzi, Claudio. 1978. Recent alluvial history in the catchment of the Arabo-Persian Gulf. In *The environmental history of the Near and Middle East since the last Ice Age*, ed. William C. Brice, 255-61. London: Academic Press.

Wagenitz, G. 1984. Studies in the flora of Arabia VII: Centaurea in the Arabian Peninsula. *Notes from the Royal Botanic Garden Edinburgh* 41(3):457-466.

Wallén, C. C. 1966. Arid zone meteorology. In *Arid lands: a geographical appraisal*, ed. E. S. Hills, 31-51. London: Methuen.

Walter, Heinrich. 1979. *Vegetation of the earth and ecological systems of the geo-biosphere*. 2nd ed. (translation of *Vegetationszonen und Klima*, 3rd ed.). New York: Springer-Verlag.

Western, A. R. 1989. The flora of the United Arab Emirates, an introduction. N.p. United Arab Emirates University.

Whybrow, Peter J., and H. A. McClure. 1981. Fossil mangrove roots and palaeoenvironments of the Miocene of the eastern Arabian Peninsula. *Palaeogeography, Palaeoclimatology, Palaeoecology* 32:213-25.

Wickens, G. E. 1982. Studies in the flora of Arabia: III, A biographical index of plant collectors in the Arabian Peninsula (including Socotra). *Notes from the Royal Botanic Garden Edinburgh* 40(2):301-330.

Wiklund, Annette. 1986. The genus *Rhanterium* (Asteraceae: Inuleae). *Botanical journal of the Linnean Society* 93:231-246.

Zohary, Michael. 1940. On the Ghada tree of northern Arabia and the Syrian Desert. *Palestine Journal of Botany, Jerusalem Series* 1:413-16.

———— 1957. A contribution to the flora of Saudi Arabia. *The Journal of the Linnean Society of London, Botany* 55:632-43.

———— 1962. *Plant life of Palestine*. New York: Ronald Press.

———— 1973. *Geobotanical foundations of the Middle East*. 2 vols. Stuttgart: Gustav Fischer Verlag.

Zohary, Michael, and N. Feinbrun-Dothan. 1966-1986. *Flora Palaestina*. Part 1 (1966), Equisetaceae to Moringaceae; part 2 (1972), Platanaceae to Umbelliferae; part 3 (1978), Ericaceae to Compositae; part 4 (1986), Alismataceae to Orchidaceae. Plates in separate vols with each part. Jerusalem.

فهرس عربي بالأسماء الدارجة للنباتات

الأرقام تشير الى الصفحات التي ترد فيها الأسماء الدارجة في النص . راجع الأسماء العلمية في الفهرس العام حيث تجد أرقام الصفحات الأخرى التي ترد فيها أنواع النباتات . العلاقة بالأسماء بين قوسين غير مؤكدة وهي مبنية على مراجع مفردة .

ARABIC INDEX AND GLOSSARY OF VERNACULAR PLANT NAMES

Page numbers refer only to vernacular names in the text. See the general index, under scientific names, for further references to the plant species concerned. Name attributions in parentheses are not in general use and are based on single, unconfirmed records.

UMM AS-SUWAYQ   أُمُ السُوَيق
Linaria tenuis, 274

UMM AS-SUWAYQAH   أُمُ السُوَيقة
Helianthemum kahiricum, 117
Helianthemum lippii, 118

UMM AL-QURAYN   أُمُ القُرَين
Hippocrepis bicontorta, 189

UMM QUTAYNAH   أُمُ قُطَينة
Sophora gibbosa, 194

BĀBŪNAJ   بابُونَج
Matricaria aurea, 300

BĀSŪL   باصول
Cistanche tubulosa, 278

BAKHATARĪ   بَخَترِيّ
Roemeria hybrida, 50

BARBĪR   بربير
Portulaca oleracea, 71

BARDĪ   بَرْدِيّ
Typha domingensis, 395

BIRKĀN, BARUKĀN   بِرْكان ، بُرُكان
Anastatica hierochuntica, 144
Limeum arabicum, 54

BARWAQ   بَرْوَق
Asphodelus tenuifolius, 399

BASBĀS   بَسْباس
Anisosciadium lanatum, 229

BAŞAL   بَصَل
Allium sphaerocephalum, 404

BU'AYTHIRĀN   بُعَيْثِران
Artemisia judaica, 304

BAQL   بَقْل
Portulaca oleracea, 71
Tribulus terrestris, 218

BALLAH   بَلّة
Silene villosa, 62

BUWAYDĀ'   بُوَيْضاء
Paronychia arabica, 69

BAYRAQ   بَيرَق
Asphodelus tenuifolius, 399

ABŪ THURAYB   أبُو ثُرَيب
Frankenia pulverulenta, 123

ABŪ KHAWĀTĪM   أبُو خَواتيم
Astragalus annularis, 184

ABŪ NASHR   أبُو نَشْر
Galium ceratopodum, 282

ATHL   أثْل
Tamarix aphylla, 119, **120**

IDHKHIR   إذْخِر
Cymbopogon commutatus, 339

UDHUN AL-ḤIMĀR   أُذُن الحِمار
Astragalus kahiricus, 186

ARĀK   أراك
Salvadora persica, 198

ARTĀ   أرْطَى
Calligonum comosum, 107
Calligonum tetrapterum, 108

ARQĀ'   أرْقاء
Helianthemum spp., 117
Helianthemum lippii, 118
Helianthemum salicifolium, 119

URAYNIBAH   أُرَيْنبة
Bassia muricata, 78

USHNĀN   أُشْنان
Seidlitzia rosmarinus, 85

AŞĀBI' AL-'ARŪS   أصابِع العَرُوس
Astragalus annularis, 184

ALĀL   ألال
Taverniera spartea, 191

UMM UDHN   أُمُ أُذن
Ononis serrata, 174

UMM AT-TURĀB   أُمُ التُّراب
Fagonia glutinosa, 211

UMM THURAYB   أُمُ ثُرَيب
Frankenia pulverulenta, 123

UMM ATH-THURAYB   أُمُ الثُّرَيب
Hypecoum geslinii, 52
Hypecoum pendulum, 52

UMM AR-RUWAYS   أُمُ الرُوَيس
Scabiosa palaestina, 284

INDEX AND GLOSSARY OF VERNACULAR PLANT NAMES

Page references here indicate only vernacular name occurrences, including English meanings of many names, tribal sources of names, and some other related data. See the general index, under scientific names, for references to other kinds of information about the plant species concerned. Vernacular attributions to names in parentheses are believed to differ from general usage and are for the most part based on unconfirmed single records; they should be taken with some caution.

Notes on vernacular name sources and transliteration may be found on pp. 36–37. For non-Arabic speakers the following additional remarks are intended as a simplified guide for approximating east Saudi Arabian colloquial pronunciations and for identifying in the index names heard in the field: Short vowels resemble their English counterparts; long vowels, indicated by macrons, are Ā (as in 'far'), Ī (as in 'machine'), and Ū (as in 'flute'). The diphthong AW is often sounded as long 'ō'. DH is sounded as th in 'the', KH as ch in Scottish 'loch'; Q usually as g in 'got'; the emphatics Ḍ and Ẓ are sounded alike — a heavy th as in 'the' but with tongue lowered and tensed. GH is a back fricative akin to the guttural French 'r'. The consonant denoted by the raised inverted comma (ʻ) is a peculiar throat constriction better ignored in the absence of expert demonstration; apostrophes in the names may be ignored entirely in colloquial speech. Ḥ is an 'h' with throat aspiration. Ṣ and Ṭ resemble their English counterparts but with tongue lowered and tensed; like other emphatics they tend to broaden associated vowels. Doubled consonants must be pronounced doubled. Stress usually falls on syllables with long vowels or diphthongs, and at doubled consonants; otherwise it is generally on the first or second syllable.

ʻABAL
  Calligonum comosum, 31, 32, **107**
  Calligonum crinitum, 108
  Calligonum tetrapterum, 108
ABŪ KHAWĀTĪM
  Astragalus annularis, 184
ABŪ NASHR
  Galium ceratopodum, 282
ABŪ THURAYB
  Frankenia pulverulenta, 123
ʻADĀM
  Ephedra alata, 47
ʻĀDHIR
  Artemisia monosperma, 31, **305**
ʻAḌĪD
  Launaea mucronata, 322
ʻAḌRIS
  Convolvulus oxyphyllus, 248
ALĀL
  Taverniera spartea, 191
ʻALANDĀ
  Ephedra alata, 47
  (Dipterygium glaucum, 127)
ʻALQĀ
  Dipterygium glaucum, 127
  Ochradenus baccatus, 157
  Scrophularia deserti, 275
  Scrophularia hypericifolia, 275
  (Fagonia bruguieri, 211)
  (Taverniera spartea, 191)
ʻAMBAṢIṢ
  Emex spinosa, 105
ʻANDAB
  Cyperus conglomeratus agg., 392
ʻANṢALĀN
  Dipcadi erythraeum, 404

ʻANTAYR
  Glossonema varians, 236
ʻAQRABĀN
  Phragmites australis, 361
ʻĀQŪL
  Alhagi maurorum, 192
ʻARĀD
  Cornulaca monacantha, 101
  Salsola cyclophylla, 87
ARĀK
  Salvadora persica, 198
ʻARFAJ
  Rhanterium epapposum, 29, 294, **295**
ARQĀ'
  Helianthemum lippii, 118
  Helianthemum salicifolium, 119
  Helianthemum spp., 117
ARṬĀ
  Calligonum comosum, 107
  Calligonum tetrapterum, 108
AṢĀBIʻ AL-ʻARŪS
  Astragalus annularis, 184
ʻAṢAL
  Suaeda monoica, 83
ʻAṢANṢAL
  Dipcadi erythraeum, 404
ATHL
  Tamarix aphylla, 119, **120**
ʻAWSAJ
  Lycium depressum, 240
  Lycium shawii, 240
  (Prosopis farcta, 167)
ʻAWSHAZ
  Lycium shawii, 240

BĀBŪNAJ
  Matricaria aurea, 300

GENERAL INDEX

Genera, species and subspecific taxa described in this book as part of the east Saudi Arabian flora are in boldface, as are main numbers in multiple page references. Synonyms are in italics. Roman type is used for all other plant names, suprageneric taxa, and general subjects. For Arabic vernacular plant names, see the separate vernacular name indexes.

Milton Keynes UK
Ingram Content Group UK Ltd.
UKHW030901141024
449569UK00025B/1275